Analytical Methods for Environmental Contaminants of Emerging Concern

Analytical Methods for Environmental Contaminants of Emerging Concern

Edited by

Núria Fontanals and Rosa Maria Marcé
Universitat Rovira i Virgili, Spain

WILEY

This edition first published 2022

Registered Offices
John Wiley & Sons, Inc., 111 River Street, Hoboken, NJ 07030, USA
John Wiley & Sons Ltd, The Atrium, Southern Gate, Chichester, West Sussex, PO19 8SQ, UK

Editorial Office
The Atrium, Southern Gate, Chichester, West Sussex, PO19 8SQ, UK

For details of our global editorial offices, customer services, and more information about Wiley products visit us at www.wiley.com.

Wiley also publishes its books in a variety of electronic formats and by print-on-demand. Some content that appears in standard print versions of this book may not be available in other formats.

A catalogue record for this book is available from the Library of Congress

Hardback ISBN: 9781119763864; ePub ISBN: 9781119763888; ePDF ISBN: 9781119763871;
Obook ISBN: 9781119763895

Cover Images: © Damsea/Shutterstock; Cover image courtesy of Núria Fontanals
Cover Design: Wiley

Set in 9.5/12.5pt STIXTwoText by Integra Software Services Pvt. Ltd, Pondicherry, India
Printed and bound by CPI Group (UK) Ltd, Croydon, CR0 4YY

C9781119763864_1304022

Contents

Contributors *xi*
Preface *xv*

1 Pesticides *1*
Irene Domínguez, Rosalía López Ruiz, Antonia Garrido Frenich, and Roberto Romero González
1.1 Overview of Pesticides *1*
1.1.1 Properties *1*
1.1.2 Legislation *2*
1.1.3 Reported or Potential Metabolites and/or Transformation Products *3*
1.1.4 Occurrence in the Environment *4*
1.2 Sample Preparation and Collection *9*
1.2.1 Protocols for Collecting and Preparing Samples *9*
1.2.2 Sample Extraction and Clean-up *10*
1.3 Determination of Pesticides *20*
1.3.1 Development of the Instrumental Method *20*
1.3.1.1 Chromatography *20*
1.3.1.2 Detection *22*
1.3.2 Figures of Merit *24*
1.3.3 Hints and Tips *24*
1.4 Future Directions and Challenges *25*
Acknowledgments *26*
Bibliography *26*

2 Pharmaceuticals *37*
Monika Paszkiewicz, Hanna Lis, Magda Caban, Anna Białk-Bielińska, and Piotr Stepnowski
2.1 Overview of Pharmaceuticals *37*
2.1.1 Properties *37*
2.1.2 Reported or Potential Metabolites and/or Transformation Products *37*
2.1.3 Occurrence *39*
2.1.4 Legislation *44*

2.2 Sampling and Sample Preparation *46*
2.2.1 Solid Samples *46*
2.2.2 Water Samples *47*
2.3 Analytical Techniques for the Determination of Pharmaceuticals *51*
2.3.1 Gas Chromatography and Gas Chromatography Coupled to Mass Spectrometry *51*
2.3.2 Liquid Chromatography and Liquid Chromatography Coupled to Mass Spectrometry *55*
2.4 Conclusion and Future Trends *60*
References *60*

3 Personal Care Products *71*
Maria Llompart, Maria Celeiro, and Thierry Dagnac
3.1 Overview of Personal Care Products *71*
3.1.1 Properties *71*
3.1.2 Legislation *72*
3.1.3 Transformation Products *73*
3.1.4 Occurrence in the Environment *73*
3.2 Sample Preparation for PCPs in the Aquatic Environment *74*
3.2.1 Sorbent-based Methodologies *75*
3.2.1.1 Solid-phase Extraction *75*
3.2.1.2 Fabric Phase Sorptive Extraction *99*
3.2.1.3 Stir-bar Sorptive Extraction *100*
3.2.1.4 Solid-phase Microextraction *102*
3.2.2 Liquid-based Extraction Techniques *104*
3.2.2.1 Microextraction Liquid Phase Approaches: DLLME, SDME, USAEME *104*
3.3 Determination of Personal Care Products *107*
3.4 Future Directions and Challenges *108*
Acknowledgements *108*
References *109*

4 New Psychoactive Substances *127*
Noelia Salgueiro-González, Ettore Zuccato, and Sara Castiglioni
4.1 Overview of New Psychoactive Substances *127*
4.1.1 Properties *127*
4.1.2 NPS Market, Dynamics and International Control *130*
4.1.3 Potential Metabolites and/or Transformation Products *131*
4.1.4 Occurrence in the Environment *132*
4.2 Sample Preparation and Collection *133*
4.2.1 Urban Wastewater *133*
4.2.1.1 Protocols for Collecting and Preparing Samples *133*
4.2.1.2 Extraction Procedures and Clean-up *134*
4.2.2 Other Environmental Matrices *136*
4.3 Determination of New Psychoactive Substances *139*
4.3.1 Development of the Instrumental Method *140*

4.3.1.1 Chromatographic Separation *140*
4.3.1.2 Detection *140*
4.3.2 Figures of Merit *142*
4.3.3 Hits and Tips *142*
4.4 Future Direction and Challenges *143*
Acknowledgments *144*
References *144*

5 Artificial Sweeteners *151*
Konstatinos Vasilatos, Maria-Christina Nika, Georgios Gkotsis, and Nikolaos S. Thomaidis
5.1 Overview of Artificial Sweeteners *151*
5.1.1 Properties *151*
5.1.2 Legislation and Environmental Risk Assessment *154*
5.1.3 Reported or Potential Metabolites and/or Transformation Products *155*
5.1.4 Occurrence in the Environment *157*
5.2 Sample Preparation and Collection *167*
5.2.1 Protocols for Collecting and Preparing Samples *167*
5.2.2 Sample Extraction and Clean-up *167*
5.3 Determination of Artificial Sweeteners *172*
5.3.1 Development of the Instrumental Method *172*
5.3.1.1 Chromatography *172*
5.3.1.2 Detection *175*
5.3.2 Figures of Merit *177*
5.3.3 Hints and Tips *178*
5.4 Future Directions and Challenges *180*
References *181*

6 Perfluorinated Substances *187*
Julian Campo and Yolanda Picó
6.1 Overview of Perfluoroalkyl Substances *187*
6.1.1 Properties *190*
6.1.2 Legislation *190*
6.1.3 Reported or Potential Metabolites and/or Transformation Products *191*
6.1.4 Occurrence in the Environment *193*
6.2 Sample Preparation and Collection *197*
6.2.1 Protocols for Collecting and Preparing Samples *197*
6.2.2 Sample Extraction and Clean-up *199*
6.3 Determination of PFASs *201*
6.3.1 Development of the Instrumental Method *201*
6.3.1.1 Chromatography-Mass Spectrometry *201*
6.3.1.2 Biosensors *206*
6.3.2 Figures of Merit *207*
6.3.3 Hints and Tips *209*
6.4 Future Directions and Challenges *210*
References *212*

7 **High Production Volume Chemicals** *223*
Óscar Castro, Eva Pocurull, Francesc Borrull, Rosa Maria Marcé, and Núria Fontanals
7.1 Overview of High Production Volume Chemicals *223*
7.1.1 Properties *223*
7.1.2 Legislation *226*
7.1.3 Reported or Potential Metabolites and/or Transformation Products *227*
7.1.4 Occurrence *228*
7.2 Sample Preparation and Collection *231*
7.2.1 Protocols for Collecting and Preparing Samples *231*
7.2.1.1 Water *231*
7.2.1.2 Air and Dust *231*
7.2.1.3 Soil, Sediments, and Sludge *232*
7.2.1.4 Biota *232*
7.2.2 Sample Extraction and Clean-Up *233*
7.2.2.1 Water *233*
7.2.2.2 Air and Dust *238*
7.2.2.3 Soil, Sediments, and Sludge *244*
7.2.2.4 Biota *247*
7.3 Determination of High Production Volume Chemicals *251*
7.3.1 Development of the Instrumental Method *251*
7.3.2 Figures of Merit *253*
7.3.3 Hints and Tips *253*
7.4 Future Directions and Challenges *253*
Acknowledgments *254*
References *254*

8 **Musk Fragrances** *263*
Irene Aparicio, Julia Martín, Juan Luis Santos, and Esteban Alonso
8.1 Overview of Musk Fragrances *263*
8.1.1 Properties *263*
8.1.2 Legislation *267*
8.1.3 Reported or Potential Metabolites and/or Transformation Products *267*
8.1.4 Occurrence in the Environment *268*
8.1.4.1 Occurrence in Wastewater and Sewage Sludge *268*
8.1.4.2 Occurrence in Surface Water, Soils, Sediments and Air *269*
8.1.4.3 Occurrence in Biota *269*
8.2 Sample Preparation and Collection *270*
8.2.1 Protocols for Collecting and Preparing Samples *270*
8.2.1.1 Air Samples *270*
8.2.1.2 Water Samples *270*
8.2.1.3 Sludge, Soil and Sediment Samples *271*
8.2.1.4 Biota *271*
8.2.2 Sample Extraction and Clean-up *271*

8.2.2.1 Air Samples *271*
8.2.2.2 Water Samples *271*
8.2.2.3 Sludge, Soil and Sediment Samples *276*
8.2.2.4 Biota *276*
8.3 Determination of Musk Fragrances *279*
8.3.1 Chromatography *279*
8.3.2 Detection *280*
8.4 Future Directions and Challenges *280*
References *281*

9 Disinfection Byproducts in Water *287*
Cristina Postigo, Joshua M. Allen, Amy A. Cuthbertson, María José Farré, and Susana Y. Kimura
9.1 Overview of Main DBP Classes *287*
9.1.1 Properties *288*
9.1.2 Legislation *298*
9.1.3 Potential Metabolites and/or Transformation Products *308*
9.1.4 Occurrence in the Environment *310*
9.2 Sample Preparation and Collection *313*
9.2.1 Protocols for Collecting and Preparing Samples *313*
9.2.2 Sample Extraction and Clean-up *314*
9.3 Determination of DBPs *320*
9.3.1 Development of the Instrumental Method *320*
9.3.1.1 Chromatography *320*
9.3.1.2 Detection *328*
9.3.2 Figures of Merit *330*
9.3.2.1 Linearity *330*
9.3.2.2 Precision and Accuracy *331*
9.3.2.3 Sensitivity *332*
9.3.3 Hints and Tips *333*
9.4 Future Directions and Challenges *335*
Acknowledgements *336*
References *336*

10 Microplastics *353*
Marta Llorca and Marinella Farré
10.1 Overview of Micro- and Nanoplastics *353*
10.1.1 Properties *353*
10.1.2 Legislation *353*
10.1.3 Origin and Distribution *354*
10.1.4 Occurrence in the Environment *355*
10.1.4.1 Water Systems *355*
10.1.4.2 Sediments *358*
10.1.4.3 Biota *359*
10.2 Sample Preparation and Collection *360*

10.2.1 Protocols for Collecting and Preparing Samples *360*
10.2.1.1 Water *360*
10.2.1.2 Sediment *360*
10.2.1.3 Biota *360*
10.2.2 Sample Extraction and Clean-up *361*
10.2.2.1 Separation *361*
10.2.2.2 Matrix Removal by Digestion *362*
10.3 Determination of MNPLs *362*
10.3.1 Physical Characterization *362*
10.3.2 Chemical Characterization *362*
10.4 Future Directions and Challenges *363*
Acknowledgments *363*
References *363*

Index *375*

Contributors

Joshua M. Allen
Department of Chemistry
High Point University
North Carolina, USA

Esteban Alonso
Departamento de Química Analítica
Universidad de Sevilla
Sevilla, Spain

Irene Aparicio
Departamento de Química Analítica
Universidad de Sevilla
Sevilla, Spain

Anna Białk-Bielińska
Department of Environmental Analysis
University of Gdansk
Gdansk, Poland

Francesc Borrull
Department of Analytical Chemistry and Organic Chemistry
Universitat Rovira i Virgili
Tarragona, Spain

Magda Caban
Department of Environmental Analysis
University of Gdansk
Gdansk, Poland

Julian Campo
Desertification Research Center (CIDE)
Environmental and Food Safety Research
University of Valencia
Moncada, Valencia, Spain

Sara Castiglioni
Department of Environmental Health Science
Istituto di Ricerche Farmacologiche Mario Negri – IRCCS
Milan, Italy

Óscar Castro
Department of Analytical Chemistry and Organic Chemistry
Universitat Rovira i Virgili
Tarragona, Spain

Maria Celeiro
Department of Analytical Chemistry
Nutrition and Food Science
Universidade de Santiago de Compostela
Santiago de Compostela, Spain

Amy A. Cuthbertson
Department of Civil and Environmental Engineering
University of California
Berkeley, CA, USA

Thierry Dagnac
Agronomic and Agrarian Research Centre (AGACAL-CIAM)
Unit of Organic Contaminants
A Coruña, Spain

Irene Domínguez
Department of Chemistry and Physics
University of Almeria
Almeria, Spain

María José Farré
Catalan Institute for Water Research (ICRA)
University of Girona
Girona, Spain

Marinella Farré
Department of Environmental Chemistry
Institute of Environmental Assessment and Water Research (IDAEA)-CSIC
Barcelona, Spain

Núria Fontanals
Department of Analytical Chemistry and Organic Chemistry
Universitat Rovira i Virgili
Tarragona, Spain

Antonia Garrido Frenich
Department of Chemistry and Physics
University of Almeria
Almeria, Spain

Georgios Gkotsis
Department of Chemistry
National and Kapodistrian University of Athens
Athens, Greece

Susana Y. Kimura
Department of Chemistry
University of Calgary
Calgary, AB, Canada

Hanna Lis
Department of Environmental Analysis
University of Gdansk
Gdansk, Poland

Maria Llompart
Department of Analytical Chemistry, Nutrition and Food Science
Universidade de Santiago de Compostela
Santiago de Compostela, Spain

Marta Llorca
Department of Environmental Chemistry
Institute of Environmental Assessment and Water Research (IDAEA)-CSIC
Barcelona, Spain

Rosalía López Ruiz
Department of Chemistry and Physics
University of Almeria
Almeria, Spain

Rosa Maria Marcé
Department of Analytical Chemistry and Organic Chemistry
Universitat Rovira i Virgili
Tarragona, Spain

Julia Martín
Departamento de Química Analítica
Universidad de Sevilla
Sevilla, Spain

Maria-Christina Nika
Department of Chemistry
National and Kapodistrian University of Athens
Athens, Greece

Monika Paszkiewicz
Department of Environmental Analysis
University of Gdansk
Gdansk, Poland

Yolanda Picó
Desertification Research Center (CIDE)
Environmental and Food Safety Research
University of Valencia
Moncada, Valencia, Spain

Eva Pocurull
Department of Analytical Chemistry and Organic Chemistry
Universitat Rovira i Virgili
Tarragona, Spain

Cristina Postigo
Department of Environmental Chemistry
Institute of Environmental Assessment and Water Research (IDAEA)-CSIC
Barcelona, Spain

Roberto Romero González
Department of Chemistry and Physics
University of Almeria
Almeria, Spain

Noelia Salgueiro-González
Department of Environmental Health Science
Istituto di Ricerche Farmacologiche Mario Negri – IRCCS
Milan, Italy

Juan Luis Santos
Departamento de Química Analítica
Escuela Politécnica Superior
Universidad de Sevilla
Sevilla, Spain

Piotr Stepnowski
Department of Environmental Analysis
University of Gdansk
Gdansk, Poland

Nikolaos S. Thomaidis
Department of Chemistry
National and Kapodistrian University of Athens
Athens, Greece

Konstatinos Vasilatos
Department of Chemistry
National and Kapodistrian University of Athens
Athens, Greece

Ettore Zuccato
Department of Environmental Health Science
Istituto di Ricerche Farmacologiche Mario Negri – IRCCS
Milan, Italy

Preface

The occurrence of organic compounds of emerging concern (CECs) in the environment has given rise to growing unease. Although most CECs have been present in the environment for years, some of them have only recently been identified. Thus, some of these compounds remain unregulated and their toxicity and risk unstudied. They occur in different environmental compartments at trace levels (low ppb or ppt). Moreover, CECs belong to different families with several origins and have a broad range of physical-chemical properties. To respond to this situation, sensitive and selective analytical methods have been developed to determine different types of CECs in a broad range of environmental samples. Most of these analytical methods comprise sample preparation followed by chromatographic separation coupled to mass spectrometry. The specific requirements at each step of the analytical method are linked to the properties of the contaminants and sample matrix.

The aim of this book was to compile the analytical methods available to determine CECs in environmental samples. It is organized by families of CECs, with each chapter presenting complete information for each family. The chapters are dedicated to the following families of CECs: pesticides (Chapter 1); pharmaceuticals (Chapter 2); personal care products (Chapter 3); novel psychoactive substances (Chapter 4); artificial sweeteners (Chapter 5); perfluorinated substances (Chapter 6); high production volume chemicals (Chapter 7); musk fragrances (Chapter 8); disinfection byproducts (Chapter 9); and microplastics (Chapter 10). The relevant environmental samples are covered in each chapter (i.e., family of CECs): different types of water (drinking, tap, surface, and sewage), solids (sediment, soil, sludge, dust, as well as biota), and atmospheric samples (air and particulate matter). With a few exceptions, the content of each chapter is organized into the same sections – an overview, sample preparation and collection, determination, and future directions. In each chapter, different analytical methods are proposed through examples of cutting-edge research studies that are summarized in the form of tables to provide an overview of the state of the art; however, the common methods currently employed are also described so that the reader can identify the methods typically used. Thus, this book provides a detailed description of the different approaches for determining each group of CECs in environmental samples with distinctive, critical information about the state of the art for their determination in the environment. The editors truly hope that this book will serve to help researchers, industrial experts, students, decision-makers, and interested members of society learn about the approaches available to monitor each family of CECs, covering most of the CECs occurring in the environment.

We would like to acknowledge the authors for their invaluable contributions. All of the authors of this book are experts and skilled professionals in environmental trace analysis and, in particular, in each of the families of CECs addressed in each specific chapter. Without their effort and dedication, this book would have not been possible. We also sincerely thank our readers for their interest in this book. Thanks for reading and we hope you enjoy it!

Núria Fontanals and Rosa Maria Marcé
October 2021

1

Pesticides

Irene Domínguez, Rosalía López Ruiz, Antonia Garrido Frenich, and Roberto Romero González

Department of Chemistry and Physics, Andalusian Center for the Assessment and Monitoring of Global Change (CAESCG), University of Almeria, Agrifood Campus of International Excellence, Almeria, Spain

1.1 Overview of Pesticides

There are thousands of chemicals that can enter the environment and pesticides are among the most significant. They have been used in the last decades in several areas, but agricultural activity is the main source of the impact of pesticides in the environment and therefore they can be present in soil, water, crops as well as in the atmosphere [1].

1.1.1 Properties

Pesticides are a class of chemicals used to limit, inhibit or prevent the growth of harmful animals, insects, weeds or fungi [2]. They can be classified according to different criteria, such as target organism, origin or chemical structure, but the most common is considering the target organism, being herbicides, fungicides, insecticides, etc., among others [3]. There are more than 800 active components and they are available in different accessible products [4]. Despite the benefits of the use of these compounds, they can be toxic to humans and many of them have been classified as endocrine disruptors, and carcinogenic effects have also been reported [5].

The widespread use of pesticides in combination with their physico-chemical properties, such as water solubility, octanol/water partition coefficient, volatility and stability against degradation by abiotic and biotic factors, are the reasons for their distribution and occurrence in different environmental matrices such as water, soil, air and biota [6] by physical processes as sedimentation leaching, sorption and volatilization.

Once pesticides are in the environment, they can be transformed by biotic or abiotic process [7], increasing the number of potential transformation products (TPs) that can be detected, and most of them are still unknown [8]. In this sense, TPs could also have environmental concern and so in addition to the parent compounds they should also be monitored in order to get a comprehensive overview of the environmental fate of pesticides.

Analytical Methods for Environmental Contaminants of Emerging Concern, First Edition. Edited by Núria Fontanals and Rosa Maria Marcé.

1.1.2 Legislation

The presence of these pollutants poses a potential risk for the environment and human health and, therefore, international organizations have set legal limits regarding the presence of pesticides in water and other environmental matrices, for controlling and preventing contamination of environmental ecosystems.

For instance, in Europe, the Water Framework Directive (WFD) is intended to protect transitional waters, inland surface waters, coastal waters and groundwater. Strategies against the chemical pollution of surface waters led to the Directive 2008/105/EC [9], establishing concentration limits of 33 priority substances and 8 other pollutants, including some pesticides such as simazine and trifluralin. Priority substances are considered to pose a significant risk to or via the aquatic environment, so environmental quality standards (EQSs) were set for each of them. Then, amending Directive 2013/39/EU [10] introduced 12 new compounds to the list and the need to establish an additional list of potential water pollutants (Watch List) that should be carefully monitored to support future reviews of the priority substances list. Currently, among the priority substances are 24 pesticides with Annual Average EQS (AA-EQS) values ranging from 1×10^{-8} μg l^{-1} for heptachlor and heptachlor epoxide to 1 μg l^{-1} for simazine. In 2020, the European Union (EU) established a new Watch List of substances, including azole compounds and providing maximum acceptable method detection limits for them from 29 to 199 ng l^{-1} [11]. Additionally, the Drinking Water Directive 98/83/EC, amended by EU 2015/1787 [12], set special quality requirements for water for human consumption. It set concentration limits for a range of hazardous substances, including pesticides, establishing a general maximum individual concentration of 0.1 μg l^{-1} for individual pesticides (0.030 μg l^{-1} in the case of aldrin, dieldrin, heptachlor and heptachlor epoxide) and 0.5 μg l^{-1} for the sum of all individual pesticides and relevant metabolites/TPs detected. The same values, 0.1 and 0.5 μg l^{-1}, for individual and total pesticides respectively, are established as groundwater quality standards in Directive 2006/118/EC [13] on the protection of groundwater against pollution and deterioration.

In the same way, the Clean Water Act (CWA) in the United States (US) establishes the basic structure for regulating quality standards for surface waters discharges of pollutants into the waters. In addition, the Safe Drinking Water Act (SDWA) was aimed at protecting drinking water and its sources (rivers, lakes, reservoirs, springs and groundwater wells). SDWA authorizes the US Environmental Protection Agency (US EPA) to set national health-based standards for drinking water to protect against contaminants, such as pesticides, that may be found in drinking water [14–16]. In this case, the proposed substance priority list is based on a combination of their frequency, toxicity and potential for human exposure at National Priorities List (NPL) sites, setting criterion maximum concentration (CMC) values for each of the pollutants listed. Aldrin, dieldrin, heptachlor and heptachlor epoxide show the lowest CMC values, between 7.7×10^{-7} and 3.2×10^{-5} μg l^{-1}.

Whereas different countries have set pesticide regulation in water matrices, regulation in soils is scarce. For instance, Spain set generic reference levels for a limited number of substances (< 60), some of them considered as persistent organic contaminants, such as dichloro-diphenyl-trichloroethane (DDT) or dichloro-diphenyl-dichloroethane (DDE), whose reference levels were 0.2 mg kg^{-1} and 0.6 mg kg^{-1} respectively [17]. These reference levels, in terms of human protection, are the maximum concentration of a substance in the soil that guarantee that contamination does not pose an unacceptable risk to humans. In addition to complying with generic reference levels, it is necessary to determine through toxicological tests that these substances do not present a serious risk to the ecosystem.

1.1.3 Reported or Potential Metabolites and/or Transformation Products

Pesticides in the environment may experience different chemical reactions, leading to the appearance of TPs and metabolites. These compounds have potentially harmful impacts on organisms, even more than their precursors [18], making their monitoring essential. However, because of the great variety of TPs, it is difficult to carry out a comprehensive analysis of their presence, and in consequence, a risk assessment evaluation.

The metabolic/transformation pathways of pesticides can be affected by biological or/and physico-chemical factors in the environment [19]. Hydrolysis is an important degradation mode of pesticides; however, multiple TPs may be produced from different processes, even after hydrolysis [20, 21].

It was noted that the number of substances that must be considered for environmental risk significantly multiplied by a factor of 7.5, just when the precursor compounds were subjected to a photolysis process [22]. This has been observed for terbutryn, mecoprop, penconazole, boscalid diuron and octhilinone pesticides in Figure 1.1, where the number of compounds that should be monitored in environmental samples considerably increase because of the presence of TPs. After evaluating genotoxicity of the proposed TPs, it was suggested that the number of substances that pose a risk onto the aquatic environment increased by a factor of >4. This fact, together with the high incidence of TPs and metabolites in natural waters, constitutes a major concern that needs to be addressed from an analytical and legislative point of view.

Several studies have revealed the presence of TPs and metabolites in waters at higher concentrations than the parent compounds [23, 24]. The physico-chemical properties (higher mobility and polarity) of the TPs and metabolites might facilitate the migration between surface water and groundwater. Since groundwater is the greatest source of freshwater in the world, the occurrence of some relevant metabolites and/or TPs led to the restriction in the use of certain pesticides, as was recently the case for chlorothalonil and previously simazine and atrazine, among others. Most of the TPs/metabolites found in natural waters are related to acetanilide and triazine herbicides [25]. Such is the case for ethanesulfonic acid (ESA) and oxanilic acid (OA), degradation products of alachlor, metolachlor, as well as acetochlor, and atrazine-desethyl (DEA), atrazine-desisopropyl (DIA), terbumeton-desethyl (TED), terbuthylazine-desethyl (TD) and terbuthylazine-2-hydroxy (T2H). Different analysis has also revealed the occurrence of 2,6-dichlorobenzamide (BAM) from dichlobenil, aminomethyl phosphonic acid (AMPA) from glyphosate, desphenyl chloridazon and methyldesphenyl chloridazon from the herbicide chloridazon and N,N-dimethylsulfamide (DMS) formed from the fungicide tolylfluanid [23, 25–27].

Metabolites were also detected in soils, especially when dissipation studies have been carried out. For instance, nine metabolites of famoxadone were detected in soil samples [28], with IN-JS940 the metabolite detected at the highest percentage in relation to the parent compound, as can be observed in Figure 1.2. Therefore, risk assessment is needed to evaluate

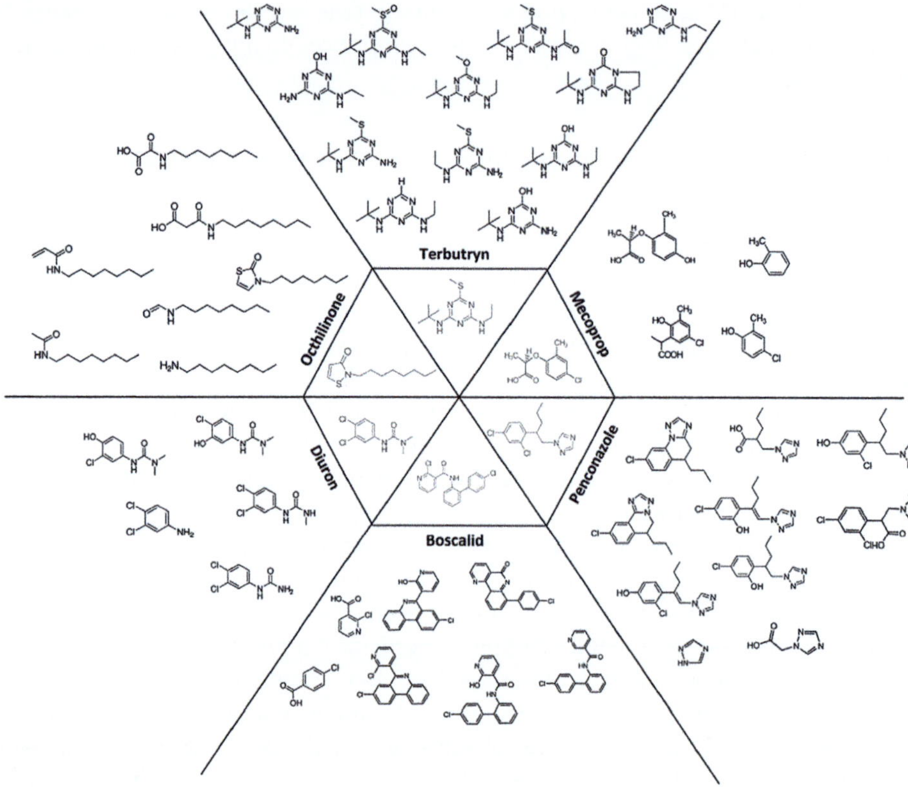

Figure 1.1 Forty-five TPs originating from six pesticidal parent compounds. Illustration of the multiplication of known substances that should be further investigates by an environmental risk assessment. Source [22]. Reproduced with permission of Elsevier B.V.

potential hazards to the fauna and flora. Tiwari et al. [29] evaluated the presence of endosulfan and chlorpyrifos metabolites in soils because of the higher toxicity of some of these compounds as chlorpyrifos oxon. They determined that metabolite concentrations increased throughout the study when the concentration of the parent molecule decreased. Moreover, it was observed that concentration of metabolites was higher in soil matrices than in water. In the same way, when 2,4-dichlorophenoxyacetic acid (2,4-D) is applied on crops or on soil, it will undergo chemical, biological and physical degradation processes depending on the environmental factors, which will determine the metabolites formed. For example, 2,4-D DMA (2,4-D dimethylamine salts) is dissociated to 2,4-D acid after its application on the soil [30], so in addition to the parent compound, different TPs should be monitored.

1.1.4 Occurrence in the Environment

Pesticides and their TPs/metabolites are widely distributed in the environment and they can be detected in water, soil, sediments, aquatic biota and air [31], as can be observed in Figure 1.3, because of surface runoff from arable lands, leaching from drainage systems, volatilization, etc. They can have toxic effects in population living close to these areas [32]. For instance, when pesticides are applied in agricultural areas, approximately

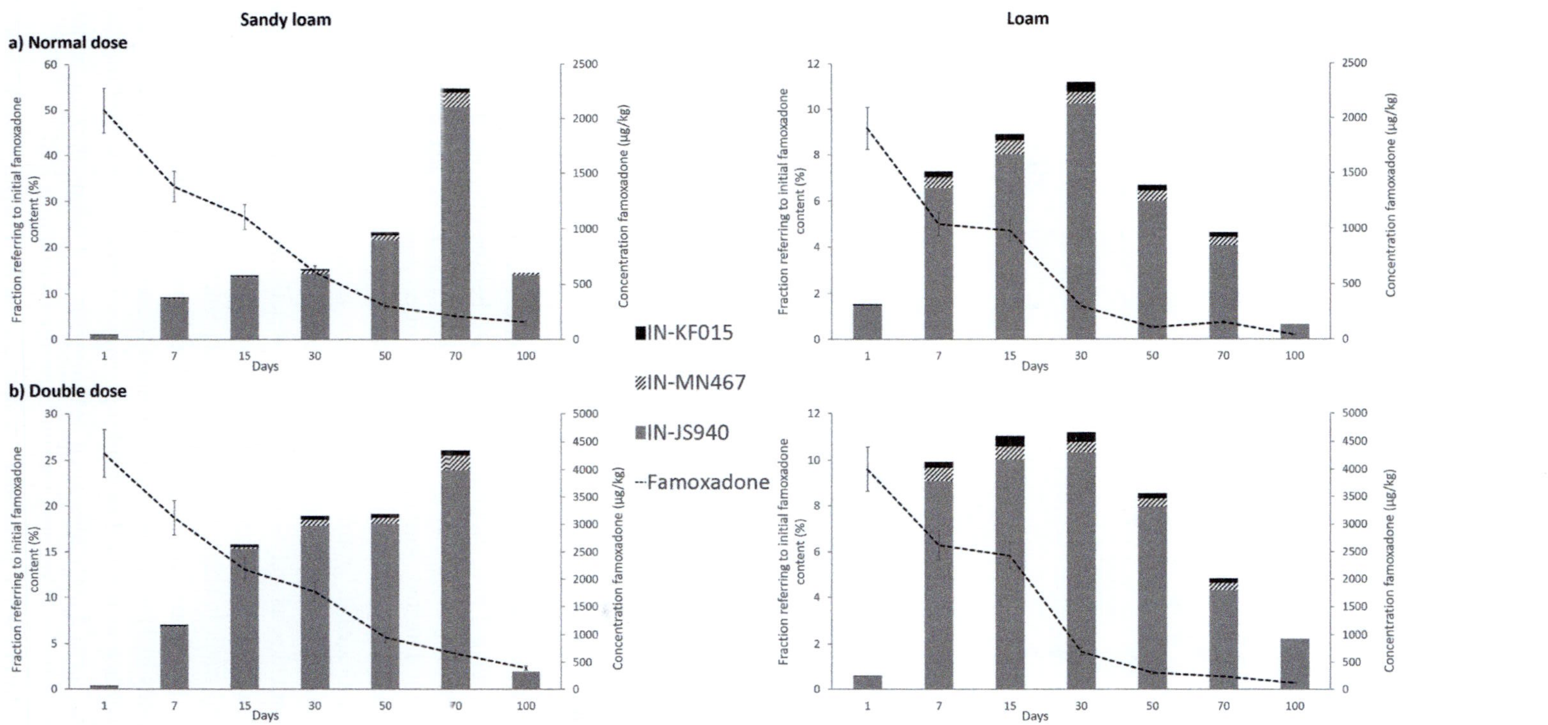

Figure 1.2 Metabolite behavior according to the concentration of famoxadone during monitoring period (100 day) for soil experiments at: (a) normal dose (2.4 mg g^{-1} soil) and (b) double dose (4.8 mg g^{-1} soil). Source [28]. Reproduced with permission of Elsevier B.V.

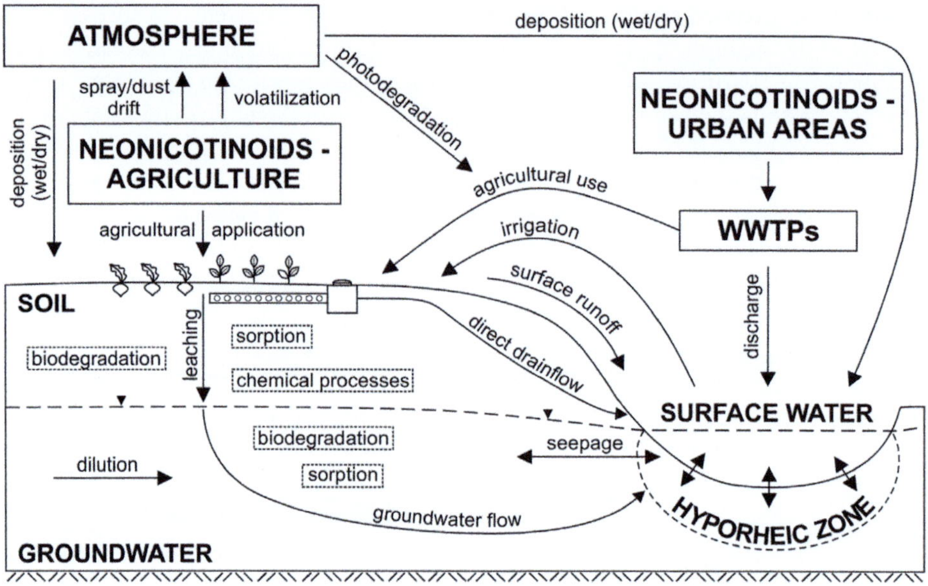

Figure 1.3 Neonicotinoid insecticides in the environment: sources, pathways, receptors and related process. Source [31]. Reproduced with permission of Elsevier B.V.

20–30% of the amount is lost due to the spray drift process, whereas another significant fraction is placed into the soil or surface waters [33].

Pesticides have been widely detected in the aquatic media. Several scientific papers and technical reports have revealed the presence of pesticides in surface water, groundwater, drinking water as well as treated wastewater which is intended to be discharged in surface waters [34].

In 2018 the European Environment Agency (EEA) indicated that chemical status of surface waters, related to pesticides, has improved between the 1st and 2nd River Basin Management Plants (RBMPs) assessments [35]. However, pesticides listed as priority substances were still detected. Among them, isoproturon and hexachlorocyclohexane were the most frequently reported, followed by endosulfan and chlorpyrifos. Furthermore, the presence of pesticides in groundwater is a failure to achieve good chemical status. As reported by EAA, several groundwater bodies exceed the permitted concentrations set by EU of total pesticides, as well as the level set for desethylatrazine, atrazine and simazine.

Herbicides such as atrazine and fungicides such as metalaxyl have been detected in aquifers for drinking water in US [25]. The most found pesticide in drinking water in Ireland was 2-methyl-4-chlorophenoxyacetic acid (MCPA), widely used for rush control in grassland [36].

A recent study in The Netherlands, intended to assess the occurrence of pesticides in ground and surface water used as drinking water sources, has revealed the detection of 15 recently authorized pesticides, such as fluopyram and thiamethoxam, which demonstrates the importance of keeping routine monitoring methods [36].

Non-agricultural uses of pesticides are also common in urban environments, such as indoor uses or applications in gardens, roads and sealed areas, among others. As a result, triazine herbicides (atrazine, simazine and prometon) and organophosphate insecticides (chlorpyriphos and diazinon) are frequently detected in U.S. urban surface waters [34].

Pesticide occurrence in soils was inversely related to the occurrence in surface water, and one of the main factors is the hydrophobicity of the compounds, expressed by the n-octanol/water partition coefficient (K_{ow}). Thus, pesticides with log $K_{ow} > 3$ were more detectable in soils than in water. In a study performed in Argentine ecosystem, it was observed that the most detected herbicide residues (>30% of detection frequency) were acetochlor, atrazine and its metabolite (hydroxyatrazine), flurochloridone, glyphosate and its metabolite (aminomethylphosphonic acid, AMPA) and metolachlor, whereas the most frequently detected insecticides were chlorpyrifos and imidacloprid [37]. In this study, glyphosate and its metabolite AMPA were also found in all the environmental samples (soil, sediments and surface water), detecting AMPA up to 713 mg kg^{-1}, glyphosate at 32 mg kg^{-1} and 5 mg l^{-1} in sediments and water respectively.

Another group of pesticides really known for their persistence are 2,4-D-based herbicides. The two chlorine that are present in their molecules confers persistence with an estimate half-life between 7 and 312 days, which depends on the environmental conditions [38]. Several studies have reported the presence of 2,4-D and derivates in the environment because of different activities, such as agricultural activities and rain and irrigation water. Discharges from manufacturing plants, leaching and accidental spills [39] represent an important source of the aforementioned herbicide.

The key properties that determine the presence of pesticides and their accumulation in biota are hydrophobicity and persistence. Thus, if water solubility is <1 mg l^{-1} or log K_{ow} is >3, they have the potential to accumulate in biota and it may be an indicator of contamination [40]. For instance, earthworms are used as indicators to the response of pesticides [41], but fish are considered as major reservoirs of pesticides and their concentration in fish tissues has been used as an indicator of bioaccumulation [42]. High concentrations of organochlorine pesticides (OCPs) have been found in fish tissues from aquatic environments in Africa [43], detecting DDTs, lindane, mirex and endosulfan at concentrations up to 100 μg kg^{-1}. Other studies, developed in South-America (southeastern Brazil and the coast of the Argentinean Pampas), showed that lindane and endosulfan levels could cause long-term or short-term damage to biota [44]. Additionally, it was observed that OCPs were highly bioaccumulated in soil mesofauna (up to 260 μg g^{-1}) [45].

In the last years, several river basins were monitored in the Iberian Peninsula, and in addition to water and sediment, biota was also analyzed. Thus, 50 pesticides were monitored in Guadalquivir river basin, but none of them were detected in biota, although there were detected in water and sediments, especially organophosphorus and triazines at concentrations up to 13 ng l^{-1} in water and 13.2 ng g^{-1} dry weight (dw) in sediment [46]. However, azinphos-ethyl, chlorpyriphos, diazinon, dimethoate and ethion were detected in different species of fish from Jucar river basin, observing that the maximum average concentration was found in European eels (up to 0.024 ng kg^{-1}), whereas in water, dichlofenthion, imazalil pyriproxyfen and prochloraz, commonly used in farming activities, were mainly detected [47]. In biota from Llobregat river basin, chlorpyrifos and azinphos-ethyl were detected at 44.75 ng g^{-1} dw and 105.81 ng g^{-1} dw, indicating possible bioaccumulation. Nevertheless, authors concluded that these values do not represent a high risk to biota. On the other hand, triazines, organophosphorus and neonicotinoids were mainly detected in water, at concentrations >600 ng l^{-1}, as can be observed in Figure 1.4, with chlorpyrifos the compound most widely detected in soils, at concentrations up to 130 ng g^{-1} dw [48]. Additionally, the Ebro river basin was also monitored, detecting imazalil and diuron at the highest concentrations in water (410 and 150 ng l^{-1} respectively), chlorpyrifos, diazinon

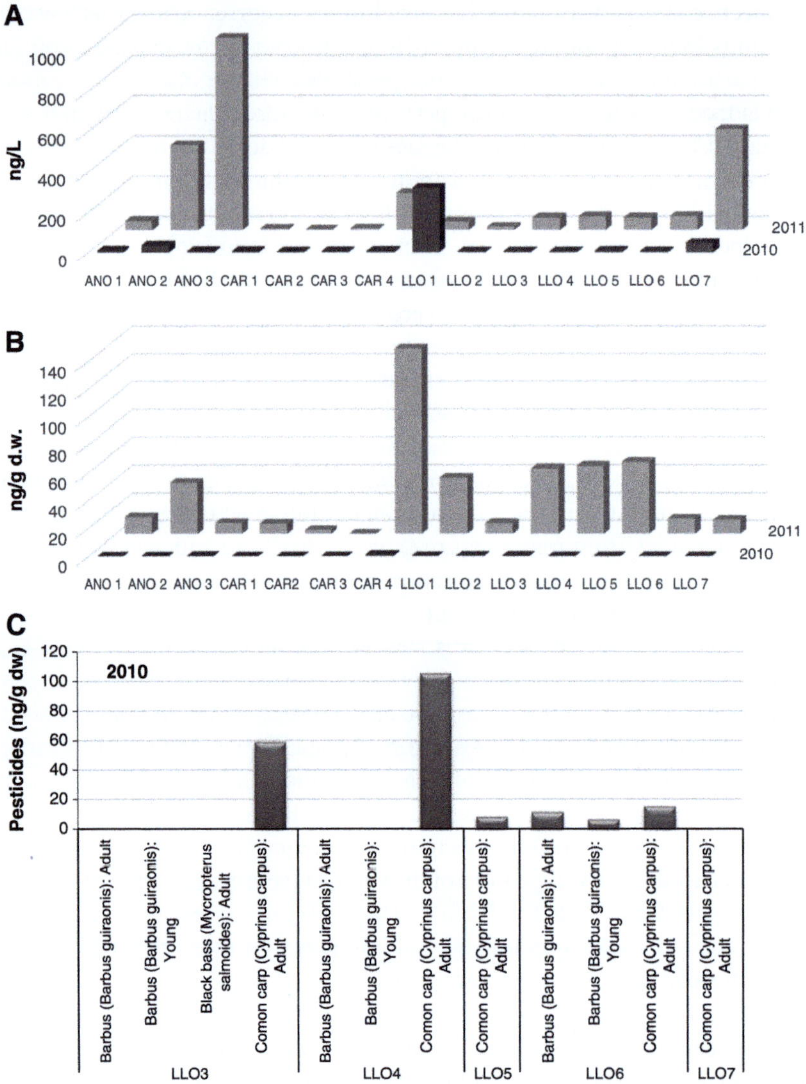

Figure 1.4 Pesticide families detected in (A) water, (B) sediment and (C) fish samples from Llobregat basin in 2010 and 2011 according to the sampling point. Source [48]. Reproduced with permission of Elsevier B.V.

and diclofenthion in sediments, whereas the only compound detected in biota was chlorpyrifos, which was detected at concentrations up to 840.2 ng g^{-1} [49].

The presence of pesticides in air depends on the specific agricultural areas, as well as those pesticides that can be transported from nearby places, observing a relationship between the concentration of some of the compounds detected in air and diseases and mortality [50].

Several studies were performed to monitor the presence of pesticides in outdoor and indoor air [51], bearing in mind that volatile pesticides, such as OCPs, can largely remain in the atmosphere after volatilizing from contaminated soils and water. For instance, p,p'-DDT and p,p'-DDD were detected in African air at concentrations of 47 pg m^{-3} and

12 pg m^{-3} respectively, whereas other compounds as dieldrin, hexachlorobenzene (HCB), aldrin, lindane and chlordane were also detected but at lower concentrations [52].

In a study developed in Vietnam, 452 pesticides were included in a database, detecting 18 (12 insecticides, 4 herbicides and 2 fungicides) in the collected air samples, and the total concentration ranged from 3.35 to 89.0 ng m^{-3}, with the most detected pesticides being permethrin, carbofuran, fenobucarb and chlorpyrifos as well as metolachlor [53]. Some prohibited pesticides in the EU, such as chlorpyrifos, permethrin, deltamethrin, cypermethrin and carbofuran, were also detected [53]. In the same country, 26 pesticides (13 pesticides, 7 fungicides and 6 herbicides) were detected in the air samples, at higher concentrations, ranging from 43 to 370 ng m^{-3}. Permethrins, chlorpyrifos and propiconazole were detected and this may result from their widespread use for both agricultural and domestic purposes in rural areas [54].

Due to their high persistence, OCPs were monitored in addition to current-use pesticides (CUPs) in two National Parks in the Rio de Janeiro State. The highest concentrations of endosulfan (up to 3202 pg m^{-3}), cypermethrin (881 pg m^{-3}) and chlorpyrifos (270 pg m^{-3}) indicated background air levels of OCPs. On the other hand, CUPs seemed to behave like pseudo-persistent organic pollutants (POPs) although it was believed that they are not persistent in the environment [55].

In another study, carbendazim, metalaxyl, myclobutanil and terbuthylazine were detected in air from rural areas of Valencia (Spain) at concentrations ranging from 16 to 174 pg m^{-3} [56].

Bearing in mind that pesticides can easily adhere to the particulate matter, they have also been analyzed in this matter. Thus, 40 CUPs were determined in PM10 of Valencia Region (Spain) observing that omethoate was detected at the highest average level (141 pg m^{-3}) [57]. A similar study was performed in the "Todos los Santos Bay" region (Brazil), where 13 pesticides were analyzed in PM2.5 samples, and concentrations ranged from 20 to 315 pg m^{-3}, being carbofuran, malathion and permethrin, the compound most widely detected [58].

1.2 Sample Preparation and Collection

In order to ensure accurate pesticide determination in environmental samples, sample collection and preparation should be carefully designed to minimize potential errors during these steps and therefore error propagation will be limited.

1.2.1 Protocols for Collecting and Preparing Samples

Sample collection is critical to get representative samples as well as to avoid any modification of the initial chemical composition of the sample.

For instance, a guidance on sampling water techniques can be found in the ISO Standard on Water Quality – Sampling 5667. The selection of sampling point, which should cover the area of surveillance, must be subject to local conditions, such as water homogeneity and vertical and lateral mixing.

The sample containers as well as transport and storage arrangements should not lead to changes in the relevant chemical status. Therefore, according to sampling for synthetic organic compounds, water samples must be stored in glass, polytetrafluoroethylene

(PTFE) or stainless-steel containers, and they should be analyzed within 24 hours and stored in the dark at 1–5°C.

On the other hand, the development and application of passive sampling techniques are highly recommended [59]. Those techniques allow for the accumulation of pesticides by passive diffusion onto a liquid or solid absorbent showing affinity for a certain type of substance. The semipermeable membrane device (SPMD) and the polar organic chemical integrative sampler (POCIS) are the most common passive samplers for organic pollutants. SPMD based on triolein sorbent can be applied for neutral organic chemicals with a log $K_{ow} > 3$ (lipophilic pesticides), while POCIS, which uses Oasis HLB phase, is intended for the sorption of more water-soluble organic chemicals with $K_{ow} < 3$ (polar pesticides). To ensure the monitoring of a high number of pollutants, different types of passive samplers could be used together [6]. In this sense, an interlaboratory study on passive sampling of emerging water pollutants showed low interlaboratory variability in the analysis of replicate samplers when POCIS was used for the monitoring of 7 polar pesticides. The same study established a series of recommendations to take into consideration, especially when passive samplers are combined with liquid chromatographic-mass spectrometric (LC-MS) methods [60].

In the same way, soil samples should be collected from growing fields using a grid pattern uniformly distributed. For instance, a 3 × 3 grid is commonly used for smaller fields, whereas 5 × 5 or even larger grids are used for very large fields and a "W" or "Z" pattern is commonly used. Each sample site represents one portion of the total sample, and then a composite sample can be formed, ensuring homogeneity. Normally, samples are collected at a 15 cm depth. Moreover, additional steps such as removing litter, plant roots and big stones from the soil samples could be needed [61] and sometimes soils should be dried with [62] or without heat [63], homogenized and finally stored at −18°C until analysis [64]. Finally, an exhaustive characterization of the soils according to several parameters, such as pH, percentage of organic matter, carbon monoxide, sand, silt, clay and grit [28, 65], is advisable to provide useful information and set a relationship between pesticide presence and physico-chemical characteristics of the soil.

In relation to air sampling, other important criteria should be considered as materials do not react with target compounds; it must be located in a place where free air masses can reach; the sampler should be protected from rainfall, dust or other sources of contamination as well as requiring low maintenance [33]. Furthermore, it should be planned bearing in mind whether only gas phase, particulate matter or both of them are going to be collected, among other factors.

Finally, for biota sampling, special care should be taken with small biota samples, which are more sensitive to contamination, and degradation of loss of analytes. Long-term storage should be performed in darkness and low temperature, and before sample treatment, dry or wet homogenization is needed [66].

1.2.2 Sample Extraction and Clean-up

For the monitoring of ultra-trace levels of pesticides in water, extraction and concentration steps are required prior to the analytical determination. Liquid-liquid extraction (LLE) allows for the detection of a large range of non-polar pesticides while requiring minimal instrumentation, being at the same time both a simple and a precise technique. Although less and less, this technique is still used for the extraction of pesticides from water samples prior to gas chromatography (GC) analysis [67, 68]. However, one major drawback is the large solvent volumes, usually dichloromethane, required in LLE. Therefore, solid phase extraction (SPE) has become the most common extraction technique [69]. Indeed, it is the most powerful sampling and enrichment approach for complex mixtures of known and unknown contaminants, and different sorbent phases can be used, allowing for the extraction of a wide range of pesticides with different physico-chemical properties.

As shown in Table 1.1, the polymeric reversed phase sorbent, Oasis HLB, is commonly used for the extraction of pesticides in water samples providing quantitative recoveries in most cases [18, 23, 70–72]. Furthermore, other sorbents have been successfully applied for the analysis of pesticides and their TPs in natural waters, such as a mixture of hydrophilic–lipophilic balance, weak anion and cation exchange sorbents (2 : 1 : 1, w/w/w) [73], and Strata-X reversed in combination with the mixture of Strata-X mixed-mode (AW and CW) and Isolute ENV + [74].

On-site integrated large-volume SPE has also been proven to be a promising tool for the monitoring of pollutants, including pesticides, in water sources [75]. On the other hand, the possibility of using on-line SPE systems allows for minimizing sample manipulation and cross-contaminations as well as improving sample throughput [76]. Hence the development of fully automated methods, based on the combination of on-line SPE and LC-MS, has been given much attention in the last few years, being applied for the monitoring of ca. 100 pesticides in natural and drinking waters [24, 77], as well as 51 pesticides, covering highly polar compounds, in surface and groundwaters [78].

Simple and miniaturized sample preparation techniques have been considered in recent years as optimal alternatives [79]. Among them, solid phase microextraction (SPME) is the most used technique, although the application of QuEChERS (Quick, Easy, Cheap, Effective, Rugged and Safe)-based protocols [21, 80], stir bar sorptive extraction (SBSE) [77], as well as liquid-phase microextraction (LPME) [81, 82], has also been suitable for the extraction of pesticides from water matrices.

SPME is a simple, sensitive, rapid and solvent-free technique in which the organic compounds are adsorbed/absorbed (depending on fiber coating) directly from the aqueous sample into the fiber and then thermally desorbed at the injection port of the GC, considerably simplifying the analysis procedure. In this sense, the availability of SPME devices in latest GC equipment leads to the complete automatization of the analytical process, allowing for improving data quality, the productivity of staff and instruments, and increasing the sample throughput [83]. This has been demonstrated in recent methodologies involving the on-line combination of SPME and GC coupled to high-resolution mass spectrometry (HRMS) allowing for the determination of priority substances, including pesticides, in surface and wastewaters [84, 85] providing limits of quantification (LOQs) at ng l^{-1} levels. Novel SPME sorbents, such as magnetic deep eutectic solvent (DES)-based polymeric hydrogel [86] and carbon nanomaterials [87, 88], have been successfully applied for the monitoring of pesticides in different water resources as can be seen in Table 1.1.

On the other hand, vacuum-assisted evaporative concentration has been effective for the monitoring of organic micropollutants [89], including pesticides such as chlorothalonil and TPs (sulphonic acids and phenols) by LC-Orbitrap-MS [90]. However, sometimes these extraction procedures are avoided, and direct injection of water samples, after filtration, can be considered when using LC. In fact, its combination with tandem

Table 1.1 Overview of analytical methods applied to monitor pesticides in environmental waters.[a]

Pesticides	Matrix	Extraction technique	Determination technique	Recovery (%)	LOQ ($\mu g\ l^{-1}$)	Reference
4 OCP and 2 OPPs	Surface water	LLE (dichloromethane)	GC-FID	80–90	0.002	[68]
296 pesticides + 156 pharmaceuticals, 18 consumer products, 10 industrial chemicals and 4 others	Coastal waters	SPE (Oasis HLB and SpePak cartridges)	LC-QTOF-MS	70–130	0.00002–0.300[b]	[70]
14 pesticides and TPs	Surface water and drinking water	SPE (Oasis HLB cartridge)	UHPLC-QTrap-MS/MS	85–105	0.01–0.1	[71]
19 acidic herbicides + metabolites	River water	SPE (Oasis HLB cartridge)	LC-QqQ-MS/MS	64–111	0.004–0.022	[72]
6 neonicotinoids and metabolites	Drinking water	SPE (Oasis HLB cartridge)	LC-DAD-QqQ-MS/MS & QTOF-MS	57–120	0.000057–0.000488[b]	[18]
ca. 500 pesticides and TPs	Surface water and groundwater	SPE (Oasis HLB cartridge)	UHPLC-QTOF-MS	—	—	[23]
8 pesticides + TPs	Surface water	Mix mode SPE: (HLB: WAX: WCX, 2 : 1: 1)	LC-QqQ-MS/MS	43–141	0.00002–0.0056[b]	[73]
125 pesticides and metabolites + 130 pharmaceuticals and metabolites + 42 antibiotics and metabolites + 63 others	Surface and marine water	SPE (Strata-X and the mixture Strata-X-AW: Strata-X-CW: Isolute ENV + (1 : 1 : 1.5))	LC-LTQ-Orbitrap-MS	83–93	—	[74]

251 contaminants (pesticides, pharmaceuticals or industrial chemicals and their transformation products)	Surface water	Onsite integrative large-volume SPE (HR-X sorbent)	UHPLC-LTQ-Orbitrap MS	60–123	—	[75]
96 including pesticides and TPs	Surface water, groundwater and drinking water	On-line SPE	UHPLC-QqQ-MS/MS	—	0.005–0.025	[24]
51 pesticides	Surface water and groundwater	On-line SPE (Prospekt-2-system)	LC-QqQ-MS/MS	80–125	0.010	[78]
8 pesticides	Surface water and groundwater	QuEChERS (Acetonitrile, $MgSO_4$ and NaCl)	GC-Q-MS	85–103	0.95–13.69	[80]
Cyflumetofen + 2 metabolites	Surface water	QuEChERS (Acetonitrile, $MgSO_4$ and NaCl)	UHPLC-QqQ-MS/MS	79–118	0.7–9.8	[21]
102 pesticides	Surface water and groundwater	SBSE (PDMS) (GC) On-line SPE (LC)	GC-Q-MS (27) UHPLC-QqQ-MS/MS (75)	—	0.015–0.025 (GC) 0.005–0.025 (LC)	[77]
10 pesticides	Surface water	HF-LPME	GC-Q-MS	85–115	0.14–1.69	[82]
14 pesticides + 16 PAHs + 26 PCBs + 6 BDEs	Surface water	On-Line SPME (DI, PA fiber)	GC-DFS-HRMS	87–116	0.0001–0.050	[84]
16 pesticides	Surface water, marine water and groundwater	Magnetic SPME with a magnetic DES-based polymeric hydrogel	GC-μECD	61–120	0.006–0.399	[86]
24 pesticides	Surface water	SPME (Novel carbon nanomaterial sorbent)	GC-Q-MS	70–123	0.0007–3.7320	[87]
18 chiral pesticides	Surface water and influent and effluent wastewater	Magnetic SPME (Amino modified multiwalled carbon nanotubes)	LC-QqQ-MS/MS	83–105	0.00035–0.00204	[88]

(Continued)

Table 1.1 (Continued)

Pesticides	Matrix	Extraction technique	Determination technique	Recovery (%)	LOQ (μg l^{-1})	Reference
Chlorothalonil + 6 TPs	Surface water and groundwater	Vacuum-assisted evaporative concentration	LC-Orbitrap-MS	85–110	0.0002–0.010	[90]
215 pesticides and TPs (Method SH2437) 30 pesticides (Method LC9045) 3 herbicides (glyphosate, AMPA and glufosinate) (Method GLYPH)	Groundwater	SH2437 &: LC9045: Direct injection GLYPH: derivatization with FMOC prior to on-line SPE	LC-QqQ-MS/MS	78–114	0.001–1.350[b] 0.001–0.028[b] 0.020[b]	[25]
150 pesticide metabolites	Surface water and groundwater	Direct injection	LC-QqQ-MS/MS	—	0.003–2.000	[27]
16 polar pesticides + pharmaceuticals	Groundwater	Extraction from passive sampler (POCIS): acetone:methanol	UHPLC-QqQ-MS/MS	42–116	0.00003–0.00135[b]	[59]

[a]Abbreviations: BDEs: Brominated diphenyl ethers; DAD: Diode array detector; DES: Deep eutectic solvent; DFS-HRMS: Magnetic sector high resolution mass spectrometry: DI: Direct injection; FID: Flame ionization detection; FMOC: 9-florenylmethyl-chloroformate; GC: Gas chromatography; HF-LPME: Hollow fiber-liquid phase microextraction; HLB: Hydrophilic-lipophilic balanced; LC: Liquid chromatography; LLE: Liquid-liquid extraction; LOQ: Limit of quantification; LTQ: Linear ion trap; MS: Mass spectrometry; MS/MS: Tandem mass spectrometry; μECD: Micro electron capture detector; OCPs: Organochlorine pesticides; OPPs: Organophosphorus pesticides; PA: Polyacrilate; PAHs: Polycyclic aromatic hydrocarbons; PCBs: Polychlorinated biphenyls; PDMS: Polydimethylsiloxane; POCIS: Polar organic chemical integrative samples; Q: Single quadrupole; QqQ: Triple quadrupole; QTOF: Quadrupole time of flight; QTRAP: Hybrid triple quadrupole-linear ion trap; SBSE: Stir bar sorptive extraction; SPE: Solid phase extraction; SPME: Solid phase microextraction; TPs: Transformation products; UHPLC: Ultra-high-performance liquid chromatography; WAX: Weak anion exchange; WCX: Weak cation exchange.
[b]Limit of detection.

MS (MS/MS) has allowed for the determination of pesticides at ultra-trace levels in surface and groundwaters [25, 27, 91].

In relation to solid samples, such as soils, the most common methods were based on solid-liquid extraction (SLE), pressurized liquid extraction (PLE) and QuEChERS (see Table 1.2). In SLE methods, solvent mixtures such as acetonitrile and water have been

Table 1.2 Overview of analytical methods applied to determine pesticides in soil matrices.[a]

Pesticides	Extraction	Determination technique	Recovery (%)	LOQ (μg kg^{-1})	Reference
Famoxadone and metabolites	SLE: Water/ Acetonitrile 1% acetic acid (50 : 50, v/v)	LC-Orbitrap-MS	72–113	20	[28]
OCPs, OPPs, pyrethroids (58)	SLE: Water/ Acetonitrile 1% acetic acid. Extraction salts: $MgSO_4$ and sodium acetate Clean-up: $MgSO_4$, PSA & C18	GC-QqQ-MS/MS	69–119	100–5000	[61]
30-multicalss	SLE: Methanol Clean-up: SPE (Oasis HLB)	LC-QTRAP-MS/MS	70–120	1–10	[93]
18-multiclass	SLE: Water/ Acetonitrile (1 : 5, v/v)	LC-QqQ-MS/MS	50–120	50	[62]
Fenamidone, propamocarb	SLE: Methanol or Water Clean-up: $MgSO_4$ & PSA	LC-QqQ-MS/MS	77–108	0.4–2	[100]
Oxanilic and sulfonic acid metabolites of acetochlor	SLE: Acetonitrile / Water (60 : 40, v/v)	LC-QqQ-MS/MS	91–120	1–2	[92]
Endosulfan, chlorpyrifos and their metabolites	SLE: Ethyl acetate with a 1 : 5 (w/v) soil-to-solvent ratio	GC-IT-MS	76–95	10–50[b]	[29]
Acetochlor and Propisochlor	SLE: Methanol/water (1 : 1 v/v) Clean-up: PSA	GC-ECD	80–116	10	[63]
10 OCPs	SLE: Water/ Acetonitrile Clean-up: $MgSO_4$	GC-QqQ-MS/MS	70–115	2–40	[106]
10 OCPs and metabolites	QuEChERS. Clean-up: Sulfuric acid and florisil	GC-IRMS	60–100	500	[115]

(Continued)

Table 1.2 (Continued)

Pesticides	Extraction	Determination technique	Recovery (%)	LOQ (μg kg^{-1})	Reference
218	QuEChERS using Acetonitrile 2.5% formic acid. Extraction salts: $MgSO_4$ & sodium acetate	LC-QqQ-MS/MS GC-QqQ-MS/MS	70–120	0.5–20	[99]
Pydiflumetofen enantiomers	QuEChERS Clean-up: $MgSO_4$ & C18	UHPLC-QqQ-MS/MS	84–103	5	[97]
Pyrethroid pesticide metabolite	QuEChERS Clean-up: d-SPE	GC-IT-MS	70–94	13	[64]
32-multiclass	UAE: 40 mL of methanol–water (4 : 1 v/v). 20 minutes	LC-Q-MS	60–110	1.5–5.0	[107]
25-Triazines, phenylureas, phenoxy acid pesticides	PLE: Dichloromethane – acetone (1 : 1, v/v) and Acetonitrile–water (2 : 1, v/v)	LC-QqQ-MS/MS	65–120	0.1–3	[95]
51 Fungicides and insecticides	PLE: Methanol:acetonitrile (70 : 30, v/v) Methanol: Acetonitrile:formic acid (65 : 30:5, v/v)	LC-QqQ-MS/MS	57–136	0.3–8.5	[96]
9 OPCs & PAHs	MAE: Hexane/water (3 : 2 v/v)	GC-QqQ-MS/MS	–	–	[104]
8 Pesticides and metabolites	DLLME	LC-FD	70–120	0.07–80	[114]
8 Chiral pesticides	MSPD-DLLME	LC-QqQ-MS/MS	87–104	0.2–1.5	[65]
5 OPPs	Deep eutectic solvent embedded sponge	LC-UV-Vis	–	–	[113]

[a]Abbreviations: DLLE: Dispersive liquid-liquid microextraction; d-SPE: Dispersive solid phase extraction; ECD: Electron capture detector; FD: Fluorescence detection; GC: Gas chromatography; HLB: Hydrophilic-lipophilic balanced; IRMS: Isotope ratio mass spectrometry; IT: Ion trap; LC: Liquid chromatography; MAE: Microwave-assisted extraction; MS: Mass spectrometry; MSPD: Matrix solid phase dispersion; MS/MS: Tandem mass spectrometry; OCPs: Organochlorine pesticides; OPPs: Organophosphorus pesticides; PAHs: Polycyclic aromatic hydrocarbons; PCBs: Polychlorinated biphenyls; PLE: Pressurized liquid extraction; PSA: Primary secondary amine; Q: Single quadrupole; QqQ: Triple quadrupole; QTRAP: Hybrid triple quadrupole-linear ion trap; SLE: Solid-liquid extraction; SPE: solid phase extraction. UAE: Ultrasonic-assisted extraction; UHPLC: Ultra-high-performance liquid chromatography; UV-Vis: Ultraviolet-visible detection.

[b]Instrumental method (μg l^{-1}).

widely used. For example, a mixture of water/acetonitrile (10 : 90, *v/v*) was utilized to monitor pesticide residues in soils from Argentina [62], or a mixture of water/acetonitrile (40 : 60, *v/v*) was used for the extraction of oxanilic and sulfonic acids metabolites [92]. Hu et al. employed methanol:water (50 : 50, *v/v*) instead of acetonitrile for the extraction of acetochlor and propisochlor in soils from Beijing (China) [63]. Colazzo et al. tested different solvents and mixtures, such as acetonitrile, methanol, water and methanol/water, choosing methanol as the best option (recoveries ranged from 45 to 90%) to determine pesticide residues in paddy fields and sugar cane from Uruguay [93].

The PLE method is based on the use of a solvent that is applied at high pressure and temperature through a solid or semisolid sample (e.g. soils) to effectively extract the analytes, being faster than conventional SLE. The selection of optimum experimental parameters, such as extraction temperature, flush volume and preheat time, can allow the extraction of a large number of pesticides within a wide polarity range in only one step [94]. Two different extraction solvents, dichloromethane:acetone (1 : 1, *v/v*) and acetonitrile:water (2 : 1, *v/v*), were used for the determination of triazines, phenylureas and phenoxy acid pesticides [95]. Fungicides and insecticides were also extracted using PLE, applying two extraction solvents, methanol:acetonitrile (70 : 30, *v/v*) and methanol:acetonitrile:formic acid (65 : 30:5, *v/v*), obtaining recoveries from 57% to 136% [96].

The QuEChERS approach has been widely used to extract pesticides from soils [3] and several modifications were carried out to improve sample extraction. One of them was matrix hydration, which consisted of the addition of water before solvent addition [97]. This approach was used for the extraction of pydiflumetofen enantiomers, obtaining recoveries between 84–103% or for the determination of afidopyropen and its metabolite residues in a cotton field with acceptable recoveries (85–100%) [98]. Another modification was the acidification of the solvent to improve the extraction of target analytes. For instance, acetonitrile, acidified with 2.5% formic acid, provided acceptable recoveries (70–120%) for the simultaneous monitoring of 218 pesticide residues in clay loam soil [99]. The clean-up step was not commonly used in soil samples and only a few studies employed dispersive solid phase extraction (d-SPE). The sorbents and salts commonly used were primary secondary amine (PSA) and anhydrous magnesium sulphate ($MgSO_4$) [98, 100]. Other sorbents, such as C18, were also used in combination with PSA [61] or with $MgSO_4$ [97].

Finally, SPE, using OASIS HLB cartridges, was also applied to concentrate the extract before chromatographic analysis [93].

Tissue analysis is more challenging than water or soil analysis due to the complexity of those matrices. Thus, for the extraction of pesticides from biota, different extraction techniques can be applied such as SPME [40], PLE, SPE, ultrasonic-assisted extraction, dispersive liquid-liquid extraction and SBSE [101], adding a freezing-lipid filtration in fatty materials. Other procedures such as QuEChERS have been widely used in the last few years, using in the clean-up step based on d-SPE a mixture of sorbents that includes $MgSO_4$, PSA, C18 and graphitized carbon black (GCB) [46, 47, 49], as indicated in Table 1.3.

Finally, pesticides are usually extracted from air using active or passive samplers (see Table 1.4). The active ones are mainly based on SPE by pumping high volumes of air

Table 1.3 Overview of analytical methods applied to determine pesticides in biota.[a]

Pesticides	Matrix	Extraction technique	Determination technique	Recovery (%)	LOQ (ng g^{-1})	Reference
50	Fish	QuEChERS	LC-QqQ	—	—	[49]
50 (TPs)	Fish	QuEChERS Clean-up: d-SPE ($MgSO_4$ + C_{18} + active coal)	LC-QqQ	58–140	0.90–11.25	[46]
40 (TPs)	Fish	QuEChERS Clean-up: d-SPE ($MgSO_4$ + C_{18} + PSA + activated charcoal)	LC-QqQ	58–145	0.01–1[b]	[47]
50	Fish	QuEChERS Clean-up: d-SPE ($MgSO_4$ + C_{18} + PSA + active coal)	LC-QqQ	58–140	0.90–11.25[1]	[48]
10 OCPs	Fish	PLE: dichloromethane:hexane 1 : 1 v/v. Clean-up: silica gel & basic alumina	GC-Q-MS	50–110	0.0055–0.0300 dw [b]	[105]
OCPs	Several species	Soxhlet: Hexane:acetone 3 : 1 v/v. Clean-up: acid silica	GC-Q-MS	—	1–4	[103]
16 OCPs	Soil & microorganisms	Soxhlet: Hexane:dichloromethane 3 : 2 v/v. Clean-up: GPC Hexane:dichloromethane 1 : 1 v/v	GC-ECD	—	—	[45]

[a]Abbreviations: d-SPE: Dispersive solid phase extraction; dw: dry weight; ECD: Electron capture detector; GPC: Gel permeation chromatography; LC: Liquid chromatography; MS: Mass spectrometry; MS/MS: Tandem mass spectrometry; OCPs: Organochlorine pesticides; PLE: Pressurized liquid extraction; PSA: Primary secondary amine; Q: Single quadrupole; QqQ: Triple quadrupole; TPs: Transformation products.

[b]Limit of detection.

Table 1.4 Overview of analytical methods applied to determine pesticides in air matrices.

Pesticides	Matrix	Extraction technique	Determination technique	Recovery (%)	LOQ (pg m^{-3})	Reference
20	Indoor air	Passive Sampling (VERAM) MAE: acetone:hexane, 1 : 1 v/v Clean-up: Alumina-C18 Cartridges	GC-Q-MS	59–174	1–10[b]	[51]
OCPs	Air samples	PAS: PUF. Soxhlet: acetone	GC-Q-MS	82–126	—	[54]
34 (OCPs & CUPs)	Air samples	PUF. Soxhlet: acetone + petroleum ether. Clean-up: silica column	GC-Q-MS	86–102	0.1–90.7[c]	[55]
40 & TPs	PM10 (remote, urban and rural areas)	MAE: ethyl acetate	LC-QqQ-MS/MS	—	6.5–32.5	[57]
35	PM10	MAE: Ethyl acetate	UHPLC-Orbitrap	73–116	2.6–75	[56]
13	PM2.5	UAE: 18% of Acetonitrile in dichloromethane	GC-Q-MS	70.2–124 (ethion 31.2–63.0)	7.5–60	[58]
452 (OCPs & CUPs)	Air particles	UAE: dichloromethane	GC-QqQ-MS/MS	—	—	[53]

[a]Abbreviations: CUPs: Current-use pesticides; GC: Gas chromatography; LC: Liquid chromatography; MAE: Microwave-assisted extraction; MS: Mass spectrometry; MS/MS: Tandem mass spectrometry; OCPs: Organochlorine pesticides; QqQ: Triple quadrupole; TPs: Transformation products; UAE: Ultrasound-assisted extraction; PAS. Passive air sampler; PUF: Polyurethane foam; UHPLC: Ultra-high-performance liquid chromatography.
[b]Limit of detection provided as ng $VERAM^{-1}$.
[c]Limit of detection.

through different sorbents such as Tenax TA, Carbotrap or polyurethane foam filter (PUF) [54, 102]. Passive samplers are based on diffusion through a well-defined barrier or membrane (e.g. Radiello) PUF disks or semipermeable membrane devices [33]. Then compounds are desorbed from the passive sampler using Soxhlet extraction [55, 103], microwave-assisted extraction [104], PLE [105] or ultrasound-assisted extraction [33]. In this sense, different solvents can be used for the extraction of pesticides from filters or membranes, such as ethyl acetate [57], acetone or petroleum ether [55]. Additionally, miniaturized methods can also be applied, and, for instance, pesticides were extracted from PM2.5 using a miniaturized device with 500 μL of 18% acetonitrile in dichloromethane followed by sonication and injection into GC-MS [58].

1.3 Determination of Pesticides

Current analytical methods for pesticide residue analysis are mainly based on the use of chromatographic techniques (GC or LC), depending on their polarity and volatility among other factors, coupled to MS analyzers. As an example, Figure 1.5 shows the chromatograms of a blank soil sample spiked with 218 pesticides by LC-MS/MS and GC-MS/MS, observing the potentiality of these techniques to analyze a wide range of compounds.

1.3.1 Development of the Instrumental Method

1.3.1.1 Chromatography

GC enables the detection of a high number of volatile or semivolatile and thermally stable pesticides, allowing the determination of persistent pesticides in environmental matrices such as water, soil, biota and air (Tables 1.1–1.4). Several multiresidue methods have been developed and, for instance, 11 OCPs were determined in soils using GC with a DB-5-MS capillary column with 5% phenyl stationary phase and 95% of methylpolysiloxane [106]. Although this is the stationary phase most commonly used, other stationary phases, such as DB-XLB and DB-35-MS, were also tested for the determination of endosulfan, chlorpyrifos and their metabolites [29], obtaining the best results with DB-XLB (larger differences in the retention time of the compounds). ZB-5MSi was also tested to determine pyrethroid pesticide metabolites in soil samples [64] and a HP-5-MS UI column to analyze 58 pesticides in soil samples [61]. Additionally, pesticides can simultaneously be analyzed with other pollutants such as polycyclic aromatic hydrocarbons (PAHs), brominated diphenyl ethers (BDEs) and polychlorinated biphenyls (PCBs) in environmental matrices, such as surface waters [84], accomplishing this with the Environmental Quality Standards (EQS) fixed by Directive 2013/39/EC [10].

For non-GC amenable compounds, although chemical derivatization can be carried out, LC is recommended and liquid chromatography (LC) and, increasingly, ultra-high-performance liquid chromatography (UHPLC) are commonly used for the analysis of pesticides, using different particle sizes as 5 μm [107], 3 μm [62] or even lower (1.8 μm) [28]. For LC-amenable pesticides, conventional C18 columns are commonly used [57], with generic mobile phases such as methanol and water 0.1%

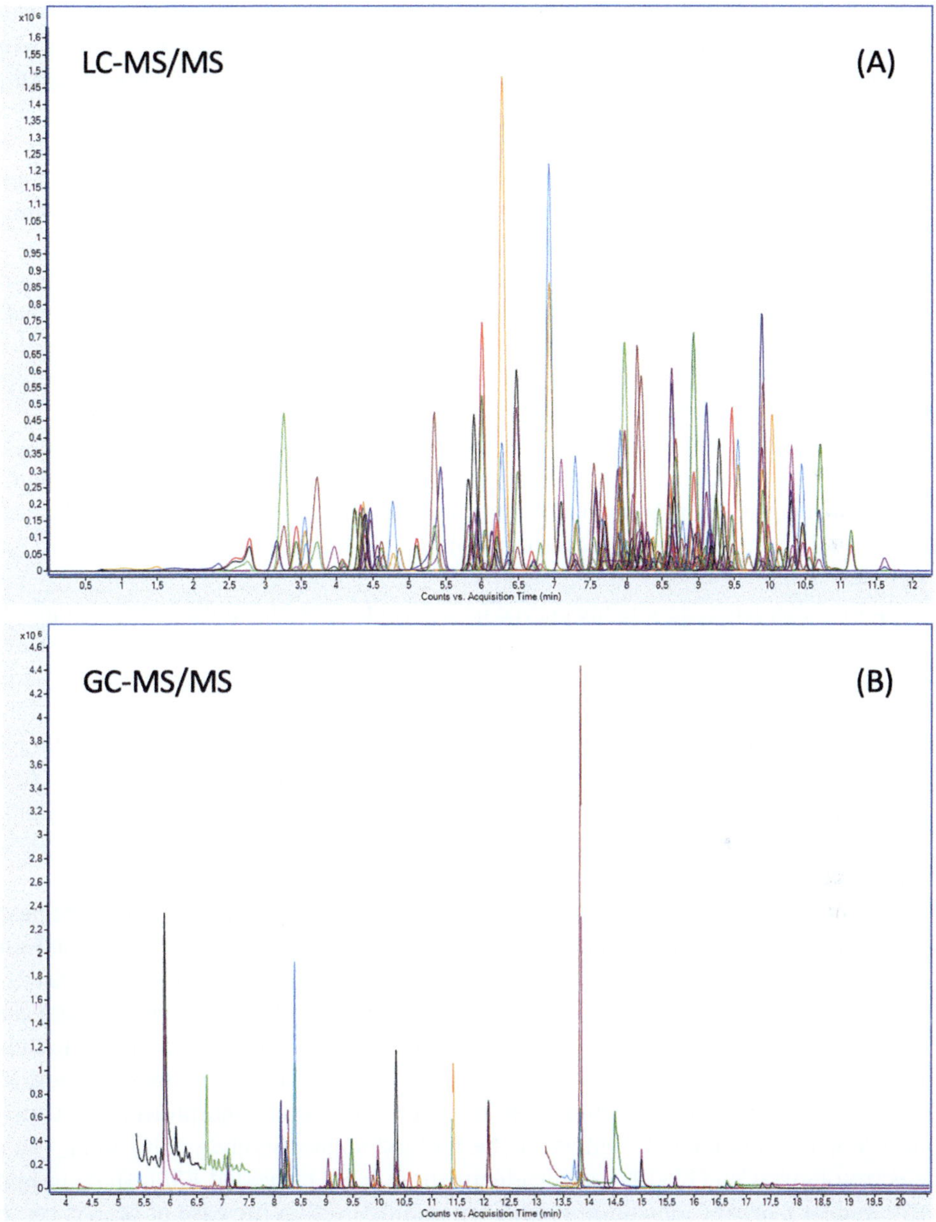

Figure 1.5 Chromatograms of the targeted pesticides by LC-MS/MS (A) and GC–MS/MS (B) of a blank soil sample spiked at 20 ng g^{-1}. Source [99]. Reproduced with permission of Elsevier B.V.

formic acid 4 mm ammonium formate, allowing, for instance, the determination of parent pesticides (famoxadone) and metabolites [28]. Moreover, hydrophilic interaction liquid chromatography (HILIC) has been proposed as a promising tool for highly polar pesticides in fresh water [72] whereas a mixed-mode stationary phase is an interesting alternative [108], allowing the separation of a wide range of compounds such as polar, medium and non-polar ones. Thus, a mixed column such as XDB-C18 was used to determine 30 pesticides (herbicides, fungicides and insecticides), using as mobile phase acetonitrile and water 0.1% formic acid [93], or an EC-C18 column, using mobile phases that consisted of methanol 0.1% formic acid and 2 mm ammonium acetate, and water 0.1% formic acid and 2 mm ammonium acetate, allowing the determination of 218 pesticides [99].

As can be observed in Tables 1.1–1.4, different multiresidue and multiclass methods based on LC-MS have been reported in recent years. Robust and reliable analytical methods have been developed, allowing for the monitoring of ca. 500 pesticides in surface and groundwaters [23], 215 pesticides and TPs in groundwater [25], as well as 251 emerging contaminants in surface water [75].

In the last years, chiral columns were also employed to determine enantiomers of pesticides [65], using different specific stationary phases, such as Chiralcel OD-RH column and a mobile phase consisting of acetonitrile and water 0.1% formic acid, or a Chiralcel OD-3 [97], that allows the determination of pydiflumetofen enantiomers in soil samples.

On the other hand, despite the proved potential of supercritical fluid chromatography (SFC) [108–110], its application in the analysis of environmental samples is still scarce. For instance, SPE was combined with SFC for the analysis of 8 emerging contaminants, including the diuron herbicide, in river water [111]. Ionic chromatography (IC) coupled with MS has tested for the analysis of polar pesticides, such as glyphosate, AMPA, glufosinate and MPPA in surface water [112], being an interesting alternative to HILIC or approaches.

1.3.1.2 Detection

Different classical detectors have been applied for the determination of target pesticides in environmental samples, such as flame ionization detector (FID) [68] and electron capture detector (ECD) [45, 63, 86] in combination with GC. Likewise, LC coupled to a UV-Vis detector was used for the analysis of organophosphorus pesticides in soils samples [113], or to a fluorescence detector, which was utilized to monitor pesticides and metabolites in soils [114]. However, in the last two decades, MS has become the main detection system in environmental monitoring, as can be observed in Tables 1.1–1.4. Most of the methodologies involving GC make use of electron ionization (EI) [99], while electrospray (ESI) or heated ESI (HESI) in positive and/or negative ionization modes are mainly used in the case of LC. As can be seen from Tables 1.1–1.4, triple quadrupole (QqQ) analyzer is the most widely used in combination with both GC and LC.

QqQ, operating in selected reaction monitoring (SRM) or multiple reaction monitoring modes (MRM), provides high selectivity and sensitivity for target analytes. Thus, QqQ and ESI $\pm$ was used to determine 218 pesticides in soils, or to analyze propamocarb and fenamidone in soils [100]. Zhao et al. [65] determined LOQ lower than the previous works (0.22 to 1.54 $\mu g\ kg^{-1}$) for the analysis of eight chiral pesticides (diniconazole, metalaxyl, paclobutrazol, epoxiconazole, myclobutanil, hexaconazole, napropamide and isocarbophos) in soils and river sediments, employing QqQ and ESI source.

Other types of MS analyzers coupled to LC have also been used. The single quadrupole MS was employed by Belmonte Vega et al. [107] to determine pesticides in environmental matrices, or the triple quadrupole-linear ion trap mass spectrometer (QTRAP), which was used to detect herbicides, insecticides and fungicides [93], achieving LOQ of 1–10 $\mu g\ kg^{-1}$.

However, other types of analyzers coupled to LC have been employed. In GC, ion trap mass spectrometer detector was employed to detect endosulfan, chlorpyrifos and their metabolites in soil matrices [29] and isotope ratio mass spectrometer (IRMS) was used to determine OCPs and metabolites [115]. Additionally, the double-focusing magnetic sector high-resolution mass spectrometer is extremely sensitive when multiple ion detection (MID) is used as the acquisition mode. It has been utilized in combination with SPME and GC for the fully automated determination of priority substances at ultra-trace levels, including pesticides, in surface water [84], as well as more complex matrices such as treated and non-treated wastewaters [85].

Although QqQ was still used to monitor the presence of pesticides in environmental samples, when a large number of compounds should be monitored, the sensitivity of these analyzers decreased by the simultaneous monitoring of a huge number of transitions. Therefore, HRMS analyzers are also used in this field, because the number of compounds to be simultaneously analyzed is theoretically unlimited and other strategies can be applied [56]. Additionally, good chromatographic peaks and resolution are still important. The use of HRMS allows that customized database can be built, including CUPs, TPs, banished pesticides and other pollutants, increasing the scope of the analysis.

HRMS enables the analysis of target, post-target (or suspected) and non-target analytes. The high sensitivity of certain HRMS analyzers, when operating in full scan mode, combined with high resolving power (>30 000 full width at half maximum, FWHM) and accurate mass measurement (1–5 ppm), allows for retrospective analysis.

While most of the HRMS applications have been aimed at targeted pesticide analysis, there are already some studies aimed at the identification of non-target analytes including new contaminants such as metabolites or TPS for a more exhaustive monitoring of water quality.

Some examples of the successful application of HRMS analyzers, specifically Q-Time of Flight (TOF) systems, in combination with LC, in target screenings allowed for the monitoring of 474 emerging contaminants, including 296 pesticides, in coastal waters [70] and ca. 500 pesticides and TPs in surface and groundwater [23]. Likewise, and as shown in Table 1.1, the potential of Orbitrap analyzers has also been employed in methods involving emerging contaminants belonging to different families and classes.

Most of the non-target screenings have been applied to wastewater samples [116], which are considered an important source of contamination of natural waters when discharged into rivers and coastal compartments. This approach was also applied in soils, detecting suspected metabolites of famoxadone in samples [28].

Nevertheless, a recent study, making use of UHPLC-Q-Orbitrap MS and software Compound Discoverer 3.0 to perform non-target analysis of pollutants in surface water, has revealed that pesticides and plastic additives were the two main types of pollutants found in Dianshan Lake in China. Once the pollutants were identified, a target screening allowed for their confirmation and quantification [117].

Moreover, HRMS is a powerful tool for the identification of unknown TPs and metabolites. In fact, novel TPs resulting from the chlorination of clothianidin, imidacloprid, desnitro-imidacloprid, imidacloprid-urea and hydrolysis products of thiamethoxam were identified in drinking water by using a HPLC-TOF MS system [18].

1.3.2 Figures of Merit

The use of the analytical methods described in previous sections provides sensitive and reliable methods that allow for low LOQs and suitable recoveries, as it can be observed in Tables 1.1–1.4. Thus, despite the lower EQS set by EU legislation for some pesticides, lower LOQs can be achieved and, for instance, concentrations down to 0.0001 $\mu g\ l^{-1}$ can be quantified (see Table 1.1), combining preconcentration techniques, such as SPE or SPME and sensitive analyzers. For instance, 51 pesticides, covering highly polar compounds, were determined in surface and groundwaters by using on-line SPE in combination with (U)HPLC and tandem mass spectrometry, providing LOQ values in the range of 0.005–0.025 $\mu g\ l^{-1}$ [78]. Higher limits were obtained in soils, allowing the quantification at $\mu g\ kg^{-1}$ or even lower (see Table 1.2), as well as in biota, where concentrations below $\mu g\ kg^{-1}$ ($ng\ g^{-1}$) can be detected (Table 1.3). Additionally, the use of passive sampling, such as in air, allows for achieving adequate recovery and precision values (Table 1.4) and low LOQs (< 6.5 $pg\ m^{-3}$) for most of the compounds [56].

In terms of recovery, most of the current analytical methods provided recoveries between 70% and 120%, although when multiresidue methods are developed, recoveries < 70% or > 120% are sometimes achieved. In the case of lower recoveries, if precision is < 20%, correction factors can be used for quantification purposes. In this sense, the use of generic methods, such as QuEChERS, which has been widely used in soils (see Table 1.2), provides acceptable recoveries (70–120%) for the simultaneous environmental monitoring of 218 pesticide residues in clay loam soil [99], or the quantitative extraction of parent pesticides, such as afidopyropen, and metabolites (recovery values: 85–100%) [98].

1.3.3 Hints and Tips

There are several hints and tips that should be considered in order to fulfill current legislation as well as to ensure reliable results:

- Sample collection: Apply a statistical-based sampling procedure to ensure spatial and/or time variation and collect representative samples. Avoid compound degradation, selecting suitable storage and transport conditions. For air or water sampling, in addition to active sampling procedures, passive sampling is an interesting alternative, which is easy to deploy and minimum maintenance is required.
- Extraction: Use sensitive and/or generic extraction procedures. Thus, for aqueous samples, SPE or SPME, selecting suitable sorbents, are the most widely extraction techniques to ensure lower LOQs as well as suitable recovery. For solid samples, such

as biota and soils, although PLE, combining different sorbents depending on the target compounds, is a good option, nowadays, QuEChERS approach is the first choice, because of its versatility, wide range of application and simplicity. Depending on the complexity of the matrix, an additional clean-up step could be necessary. The final extract can be injected in both GC and LC equipment or it can easily be evaporated and redissolved in a GC solvent as ethyl acetate.

- Analytical determination: Select LC or GC depending on the characteristics of the monitored compounds (GC for non-polar compounds and LC for polar pesticides). Either LC or GC should be coupled to QqQ, providing reliable, sensitive, fast and wide-scope analysis.
- LRMS vs HRMS: Use QqQ (LRMS analyzer) for targeted analysis of pesticide residues in environmental samples, whereas HRMS ((Q)-Orbitrap or (Q)-TOF) must be utilized if suspect or unknown analysis is also required.
- Data treatment: Well-set workflows and software are available for targeted analysis whereas when suspect or unknown analysis is performed, large raw data files are obtained, and high expertise and suitable software is needed for data interpretation and pesticide identification.

1.4 Future Directions and Challenges

Nowadays, in order to increase sample throughput and minimize the number of analyses that should be performed to get a comprehensive view of the presence of pesticides in the environment, current analytical methods are monitoring a higher number of pesticides. However, they have different physico-chemical properties (from non-polar to highly polar pesticides, and from low to high volatility) and this has hampered the development of an all-in-one approach, where a wide-scope analysis can be performed, analyzing pesticides and TPs. Therefore, the combination of different platforms (GC-MS & LC-MS) is a suitable approach to minimize the number of analyses to be carried out, covering a wide-scope of pesticides, although several compounds will still require specific methods to be detected (e.g. highly polar pesticides or metabolites).

Simplification of sample treatment is still necessary in order to minimize sample handling and therefore, error or bias associated to this step. Thus, automated methodologies that allow for monitoring of many priority substances and specific pollutants at concentrations below EQSs (e.g. water samples) are still demanding, because they are fast and cost-effective

One of the main challenges of environmental analytical chemistry is that there is a huge number of pesticides and/or TPs present in the environment, but many of them are still unknown because, for instance, the dissipation of the parent compounds has not been evaluated. Therefore, a targeted approach should be combined with suspect and unknown analyses to comprehensively assess chemical pollution related to pesticides. Special interest should be paid to the identification of new metabolites and/or TPs, and other emerging contaminants, and therefore the faster acquisition rate in HRMS analyzers and the development of efficient and user-friendly software are still needed to carry out a deconvolution process for the generation of clean mass spectra of coeluting peaks as well as an accurate putative identification. In this sense, the development of databases that include detected transformation products are needed to make this process much easier and this will facilitate the inclusion of TPs in future legislation.

Acknowledgments

Authors gratefully acknowledge to the Spanish Ministry of Economy and Competitiveness (MINECO), Spain and FEDER-EU (project ref. PID2019-106201RB-I00) for financial support. ID thanks the University of Almeria for the Hipatia postdoctoral contract. RLR acknowledges "Plan Propio of Investigation" of University of Almería, cofunded by CAJAMAR and the Operational Program Funds Europeans of Regional Development of Andalusia (2014–2020) (FEDER), for financial support.

Bibliography

1 Carvalho, F.P. (2017). Pesticides, environment, and food safety. *Food Energy Secur.* 6: 48–60. doi: 10.1002/fes3.108.

2 Pietrzak, D., Kania, J., Malina, G., Kmiecik, E., and Wator, K. (2019). Pesticides from the EU first and second watch lists in the water environment. *Clean Soil Air Water* 47: 1–13. doi: 10.1002/clen.201800376.

3 Samsidar, A., Siddiquee, S., and Shaarani, S.M. (2018). A review of extraction, analytical and advanced methods for determination of pesticides in environment and foodstuffs. *Trends Food Sci. Technol.* 71: 188–201. doi: 10.1016/j.tifs.2017.11.011.

4 Nasiri, M., Ahmadzadeh, H., and Amiri, A. (2020). Sample preparation and extraction methods for pesticides in aquatic environments: a review. *Trends Anal. Chem.* 123: 115772. doi: 10.1016/j.trac.2019.115772.

5 Choudri, B.S., Charabi, Y., Al-Nasiri, N., and Al-Awadhi, T. (2020). Pesticides and herbicides. *Water Environ. Res.* 92: 1425–1432. doi: 10.1002/wer.1380.

6 Ochoa, V. and Maestroni, B. (2018). Pesticides in water, soil, and sediments. In: *Integrated Analytical Approaches for Pesticide Management*, 1st ed. (ed. B. Maestroni and A. Cannavan), 133–147. Amsterdam: Elsevier. doi: 10.1016/B978-0-12-816155-5.00009-9.

7 Ramakrishnan, B., Venkateswarlu, K., Sethunathan, N., and Megharaj, M. (2019). Local applications but global implications: can pesticides drive microorganisms to develop antimicrobial resistance? *Sci. Total Environ.* 654: 177–189. doi: 10.1016/j.scitotenv.2018.11.041.

8 Hernández, F., Bakker, J., Bijlsma, L., de Boer, J., Botero-Coy, A.M., Bruinen de Bruin, Y., Fischer, S., Hollander, J., Kasprzyk-Hordern, B., Lamoree, M., López, F.J., Ter Laak, T.L., van Leerdam, J.A., Sancho, J.V., Schymanski, E.L., de Voogt, P., and Hogendoorn, E.A. (2019). The role of analytical chemistry in exposure science: focus on the aquatic environment. *Chemosphere* 222: 564–583. doi: 10.1016/j.chemosphere.2019.01.118.

9 European Union (2008). Directive 2008/105/EC of the European Parliament and of the Council of 16 December 2008 on environmental quality standards in the field of water policy, amending and subsequently repealing Council Directives 82/176/EEC, 83/513/EEC, 84/156/

EEC, 84/491/EEC. 86/280/EEC and amending Directive 2000/60/EC of the European Parliament and of the Council. *Official Journal of the European Union* L348: 84–97.

10 European Union (2013). Directive 2013/39/EU of the European Parliament and of the Council of 12 August 2013 amending Directives 2000/60/EC and 2008/105/EC as regards priority substances in the field of water policy. *Official Journal of the European Union* L226: 1–17.

11 European Union (2020). Commission implementing Decision (EU) 2020/1161 of 4 August 2020 establishing a watch list of substances for Union-wide monitoring in the field of water policy pursuant to Directive 2008/105/EC of the European Parliament and of the Council. *Official Journal of the European Union* L257: 32–35.

12 European Union (2015). Commission Directive (EU) 2015/1787 of 6 October 2015 amending Annexes II and III to Council Directive 98/83/EC on the quality of water intended for human consumption. *Official Journal of the European Union* L260: 6–17.

13 European Union (2006). Directive 2006/118/EC of the European parliament and of the Council of 12 December 2006 on the protection of groundwater against pollution and deterioration. *Official Journal of the European Union* L372: 19–31.

14 US EPA (2020). Water quality criteria. https://www.epa.gov/wqc. Accessed 14 December 2021.

15 US EPA (2020). National primary drinking water regulations. https://www.epa.gov/ground-water-and-drinking-water/national-primary-drinking-water-regulations#Organic. Accessed 14 December 2021.

16 US EPA (2020) Main water legislation. https://www.epa.ie/water/waterleg. Accessed 14 December 2021.

17 Spanish Ministry of Presidency (2005). Royal Decree 9/2005 of 14 January which establishes a list of potentially soil contaminating activities and criteria and standards for declaring that sites are contaminated. *Official Gazette of the Spanish State* 15: 1833–1843.

18 Wong, K.L.K., Webb, D.T., Nagorzanski, M.R., Kolpin, D.W., Hladik, M.L., Cwiertny, D.M., and LeFevre, G.H. (2019). Chlorinated byproducts of neonicotinoids and their metabolites: an unrecognized human exposure potential? *Environ. Sci. Technol. Lett.* 6: 98–105. doi: 10.1021/acs.estlett.8b00706.

19 Kiss, A. and Virág, D. (2009). Photostability and photodegradation pathways of distinctive pesticides. *J. Environ. Qual.* 38: 157–163. doi: 10.2134/jeq2007.0504.

20 Ibáñez, M., Sancho, J.V., Pozo, Ó.J., and Hernández, F. (2006). Use of liquid chromatography quadrupole time-of-flight mass spectrometry in the elucidation of transformation products and metabolites of pesticides. Diazinon as a case study. *Anal. Bioanal. Chem.* 384: 448–457. doi: 10.1007/s00216-005-0167-6.

21 Li, M., Wang, R., Kong, Z., Gao, T., Wang, F., and Fan, B. (2020). Cyflumetofen degradation in different aquatic environments and identification of hydrolytic products. *J. Environ. Chem. Eng.* 8: 104512. doi: 10.1016/j.jece.2020.104512.

22 Hensen, B., Olsson, O., and Kümmerer, K. (2020). A strategy for an initial assessment of the ecotoxicological effects of transformation products of pesticides in aquatic systems following a tiered approach. *Environ. Int.* 137: 105533. doi: 10.1016/j.envint.2020.105533.

23 Fonseca, E., Renau-Pruñonosa, A., Ibáñez, M., Gracia-Lor, E., Estrela, T., Jiménez, S., Pérez-Martín, M.A., González, F., Hernández, F., and Morell, I. (2019). Investigation of pesticides and their transformation products in the Júcar River Hydrographical Basin (Spain) by wide-scope high-resolution mass spectrometry screening. *Environ. Res.* 177: 108570. doi: 10.1016/j.envres.2019.108570.

24 Quintana, J., de La Cal, A., and Boleda, M.R. (2019). Monitoring the complex occurrence of pesticides in the Llobregat basin, natural and drinking waters in Barcelona metropolitan area (Catalonia, NE Spain) by a validated multi-residue online analytical method. *Sci. Total Environ.* 692: 952–965. doi: 10.1016/j.scitotenv.2019.07.317.

25 Fisher, I.J., Phillips, P.J., Bayraktar, B.N., Chen, S., McCarthy, B.A., and Sandstrom, M.W. (2021). Pesticides and their degradates in groundwater reflect past use and current management strategies, Long Island, New York, USA. *Sci. Total Environ.* 752: 141895. doi: 10.1016/j.scitotenv.2020.141895.

26 Kiefer, K., Müller, A., Singer, H., and Hollander, J. (2019). New relevant pesticide transformation products in groundwater detected using target and suspect screening for agricultural and urban micropollutants with LC-HRMS. *Water Res.* 165: 114972. doi: 10.1016/j.watres.2019.114972.

27 Reemtsma, T., Alder, L., and Banasiak, U. (2013). Emerging pesticide metabolites in groundwater and surface water as determined by the application of a multimethod for 150 pesticide metabolites. *Water Res.* 47: 5535–5545. doi: 10.1016/j.watres.2013.06.031.

28 Lopez-Ruiz, R., Romero-González, R., and Garrido-Frenich, A. (2019). Residues and dissipation kinetics of famoxadone and its metabolites in environmental water and soil samples under different conditions. *Environ. Pollut.* 252: 163–170. doi: 10.1016/j.envpol.2019.05.123.

29 Tiwari, M.K. and Guha, S. (2013). Simultaneous analysis of endosulfan, chlorpyrifos, and their metabolites in natural soil and water samples using gas chromatography-tandem mass spectrometry. *Environ. Monit. Assess.* 185: 8451–8463. doi: 10.1007/s10661-013-3186-3.

30 Peterson, M.A., McMaster, S.A., Riechers, D.E., Skelton, J., and Stahlman, P.W. (2016). 2,4-D past, present, and future: a review. *Weed Technol.* 30: 303–345. doi: 10.1614/wt-d-15-00131.1.

31 Pietrzak, D., Kania, J., Kmiecik, E., Malina, G., and Wator, K. (2020). Fate of selected neonicotinoid insecticides in soil–water systems: current state of the art and knowledge gaps. *Chemosphere* 255 (126981). doi: 10.1016/j.chemosphere.2020.126981.

32 Dereumeaux, C., Fillol, C., Quenel, P., and Denys, S. (2020). Pesticide exposures for residents living close to agricultural lands: a review. *Environ. Int.* 134: 105210. doi: 10.1016/j.envint.2019.105210.

33 Nascimento, M.M., Da Rocha, G.O., and De Andrade, J.B. (2018). Pesticides in the atmospheric environment: an overview on their determination methodologies. *Anal. Methods.* 10: 4484–4504. doi: 10.1039/c8ay01327f.

34 Stehle, S., Bline, A., Bub, S., Petschick, L.L., Wolfram, J., and Schulz, R. (2019). Aquatic pesticide exposure in the U.S. as a result of non-agricultural uses. *Environ. Int.* 133: 105234. doi: 10.1016/j.envint.2019.105234.

35 European Environment Agency (2018). European waters. Assessment of status and pressures 2018. EEA Report No 7/2018. Available at: https://www.eea.europa.eu/publications/state-of-water. Accessed 14 December 2021.

36 Sjerps, R.M.A., Kooij, P.J.F., van Loon, A., and Van Wezel, A.P. (2019). Occurrence of pesticides in Dutch drinking water sources. *Chemosphere* 235: 510–518. doi: 10.1016/j.chemosphere.2019.06.207.

37 Pérez, D.J., Iturburu, F.G., Calderon, G., Oyesqui, L.A.E., De Gerónimo, E., and Aparicio, V.C. (2021). Ecological risk assessment of current-use pesticides and biocides in soils, sediments and surface water of a mixed land-use basin of the Pampas region, Argentina. *Chemosphere* 263: 128061. doi: 10.1016/j.chemosphere.2020.128061.

38 Ordaz-Guillén, Y., Galíndez-Mayer, C.J., Ruiz-Ordaz, N., Juárez-Ramírez, C., Santoyo-Tepole, F., and Ramos-Monroy, O. (2014). Evaluating the degradation of the herbicides picloram and 2,4-D in a compartmentalized reactive biobarrier with internal liquid recirculation. *Environ. Sci. Pollut. Res.* 21: 8765–8773. doi: 10.1007/s11356-014-2809-8.

39 Magnoli, K., Carranza, C.S., Aluffi, M.E., Magnoli, C.E., and Barberis, C.L. (2020). Herbicides based on 2,4-D: its behavior in agricultural environments and microbial biodegradation aspects. A review. *Environ. Sci. Pollut. Res.* 27: 38501–38512. doi: 10.1007/s11356-020-10370-6.

40 Andreu, V. and Picó, Y. (2012). Determination of currently used pesticides in biota. *Anal. Bioanal. Chem.* 404: 2659–2681. doi: 10.1007/s00216-012-6331-x.

41 Pelosi, C., Barot, S., Capowiez, Y., Hedde, M., and Vandenbulcke, F. (2014). Pesticides and earthworms: a review. *Agron. Sustain. Dev.* 34: 199–228. doi: 10.1007/s13593-013-0151-z.

42 Schäfer, S., Buchmeier, G., Claus, E., Duester, L., Heininger, P., Körner, A., Mayer, P., Paschke, A., Rauert, C., Reifferscheid, G., Rüdel, H., Schlechtriem, C., Schröter-Kermani, C., Schudoma, D., Smedes, F., Steffen, D., and Vietoris, F. (2015). Bioaccumulation in aquatic systems: methodological approaches, monitoring and assessment. *Environ. Sci. Eur.* 27: 5. doi: 10.1186/s12302-014-0036-z.

43 Olisah, C., Okoh, O.O., and Okoh, A.I. (2020). Occurrence of organochlorine pesticide residues in biological and environmental matrices in Africa: a two-decade review. *Heliyon* 6: e03518. doi: 10.1016/j.heliyon.2020.e03518.

44 Girones, L., Oliva, A.L., Marcovecchio, J.E., and Arias, A.H. (2020). Spatial distribution and ecological risk assessment of residual organochlorine pesticides (OCPs) in South American marine environments. *Curr. Environ. Heal. Reports* 7: 147–160. doi: 10.1007/s40572-020-00272-7.

45 Lupi, L., Bedmar, F., Wunderlin, D.A., and Miglioranza, K.S.B. (2019). Levels of organochlorine pesticides in soils, mesofauna and streamwater from an agricultural watershed in Argentina. *Environ. Earth Sci.* 78: 1–9. doi: 10.1007/s12665-019-8579-3.

46 Masiá, A., Campo, J., Vázquez-Roig, P., Blasco, C., and Picó, Y. (2013). Screening of currently used pesticides in water, sediments and biota of the Guadalquivir River Basin (Spain). *J. Hazard. Mater.* 263: 95–104. doi: 10.1016/j.jhazmat.2013.09.035.

47 Belenguer, V., Martinez-Capel, F., Masiá, A., and Picó, Y. (2014). Patterns of presence and concentration of pesticides in fish and waters of the Júcar River (eastern Spain). *J. Hazard. Mater.* 265: 271–279. doi: 10.1016/j.jhazmat.2013.11.016.

48 Masiá, A., Campo, J., Navarro-Ortega, A., Barceló, D., and Picó, Y. (2015). Pesticide monitoring in the basin of Llobregat River (Catalonia, Spain) and comparison with historical data. *Sci. Total Environ.* 503–504: 58–68. doi: 10.1016/j.scitotenv.2014.06.095.

49 Ccanccapa, A., Masiá, A., Navarro-Ortega, A., Picó, Y., and Barceló, D. (2016). Pesticides in the Ebro River basin: occurrence and risk assessment. *Environ. Pollut.* 211: 414–424. doi: 10.1016/j.envpol.2015.12.059.

50 Maceira, A., Marcé, R.M., and Borrull, F. (2020). Analytical methods for determining organic compounds present in the particulate matter from outdoor air. *Trends Anal. Chem.* 122: 115707. doi: 10.1016/j.trac.2019.115707.

51 Sanjuán-Herráez, D., Rodríguez-Carrasco, Y., Juan-Peiró, L., Pastor, A., and de La Guardia, M. (2011). Determination of indoor air quality of a phytosanitary plant. *Anal. Chim. Acta* 694: 67–74. doi: 10.1016/j.aca.2011.03.039.

52 Batterman, S.A., Chernyak, S.M., Gounden, Y., Matooane, M., and Naidoo, R.N. (2008). Organochlorine pesticides in ambient air in Durban, South Africa. *Sci. Total Environ.* 397: 119–130. doi: 10.1016/j.scitotenv.2008.02.033.

53 Duong, H.T., Kadokami, K., Trinh, H.T., Phan, T.Q., Le, G.T., Nguyen, D.T., Nguyen, T.T., and Nguyen, D.T. (2019). Target screening analysis of 970 semi-volatile organic compounds adsorbed on atmospheric particulate matter in Hanoi, Vietnam. *Chemosphere* 219: 784–795. doi: 10.1016/j.chemosphere.2018.12.096.

54 Anh, H.Q., Tomioka, K., Tue, N.M., Tuyen, L.H., Chi, N.K., Minh, T.B., Viet, P.H., and Takahashi, S. (2019). A preliminary investigation of 942 organic micro-pollutants in the T atmosphere in waste processing and urban areas, northern Vietnam: levels, potential sources, and risk assessment. *Ecotoxicol. Environ. Saf.* 167: 354–364. doi: 10.1016/j.ecoenv.2018.10.026.

55 Guida, Y.D.S., Meire, R.O., Torres, J.P.M., and Malm, O. (2018). Air contamination by legacy and current-use pesticides in Brazilian mountains: an overview of national regulations by monitoring pollutant presence in pristine areas. *Environ. Pollut.* 242: 19–30. doi: 10.1016/j.envpol.2018.06.061.

56 Coscollà, C., León, N., Pastor, A., and Yusà, V. (2014). Combined target and post-run target strategy for a comprehensive analysis of pesticides in ambient air using liquid chromatography-Orbitrap high resolution mass spectrometry. *J. Chromatogr. A* 1368: 132–142. doi: 10.1016/j.chroma.2014.09.067.

57 Coscollá, C., Hart, E., Pastor, A., and Yusà, V. (2013). LC-MS characterization of contemporary pesticides in PM10 of Valencia Region, Spain. *Atmos. Environ.* 77: 394–403. doi: 10.1016/j.atmosenv.2013.05.022.

58 Nascimento, M.M., Da Rocha, G.O., and De Andrade, J.B. (2017). Pesticides in fine airborne particles: from a green analysis method to atmospheric characterization and risk assessment. *Sci. Rep.* 7: 1–11. doi: 10.1038/s41598-017-02518-1.

59 Pinasseau, L., Wiest, L., Volatier, L., Mermillod-Blondin, F., and Vulliet, E. (2020). Emerging polar pollutants in groundwater: potential impact of urban stormwater infiltration practices. *Environ. Pollut.* 266: 115387. doi: 10.1016/j.envpol.2020.115387.

60 Vrana, B., Smedes, F., Prokeš, R., Loos, R., Mazzella, N., Miege, C., Budzinski, H., Vermeirssen, E., Ocelka, T., Gravell, A., and Kaserzon, S. (2016). An interlaboratory study on passive sampling of emerging water pollutants. *Trends Anal. Chem.* 76: 153–165. doi: 10.1016/j.trac.2015.10.013.

61 Yu, Y., Liu, X., He, Z., Wang, L., Luo, M., Peng, Y., and Zhou, Q. (2016). Development of a multi-residue method for 58 pesticides in soil using QuEChERS and gas chromatography-tandem mass spectrometry. *Anal. Methods* 8: 2463–2470. doi: 10.1039/c6ay00337k.

62 Gerónimo, E., Botero-Coy, A.M., Marín, J.M., Aparicio, V.C., Costa, J.L., Sancho, J.V., and Hernández, F. (2015). A simple and rapid analytical methodology based on liquid chromatography-tandem mass spectrometry for monitoring pesticide residues in soils from Argentina. *Anal. Methods.* 7: 9504–9512. doi: 10.1039/c5ay01582k.

63 Hu, J.Y., Zhen, Z.H., and Deng, Z.B. (2011). Simultaneous determination of acetochlor and propisochlor residues in corn and soil by solid phase extraction and gas chromatography with electron capture detection. *Bull. Environ. Contam. Toxicol.* 86: 95–100. doi: 10.1007/s00128-010-0130-x.

64 Bragança, I., Lemos, P.C., Delerue-Matos, C., and Domingues, V.F. (2019). Pyrethroid pesticide metabolite, 3-PBA, in soils: method development and application to real agricultural soils. *Environ. Sci. Pollut. Res.* 26: 2987–2997. doi: 10.1007/s11356-018-3690-7.

65 Zhao, P., Zhao, J., Lei, S., Guo, X., and Zhao, L. (2018). Simultaneous enantiomeric analysis of eight pesticides in soils and river sediments by chiral liquid chromatography-tandem mass spectrometry. *Chemosphere* 204: 210–219. doi: 10.1016/j.chemosphere.2018.03.204.

66 Sørensen, L., Silva, M.S., Meier, S., and Booth, A.M. (2015). Advances in miniaturization and increasing sensitivity in analysis of organic contaminants in marine biota samples. *Trends Environ. Anal. Chem.* 6–7: 39–47. doi: 10.1016/j.teac.2015.03.001.

67 Lundqvist, J., Von Brömssen, C., Rosenmai, A.K., Ohlsson, A., Le Godec, T., Jonsson, O., Kreuger, J., and Oskarsson, A. (2019). Assessment of pesticides in surface water samples from Swedish agricultural areas by integrated bioanalysis and chemical analysis. *Environ. Sci. Eur.* 31: 53. doi: 10.1186/s12302-019-0241-x.

68 Hashmi, T.A., Qureshi, R., Tipre, D., and Menon, S. (2019). Investigation of pesticide residues in water, sediments and fish samples from Tapi River, India as a case study and its forensic significance. *Environ. Forensics.* 21: 1–10. doi: 10.1080/15275922.2019.1693441.

69 Ruiz-Gil, L., Romero-González, R., Garrido Frenich, A., and Martínez Vidal, J.L. (2008). Determination of pesticides in water samples by solid phase extraction and gas chromatography tandem mass spectrometry. *J. Sep. Sci.* 31: 151–161. doi: 10.1002/jssc.200700299.

70 Xie, H., Chen, J., Huang, Y., Zhang, R., Chen, C.E., Li, X., and Kadokami, K. (2020). Screening of 484 trace organic contaminants in coastal waters around the Liaodong Peninsula, China: occurrence, distribution, and ecological risk. *Environ. Pollut.* 267: 115436. doi: 10.1016/j.envpol.2020.115436.

71 Wan, Y., Tran, T.M., Nguyen, V.T., Wang, A., Wang, J., and Kannan, K. (2021). Neonicotinoids, fipronil, chlorpyrifos, carbendazim, chlorotriazines, chlorophenoxy herbicides, bentazon, and selected pesticide transformation products in surface water and drinking water from northern Vietnam. *Sci. Total Environ.* 750: 141507. doi: 10.1016/j.scitotenv.2020.141507.

72 Fauvelle, V., Mazzella, N., Morin, S., Moreira, S., Delest, B., and Budzinski, H. (2015). Hydrophilic interaction liquid chromatography coupled with tandem mass spectrometry for acidic herbicides and metabolites analysis in fresh water. *Environ. Sci. Pollut. Res.* 22: 3988–3996. doi: 10.1007/s11356-014-2876-x.

73 Tan, B., Xiong, J., Li, H., and You, J. (2020). Simultaneous analysis of current-use pesticides and their transformation products in water using mixture-sorbent solid phase extraction and high-performance liquid chromatography–tandem mass spectrometry. *J. Sep. Sci.* 43: 2409–2418. doi: 10.1002/jssc.202000115.

74 Čelić, M., Jaén-Gil, A., Briceño-Guevara, S., Rodríguez-Mozaz, S., Gros, M., and Petrovic, M. (2021). Extended suspect screening to identify contaminants of emerging concern in riverine and coastal ecosystems and assessment of environmental risks. *J. Hazard. Mater.* 404: 124102. doi: 10.1016/j.jhazmat.2020.124102.

75 Schulze, T., Ahel, M., Ahlheim, J., Ait-Aissa, S., Brion, F., Di Paolo, C., Fro ment, J., Hidasi, A.O., Hollender, J., Hollert, H., Hu, M., Klob, A., Koprivica, S., Krauss, M., Muz, M., Oswald, P., Petre, M., Schollée, J.E., Seiler, T.B., Shao, Y., Slobod nik, J., Sonavane, M., Suter, M.J.F., Tollefsen, K.E., Tousova, Z., Walz, J.H., and Brack, W. (2017). Assessment of a novel device for onsite integrative large-volume solid phase extraction of water samples to enable a comprehensive chemical and effect-based analysis. *Sci. Total Environ.* 581–582: 350–358. doi: 10.1016/j.scitotenv.2016.12.140.

76 Hurtado-Sánchez, M.C., Romero-González, R., Rodríguez-Cáceres, M.I., Durán-Merás, I., and Garrido Frenich, A. (2013). Rapid and sensitive on-line solid phase extraction ultra-high-performance liquid chromatography-electrospray-tandem mass spectrometry analysis of pesticides in surface waters. *J. Chromatogr. A* 1305: 193–202. doi: 10.1016/j.chroma.2013.07.045.

77 Postigo, C., Ginebreda, A., Barbieri, M.V., Barceló, D., Martín-Alonso, J., Cal, A., Boleda, M.R., Otero, N., Carrey, R., Solà, V., Queralt, E., Isla, E., Casanovas, A., Frances, G., and López de Alda, M. (2021). Investigative monitoring of pesticide and nitrogen pollution sources in a complex multi-stressed catchment: the lower Llobregat River basin case study (Barcelona, Spain). *Sci. Total Environ.* 755: 142377. doi: 10.1016/j.scitotenv.2020.142377.

78 Barbieri, M.V., Monllor-Alcaraz, L.S., Postigo, C., and López de Alda, M. (2020). Improved fully automated method for the determination of medium to highly polar pesticides in surface and groundwater and application in two distinct agriculture-impacted areas. *Sci. Total Environ.* 745: 140650. doi: 10.1016/j.scitotenv.2020.140650.

79 Domínguez, I., Romero-González, R., Arrebola Liébanas, F.J., Martínez Vidal, J.L., and Garrido Frenich, A. (2016). Automated and semi-automated extraction methods for GC–MS determination of pesticides in environmental samples. *Trends Environ. Anal. Chem.* 12: 1–12. doi: 10.1016/j.teac.2016.09.001.

80 Abdel Ghani, S.B. and Hanafi, A.H. (2016). QuEChERS method combined with GC–MS for pesticide residues determination in water. *J. Anal. Chem.* 71: 508–512. doi: 10.1134/S1061934816050117.

81 Garrido Frenich, A., Romero-González, R., Martínez Vidal, J.L., Martínez Ocaña, R., and Baquero Feria, P. (2011). Comparison of solid phase microextraction and hollow fiber liquid phase microextraction for the determination of pesticides in aqueous samples by gas chromatography triple quadrupole tandem mass spectrometry. *Anal. Bioanal. Chem.* 399: 2043–2059. doi: 10.1007/s00216-010-4236-0.

82 Menezes, H.C., Paulo, B.P., Paiva, M.J.N., and Cardeal, Z.L. (2016). A simple and quick method for the determination of pesticides in environmental water by HF-LPME-GC/MS. *J. Anal. Methods Chem.* 2016: 1–11. doi: 10.1155/2016/7058709.

83 Tankiewicz, M., Fenik, J., and Biziuk, M. (2011). Solventless and solvent-minimized sample preparation techniques for determining currently used pesticides in water samples: a review. *Talanta* 86: 8–22. doi: 10.1016/j.talanta.2011.08.056.

84 Domínguez, I., Arrebola, F.J., Romero-González, R., Nieto-García, A., Martínez Vidal, J.L., and Garrido Frenich, A. (2017). Solid phase microextraction and gas chromatography coupled to magnetic sector high resolution mass spectrometry for the ultra-trace determination of contaminants in surface water. *J. Chromatogr. A* 1518: 15–24. doi: 10.1016/j.chroma.2017.08.061.

85 Domínguez, I., Arrebola, F.J., Gavara, R., Martínez Vidal, J.L., and Garrido Frenich, A. (2018). Automated and simultaneous determination of priority substances and polychlorinated biphenyls in wastewater using headspace solid phase microextraction and high resolution mass spectrometry. *Anal. Chim. Acta* 1002: 39–49. doi: 10.1016/j.aca.2017.11.056.

86 Jamshidi, F., Nouri, N., Sereshti, H., and Aliabadi, M.H.S. (2020). Synthesis of magnetic poly (acrylic acid-menthol deep eutectic solvent) hydrogel: application for extraction of pesticides. *J. Mol. Liq.* 318: 114073. doi: 10.1016/j.molliq.2020.114073.

87 Valenzuela, E.F., de Paula, F.G.F., Teixeira, A.P.C., Menezes, H.C., and Cardeal, Z.L. (2020). A new carbon nanomaterial solid-phase microextraction to pre-concentrate and extract pesticides in environmental water. *Talanta* 217: 121011. doi: 10.1016/j.talanta.2020.121011.

88 Zhao, P., Wang, Z., Li, K., Guo, X., and Zhao, L. (2018). Multi-residue enantiomeric analysis of 18 chiral pesticides in water, soil and river sediment using magnetic solid-phase extraction based on amino modified multiwalled carbon nanotubes and chiral liquid chromatography coupled with tandem mass spectrometry. *J. Chromatogr. A* 1568: 8–21. doi: 10.1016/j.chroma.2018.07.022.

89 Mechelke, J., Longrée, P., Singer, H., and Hollender, J. (2019). Vacuum-assisted evaporative concentration combined with LC-HRMS/MS for ultra-trace-level screening of organic micropollutants in environmental water samples. *Anal. Bioanal. Chem.* 2555–2567. doi: 10.1007/s00216-019-01696-3.

90 Kiefer, K., Bader, T., Minas, N., Salhi, E., Janssen, E.M.L., Von Gunten, U., and Hollender, J. (2020). Chlorothalonil transformation products in drinking water resources: widespread and challenging to abate. *Water Res.* 183: 116066. doi: 10.1016/j.watres.2020.116066.

91 Reemtsma, T., Alder, L., and Banasiak, U. (2013). A multimethod for the determination of 150 pesticide metabolites in surface water and groundwater using direct injection liquid chromatography-mass spectrometry. *J. Chromatogr. A* 1271: 95–104. doi: 10.1016/j.chroma.2012.11.023.

92 Dagnac, T., Jeannot, R., Mouvet, C., and Baran, N. (2002). Determination of oxanilic and sulfonic acid metabolites of acetochlor in soils by liquid chromatography-electrospray ionisation mass spectrometry. *J. Chromatogr. A* 957: 69–77. doi: 10.1016/S0021-9673(02)00310-2.

93 Colazzo, M., Pareja, L., Cesio, M.V., and Heinzen, H. (2018). Multi-residue method for trace pesticide analysis in soils by LC-QQQ-MS/MS and its application to real samples. *Int. J. Environ. Anal. Chem.* 98: 1292–1308. doi: 10.1080/03067319.2018.1551530.

94 Martínez Vidal, J.L., Padilla Sánchez, J.A., Plaza-Bolaños, P., Garrido Frenich, A., and Romero-González, R. (2010). Use of pressurized liquid extraction for the simultaneous analysis of 28 polar and 94 non-polar pesticides in agricultural soils by GC/QqQ-MS/MS and UPLC/QqQ-MS/MS. *J. AOAC Int.* 93: 1715–1731. doi: 10.1093/jaoac/93.6.1715.

95 Homazava, N., Gachet Aquillon, C., Vermeirssen, E., and Werner, I. (2014). Simultaneous multi-residue pesticide analysis in soil samples with ultra-high-performance liquid chromatography–tandem mass spectrometry using QuEChERS and pressurised liquid extraction methods. *Int. J. Environ. Anal. Chem.* 94: 1085–1099. doi: 10.1080/03067319.2014.954558.

96 Pérez-Mayán, L., Ramil, M., Cela, R., and Rodríguez, I. (2020). Multiresidue procedure to assess the occurrence and dissipation of fungicides and insecticides in vineyard soils from Northwest Spain. *Chemosphere* 261: 127696. doi: 10.1016/j.chemosphere.2020.127696.

97 Wu, X., Dong, F., Xu, J., Liu, X., Wu, X., and Zheng, Y. (2020). Enantioselective separation and dissipation of pydiflumetofen enantiomers in grape and soil by supercritical fluid chromatography–tandem mass spectrometry. *J. Sep. Sci.* 43: 2217–2227. doi: 10.1002/jssc.201901332.

98 Hou, X., Qiao, T., Zhao, Y., and Liu, D. (2019). Dissipation and safety evaluation of afidopyropen and its metabolite residues in supervised cotton field. *Ecotoxicol. Environ. Saf.* 180: 227–233. doi: 10.1016/j.ecoenv.2019.04.089.

99 Acosta-Dacal, A., Rial-Berriel, C., Díaz-Díaz, R., Bernal-Suárez, M.M., and Luzardo, O.P. (2021). Optimization and validation of a QuEChERS-based method for the simultaneous environmental monitoring of 218 pesticide residues in clay loam soil. *Sci. Total Environ.* 753: 142015. doi: 10.1016/j.scitotenv.2020.142015.

100 Chen, K., Li, S., Hu, M., Xu, J., Wu, X., Dong, F., Zheng, Y., and Liu, X. (2017). Dissipation dynamics of fenamidone and propamocarb hydrochloride in pepper, soil and residue analysis in vegetables by ultra-performance liquid chromatography coupled with tandem mass spectrometry. *Int. J. Environ. Anal. Chem.* 97: 134–144. doi: 10.1080/03067319.2017.1291807.

101 Ismail, N.A.H., Wee, S.Y., and Aris, A.Z. (2017). Multi-class of endocrine disrupting compounds in aquaculture ecosystems and health impacts in exposed biota. *Chemosphere* 188: 375–388. doi: 10.1016/j.chemosphere.2017.08.150.

102 Egea Gonzalez, F.J., Mena Granero, A., Glass, C.R., Garrido Frenich, A., and Martinez Vidal, J.L. (2004). Screening method for pesticides in air by gas chromatography/tandem mass spectrometry. *Rapid Commun. Mass Spectrom.* 18: 537–543. doi: 10.1002/rcm.1359.

103 Van Ael, E., Covaci, A., Blust, R., and Bervoets, L. (2012). Persistent organic pollutants in the Scheldt estuary: environmental distribution and bioaccumulation. *Environ. Int.* 48: 17–27. doi: 10.1016/j.envint.2012.06.017.

104 Miyawaki, T., Tobiishi, K., Takenaka, S., and Kadokami, K. (2018). A rapid method, combining microwave-assisted extraction and gas chromatography-mass spectrometry with a database, for determining organochlorine pesticides and polycyclic aromatic hydrocarbons in soils and sediments. *Soil Sediment Contam.* 27: 31–45. doi: 10.1080/15320383.2017.1360245.

105 Zhao, X., Cui, T., Guo, R., Liu, Y., Wang, X., An, Y., Qiao, X., and Zheng, B. (2019). A clean-up method for determination of multi-classes of persistent organic pollutants in sediment and biota samples with an aliquot sample. *Anal. Chim. Acta* 1047: 71–80. doi: 10.1016/j.aca.2018.10.011.

106 Mesquita, T.C.R., Santos, R.R., Cacique, A.P., De Sá, L.J., Silvério, F.O., and Pinho, G.P. (2018). Easy and fast extraction methods to determine organochlorine pesticides in sewage sludge, soil, and water samples based at low temperature. *J. Environ. Sci. Heal. – Part B Pestic. Food Contam. Agric. Wastes.* 53: 199–206. doi: 10.1080/03601234.2017.1405626.

107 Belmonte Vega, A., Garrido Frenich, A., and Martínez Vidal, J.L. (2005). Monitoring of pesticides in agricultural water and soil samples from Andalusia by liquid chromatography coupled to mass spectrometry. *Anal. Chim. Acta* 538: 117–127. doi: 10.1016/j.aca.2005.02.003.

108 Knoll, S., Rösch, T., and Huhn, C. (2020). Trends in sample preparation and separation methods for the analysis of very polar and ionic compounds in environmental water and biota samples. *Anal. Bioanal. Chem.* 412: 6149–6165. doi: 10.1007/s00216-020-02811-5.

109 Deng, H., Ji, Y., Tang, S., Yang, F., Tang, G., Shi, H., and Kee Lee, H. (2020). Application of chiral and achiral supercritical fluid chromatography in pesticide analysis: a review. *J. Chromatogr. A* 1634: 461684. doi: 10.1016/j.chroma.2020.461684.

110 Bieber, S., Greco, G., Grosse, S., and Letzel, T. (2017). RPLC-HILIC and SFC with mass spectrometry: polarity-extended organic molecule screening in environmental (water) samples. *Anal. Chem.* 89: 7907–7914. doi: 10.1021/acs.analchem.7b00859.

111 Salvatierra-Stamp, V.D.C., Ceballos-Magaña, S.G., Gonzalez, J., Ibarra-Galván, V., and Muñiz-Valencia, R. (2015). Analytical method development for the determination of emerging contaminants in water using supercritical-fluid chromatography coupled with diode-array detection. *Anal. Bioanal. Chem.* 407: 4219–4226. doi: 10.1007/s00216-015-8581-x.

112 Geerdink, R.B., Hassing, M., Ayarza, N., Bruggink, C., Wielheesen, M., Claassen, J., and Epema, O.J. (2020). Analysis of glyphosate, AMPA, glufosinate and MPPA with ion chromatography tandem mass spectrometry using a membrane suppressor in the ammonium form application to surface water of low to moderate salinity. *Anal. Chim. Acta* 1133: 66–76. doi: 10.1016/j.aca.2020.05.058.

113 Gissawong, N., Mukdasai, S., Boonchiangma, S., Sansuk, S., and Srijaranai, S. (2020). A rapid and simple method for the removal of dyes and organophosphorus pesticides from water and soil samples using deep eutectic solvent embedded sponge. *Chemosphere* 260: 127590. doi: 10.1016/j.chemosphere.2020.127590.

114 Asensio-Ramos, M., Hernández-Borges, J., Ravelo-Pérez, L.M., Alfonso, M.M., Palenzuela, J.A., and Rodríguez-Delgado, M.A. (2012). Dispersive liquid-liquid microextraction of pesticides and metabolites from soils using 1,3-dipentylimidazolium hexafluorophosphate ionic liquid as an alternative extraction solvent. *Electrophoresis* 33: 1449–1457. doi: 10.1002/elps.201100522.

115 Ivdra, N., Herrero-Martín, S., and Fischer, A. (2014). Validation of user- and environmentally friendly extraction and clean-up methods for compound-specific stable carbon isotope analysis of organochlorine pesticides and their metabolites in soils. *J. Chromatogr. A* 1355: 36–45. doi: 10.1016/j.chroma.2014.06.014.

116 Domínguez, I., Arrebola, F.J., Martínez Vidal, J.L., and Garrido Frenich, A. (2020). Assessment of wastewater pollution by gas chromatography and high resolution

Orbitrap mass spectrometry. *J. Chromatogr. A* 1619: 460964. doi: 10.1016/j.chroma.2020.460964.

117 Meng, D., Fan, D., Gu, W., Wang, Z., Chen, Y., Bu, H., and Liu, J. (2020). Development of an integral strategy for non-target and target analysis of site-specific potential contaminants in surface water: a case study of Dianshan Lake, China. *Chemosphere* 243: 125367. doi: 10.1016/j.chemosphere.2019.125367.

2

Pharmaceuticals

Monika Paszkiewicz, Hanna Lis, Magda Caban, Anna Białk-Bielińska, and Piotr Stepnowski

Department of Environmental Analysis, Faculty of Chemistry, University of Gdansk, Gdansk, Poland

2.1 Overview of Pharmaceuticals

2.1.1 Properties

It has already been more than 20 years since the problem of pharmaceutical residues in the environment gained huge scientific attention all over the world [1, 2]. In general, pharmaceuticals are chemicals that have been designed to be biologically active at very low doses for therapeutic, prophylactic and preventative purposes. Therefore, they may be classified based on their mode of action or medical purposes. From the chemical point of view, this group of substances is usually referred to as polar and ionizable compounds. However, it should be noted that this is a class of chemicals with a wide range of physico-chemical properties (Figure 2.1) [1]. Based on the selected properties presented in Figure 2.1, pharmaceuticals are compounds with either a very simple chemical structure or a very complicated one, that can be furthermore classified as weak acids/bases or amphoteric compounds [1]. Their structure and hence properties such as acidity and hydrophilicity/lipophilicity influence their fate in the environment. To prove this the estimated adsorption coefficients have also been presented in Figure 2.1, highlighting that some of these chemicals might be recognized as very mobile while others present rather strong sorption potential to soil and sediments [3, 4]. This also explains why pharmaceuticals have been detected in many different environmental compartments [1–3, 5–10].

2.1.2 Reported or Potential Metabolites and/or Transformation Products

During the last two decades many studies have been performed in order to evaluate their sources, occurrence, fate and effects on human health and other organisms [1, 2, 4–10]. Many different sources of their presence have been pointed out, including their inefficient removal in the wastewater treatment plants, husbandry, aquacultures, inappropriate drugs disposal, pharmaceutical industry etc. [1, 2, 4–10]. In general, after their

Analytical Methods for Environmental Contaminants of Emerging Concern, First Edition. Edited by Núria Fontanals and Rosa Maria Marcé.

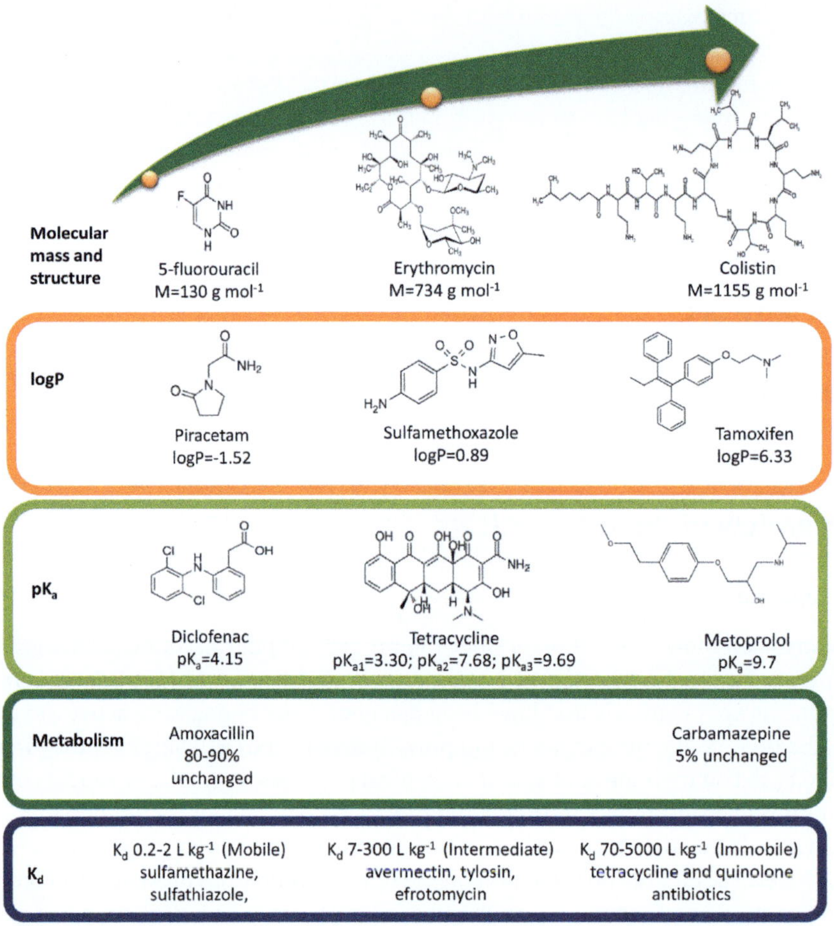

Figure 2.1 Selected properties of pharmaceuticals.

application in human medicine or in veterinary use, pharmaceuticals undergo different processes leading to the excretion of their metabolites, which should usually be less active and more polar than native forms [1, 2]. In some cases, however, metabolism may lead to more active compounds, such as metabolites of carbamazepine, diclofenac, acetaminophen [11–14] or tamoxifen (hydroxyl-desmethyltamoxifen and 4-hydroxytamoxifen), which (these last two) can be up to 100 times more potent and active than the parent compound [15]. This in turn leads to the occurrence in the environment not only of their parent forms but also of their metabolites [11–14, 16]. However, it must also be taken into account that pharmaceuticals and their metabolites may undergo many different transformation processes in the waste water treatment plant, drinking water treatment processes or being present in the environment (such as hydrolysis, photodegradation and biodegradation), which can also lead to the production of many new additional degradation products of pharmaceuticals [11]. For example, glucuronide- and sulfate-conjugates (phase II metabolites) have been observed to be readily deconjugated during biological wastewater treatment to the active form of the pharmaceutical or its corresponding phase I metabolite [11]. All of these compounds, including metabolites and degradation products, are usually referred to as "transformation products (TPs)" in the scientific literature [11, 12, 14].

It must also be highlighted that the environmental fate of TPs can be different from the parent compound, meaning that the environmental compartments that are exposed to the parent compound may not be exposed to a TP [11]. While some of the TPs are known and have already been detected in the aquatic environment in concentrations higher than their parent compounds, the majority of TPs present there have not been identified yet [11]. The current state of knowledge on the presence of TPs in the environment has been quite recently presented [11–14, 16]. Based on these data, selected TPs have been presented in Table 2.1. Taking into consideration these data, it might be concluded that human and the environment are exposed to a highly variable and still not fully known cocktail of pharmaceuticals and their TPs.

2.1.3 Occurrence

In 2019 the German Environmental Agency (Umwelt Bundesamt) published the Final report on *The database "Pharmaceuticals in the Environment" – Update and new analysis* (Texte 67/2019). This is a summary of the state of knowledge on this problem [17]. It is based on the data published all over the world in 1,519 publications from peer-reviewed journals and 240 review papers from 1987 up to 2016. The analysis of these data revealed that pharmaceuticals and their transformation products have been detected in 75 countries, and in 54 different matrices worldwide. It was proved that 771 active pharmaceutical substances and their TPs have been measured globally, and 596 substances in the European Union. It must also be noted that most of these compounds have been detected in the effluents of wastewater treatment plants. In contrast, 528 substances were determined in surface water, groundwater and drinking water all over the world. However, 19 of them have been detected in all five United Nation regions [17]: 17-α-ethynylestradiol, 17-β-estradiol, acetylsalicylic acid, carbamazepine, ciprofloxacin, clofibric acid, diclofenac, estriol, estrone, ibuprofen, indomethacin, ketoprofen, naproxen, paracetamol, sulfamethazine, sulfamethoxazole, triclocarban, triclosan and trimethoprim.

In comparison with the previous version of these data five substances (indomethacin, ketoprofen, sulfamethazine, triclocarban and triclosan) have now been detected in all five UN regions [17]. Even though there have been more than 50 different matrices analysed, it has been proved that pharmaceutical residues were mostly found in groundwater, surface water and drinking water, as well as soil and sediments [17]. Moreover, it was also observed that the most commonly detected pharmaceuticals in surface water or wastewater worldwide belong to the group of antibiotics and analgesic/anti-inflammatories [1]. Usually, more than 40% of the analysed samples were positive for at least one target, with concentrations for half of them $<0.1\ \mu g\ L^{-1}$ [1]. Furthermore, most of the published data on the Maximal Environmental Concentrations (MECs) come from China, USA, Spain and Germany [17].

In general, their concentrations are in the range of $ng\ L^{-1}$ up to $\mu g\ L^{-1}$ in the water bodies (wastewater, surface water, ground water and drinking water) and from $\mu g\ kg^{-1}$ up to $mg\ kg^{-1}$ in sludge/soils or sediments. In Figure 2.2 the range of pharmaceutical concentrations in different environmental samples has been presented based on the previously reviewed data in [1] and [17].

Table 2.1 Pharmaceuticals and their selected transformation products [12–14].

Pharmaceutical (native form)	Therapeutic group	Excretion of parent form [%]	Main metabolites/ Transformation products	Justification	Additional comments
Carbamazepine (CBZ)	Antidepressant/ antiepileptic	5%	carbamazepine-10,11-epoxide (CBZ-Ep)	Active metabolite; however, unstable; can be transformed to Di-OH-CBZ, hence the concentration in the environment is usually lower than for Di-OH-CBZ; detected in the environment up to 4000 ng L^{-1}. It is also a transformation product of CBZ in the process of water chlorination. Toxicity to *V. fischeri* is similar to the toxicity of parent form.	There were no losses in the water treatment processes in terms of the content of CBZ metabolites. In some cases, the amount of 2-OH-CBZ and Di-OH-CBZ increased after the purification process, which was most likely due to the fact that some CBZ and its metabolites are also excreted in the form of glucuronides.
			10,11-dihydro-trans-10,11-dihydroxy-carbamazepine (Di-OH-CBZ)	Not active, concentrations at the level of 4000 ng L^{-1}, concentrations higher than for the native form. It is also a transformation product of CBZ in the process of water chlorination. More toxic to *V. fischeri* than the native substance.	
			2-hydroxycarbamazepine (2-OH-CBZ)	More toxic to *V. fischeri* than the native substance; detected in the environmental samples.	
			3-hydroxycarbamazepine (3-OH-CBZ)	More toxic to *V. fischeri* than the native substance; detected in the environmental samples.	
			10-hydroxycarbamazepine (10-OH-CBZ)	Often detected in environmental samples, more often than other metabolites, although the percentage of this metabolite is low, probably it is a TP of CBZ in WWTP purification processes.	

Metopropol (MTP)	β-blocker	<10%	metoprolol acid (MTPA), 60–65%	Detected in the environmental samples at concentrations 10 times higher than for the native form; this substance is also a product of the transformation of other beta-blockers, including atenolol, in the biodegradation process, hence its concentration is almost 10 times higher in effluents than in influents; acute toxicity to *V. fischeri* has not been demonstrated.
			α-hydroxymetoprolol (α-HMTP)	Detected, but only in wastewater, which is suspected to be due to the biodegradability of MTP in the water treatment process; acute toxicity to *V. fischeri* has not been demonstrated.
			O-desmethylmetoprolol (*O*-DMTP)	MTP biodegradation product; however, it is rapidly biodegradable and therefore not detected in environmental samples.
Sulfamethoxazole (SMX)	Antibiotic	20%	*N*4-acetylo-sulfamethoxazole (*N*4-Ac-SMX), 50–70%	Main metabolite of SMX. Excreted in 50–70%. Detected in the environmental samples.
Ciprofloxacin (CIP)		40–50%		CIP is the metabolite of enrofloxacin; metabolites of CIP are less active than CIP itself, moreover they are excreted in smaller amount that native form.
Metronidazol (METR)		20%	hydroxymetronidazol (METR-OH), main metabolite	Active; in treated waters discharged from hospital treatment plants at a concentration of 11 $\mu g\ L^{-1}$; although the concentrations in WWTP wastewater are lower (160 $ng\ L^{-1}$).

(*Continued*)

Table 2.1 (Continued)

Pharmaceutical (native form)	Therapeutic group	Excretion of parent form [%]	Main metabolites/ Transformation products	Justification	Additional comments
Ibuprofen (IBU)	NSAIDs	10–15%	2-hydroxy-ibuprofen (2-OH-IBU), 26%	Concentration in influents 6840 ng L^{-1} while in the effluents 1130 ng L^{-1}; hydroxyl derivatives more toxic to *V. fischeri* than the native form.	It was found that these compounds are also produced as TPs in the biodegradation process of IBU. CX-IBU is removed by biodegradation, but OH-IBU is stable, therefore it is the dominant TP in purified waters, although there are studies where it has been found that both can be biodegradable – it depends on the test conditions.
			carboxy-ibuprofen (CX-IBU), 43%	Concentration in influents 38.4 μg L^{-1} while in the effluents 10.6 μg L^{-1}.	
			1-hydroxy-ibuprofen (1-OH-IBU)	The percentage of this metabolite is lower than for others; this is reflected in its concentrations detected in wastewaters, lower than for 2-OH-IBU; however, hydroxyl derivatives are more toxic than the native form to *V. fischeri*.	
Diclofenac (DIC)		5–10%	4-hydroxydiclofenac (4-OH-DCF), main metabolite, 16%	Concentration in WWTP effluents at the level of 1600 ng L^{-1}.	
			5-hydroxydiclofenac (5-OH-DCF), 6,1%	Concentration in WWTP effluents at the level of 860 ng L^{-1}.	
			4-hydroxydiclofenac dehydrate (4-OH-DCF-H_2O)	Concentration in WWTP effluents at the level of 660 ng L^{-1}.	
Ketoprofen (KET)		20%	glucuronide-KET 80%	Excreted mainly in the form of glucuronides, which are unstable, which leads to the formation of the native form.	

Naproxen (NPX)		<1%	6-*O*-desmethyl-naproxen (6-*O*-DMNPX) (<1%)	It is a metabolites, but it is also produced during the biodegradation process in WWTP.
			NPX and 6-*O*-DMNPX excreted mainly as coniugates (66–92%)	
Tramadol (TRA)	Opioid painkiller	30%	*O*-desmethyl-tramadol (*O*-DMTRA)	Active metabolite.
Fluoxatine (FLX)	Antidepressant	11%	Norfluoxetine (NOR-FLX)	
Venlafaxine (VNF)		5%	*O*-desmethylvenlafaxine (*O*-DM-VNF)	
Estron (E1)	Steroid hormones		2-hydroxyestrone (2-OH-E1)	Detected at very low concentrations (max. 14 ng L^{-1}).
			4-hydroxyestrone (4-OH-E1)	Detected at very low concentrations (max. 14 ngL^{-1}).
Tamoxifen (TAM)	Anticancer drug	<30%	Hydroxytamoxifen (OH-TAM)	Active metabolite.
			endoxifen	Active metabolite.
Cyclophosphamide (CF)/ ifosfamide (IF)			carboxyphosphamide	
			4-OH-cyclophosphamide/ 4-OH-ifosfamide	Active metabolites, but unstable.
Methothrexate (MET)			7-hydroxymethothrexate (7-OH-MET)	Active metabolite and TPs – it is produces during the MET biodegradation; however, it is stable.
Fenbendazole (FEN)	Antiparasites		fenbendazole sulfoxide (FEN-SO) oxfendazol (OXFEN)	Active metabolite.

Figure 2.2 Pharmaceuticals occurrence in the environment. Average range of concentration based on the data presented in *[1], examples of MECs based on the data in **[17] (applied abbreviations: E2 – estradiol, EE2 – 17-α-estradiol, DIC – diclofenac, AZ – azithromycin, CL – clarithromycin, ER – erythromycin, CIP – ciprofloxacin, AMX – amoxicillin, IBU – ibuprofen, TMP – trimethoprim).

2.1.4 Legislation

One of the biggest concerns resulting from the presence of pharmaceutical residues in the environment is the lack of sufficient data to evaluate their real risk to human health and other organisms in the ecosystem. This problem is getting even more complicated if the TPs are taken into account, which in many cases have not yet been sufficiently identified or their (eco) toxicity has not been determined. For all the mentioned reasons, the scientific community and regulatory and authority agencies started to pay special attention to the problem of the presence of pharmaceuticals and their TPs in the environment by introducing specific guidelines on their risk assessment and water policy actions (EU Watch List), as well as supporting and founding research projects concerning this problem. Currently, one of the most important regulations is the European Watch List (WL) for emerging water pollutants, which is aimed at increasing monitoring for substances with a significant risk for human health and the environment in order to perform a future EU-wide risk assessment. Several pharmaceuticals have been listed on the WL. However, it must also be stated that such WL is updated, and it includes chemicals that need to be monitored by EU Member States in surface water at least once per year for 4 years. The first Watch List was established by the Commission Implementing Decision (EU) 2015/4951 in 2015 [18], and it included such pharmaceuticals as:

- (1st WL) 17-β-estradiol, 17-α-ethynylestradiol, diclofenac, erythromycin, clarithromycin, azithromycin.

It was updated in 2018 by Commission Implementing Decision (EU) 2018/840 [19]. During this update, the Commission removed diclofenac (due to its strong monitoring background) from the WL; however, two antibiotics (amoxicillin and ciprofloxacin) were added to the list:

- (2nd WL) 17-β-estradiol, 17-α-ethynylestradiol, erythromycin, clarithromycin, azithromycin, amoxicillin, ciprofloxacin.

The selection of candidate substances was based on hazard properties, the availability of reliable safety thresholds (such as the contribution to antimicrobial resistance) and the availability of proper analytical methods for their monitoring in the environment.

Finally, in 2020 the 3rd WL was adopted by the Commission Implementing Decision (EU) 2020/1161 [20]. In terms of pharmaceuticals, it includes only two compounds from the 2nd WL and few new groups of pharmaceuticals such as two antibiotics, which are often prescribed together to overcome antimicrobial resistance, an antidepressant pharmaceutical and its metabolite, and a group of three azole pharmaceuticals:

- (3rd WL) amoxicillin, ciprofloxacin, sulfamethoxazole, trimethoprim, venlafaxine, *O*-desmethylvenlafaxine, clotrimazole, fluconazole, miconazole.

Moreover, there are already existing regulations regarding the Environmental Risk Assessment (ERA) pharmaceuticals such as the Committee for Medicinal Products for Human Use (CHMP, 2006 [21]) and the Committee for Medicinal Products for Veterinary Use (CVMP, 2000, 2004 [22]). According to these documents the ERA process is different for veterinary and human medicines; however, it usually starts with an initial exposure assessment (Phase I) that is based on a calculation of the predicted/measured environmental concentration (PEC or MEC respectively). A fate and effects analysis (phase II) is only required (with some exceptions) when the so-called action limits are exceeded in different environmental compartments. Hence, risk assessment (determined by the Risk Quotient (RQ)) is performed by calculating the ratio of the PEC (or MEC) and PNEC on non-target organisms. If $RQ < 1$, no risk is estimated, hence further testing is not recommended [23–25]. Moreover, the approaches for the ERA of pharmaceuticals are in some respect dissimilar in the U.S. and in the EU. For example, in the EU the introduced guideline excludes testing of pharmaceuticals whose $PEC_{surface\ water}$ is below the action limit of 0.01 $\mu g\ L^{-1}$, whereas in the U.S. this threshold value is 0.1 $\mu g\ L^{-1}$. It must also be highlighted that the current regulations refer mainly to the acute toxicity of only single compounds, while chronic and mixture toxicity is not obligatory. Finally, while assessing the risks posed by the residues of pharmaceuticals their TPs should also be accounted. This adds a further complexity to any chemical risk assessment. TPs may contribute significantly to the risk posed by the parent compound (a) if they are formed in a high proportion, (b) if they are more persistent/more mobile than the native forms or (c) if they are highly toxic. Current ERA guidelines refer to the issue of TPs only insofar as simulation-type degradation studies at higher tiers of the assessment, which usually include the identification of major TPs [11].

2.2 Sampling and Sample Preparation

The most commonly used method for the collection of environmental samples is grab sampling. Due to significant dilutions, the volume of samples, especially of drinking or marine water, should be sufficiently large due to the trace content of pharmaceuticals in this type of matrix. Water samples of 1000 mL are recommended for the determination of drug residues in drinking water, groundwater and treated wastewater (EPA Method 1694, December 2007, EPA-821-R-08-002). A different approach to monitoring the level of water pollution with pharmaceuticals is passive extraction, where the sampling process is carried out simultaneously with the extraction of analytes [26]. The use of passive extraction techniques has the advantage that the result obtained is a time-weighted concentration independent of the momentary and temporary variations in the concentration of the analytes. In the extraction of pharmaceuticals, the Polar Organic Chemical Integrative Sampler (POCIS) type sampler is most often used, in which the Oasis HLB sorbent is the acceptor phase (the same as in the case of SPE). In a paper by Bueno et al. [27] regarding the quantification of carbamazepine and its transformation products in French coastal waters using POCIS (Oasis HLB), the presence of 20 other compounds including β-blockers, lipid regulators, analgesics, antibiotics and antidepressants was revealed. In turn, Rimayi et al. [28] used the Chemcatcher passive sampler deployed in environmental waters over a period of 14 days. This resulted in the identification of over 200 compounds including pesticides, pharmaceuticals and personal care products, drugs of abuse and their metabolites in environmental waters. In recent years, the possibility of using other receiving phases in POCIS in order to increase the range of polarity of isolated compounds, namely Strata-X Strata XAW or Oasis WAX, Oasis MAX, molecularly imprinted polymers (MIP), carbon nanotubes and ionic liquids, has been described [29–32]. Although both standard POCIS design and prototype devices have great potential applications in water monitoring, there is still a gap between the determination of sampling frequency in the laboratory and its applicability in the field.

Extraction procedure is the second after sampling largest source of errors in determining the actual content of pollutants in the environmental samples. Thus it is important to select the proper procedure of extraction according to the selected analytes and the type of the studied matrix. The procedures used to extract pharmaceuticals depend mainly on the physical state of the matrix. Therefore the procedures can be divided into those targeting soil and sediment samples and those for water samples – mainly marine and fresh surface water, but also wastewater and drinking water.

2.2.1 Solid Samples

The determination of pharmaceuticals in soils/sediments is problematic due to the complexity of these matrices, the low concentration of targets and the lack of reference methods. These difficulties have been highlighted by numerous authors such as Białk-Bielińska [33], Brinkman [34], Pavlovic [35], Kemper [36], Buchberger [37], Tadeo [38], Babic and Mutavdzic [39] or Havens [40], who have presented reviews on the methods of determining of pharmaceuticals in soils available over the last decade. The developed methods are multi-steps with sample pre-treatment, solid-solvent extraction (assisted or not by ultrasounds or microwaves) and extract clean-up. A step-by-step summary of exemplary extraction techniques of pharmaceuticals from solid samples is presented in Table 2.2. Clean-up of extract by the solid-phase extraction (SPE) technique using HLB (hydrophilic-lipophilic adsorbent) columns is a common practice (Table 2.2). The most commonly used solvents are a mixture of acetonitrile and citric acid [41, 42], acetonitrile and EDTA-McIlvaine buffer [43, 44], and methanol with water [45–47]. In a nutshell, the extraction of pharmaceuticals from the soil can be carried out in various ways, depending on the chosen technique of quantification (GC or LC coupled with MS) and the properties of the analytes. For example, Aznar et al. [48] presented a method that allows the determination of pharmaceuticals belonging to three different therapeutic groups, in which they used ultrasound-assisted extraction (UAE). Kumirska et al. [49] used microwave-assisted extraction (MAE) to extract 13 NSAIDs and oestrogens from solid matrices such as sediment, sludge and soils. It may be concluded that both Aznar et al. and Kumirska et al. presented a rather standard approach to analysing pharmaceuticals in soil. The other methods used for this purpose are usually modifications or extensions of those mentioned above. A novel approach was characterized by the research conducted by Mijangos et al. [50], in which 11 endocrine-disrupting compounds were extracted from the soil using focused ultrasonic solid-liquid extraction (FUSLE), and the purification of the extract was performed using dispersive solid-phase extraction (dSPE). The use of FUSLE along with the simplified clean-up technique requires the use of only small samples, ranging from 0.01 to 1 g. The required amount of solvent ranges from 5 to 15 mL, and the extraction time can be several minutes or even seconds. So, it is an attractive method not only from the point of view of efficiency, but also of greenness and economy.

2.2.2 Water Samples

The high polarity of pharmaceuticals makes it unsuitable to use the standard liquid-liquid extraction technique. Furthermore, this technique is not environmentally friendly. Therefore, for the extraction of pharmaceuticals from aqueous samples the SPE technique is the most widely used. This technique allows for the simultaneous extraction of analytes from various groups and the clean-up of the sample with a high concentration factor and the low consumption of organic solvents. However, it should be mentioned that for SPE the water samples should be filtered, and it is recommended to adjust pH to accelerate the analytes adsorption. Filtration removes the fraction of target compounds adsorbed in suspended particles, and it is therefore recommended to wash the filters with methanol [51]. The extraction of pharmaceuticals from water requires a high volume of samples due to the low concentration of targets (at the level of a few ng/L of water) [33, 52] and a concentration factor higher than 1000. This means that with the extract volume equal to 1 mL, the volume of the tested water sample should be as much as 1000 mL, which is recommended for using the EPA Method 1694 (December 2007, EPA-821-R-08-002) [53–55]. In addition, with reference to the review by Białk-Bielińska et al. [33], the average volume of water sample that can be passed through the standards SPE column without significant loss of extraction rate and efficiency is 100–200 mL. This

Table 2.2 Exemplary extraction/clean-up procedures used in the analysis of pharmaceuticals in soil matrices.

Therapeutic class	Sample pre-treatment	Extraction and extract clean-up	Analytical method MQL/MDL [ng g^{-1}]	Recovery [%]	Ref
Veterinary antibiotics and hormones	Homogenization and sieving	• extraction solvent 5 mL MeOH:ACN:0.1 M EDTA:McIlvaine buffer (pH 4): 30 : 20:25 : 25, v/v/v/v/) • 3 × UAE (10 minutes) • 20 mL of supernatant diluted by H_2O to 500 mL • Adjustment to pH 2.3 and filtration • SPE (Oasis HLB column)	LC-MS/MS 2–10/0.5–3	72–121	[112]
Quinolone antibiotics	Air-drying, sieving	• extraction solvent 5 mL 50% $Mg(NO_3)_2$ in 4% aq. ammonia • 3 × UAE [10 minutes] • SPE (Oasis HLB column)	LC-MS/MS 0.004–0.011/0.001/0.003	67–88	[113]
Veterinary antianxiety medications	Air-drying, sieving	• extraction solvent 2 mL H_2O, 2 mL 2 M aq. NaOH, 10 mL EtOAc • UAE (15 minutes) • Supernatant (5 mL) dried under N_2 (50°C) • Dissolving in 1 mL MeOH • Vortex with 1 mL 1% aq. FA (pH 3.5) • LLE with 0.5 mL n-hexane	LC-HRMS No data	58.9–102.6	[114]
4 analytes including diclofenac and naproxen	Air-drying, sieving	• 2 × LLE using phosphate buffer (pH 2): MeOH (3 : 4, v/v) (shaking 260 rpm 60 minutes) • LLE using 20 mL MeOH (shaking 260 rpm 60 minutes) • Supernatant mixed with 1200 mL deionized water • SPE (Oasis HLB column)	No data	61.7–74.5	[115]

15 analytes including NSAIDs, lipid regulators, antiepileptics, β-blockers, antidepressants	Air-drying, sieving	• STEP 1Anhydration with Na_2SO_4 (60 minutes) • 8 mL ACN (2% NH_4OH) • UAE (15 minutes) • Solvent evaporation and collection in graduated tubes • Samples washed with 1 mL of ACN (2% NH_4OH) • Extracts evaporated to dryness • STEP 2UAE (15 minutes) with 8 mL ACN (2% of formic acid) • Sampled washing with acidic solvent • Extracts collected to graduated tubes from Step 1 • Evaporation to dryness and reconstitution to 1 mL with ACN	GC-MS 0.14–0.65/0.07–0.53	40.8–112.8	[48]
NSAIDs, Oestrogenic hormones	Air-drying, sieving	• 10 mL H_2O (pH 2) and 10 mL ACN • Addition of anhydrous $MgSO_4$ (4 g) and Na_2SO_4 (1 g) • Vortex 1 minutes • MAE (23 minutes) • Organic layer evaporation to dryness (N_2) • Dissolved in 100 mL H_2O (pH 2) • SPE (Oasis HLB column)	GC-MS 0.9–17.1/0.3–5.7	38.7–108.2	[49]
11 Analytes including 6 hormones	Freeze-drying, homogenization	• 10 mL acetone:n-hexane (70 : 30, v/v) solvent mixture using an ice bath, IS addition • FUSLE sonication (5 minutes) • Evaporation and extract reconstruction in ACN • dSPE (Envi-carb protocol)	LC-MS/MS/1–3.2	5–30	[116]

(*Continued*)

Table 2.2 (Continued)

Therapeutic class	Sample pre-treatment	Extraction and extract clean-up	Analytical method MQL/MDL [ng g^{-1}]	Recovery [%]	Ref
Veterinary antibiotics	Freeze-drying, homogenization	• Surrogates and 0.1 g of NaF addition (as ion exchanger) • extraction solvent 5 mL MeOH:EDTA:citrate buffer (3 : 2 : 1, v/v/v) • 3 × UAE (15 minutes) • Combined supernatant degreased by n-hexane • Dilution to 250 mL with sterilized water • Filtration and adjustment to pH 4 • Tandem SPE (SAX plus HLB columns)	UPLC-MS/MS No data	47.3–90.0	[117]

seems to be insufficient, especially for drinking water characterized by a particularly low level of pharmaceutical concentrations. To solve this problem, the commonly used SPE technique can also be performed with the use of speed extraction disks, which are characterized by a quick extraction of samples with a large volume: 80–90 mL per minute for volumes up to 2 L. According to literature reports, speedisks were used in the monitoring of pharmaceuticals in drinking water, rivers, seawater and groundwater [52, 56–59]. As already mentioned in subsection 2.2, another approach to the extraction of pharmaceuticals is passive extraction, where the sampling process is carried out simultaneously with the extraction of analytes [26, 32, 60]. Furthermore, in addition to classical sample preparation techniques, alternative solutions are increasingly being used. One of these is dSPE, whereby a solid sorbent is added directly to a liquid sample which greatly facilitates contact between the analyte and the sorbent surface. The use of dSPE makes it possible to significantly shorten the extraction time, reduce the amount of sorbent used and simplify the whole process. The extraction efficiency is also higher compared to conventional methods because the interaction between sorbent and analyte is facilitated. Various sorbents including modified and unmodified carbon nanotubes were used in dSPE for extraction of sulfonamides, β-blockers and a mixture of pharmaceuticals and their metabolites [29, 61–64]. The solid-phase microextraction [65], stir bar sorptive extraction [66] or aqueous two-phase systems [67] were also tested for pharmaceuticals.

2.3 Analytical Techniques for the Determination of Pharmaceuticals

2.3.1 Gas Chromatography and Gas Chromatography Coupled to Mass Spectrometry

Gas chromatography (GC) is a powerful technique for the separation and quantification of organics in complex matrices, such as extracts of environmental water and solids [33, 68]. The simplicity of GC operations, the high resolution of the capillary columns and the lack of liquid waste are some of the advantages compared to LC. Nevertheless, for most of the pharmaceuticals derivatization is needed to improve the volatility and thermal stability in the hot injector and capillary column. The disadvantage of derivatization for GC is that this is an additional step during the analytical protocol and derivative stability is low (for example, the silylated derivatives are fragile to moisture). The artefact and impact of environmental matrix in derivatization can be obtained [69]. The advantage of derivatization is that higher MS response/lower detection limit of targets and higher capacity of columns can be achieved (for example for oestrogens [70]). Only a few pharmaceuticals can be separated by GC without derivatization, for example tri-cyclic antidepressants and oestrogens. Most of the pharmaceuticals structures have at least one polar functional group, which increases the polarity and lowers the volatility of the molecule. With the increasing number of heteroatoms and molecular mass, the probability that a pharmaceutical can be analysed by GC is decreasing. Generally, the mass limit in GC is 1000 amu, but the practical limit is even lower – about 800 amu. Thus, pharmaceuticals such as most of the tetracyclines, sulphonamides and macrolides cannot be analysed by

GC, even if the derivatization would be successful, because of each substituent increase of the mass of the analyte. β-blockers, β-agonists, non-steroidal anti-inflammatory drugs (NSAIDs) and oestrogens can be analysed by GC after the single-step derivatization, because the carboxyl and hydroxyl groups are easily transformed into derivatives by commercially available reagents. The amine and amides can be derivatized as well, and the most efficient are the acylation reagent, such as perfluorinated acid anhydrites.

Derivatization is mostly performed by silylation, which aims to exchange the labile polar hydrogen in the active groups of analytes into non-polar alkyl-silyl groups [71]. The currently available reagent allows for a quick and reproducible reaction; nevertheless the conditions initially need to be optimized, especially for mixtures of pharmaceuticals [72]. The optimal reaction duration and temperature for most pharmaceuticals are 10–30 minutes and 60°C. MSTFA (*N*-methyl-*N*-(trimethylsilyl)trifluoroacetamide) and BSTFA (*N,O*-bis(trimethylsilyl)trifluoroacetamide) with TMCS (trimethylchlorosilane) as catalyst are the most popular reagents. Two-step derivatization could be needed for pharmaceuticals with two types of active groups. For example, the derivatization of β-blockers can be performed by two steps: 1. introduction of the trifluoroacetate group to the nitrogen atom by *N*-methyl-bis(trifluoroacetamide) (MBTFA), and 2. silylation of the hydroxyl group by any standard silylation reagent [73]. The novel reagent DIMETRIS (dimethyl(3,3,3-trifluoropropyl)silyldiethylamine) was used for β-blockers, NSAID and oestrogens analysis in environmental samples [58, 70, 74]. This reagent is used to introduce the fluorine atoms using the silylation mechanism. The alkylation with perfluorinated reagents TFAA (trifluoroacetic anhydride) or PFPAA (pentafluoropropionic anhydride) can be performed, but the removal of acidic by-products is needed. The old-fashioned technique of methylation with unstable and very flammable diazomethane was changed into silylation by trimethylsilyldiazomethane [75]. Some special solvents and additives could be needed for the derivatization of pharmaceuticals. For example, the silylation of 17-α-ethynylestradiol needs to be performed with pyridine addition to protect the ethynyl group in the hot injector [70]. For the silylation the water needs to be removed from the extract, thus additional time needs to be added to the whole protocol. Nevertheless, there are derivatization techniques for in situ derivatization, coupled with the extraction process, presented in Table 2.3.

Mass spectrometry (MS) is actually the only appropriate technique for the detection of traces of pharmaceuticals in environmental samples. Other detectors coupled to GC, such as the Electron Capture Detector (ECD), Flame Ionization Detector (FID) and Nitrogen Phosphorus Detector (NPD), are not selective enough to sufficiently assure the chromatographic signal origin. With MS more confirmation points (m/z values, fragmentation pathways, ratios of ions) are obtained compared to the use of only retention time with other detectors. Thus, in Table 2.3 presenting the selected exemplary methods of pharmaceuticals analysis, only GC/MS can be found. The mode of the mass spectra recording used was SIM (single quadrupole) and SRM/MRM (triple quadrupole), while full spectra recording for trace analysis is not recommended due to the high detection limits caused by noise. The advantage of the GC/MS is the mass library, where the most often analysed derivatives of pharmaceutical and metabolites can be found. The mass spectra of the silylated derivatives are easy to interpret, because of the repeatable fragmentation pattern related to substituent detachment [76]. The stable isotope labelled

Table 2.3 GC/MS application for determination of pharmaceuticals in the environmental samples.

Analytes	Matrix		Derivatization technique and reagent	Detection limit	Ref.
Aspirin, ibuprofen, tramadol, fluoxetine, metoprolol, naproxen, diclofenac, pindolol, estrone, β-estradiol, 17-α-ethinylestradiol, estriol	River water	SPE-GC/MS/MS (triple quadrupole)	In-port silylation by *N*-methyl-*N*-(trimethylsilyl) trifluoroacetamide (MSTFA) and *N*-tert-butyl dimethyl-*N*-methyltrifluoroacetamide (MTBSTFA)	2.71–7.31 ng/L	[118]
20 pharmaceuticals [8 NSAIDs, 5 oestrogenic hormones, 2 antiepileptic drugs, 2 β-blockers, 3 antidepressants]	Soil	UAE-GC/MS	Silylation by *N,O*-bis(trimethylsilyl) trifluoroacetamide (BSTFA) and 1% trimethylchlorosilane (TMCS) in pyridine and ethyl acetate (2 : 1:1, v/v/v)	0.3–1.7 ng/g	[119]
Ibuprofen, gemfibrozil, naproxen, ketoprofen and diclofenac	Water sample	dSPE-GC/MS	In situ trimethylphenylammonium hydroxide	1–16 ng/L	[120]
Flufenamic acid, mefenamic acid, flurbiprofen, clofibrate, ketoprofen, naproxen, tolfenamic acid, gemfibrozil	River water	SPME-GC/MS	Aqueous derivatization by tetrabutylammonium hydrogen sulfate and dimethyl sulfate	0.06–1.24 ng/L	[80]
Natural and synthetic oestrogens	Wastewater	SPE-GCxGC/TOF-MS	BSTFA + 1% TMCS + pyridine	Not specified	[121]
Paracetamol, ibuprofen, flurbiprofen, naproxen, diclofenac, 4-OH-diclofenac, 5-OH-diclofenac		SPE-GC/MS	BSTFA + 1% TMCS	2–4 ng/L	[86]

internal standards (SILISs) can be used, but caution should be taken in the choice of molecule, because of the possible low rate of separation with the target and overlapping of some m/z [77].

The separation of pharmaceutical derivatives is obtained using the standard type "5" capillary columns, (95% dimethyl-polysiloxane with 5% phenyl-polysiloxane as the stationary phase) with dimensions of 30 m length × 0.25 mm I.D. × 0.25 μm film thickness, and this is generally the most popularly used column in GC. The chiral columns can be applied for selected pharmaceutical analysis (review in [78]). The 70 eV of the EI ion source is also a standard for GC/MS. The spitless injection is used to ensure the low detection limits. The large volume injection was tested for determination of various pharmaceutical and personal care products [79], and has shown that such introduction techniques can be used only for extracts with a low mass of matrix ingredients. SPME, as the way of sample introduction into the GC/MS system, was tested for selected acidic pharmaceuticals with aqueous derivatization by dimethyl sulfate [80]. The two-dimensional techniques (GC × GC) coupled with time-of-flight mass spectrometry was used for the quantification of pharmaceuticals in environmental samples (review in [81]). Such separation technique is especially valuable for analysis of complex matrix, such as wastewater.

The extraction of environmental samples for GC analysis can be prepared by various techniques. The appropriate minimalization in the matrix composition and the sufficient concentration factor need to be achieved. Thus, the standard liquid-liquid and liquid-solid extraction is not suitable. Solid-phase extraction (SPE) is most popular for water samples, because of the high concentration factor and the exchange of the medium to organic solvent, which can be quickly evaporated and replaced into the derivatization reagent. Because of the low volume of derivatization reagent used (about 50–100 μL), the concentration factor can reach 20,000 for surface water samples for a 1–2 L volume sample. In comparison, in a case of LC/MS with about 1 mL of extract volume, a factor of 2,000 can be reached. The high concentration factor by SPE-GC/MS-based methods allows sub-traces of pharmaceuticals to be tracked in drinking and ground water [82]. The SPE is also normally applied for purification of extracts of biosolids and solids (for example for analysis of NSAIDs in mussel tissue [83]). The suppression/enhancing of the analyte signal by the matrix components ("matrix effect") during SPE-GC/MS analysis is mainly connected with impurities accumulated in the injector and the start of the capillary column, rather than the impact on EI ionization [69], which is a crucial issue in the electrospray of LC/MS. Therefore, during analysis of pharmaceuticals by GC/MS in environmental samples, special attention needs to be given to lowering the interferents in the extract and purity of the GC system.

The pharmaceuticals can be simultaneously analysed by GC/MS with pesticides, endocrine disruptors and other semi-polar compounds [84, 85], if the extraction technique allows the efficient recovery of targets in a single extraction batch. The number of analytes in a single run is actually limited to the resolution of the capillary columns, but the effective recovery, presence of impurities and actual scope of the research limited this number to a practical 10–50 compounds in a single run. The phase I metabolites of pharmaceuticals, such as hydroxy- and carboxy-metabolites of NSAIDs, can be analysed together with the natives with the same extraction and derivatization protocols [86]. The phase II metabolites such as glucuronides cannot be analysed by GC, because of low thermal stability.

2.3.2 Liquid Chromatography and Liquid Chromatography Coupled to Mass Spectrometry

High-performance LC (HPLC) is still a widely used separation system and remains an excellent choice for the comprehensive analysis of pharmaceuticals in complex pharmaceutical samples. Currently, LC separation is most commonly achieved under reversed-phase LC (RPLC) conditions, e.g. using a C18 column [7, 13, 62, 87–89]. RPLC is a suitable choice for a wide range of compounds; however, highly polar substances are not retained and to separate mixtures of polar and highly polar compounds Hydrophilic Interaction Liquid Chromatography (HILIC) and Mixed-Mode LC (MMLC) are used [90, 91]. In recent years, there has been a significant increase in the use of ultra-high-performance LC (UHPLC) due to its reduced analysis time and chromatographic resolution and improved sensitivity (Table 2.4). Generally, mobile phases consisting of methanol (MeOH), acetonitrile (ACN) and mixtures of them with different additives (i.e. formic acid (FA) or ammonium acetate (NH_4AC)) at different concentrations are used. The eluent type and composition are known to have a significant influence on obtaining reproducible retention times, satisfactory peak shapes and good ionization efficiencies [62, 87, 92–94]. In a publication by Tran et al. [87] the effect of different mobile phases on the separation of multiple classes of antimicrobials (AMX) was investigated. The addition of NH_4AC in the mobile phases significantly suppressed the signal intensity of almost all AMX agents and their isotopically labelled internal standards (ILIS). In contrast, the addition of formic acid improves the peak shapes and increases the detection sensitivity for most target compounds and their ILIS except triclosan, triclosan-d3, triclocarban, triclocarban-^{13}C6, chloramphenicol and chloramphenicol-d5, for which a slight decrease in detection sensitivity was observed. Finally, a gradient consisting of 0.1% FA in water and 0.1% FA in a mixture of MeOH and ACN were selected as suitable mobile phases for routine HPLC-MS/MS analyses for the target compounds. A similar approach was used to determine 56 antibiotics (tetracyclines, sulfonamides, β-lactams, macrolides and quinolones) in water. ACQUITY UPLC® BEH C18 column, and a mobile phase consisting of a mixture of 0.1% formic acid in water and in acetonitrile was used for the separation of analytes. The method was implemented to assess the potential presence of antibiotics in wastewater and seawater in Tunisia [95].

Analysing the available literature data, it can be stated that sporadically a UV-VIS or diode array spectrophotometric detector is used, e.g. for analysis of sulfonamides [96] or non-steroidal anti-inflammatory drugs (NSAIDs) [97], whereas LC/UPLC in combination with a mass spectrometer having high sensitivity and specificity have been well developed in recent years and have become popular for monitoring single- or multi-class drug residues in different matrices. The preferred approach for pharmaceutical analysis is the use of triple quadrupole (QqQ) or ion trap (IT), in combination with electrospray ionization (ESI) [62, 87, 88, 93, 98]. Other hybrid techniques such as quadrupole analyser in combination with time-of-flight analyser, Orbitrap and hybrid quadrupole linear ion trap are also used [28, 62, 99, 100]. In addition to ESI ionization,

Table 2.4 LC/MS application for determination of pharmaceuticals in the environmental samples.

Analytes	Matrix	Analytical method	Analysis conditions	
156 target analytes screened 120 satisfactorily determined 71 and 73 compounds quantified	Wastewater and receiving waters in South Africa	UHPLC-Orbitrap MS ESI Orbitrap analyser	Analytical column: XBridge™ C18, 100 mm × 2.1 mm, 3.5 μm Mobile phase: [A] 0.1% formic acid in water and [B] 0.1% formic acid in acetonitrile Elution programme: 0–2% B, progressed to 98% B in 15 minutes, then a hold in 2 minutes, and returned to the initial conditions	[100]
17-α-ethynylestradiol [EE2], 17-β-estradiol [E2], estrone [E1], testosterone, progesterone, diclofenac, primidone	Tropical estuarine sediments	LC-MS/MS ESI[+]/[−] modes Triple quadrupole analyser	Analytical column: Kinetex XB C18, 30 mm × 2.1 mm, 1.7 μm Mobile phase: *for negative ionization mode* [A] 0.2% NH_4OH in 95% water [95 : 5, H_2O:OS] [OS consists of 60% methanol and 40% acetonitrile] and [B] 0.2% NH_4OH in 95% OS [95 : 5, OS: H_2O]. *for positive ionization mode* [A] 95% water [95 : 5, H_2O:OS] [OS] and [B] 95% OS [95 : 5, OS: H_2O]. Elution programme: *for negative ionization mode* 0 – 3 minutes 5% B progressed to 95% in 4 minutes and then holding it constant for 2 minutes and returned to the initial conditions *for positive ionization mode* 0–3 minutes 5% B progressed to 95% in 3 minutes and then holding it constant for 2 minutes and returned to the initial conditions	[89]

Ceftazidime, meropenem, amoxicillin Ciprofloxacin, lincomycin, clindamycin, erythromycin, azithromycin, clarithromycin, tylosin, sulfamethazine, sulfamethoxazole, trimethoprim, tetracycline, minocycline chlortetracycline, oxytetracycline triclosan, triclocarban, vancomycin chloramphenicol	Environmental water samples [surface water, treated wastewater effluents]	UHPLC -MS/MS ESI [+]/[−] modes Triple quadrupole analyser	Analytical column: Poroshell 120EC-C18, 100 mm × 3.0 mm, 2.7 μm Mobile phase: [A] 0.1% formic acid in water and [B] 0.1% formic acid in a 50 : 50 [v/v] mixture of methanol and acetonitrile Elution programme: 0–1.5 minutes 10% B progressed to 90% B at 15 minutes, held until 22 minutes, and returned to the initial conditions	[87]
40 multiclass antibiotics from cephalosporin, fluoroquinolone, lincosamide, macrolide, nitroimidazole, quinolone, sulfonamide and tetracycline groups	Environmental matrices	LC-MS/MS Ion source: ESI Triple quadrupole/ linear ion trap analyser	Analytical column: Kinetex C18, 100 mm × 2.1 mm, 2.6 μm Mobile phase: [A] 0.001% formic acid in water and [B] methanol Elution programme: 0–0.01 minutes 5% B progressed to 10% B at 3.0 minutes, then to 28% B at 6.0 minutes, 70% B at 11.0 minutes, 85% B at 13 minutes, and finally returned to the initial conditions	[88]

(*Continued*)

Table 2.4 (Continued)

Analytes	Matrix	Analytical method	Analysis conditions	
acetaminophen, acetazolamide, acetylsalicylic acid, amiodarone, amoxicillin, ampicillin, atenolol, azithromycin, caffeine, carbamazepine, ciprofloxacin, diclofenac, clarithromycin, cyclophosphamide, erythromycin A, flumequine, gemfibrozil, hydrochlorothiazide, ibuprofen, josamycin, ketoprofen, lorazepam, losartan, metoprolol, metronidazole, niflumic acid, nordiazepam, 19-norethindrone, norfloxacin, ofloxacin, oxazepam, oxolinic acid, phenazone, piperacillin, roxithromycin, spiramycin, sulfadiazine, sulfamethazine, tylosine sulfamethoxazole, tetracycline, trimethoprim	Three fish species hake [*Merluccius merluccius*], red mullet [*Mullus surmuletus*], sole [*Solea solea*] and one crustacean species shrimp [*Palaemon serratus*]	UPLC-MS/MS Ion source: ESI[+]/[−] Triple quadrupole analyser	Analytical column: C18 Acquity UPLC HSS T3, 50 mm × 2.1 mm, 1.8 μm Mobile phase: *for positive ionization mode* [A] 0.1% formic acid in water and [B] acetonitrile *for negative ionization mode* [A] 0.01% formic acid and eluent [B] acetonitrile Elution programme: 0–2 minutes 2% [B] progressed to 60% at 4 minutes, then to 100% at 6 minutes, held 1 minute and returned to the initial conditions	[92]
56 antimicrobial drugs (tetracyclines, sulfonamides,β-lactams, macrolides and quinolones)	Wastewater and sea water	UPLC-MS/MS Ion source: ESI[+] Triple quadrupole analyser	Analytical column: ACQUITY UPLC®BEH C18 column [100 mm × 2.1 mm] with a 1.7 μm particle size Mobile phase: [A] 0.1% formic acid in water and [B] acetonitrile Elution programme: 0 minutes – 1% B at progressed to 99% B at 10 minutes and then returned to the initial conditions	[95]

List of abbreviations: ACN: acetonitrile; dSPE: dispersive solid-phase extraction; ESI: electrospray ionization; EtOAc: ethylacetate; FA: formic acid; FUSLE: focused ultrasonic solid-liquid extraction, HPLC: high-performance liquid chromatography; HRMS: high-resolution mass spectrometry; LCL: lowest calibrated level; LLE: liquid-liquid extraction; MAE: microwave-assisted extraction; MDL: method detection limit; MQL: method quantification limit; MRM: multiple reaction monitoring; MS: mass spectrometry; MS/MS: tandem mass spectrometry; NSAIDs: non-steroidal anti-inflammatory drugs; SIM: selected ion monitoring; SPE: solid-phase extraction; SRM: selected reaction monitoring; UAE: ultrasound-assisted extraction; UPLC: ultra-performance liquid chromatography; UV: ultraviolet light.

other techniques, e.g. atmospheric pressure chemical ionization (APCI) or atmospheric pressure photoionisation (APPI), can be used, especially for less polar compounds that are usually not ionized by ESI [101]. In MS/MS systems, fragmentation using ESI is performed in either positive or negative mode. In general, the negative mode is preferred for acidic pharmaceuticals, while positive ionization is used for neutral and basic pharmaceuticals. A series of experiments to achieve satisfactory separation and high sensitivity for ionization of pharmaceuticals in positive and negative modes were carried out by Wu et al. [102]. The use of methanol as the organic phase resulted in increased sensitivity for most compounds compared to acetonitrile. An improvement in signal shape was observed when formic acid or ammonium acetate was added to the aqueous phase. Although 0.1% formic acid in water is very often used as a mobile phase for the simultaneous determination of both acidic and basic pharmaceuticals, it was observed that reducing the FA concentration to 0.001% improved the sensitivity by up to 14-fold for some acidic drugs detected in negative mode, including naproxen, ibuprofen, gemfibrozil and diclofenac, while this change had negligible effect for compounds analysed in the ESI + mode. Several advantages of the use of MS/MS in the analysis of pharmaceuticals in environmental samples should be highlighted. Firstly, it is not necessary to achieve full separation in chromatographic analysis for selective detection, which enables shorter analysis times using shorter LC columns. Another advantage of the MS/MS method is the ability to confirm the presence of compounds based on precursor and product ions [62, 66, 87, 89, 92, 94, 103, 104]. This is achieved by multiple reaction monitoring (MRM, also called selective reaction monitoring (SRM)) of two transitions between precursor and product ions to achieve four identification points (IPs) as a minimum requirement for positive identification and confirmation criteria as defined in EU Commission Decision 2002/657/EC. The first transition of the MRM is prone to false positive identification, therefore the results of the first MRM transition are verified by a second MRM transition leading to correct identification. Therefore, most environmental pharmaceutical studies use a quantitative approach with two MRM transitions to accurately identify target pharmaceuticals. Another advantage of LC-MS/MS, which has not yet been exploited in the analysis of pharmaceuticals in the environment, is the structure determination of unknown degradation and transformation products. In addition to the undoubted advantages of the LC-MS technique, it is widely recognized that the main drawback of electrospray ionization mass spectrometry is its susceptibility to matrix components in environmental samples [87–89, 93, 94, 105–108]. Consequently, HPLC-MS/MS analysis may be subject to signal suppression or enhancement of analytes, possibly due to the presence of co-matrix components in the samples. For this reason, assessment of matrix effects is extremely important to ensure accurate and reproducible quantitative data. Various methods to reduce the influence of matrix components, including the use of isotopically labelled standards, changes in mass spectrometer operating and chromatographic conditions and modifications to the sample extraction procedure have been described [87, 89, 108–111]. If signal suppression/enhancement cannot be sufficiently minimized by the strategies described, appropriate calibration techniques can be used to determine the value of matrix effects.

2.4 Conclusion and Future Trends

Despite the undoubtedly impressive development of research methodologies that has been made, there are still numerous methodological challenges to overcome. Isolation and pre-concentration of analytes remains a key step in environmental analysis. Significant dilution of pharmaceuticals in seawater samples and condensation of contaminants in wastewater require modification of existing SPE procedures. Furthermore, most SPE sorbents are disposable, which is costly and contrary to the principles of green analytical chemistry. Therefore, the development of suitable sample preparation methods that fit within the principles of green analytical chemistry remains a necessary and challenging task. Alternative sampling methods, such as composite samples or passive samplers, can provide a more representative chemical profile. In addition, an inappropriate extraction method can affect matrix effects, which are indicated as a major problem in the analysis of trace pharmaceuticals using the LC/MS technique. Matrix effects can limit the usefulness of coupled techniques in quantitative analysis, especially when performing compound determinations in environmental, food, wastewater or biological samples. Typically, suppression or enhancement of the analyte response is accompanied by reduced precision and accuracy of subsequent measurements. Various ways to reduce the influence of matrix components, including the use of isotopically labelled standards, changes in mass spectrometer operating conditions and chromatographic conditions, and modifications to the sample extraction procedure can be used [31]. If matrix effects cannot be minimized sufficiently, appropriate calibration techniques to determine matrix effect values are currently in practice. Further work in this area should be carried out and, in particular, standardisation of methodologies should be attempted.

In recent years, the performance of LC instrumentation, in particular HRMS, has increased, and other tools, e.g. HILIC and IMS, have been recognized as promising additions for the analysis of difficult compounds. It is also necessary to carry out research primarily aimed at the identification of a broad spectrum of pharmaceuticals, including the use of non-target and suspect screening approaches. Many different data processing strategies have been developed in recent years. However, structure elucidation remains a difficult and time-consuming task. In the future, it is likely that data processing and software capabilities will increase, enabling approaches to process large amounts of data. The reproducibility and comparability of non-target and suspect screening is already being discussed and should become an important research topic in the near future.

References

1 Roig, B. and D'Aco, V. (2016). Distribution of pharmaceutical residues in the environment. In: *Pharmaceuticals in the Environment*, 1st edn (ed., R.E. Hester and R.M. Harrison), 34–69. Cambridge: Royal Society of Chemistry. doi: 10.1039/9781782622345.

2 Kümmerer, K. (2008). Pharmaceuticals in the environment – a brief summary. In: *Pharmaceuticals in the Environment Sources, Fate, Effects and Risks*, 3rd edn (ed. K. Kümmerer), 3–21. Berlin Heidelberg: Springer. doi: 10.1007/978-3-662-09259-0.

3 Tolls, J. (2001). Sorption of veterinary pharmaceuticals in soils: a review. *Environ. Sci. Technol.* 35(17): 3397–3406. doi: 10.1021/es0003021.

4 Conde-Cid, M., Núñez-Delgado, A., Fernández-Sanjurjo, M.J., Álvarez-Rodríguez, E., Fernández-Calviño, D., and Arias-Estévez, M. (2020). Tetracycline and sulfonamide antibiotics in soils: presence, fate and environmental risks. *Processes* 8(11): 1–40. doi: 10.3390/pr8111479.

5 Khan, N.A., Ahmed, S., Farooqi, I.H., Ali, I., Vambol, V., Changani, F., Yousefi, M., Vambol, S., Khan, S.U., and Khan, A.H. (2020). Occurrence, sources and conventional treatment techniques for various antibiotics present in hospital wastewaters: a critical review. *Trends Anal. Chem.* 129: 115921. doi: 10.1016/j.trac.2020.115921.

6 Wang, J., Chu, L., Wojnárovits, L., and Takács, E. (2020). Occurrence and fate of antibiotics, antibiotic resistant genes (ARGs) and antibiotic resistant bacteria (ARB) in municipal wastewater treatment plant: an overview. *Sci. Total Environ.* 744: 140997. doi: 10.1016/j.scitotenv.2020.140997.

7 Gworek, B., Kijeńska, M., Zaborowska, M., Wrzosek, J., Tokarz, L., and Chmielewski, J. (2019). Pharmaceuticals in aquatic environment. Fate and behavior, ecotoxicology and risk assessment, a review. *Acta Pol. Pharm. – Drug Res.* 76(3): 397–407. doi: 10.32383/appdr/103368.

8 Kovalakova, P., Cizmas, L., McDonald, T.J., Marsalek, B., Feng, M., and Sharma, V.K. (2020). Occurrence and toxicity of antibiotics in the aquatic environment: a review. *Chemosphere* 251: 126351. doi: 10.1016/j.chemosphere.2020.126351.

9 Carvalho, I.T. and Santos, L. (2016). Antibiotics in the aquatic environments: a review of the European scenario. *Environ. Int.* 94: 736–757. doi: 10.1016/j.envint.2016.06.025.

10 Kaczala, F. and Blum, E.S. (2015). The occurrence of veterinary pharmaceuticals in the environment: a review. *Curr. Anal. Chem.* 12(3): 169–182. doi: 10.2174/1573411012666151009193108.

11 Escher, B.I. and Fenner, K. (2011). Recent advances in environmental risk assessment of transformation products. *Environ. Sci. Technol.* 45(9): 3835–3847. doi: 10.1021/es1030799.

12 Evgenidou, E.N., Konstantinou, I.K., and Lambropoulou, D.A. (2015). Occurrence and removal of transformation products of PPCPs and illicit drugs in wastewaters: a review. *Sci. Total Environ.* 505: 905–926. doi: 10.1016/j.scitotenv.2014.10.021.

13 Celiz, M., Tso, J.D., and Aga, D.S. (2009). Pharmaceutical metabolites in the environment: analytical challenges and ecological risks. *Environ. Toxicol. Chem.* 28(12): 2473–2484. doi: 10.1897/09-173.1.

14 Yin, L., Wang, B., Yuan, H., Deng, S., Huang, J., Wang, Y., and Yu, G. (2017). Pay special attention to the transformation products of PPCPs in environment. *Emerg. Contam.* 3(2): 69–75. doi: 10.1016/j.emcon.2017.04.001.

15 Besse, J.P., Latour, J.F., and Garric, J. (2012). Anticancer drugs in surface waters. What can we say about the occurrence and environmental significance of cytotoxic, cytostatic and endocrine therapy drugs? *Environ. Int.* 39(1): 73–86. doi: 10.1016/j.envint.2011.10.002.

16 Madikizela, L.M., Ncube, S., Tutu, H., Richards, H., Newman, B., Ndungu, K., and Chimuka, L. (2020). Pharmaceuticals and their metabolites in the marine environment: sources, analytical methods and occurrence. *Trends Environ. Anal. Chem.* 28: e00104. doi: 10.1016/j.teac.2020.e00104.

17 Godoi, F.C., Prakash, S., and Bhandari, B.R., (2019). Final report The database "Pharmaceuticals in the Environment" – update and new analysis. Germany: German Environmental Agency, Umwelt Bundesamt, p. 103. Report No.: 67/2019.

18 Commission Implementing Decision (EU) 2015/495 of 20 March 2015 establishing a watch list of substances for Union-wide monitoring in the field of water policy pursuant to Directive 2008/105/EC of the European Parliament and of the Council.

19 Commission Implementing Decision (EU) 2018/840 of 5 June 2018 establishing a watch list of substances for Union-wide monitoring in the field of water policy pursuant to Directive 2008/105/EC of the European Parliament and of the Council and repealing Comm.

20 Commission Implementing Decision (EU) 2020/1161 of 4 August 2020 establishing a watch list of substances for Union-wide monitoring in the field of water policy pursuant to Directive 2008/105/EC of the European Parliament and of the Council.

21 European Medicines Agency (2006). *Guideline on the Environmental Risk Assessment of Medical Products for Human Use*. London: European Medicines Agency. Report No.: EMEA/CHMP/SWP/4447/00.

22 *Guideline on the Environmental Risk Assessment for Veterinary Medicinal Products in Support of the VICH GL6 and GL 38*. London: European Medicines Agency; 2008, 77 p. Report No.: EMEA/CVMP/ERA/418282/2005-Rev.1.

23 Ankley, G.T., Brooks, B.W., Huggett, D.B., and Sumpter, J.P. (2007). Repeating history: Pharmaceuticals in the environment. *Environ. Sci. Technol.* 15: 8211–8217. doi: 10.1021/es072658j.

24 Schmitt, H., Boucard, T., Garric, J., Jensen, J., Parrott, J., Péry, A., Römbke, J., Straub, J.O., Hutchinson, T.H., Sanchez-Argüello, P., Wennmalm, A., and Duis, K. (2010). Recommendations on the environmental risk assessment of pharmaceuticals: Effect characterization. *Integr. Environ. Assess. Manag.* 6: 588–602. doi: 10.1897/IEAM_2009-053.1.

25 Tarazona, J.V., Escher, B.I., Giltrow, E., Sumpter, J., and Knacker, T. (2010). Targeting the environmental risk assessment of pharmaceuticals: Facts and fantasies. *Integr. Environ. Assess. Manag.* 6: 603–613. doi: 10.1897/IEAM_2009-052.1.

26 Męczykowska, H., Kobylis, P., Stepnowski, P., and Caban, M. (2017). Calibration of passive samplers for the monitoring of pharmaceuticals in water-sampling rate variation. *Crit. Rev. Anal. Chem.* 47(3): 204–222. doi: 10.1080/10408347.2016.1259063.

27 Martínez Bueno, M.J., Herrera, S., Munaron, D., Boillot, C., Fenet, H., Chiron, S., and Gómez, E. (2016). POCIS passive samplers as a monitoring tool for pharmaceutical residues and their transformation products in marine environment. *Environ. Sci. Pollut. Res.* 23(6): 5019–5029. doi: 10.1007/s11356-014-3796-5.

28 Rimayi, C., Chimuka, L., Gravell, A., Fones, G.R., and Mills, G.A. (2019). Use of the Chemcatcher® passive sampler and time-of-flight mass spectrometry to screen for emerging pollutants in rivers in Gauteng Province of South Africa. *Environ. Monit. Assess.* 191(6): 388. doi: 10.1007/s10661-019-7515-z.

29 Jakubus, A., Godlewska, K., Gromelski, M., Jagiello, K., Puzyn, T., Stepnowski, P., and Paszkiewicz, M. (2019). The possibility to use multi-walled carbon nanotubes as a sorbent for dispersive solid phase extraction of selected pharmaceutical and their metabolites: Effect of extraction condition. *Microchem. J.* 146: 1113–1125. doi: 10.1016/j.microc.2019.02.051.

30 Męczykowska, H., Kobylis, P., Stepnowski, P., and Caban, M. (2017). Ionic liquids for the passive sampling of sulfonamides from water – applicability and selectivity study. *Anal. Bioanal. Chem.* 409(16): 3951–3958. doi: 10.1007/s00216-017-0342-6.

31 Caban, M., Męczykowska, H., and Stepnowski, P. (2016). Application of the PASSIL technique for the passive sampling of exemplary polar contaminants (pharmaceuticals and phenolic derivatives) from water. *Talanta* 155: 185–192. doi: 10.1016/j.talanta.2016.04.035.

32 Godlewska, K., Stepnowski, P., and Paszkiewicz, M. (2021). Pollutant analysis using passive samplers: Principles, sorbents, calibration and applications. A review. *Environ. Chem. Lett.* 19: 465–520. doi: 10.1007/s10311-020-01079-6.

33 Białk-Bielińska, A., Kumirska, J., Borecka, M., Caban, M., Paszkiewicz, M., Pazdro, K., and Stepnowski, P. (2016). Selected analytical challenges in the determination of pharmaceuticals in drinking/marine waters and soil/sediment samples. *J Pharm. Biomed. Anal.* 121: 271–296. doi: 10.1016/j.jpba.2016.01.016.

34 Stolker, A.A.M. and Brinkman, U.A.T. (2005). Analytical strategies for residue analysis of veterinary drugs and growth-promoting agents in food-producing animals – A review. *J. Chromatogr. A* 1067(1–2): 15–53. doi: 10.1016/j.chroma.2005.02.037.

35 Pavlović, D.M., Babić, S., Horvat, A.J.M., and Kaštelan-Macan, M. (2007). Sample preparation in analysis of pharmaceuticals. *Trends Anal. Chem.* 26(11): 1062–1075. doi: 10.1016/j.trac.2007.09.010.

36 Kemper, N. (2008). Veterinary antibiotics in the aquatic and terrestrial environment. *Ecol. Indic.* 8(1): 1–13. doi: 10.1016/j.ecolind.2007.06.002.

37 Buchberger, W. (2011). Current approaches to trace analysis of pharmaceuticals and personal care products in the environment. *J. Chromatogr. A* 1218(4): 603–618. doi: 10.1016/j.chroma.2010.10.040.

38 Tadeo, J.L., Sánchez-Brunete, C., Albero, B., García-Valcárcel, A.I., and Pérez, R.A. (2012). Analysis of emerging organic contaminants in environmental solid samples. *Cent. Eur. J. Chem.* 10(3): 480–520. doi: 10.2478/s11532-011-0157-9.

39 Babić, S. and Mutavdžić Pavlović, D. (2013). Analysis of PhACs in solid environmental samples (soil, sediment, and sludge). *Compr. Anal. Chem.* 62: 129–167. doi: 10.1016/B978-0-444-62657-8.00005-7.

40 Havens, S.M., Hedman, C.J., Hemming, J.D.C., Mieritz, M.G., Shafer, M.M., and Schauer, J.J. (2014). Comparison of accelerated solvent extraction, soxhlet and sonication techniques for the extraction of estrogens, androgens and progestogens from soils. *J. Agric. Chem. Environ.* 03(03): 103–120. doi: 10.4236/jacen.2014.33013.

41 Liang, X., Chen, B., Nie, X., Shi, Z., Huang, X., and Li, X. (2013). The distribution and partitioning of common antibiotics in water and sediment of the Pearl River Estuary, South China. *Chemosphere* 92(11): 1410–1416. doi: 10.1016/j.chemosphere.2013.03.044.

42 Chen, H., Liu, S., Xu, X.R., Liu, S.S., Zhou, G.J., and Sun, K.F. (2015). Antibiotics in typical marine aquaculture farms surrounding Hailing Island, South China: Occurrence, bioaccumulation and human dietary exposure. *Mar. Pollut. Bull.* 90(1–2): 181–187. doi: 10.1016/j.marpolbul.2014.10.053.

43 Na, G., Fang, X., Cai, Y., Ge, L., Zong, H., Yuan, X., Yao, Z., and Zhang, Z. (2013). Occurrence, distribution, and bioaccumulation of antibiotics in coastal environment of Dalian, China. *Mar. Pollut. Bull.* 69(1–2): 233–237. doi: 10.1016/j.marpolbul.2012.12.028.

44 Capone, D.G., Weston, D.P., Miller, V., and Shoemaker, C. (1996). Antibacterial residues in marine sediments and invertebrates following chemotherapy in aquaculture. *Aquaculture* 145(1–4): 55–75. doi: 10.1016/S0044-8486(96)01330-0.

45 Stewart, M., Olsen, G., Hickey, C.W., Ferreira, B., Jelić, A., Petrović, M., and Barcelo, D. (2014). A survey of emerging contaminants in the estuarine receiving environment around Auckland, New Zealand. *Sci. Total Environ.* 468–469: 202–210. doi: 10.1016/j.scitotenv.2013.08.039.

46 Choi, M., Furlong, E.T., Werner, S.L., Pait, A.S., Lee, I.S., and Choi, H.G. (2014). Cimetidine, acetaminophen, and 1,7-dimethylxanthine, as indicators of wastewater pollution in marine sediments from Masan Bay, Korea. *Ocean Sci. J.* 49(3): 231–240. doi: 10.1007/s12601-014-0023-8.

47 Norambuena, L., Gras, N., and Contreras, S. (2013). Development and validation of a method for the simultaneous extraction and separate measurement of oxytetracycline, florfenicol, oxolinic acid and flumequine from marine sediments. *Mar. Pollut. Bull.* 73(1): 154–160. doi: 10.1016/j.marpolbul.2013.05.027.

48 Aznar, R., Sánchez-Brunete, C., Albero, B., Rodríguez, J.A., and Tadeo, J.L. (2014). Occurrence and analysis of selected pharmaceutical compounds in soil from Spanish agricultural fields. *Environ. Sci. Pollut. Res.* 21(6): 4772–4782. doi: 10.1007/s11356-013-2438-7.

49 Kumirska, J., Migowska, N., Caban, M., Łukaszewicz, P., and Stepnowski, P. (2015). Simultaneous determination of non-steroidal anti-inflammatory drugs and oestrogenic hormones in environmental solid samples. *Sci. Total Environ.* 508: 498–505. doi: 10.1016/j.scitotenv.2014.12.020.

50 Mijangos, L., Ziarrusta, H., Prieto, A., Zugazua, O., Zuloaga, O., Olivares, M., Usobiaga, A., Paschke, A., and Etxebarria, N. (2018). Evaluation of polar organic chemical integrative and hollow fibre samplers for the determination of a wide variety of organic polar compounds in seawater. *Talanta* 185: 469–476. doi: 10.1016/j.talanta.2018.03.103.

51 Wille, K., De Brabander, H.F., Vanhaecke, L., De Wulf, E., Van Caeter, P., and Janssen, C.R. (2012). Coupled chromatographic and mass-spectrometric techniques for the analysis of emerging pollutants in the aquatic environment. *Trends Analyt. Chem.* 35: 87–108. doi: 10.1016/j.trac.2011.12.003.

52 Pazdro, K., Borecka, M., Siedlewicz, G., Białk-Bielińska, A., and Stepnowski, P. (2015). Analysis of the residues of pharmaceuticals in marine environment: State-of-the-art, analytical problems and challenges. *Curr. Anal. Chem.* 12(3): 202–226. doi: 10.2174/1573411012666151009193536.

53 Wang, C., Shi, H., Adams, C.D., Gamagedara, S., Stayton, I., and Timmons, T. (2011). Investigation of pharmaceuticals in Missouri natural and drinking water using high performance liquid chromatography-tandem mass spectrometry. *Water Res.* 45(4): 1818–1828. doi: 10.1016/j.watres.2010.11.043.

54 Ferrer, I., Zweigenbaum, J.A., and Thurman, E.M. (2010). Analysis of 70 environmental protection agency priority pharmaceuticals in water by EPA method 1694. *J. Chromatogr. A* 1217(36): 5674–5686. doi: 10.1016/j.chroma.2010.07.002.

55 Sacher, F., Lange, F.T., Brauch, H.J., and Blankenhorn, I. (2001). Pharmaceuticals in groundwaters: Analytical methods and results of a monitoring program in Baden-Württemberg, Germany. *J. Chromatogr. A* 938(1–2): 199–210. doi: 10.1016/S0021-9673(0101266-3).

56 Borecka, M., Białk-Bielińska, A., Siedlewicz, G., Kornowska, K., Kumirska, J., Stepnowski, P., and Pazdro, K. (2013). A new approach for the estimation of expanded uncertainty of results of an analytical method developed for determining antibiotics in seawater using solid-phase extraction disks and liquid chromatography coupled with tandem mass spectrometry technique. *J. Chromatogr. A* 1304: 138–146. doi: 10.1016/j.chroma.2013.07.018.

57 Borecka, M., Siedlewicz, G., Haliński, Ł.P., Sikora, K., Pazdro, K., Stepnowski, P., and Białk-Bielińska, A. (2015). Contamination of the southern Baltic Sea waters by the residues of selected pharmaceuticals: Method development and field studies. *Mar. Pollut. Bull.* 94(1–2): 62–71. doi: 10.1016/j.marpolbul.2015.03.008.

58 Caban, M., Lis, H., Kumirska, J., and Stepnowski, P. (2015). Determination of pharmaceutical residues in drinking water in Poland using a new SPE-GC-MS(SIM) method based on Speedisk extraction disks and DIMETRIS derivatization. *Sci. Total Environ.* 538: 402–411. doi: 10.1016/j.scitotenv.2015.08.076.

59 Noppe, H., De Wasch, K., Poelmans, S., Van Hoof, N., Verslycke, T., and Janssen, C.R. (2005). Development and validation of an analytical method for detection of estrogens in water. *Anal. Bioanal. Chem.* 382(1): 91–98. doi: 10.1007/s00216-005-3174-8.

60 Godlewska, K., Stepnowski, P., and Paszkiewicz, M. (2020). Application of the polar organic chemical integrative sampler for isolation of environmental micropollutants – A review. *Crit. Rev. Anal. Chem.* 50(1): 1–28. doi: 10.1080/10408347.2019.1565983.

61 Herrera-Herrera, A.V., Hernández-Borges, J., Afonso, M.M., Palenzuela, J.A., and Rodríguez-Delgado, M.Á. (2013). Comparison between magnetic and non magnetic multi-walled carbon nanotubes-dispersive solid-phase extraction combined with ultra-high performance liquid chromatography for the determination of sulfonamide antibiotics in water samples. *Talanta* 116: 695–703. doi: 10.1016/j.talanta.2013.07.060.

62 Li, J., Ren, X., Diao, Y., Chen, Y., Wang, Q., and Jin, W. (2018). Multiclass analysis of 25 veterinary drugs in milk by ultra-high performance liquid chromatography-tandem mass spectrometry. *Food Chem.* 257: 259–264. doi: 10.1016/j.foodchem.2018.02.144.

63 Jakubus, A., Gromelski, M., Jagiello, K., Puzyn, T., Stepnowski, P., and Paszkiewicz, M. (2019). Dispersive solid-phase extraction using multi-walled carbon nanotubes combined with liquid chromatography–mass spectrometry for the analysis of β-blockers: Experimental and theoretical studies. *Microchem. J.* 146: 258–269. doi: 10.1016/j.microc.2018.12.063.

64 Tsai, W.H., Huang, T.C., Huang, J.J., Hsue, Y.H., and Chuang, H.Y. (2009). Dispersive solid-phase microextraction method for sample extraction in the analysis of four tetracyclines in water and milk samples by high-performance liquid chromatography with diode-array detection. *J. Chromatogr. A* 1216(12): 2263–2269. doi: 10.1016/j.chroma.2009.01.034.

65 Vera-Candioti, L., Gil García, M.D., Martínez Galera, M., and Goicoechea, H.C. (2008). Chemometric assisted solid-phase microextraction for the determination of anti-inflammatory and antiepileptic drugs in river water by liquid chromatography-diode array detection. *J. Chromatogr. A* 1211(1–2): 22–32. doi: 10.1016/j.chroma.2008.09.093.

66 Bratkowska, D., Fontanals, N., Cormack, P.A.G., Borrull, F., and Marcé, R.M. (2012). Preparation of a polar monolithic stir bar based on methacrylic acid and divinylbenzene for the sorptive extraction of polar pharmaceuticals from complex water samples. *J. Chromatogr. A* 1225: 1–7. doi: 10.1016/j.chroma.2011.12.064.

67 Assis, R.C., Mageste, A.B., de Lemos, L.R., Orlando, R.M., and Rodrigues, G.D. (2020). Application of aqueous two-phase systems for the extraction of pharmaceutical compounds from water samples. *J. Mol. Liq.* 301: 112411. doi: 10.1016/j.molliq.2019.112411.

68 Sadkowska, J., Caban, M., Chmielewski, M., Stepnowski, P., and Kumirska, J. (2019). The use of gas chromatography for determining pharmaceutical residues in clinical, cosmetic, food and environmental samples in the light of the requirements of sustainable development. *Arch. Environ. Prot.* 45(1): 42–49. doi: 10.24425/aep.2019.124829.

69 Caban, M., Migowska, N., Stepnowski, P., Kwiatkowski, M., and Kumirska, J. (2012). Matrix effects and recovery calculations in analyses of pharmaceuticals based on the determination of β-blockers and β-agonists in environmental samples. *J. Chromatogr. A* 1258: 117–127. doi: 10.1016/j.chroma.2012.08.029.

70 Caban, M., Czerwicka, M., Łukaszewicz, P., Migowska, N., Stepnowski, P., Kwiatkowski, M., and Kumirska, J. (2013). A new silylation reagent dimethyl(3,3,3-trifluoropropyl) silyldiethylamine for the analysis of estrogenic compounds by gas chromatography-mass spectrometry. *J. Chromatogr. A* 1301: 215–224. doi: 10.1016/j.chroma.2013.05.073.

71 Evershed, R. (1993). Advances in silylation. In: *Handbook of Derivatives for Chromatography*, 2nd (ed. K. Blau and J.M. Halket), 51–108. London: Wiley.

72 Kumirska, J., Plenis, A., Łukaszewicz, P., Caban, M., Migowska, N., Białk-Bielińska, A., Czerwicka, M., and Stepnowski, P. (2013). Chemometric optimization of derivatization reactions prior to gas chromatography – mass spectrometry analysis. *J. Chromatogr. A* 1296: 164–178. doi: 10.1016/j.chroma.2013.04.079.

73 Caban, M., Stepnowski, P., Kwiatkowski, M., Migowska, N., and Kumirska, J. (2011). Determination of β-blockers and β-agonists using gas chromatography and gas chromatography – mass spectrometry – a comparative study of the derivatization step. *J. Chromatogr. A* 1218(44): 8110–8122. doi: 10.1016/j.chroma.2011.08.093.

74 Caban, M., Mioduszewska, K., Łukaszewicz, P., Migowska, N., Stepnowski, P., Kwiatkowski, M., and Kumirska, J. (2014). A new silylating reagent – dimethyl(3,3,3-trifluoropropyl)silyldiethylamine – for the derivatisation of non-steroidal anti-inflammatory drugs prior to gas chromatography-mass spectrometry analysis. *J. Chromatogr. A* 1346: 107–116. doi: 10.1016/j.chroma.2014.04.054.

75 Migowska, N., Stepnowski, P., Paszkiewicz, M., Gołębiowski, M., and Kumirska, J. (2010). Trimethylsilyldiazomethane (TMSD) as a new derivatization reagent for trace analysis of selected non-steroidal anti-inflammatory drugs (NSAIDs) by gas chromatography methods. *Anal. Bioanal. Chem.* 397(7): 3029–3034. doi: 10.1007/s00216-010-3853-y.

76 Caban, M. and Stepnowski, P. (2018). Silylation of acetaminophen by trifluoroacetamide-based silylation agents. *J. Pharm. Biomed. Anal.* 154: 433–437. doi: 10.1016/j.jpba.2018.03.037.

77 Caban, M. and Stepnowski, P. (2020). The application of isotopically labeled analogues for the determination of small organic compounds by GC/MS with selected ion monitoring. *Anal. Methods* 12(30): 3854–3864. doi: 10.1039/D0AY00723D.

78 Evans, S.E. and Kasprzyk-Hordern, B. (2014). Applications of chiral chromatography coupled with mass spectrometry in the analysis of chiral pharmaceuticals in the environment. *Trends Environ. Anal. Chem.* 1: 34–51. doi: 10.1016/j.teac.2013.11.005.

79 Guitart, C. and Readman, J.W. (2010). Critical evaluation of the determination of pharmaceuticals, personal care products, phenolic endocrine disrupters and faecal steroids by GC/MS and PTV-GC/MS in environmental waters. *Anal. Chim. Acta* 658(1): 32–40. doi: 10.1016/j.aca.2009.10.066.

80 Huang, S., Zhu, F., Jiang, R., Zhou, S., Zhu, D., Liu, H., and Ouyang, G. (2015). Determination of eight pharmaceuticals in an aqueous sample using automated derivatization solid-phase microextraction combined with gas chromatography-mass spectrometry. *Talanta* 136: 198–203. doi: 10.1016/j.talanta.2014.11.071.

81 Aspromonte, J., Wolfs, K., and Adams, E. (2019). Current application and potential use of GC × GC in the pharmaceutical and biomedical field. *J. Pharm. Biomed. Anal.* 176: 112817. doi: 10.1016/j.jpba.2019.112817.

82 Caban, M., Białk-Bielińska, A., Stepnowski, P., and Kumirska, J. (2016). Current issues in pharmaceutical residues in drinking water. *Curr. Anal. Chem.* 12(3): 1–9. doi: 10.2174/15 73411012666151009194401.

83 Świacka, K., Szaniawska, A., and Caban, M. (2019). Evaluation of bioconcentration and metabolism of diclofenac in mussels Mytilus trossulus – laboratory study. *Mar. Pollut. Bull.* 141: 249–255. doi: 10.1016/j.marpolbul.2019.02.050.

84 Rosal, R., Rodríguez, A., Perdigón-Melón, J.A., Petre, A., García-Calvo, E., Gómez, M.J., Agüera, A., and Fernández-Alba, A.R. (2010). Occurrence of emerging pollutants in urban wastewater and their removal through biological treatment followed by ozonation. *Water Res.* 44(2): 578–588. doi: 10.1016/j.watres.2009.07.004.

85 Boyd, G.R., Palmeri, J.M., Zhang, S., and Grimm, D.A. (2004). Pharmaceuticals and personal care products (PPCPs) and endocrine disrupting chemicals (EDCs) in stormwater canals and Bayou St. John in New Orleans, Louisiana, USA. *Sci. Total Environ.* 333: 137–148. doi: 10.1016/j.scitotenv.2004.03.018.

86 Kołecka, K., Gajewska, M., Stepnowski, P., and Caban, M. (2019). Spatial distribution of pharmaceuticals in conventional wastewater treatment plant with Sludge Treatment Reed Beds technology. *Sci. Total Environ.* 647: 149–157. doi: 10.1016/j.scitotenv.2018.07.439.

87 Tran, N.H., Chen, H., Do, T.V., Reinhard, M., Ngo, H.H., He, Y., and Yew-Hoong Gin, K. (2016). Simultaneous analysis of multiple classes of antimicrobials in environmental water samples using SPE coupled with UHPLC-ESI-MS/MS and isotope dilution. *Talanta* 159: 163–173. doi: 10.1016/j.talanta.2016.06.006.

88 Rashid, A., Mazhar, S.H., Zeng, Q., Kiki, C., Yu, C., and Sun, Q. (2020). Simultaneous analysis of multiclass antibiotic residues in complex environmental matrices by liquid chromatography with tandem quadrupole mass spectrometry. *J. Chromatogr. B* 1145: 122103. doi: 10.1016/j.jchromb.2020.122103.

89 Fauzan, T., Omar, T., Zaharin, A., Yuso, F., and Mustafa, S. (2017). An improved SPE-LC-MS/MS method for multiclass endocrine disrupting compound determination in tropical estuarine sediments. *Talanta* 173: 51–59. doi: 10.1016/j.talanta.2017.05.064.

90 Hemström, P. and Irgum, K. (2006). Hydrophilic interaction chromatography. *J. Sep. Sci.* 29: 1784–1821. doi: 10.1002/jssc.200600199.

91 Montes, R., Aguirre, J., Vidal, X., Rodil, R., Cela, R., and Quintana, J.B. (2017). Screening for polar chemicals in water by trifunctional mixed-mode liquid chromatography-high resolution mass spectrometry. *Environ. Sci. Technol.* 51(11): 6250–6259. doi: 10.1021/acs.est.6b05135.

92 Miossec, C., Mille, T., Lanceleur, L., and Monperrus, M. (2020). Simultaneous determination of 42 pharmaceuticals in seafood samples by solvent extraction coupled to liquid chromatography–tandem mass spectrometry. *Food Chem.* 322: 126765. doi: 10.1016/j.foodchem.2020.126765.

93 Afsa, S., Hamden, K., Martin, P.A.L., and Mansour, H.B. (2020). Occurrence of 40 pharmaceutically active compounds in hospital and urban wastewaters and their contribution to Mahdia coastal seawater contamination. *Environ. Sci. Pollut. Res. Int.* 27(2): 1941–1955. doi: 10.1007/s11356-019-06866-5.

94 Liang, Y., Liu, J., Zhong, Q., Yu, D., Yao, J., Huang, T., Zhu, M., and Zhou, T. (2019). A fully automatic cross used solid-phase extraction online coupled with ultra-high performance liquid chromatography–tandem mass spectrometry system for the trace analysis of multi-class pharmaceuticals in water samples. *J. Pharm. Biomed. Anal.* 174: 330–339. doi: 10.1016/j.jpba.2019.06.004.

95 Tahrani, L., Loco, J.V., Anthonissen, R., Verschaeve, L., Mansour, H.B., and Reyns, T. (2017). Identification and risk assessment of human and veterinary antibiotics in the wastewater treatment plants and the adjacent sea in Tunisia. *Water Sci. Technol.* 76(11–12): 3000–3021. doi: 10.2166/wst.2017.465.

96 Errayess, S.A., Lahcen, A.A., Idrissi, L., Marcoaldi, C., Chiavarini, S., and Amine, A. (2017). A sensitive method for the determination of Sulfonamides in seawater samples by solid phase extraction and UV–visible spectrophotometry. *Spectrochim. Acta – Part A Mol. Biomol. Spectrosc.* 181: 276–285. doi: 10.1016/j.saa.2017.03.061.

97 Ngubane, N.P., Naicker, D., Ncube, S., Chimuka, L., and Madikizela, L.M. (2019). Determination of naproxen, diclofenac and ibuprofen in Umgeni estuary and seawater: A case of northern Durban in KwaZulu–Natal Province of South Africa. *Reg. Stud. Mar. Sci.* 29: 100675. doi: 10.1016/j.rsma.2019.100675.

98 Omotola, E.O. and Olatunji, O.S. (2020). Quantification of selected pharmaceutical compounds in water using liquid chromatography-electrospray ionisation mass spectrometry (LC-ESI-MS). *Heliyon* 6(12): e05787. doi: 10.1016/j.heliyon.2020.e05787.

99 Gago-Ferrero, P., Bletsou, A.A., Damalas, D.E., Aalizadeh, R., Alygizakis, N.A., Singer, H.P., Hollender, J., and Thomaidis, N.S. (2020). Wide-scope target screening of >2000 emerging contaminants in wastewater samples with UPLC-Q-ToF-HRMS/MS and smart evaluation of its performance through the validation of 195 selected representative analytes. *J. Hazard. Mater.* 387: 121712. doi: 10.1016/j.jhazmat.2019.121712.

100 Mhuka, V., Dube, S., and Nindi, M.M. (2020). Occurrence of pharmaceutical and personal care products (PPCPs) in wastewater and receiving waters in South Africa using LC-Orbitrap^{TM} MS. *Emerg. Contam.* 6: 250–258. doi: 10.1016/j.emcon.2020.07.002.

101 Louw, S., Njoroge, M., Chigorimbo-Murefu, N., and Chibale, K. (2012). Comparison of electrospray ionisation, atmospheric pressure chemical ionisation and atmospheric pressure photoionisation for the identification of metabolites from labile artemisinin-based anti-malarial drugs using a QTRAP® mass spectrometer. *Rapid Commun. Mass Spectrom.* 26(20): 2431–2442. doi: 10.1002/rcm.6359.

102 Wu, X., Conkle, J.L., and Gan, J. (2012). Multi-residue determination of pharmaceutical and personal care products in vegetables. *J. Chromatogr. A* 1254: 78–86. doi: 10.1016/j.chroma.2012.07.041.

103 Kiszkiel-Taudul, I. (2021). Determination of antihistaminic pharmaceuticals in surface water samples by SPE-LC-MS/MS method. *Microchem. J.* 162: 105874. doi: 10.1016/j.microc.2020.105874.

104 Duan, Y.P., Dai, C.M., Zhang, Y.L., and Chen, L. (2013). Selective trace enrichment of acidic pharmaceuticals in real water and sediment samples based on solid-phase extraction using multi-templates molecularly imprinted polymers. *Anal. Chim. Acta* 758: 93–100. doi: 10.1016/j.aca.2012.11.010.

105 Ebele, A.J., Abdallah, M.A.E., and Harrad, S. (2017). Pharmaceuticals and personal care products (PPCPs) in the freshwater aquatic environment. *Emerg. Contam.* 3(1): 1–16. doi: 10.1016/j.emcon.2016.12.004.

106 Maranata, G.J., Surya, N.O., and Hasanah, A.N. (2021). Optimising factors affecting solid phase extraction performances of molecular imprinted polymer as recent sample preparation technique. *Heliyon* 7(1): e05934. doi: 10.1016/j.heliyon.2021.e05934.

107 Gosetti, F., Mazzucco, E., Gennaro, M.C., and Marengo, E. (2016). Contaminants in water: Non-target UHPLC/MS analysis. *Environ. Chem. Lett.* 14(1): 51–65. doi: 10.1016/S0045-6535(02)00769-5.

108 Martínez-Piernas, A.B., Plaza-Bolaños, P., García-Gómez, E., Fernández-Ibáñez, P., and Agüera, A. (2018). Determination of organic microcontaminants in agricultural soils irrigated with reclaimed wastewater: Target and suspect approaches. *Anal. Chim. Acta* 1030: 115–124. doi: 10.1016/j.aca.2018.05.049.

109 Le-Minh, N., Stuetz, R.M., and Khan, S.J. (2012). Determination of six sulfonamide antibiotics, two metabolites and trimethoprim in wastewater by isotope dilution liquid chromatography/tandem mass spectrometry. *Talanta* 89: 407–416. doi: 10.1016/j.talanta.2011.12.053.

110 Hug, C., Ulrich, N., Schulze, T., Brack, W., and Krauss, M. (2014). Identification of novel micropollutants in wastewater by a combination of suspect and nontarget screening. *Environ. Pollut.* 184: 25–32. doi: 10.1016/j.envpol.2013.07.048.

111 Zhou, W., Yang, S., and Wang, P.G. (2017). Matrix effects and application of matrix effect factor. *Bioanalysis* 9(23): 1839–1844. doi: 10.4155/bio-2017-0214.

112 Ho, Y.B., Zakaria, M.P., Latif, P.A., and Saari, N. (2014). Occurrence of veterinary antibiotics and progesterone in broiler manure and agricultural soil in Malaysia. *Sci. Total Environ.* 488–489(1): 261–267. doi: 10.1016/j.scitotenv.2014.04.109.

113 Wu, X.L., Xiang, L., Yan, Q.Y., Jiang, Y.N., Li, Y.W., Huang, X.P., Li, H., Cai, Q.Y., and Mo, C.H. (2014). Huang, Distribution and risk assessment of quinolone antibiotics in the soils from organic vegetable farms of a subtropical city, Southern China. *Sci. Total Environ.* 487(1): 399–406. doi: 10.1016/j.scitotenv.2014.04.015.

114 Choi, J.H., Lamshöft, M., Zühlke, S., Abd El-Aty, A.M., Rahman, M.M., Kim, S.W., Shim, J.H., and Spiteller, M. (2014). Analyses and decreasing patterns of veterinary antianxiety medications in soils. *J. Hazard. Mater.* 275: 154–165. doi: 10.1016/j.jhazmat.2014.05.005.

115 Dodgen, L.K., Li, J., Wu, X., Lu, Z., and Gan, J.J. (2014). Transformation and removal pathways of four common PPCP/EDCs in soil. *Environ. Pollut.* 193: 29–36. doi: 10.1016/j.envpol.2014.06.002.

116 Mijangos, L., Bizkarguenaga, E., Prieto, A., Fernández, L.A., and Zuloaga, O. (2015). Simultaneous determination of a variety of endocrine disrupting compounds in carrot, lettuce and amended soil by means of focused ultrasonic solid-liquid extraction and dispersive solid-phase extraction as simplified clean-up strategy. *J. Chromatogr. A* 1389: 8–1318. doi: 10.1016/j.chroma.2015.02.036.

117 Hou, J., Wan, W., Mao, D., Wang, C., Mu, Q., Qin, S., and Luo, Y. (2015). Occurrence and distribution of sulfonamides, tetracyclines, quinolones, macrolides, and nitrofurans in livestock manure and amended soils of Northern China. *Environ. Sci. Pollut. Res.* 22(6): 4545–4554. doi: 10.1007/s11356-014-3632-y.

118 Sharma, M.T., Krishna, M., and Mudiam, R. (2020). Estimation of measurement uncertainty for the quantitative analysis of pharmaceutical residues in river water using solid-phase extraction coupled with injector port silylation-gas chromatography-tandem mass spectrometry. *Microchem. J.* 159: 105560. doi: 10.1016/j.microc.2020.105560.

119 Kumirska, J., Łukaszewicz, P., Caban, M., Migowska, N., Plenis, A., Białk-Bielińska, A., Czewicka, M., Qi, F., and Stepnowski, P. (2019). Determination of twenty pharmaceutical contaminants in soil using ultrasound-assisted extraction with gas chromatography-mass spectrometric detection. *Chemosphere* 232: 232–242. doi: 10.1016/j.chemosphere.2019.05.164.

120 Nyi, N., Fong, S., Li, Y., and Kee, H. (2015). Graphene oxide-based dispersive solid-phase extraction combined with in situ derivatization and gas chromatography – mass spectrometry for the determination of acidic pharmaceuticals in water. *J. Chromatogr. A* 1426: 69–76. doi: 10.1016/j.chroma.2015.11.070.

121 Kopperi, M., Ruiz-Jiménez, J., Hukkinen, J.I., and Riekkola, M.L. (2013). New way to quantify multiple steroidal compounds in wastewater by comprehensive two-dimensional gas chromatography-time-of-flight mass spectrometry. *Anal. Chim. Acta* 761: 217–226. doi: 10.1016/j.aca.2012.11.059.

3

Personal Care Products

Maria Llompart[1], Maria Celeiro[1], and Thierry Dagnac[2]

[1] *CRETUS, Department of Analytical Chemistry, Nutrition and Food Science, Faculty of Chemistry, Universidade de Santiago de Compostela, Santiago de Compostela, Spain*
[2] *Agronomic and Agrarian Research Centre (AGACAL-CIAM), Unit of Organic Contaminants, A Coruña, Spain*

3.1 Overview of Personal Care Products

Personal care products (PCPs) are a diverse group of household chemical compounds employed for beauty, health, and cleaning purposes. After their use, a significant quantity of these products goes down the drain and large quantities enter the environment daily through sewage effluents from wastewater treatment plants (WWTPs), due to their incomplete or inefficient removal. PCPs include several families such as ultraviolet (UV) filters, preservatives, plasticizers, fragrance allergens, or siloxanes, among others [1].

3.1.1 Properties

UV filters are molecules formed in general by one or more benzene rings and/or carbonyl groups, allowing a high electron delocalization that provides a high molar absorptivity. Benzophenones, salicylates, cinnamates, benzotriazoles, benzimidazoles, triazines, camphor, or p-aminobenzoates derivatives are some of their chemical families. UV filters can be "fat-soluble" or "water-soluble" depending on the properties of their main chemical groups. They are incorporated not only in sunscreens as sun protectors, but also added to all kind of daily-used cosmetic products, such as moisturizers, make-up, after-shave balms, or shampoos, where mixtures of several UV filters are used to cover a broad spectrum of wavelengths [2, 3].

Preservatives are added to daily care products to avoid microbial growth. Parabens, probably the best known type of preservatives, are considered ideal because they have a wide spectrum of antimicrobial activity, highly stability to pH variation, and a low production cost. Although they are still considered as emerging pollutants in the aquatic environment, the first results demonstrating the presence of parabens in water were published in 1996

Analytical Methods for Environmental Contaminants of Emerging Concern, First Edition. Edited by Núria Fontanals and Rosa Maria Marcé.

[4]. Another preservative of high concern is triclosan (TCS), a broad-spectrum antibacterial agent present as an active ingredient in products such as soaps or toothpastes. It is considered an endocrine disruptor and its increasing presence in water as well as its toxicity to aquatic organisms are a cause of concern to human and environmental health.

Phthalates are the most important group of plasticizers, used as dispersants, binders, and emulsifiers in cosmetics and pharmaceuticals products. When used as plasticizers, phthalates are not covalently bonded to the polymer matrix and thus readily leach from plastics, especially when they are exposed to elevated temperatures, low pH, or UV light. These compounds interfere with the activity of androgens such as testosterone and dihydrotestosteron, inhibiting the enzymes involved in steroidogenesis, thereby decreasing circulating levels of androgens [5].

Fragrances, a group of chemicals included in almost every daily care product, are mainly used for their odorous properties. However, some of them have been shown to cause skin sensitivity, rashes, dermatitis, coughing, asthma attacks, migraine, etc. According to the Scientific Committee on Cosmetic Products and Non-Food Products (SCCS), 26 of these ingredients correspond to the most frequently recognized allergens.

Siloxanes have been recently identified as emerging organic contaminants [6] and are now raising concern amongst environmental scientists. They are classified into linear and cyclic compounds and consist of a structural unit of alternating Si-O bond with organic side chains. They are widely used in personal care products and cosmetics and in industrial applications such as lubricants, paints, biomedical products, and antifoaming agents. Due to their high vapour pressure, and their high lipophility, they are prone to long-range transport and bioaccumulation.

Although the environmental concentrations and toxicity of PCPs have been largely overlooked in comparison to pharmaceuticals, these compounds are considered as chemicals of emerging concern for their negative impact and their persistence in different environmental compartments [1, 7–10]. Due to their relative stability to biodegradation, several of them, e.g. UV filters or siloxanes, have been shown to accumulate in the suspended particles contained in water, sediments, or sludge, altering the endocrine system of aquatic fauna.

3.1.2 Legislation

Since PCPs are main constituents of cosmetics products, Regulation EU No 1223/2009 establishes the legal requirements in terms of banned compounds or maximum allowed concentrations in the final cosmetic product [11]. However, there is still a lack of legal requirements to control these compounds in the aquatic environment. In fact, the term "contaminants of emerging concern" (CECs) became a common term for chemicals that are currently not regulated (not submitted to a routine monitoring and/or emission control regime) while they are expected to be under scrutiny in future regulation. Most CECs are not necessarily "new" chemicals; most of them involve compounds that have entered the environment for years, but their presence has only recently begun to be investigated. For example, PCPs [12, 13] are still considered emerging pollutants due to the unknown risks associated with their presence. There is a growing concern regarding the effects these compounds may have on aquatic life since several studies suggest that some PCPs act as endocrine disruptors. In Europe, Directive 2000/60/EC [14] establishes the action framework in the field of water policy for the identification of priority substances amongst those that pose a significant risk to, or via, the aquatic environment. Annex X includes a list of priority substances in the field of water policy where the phthalate di-(2-ethylhexyl)phthalate (DEHP) is the only PCP belonging to the families covered in this chapter (UV filters, preservatives, plasticizers, fragrance allergens, and siloxanes). Later, in 2015, the 1st Surface Water European Watch List (WL) was established by Commission Implementing Decision (EU) 2015/4951, listing 33 substances, including the antioxidant butylhydroxytoluene (BHT) and the UV filter 2-ethylhexylmethoxycinnamate (2-EHMC). Although this UV filter was removed from the WL in the further update in June 2018 by the Commission Implementing Decision (EU) 2018/8402, 2-EHMC has been very recently proposed to be included again as a candidate substance for the 3rd WL [15].

3.1.3 Transformation Products

Once released into the environment, PCPs are subjected to different processes such as photodegradation, biodegradation, or chemical oxidation. Depending on the compartment in which they are present, different transformations can take place, giving rise to by-products that can show different environmental behaviour and ecotoxicological profile. Transformation products (TPs) can be more persistent or toxic than their corresponding parent compounds. For example, the TPs of the antimicrobial triclosan (TCS), with similar chemical structures to highly toxic and persistent contaminants such as dioxins, have been evaluated. Results demonstrated that the parent compound is easily degraded by sunlight and UV light, and up to 6 photoproducts were detected in real wastewater, including dichlorohydroxydiphenyl ether (2,8-DCDD) and a possible DCDD isomer or dichlorohydroxydibenzofuran [16, 17].

The occurrence of UV filter TPs, mainly photodegradation ones, have also been extensively studied in the aquatic environment due to the great ability of these PCPs to absorb solar radiation. In addition, UV filters are massively employed in sunscreens, and they can be absorbed by the human body through the skin. These compounds and their metabolites can be excreted in urine and/or faeces, constituting a continuous indirect input all over the year through wastewater treatment plants (WWTPs). The UV filters' degradation and TPs formation have been demonstrated in natural waters (river, lake, tap, seawater...) as well as in chlorine media, like swimming pools, resulting in chlorinated by-products that are often more toxic than the parent UV filters [2, 18, 19]. It is important to note that TPs are usually formed at trace levels, making necessary the development of sensitive analytical methodologies to identify and accurately monitor them when seeking a full understanding of their behaviour in the aquatic environment.

3.1.4 Occurrence in the Environment

The first review specifically oriented to PCPs in the environment appeared in 1999 by Daughton and Ternes [20]. From that, in the last two decades several revisions reporting specific topics such as the occurrence, distribution and fate, or overviews of analytical methods [2, 3, 7, 8, 10, 21–25] have been published for the simultaneous analysis of PCPs as well as for individual families.

PCP residues' occurrence has been reported in river water, lake water, seawater, and even ground or tap water [1, 5, 7, 8, 10, 24] due to their universal consumption, low human metabolic capability, and improper disposal. However, urban wastewater constitutes one of the main sources of PCPs in the aquatic environment (see Figure 3.1), since the WWTP

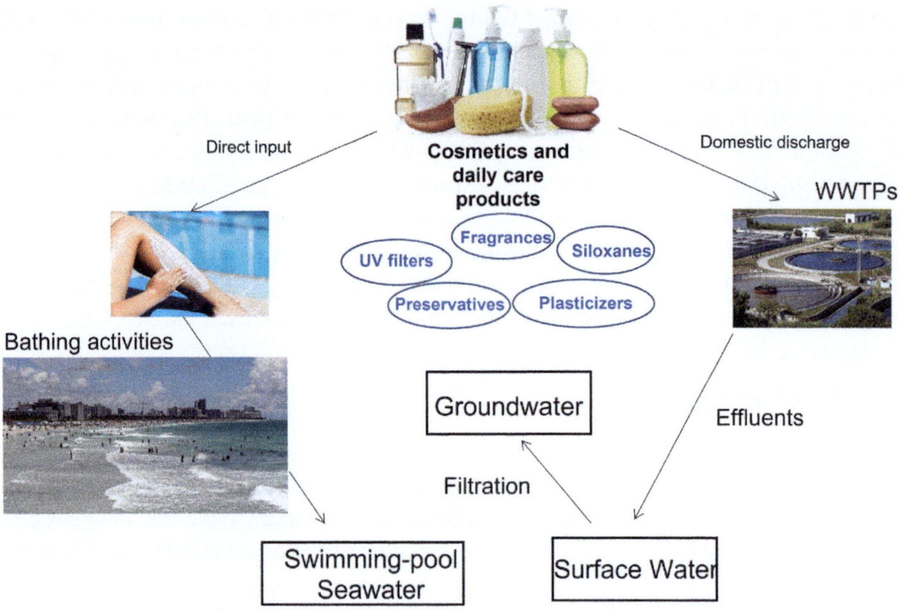

Figure 3.1 Pathways of personal care products to reach the aquatic environment.

processes do not efficiently remove these compounds. Furthermore, the by-products may become secondary pollutants and can be further modified in receiving water bodies [26]. For example, most of the parabens used in common household, daily care or cosmetic products (MeP, EtP, BuP, and BzP) have been detected in raw wastewater, with values reaching the mg L^{-1} level [2, 26]. UV filters were also detected up to mg L^{-1} concentration levels in river and recreational water such as swimming pools [2]. The occurrence of these PCPs in these waters, especially in summer, is directly related to the use of sunscreens by users of bathing areas. It is also important to note that swimming-pool water is a sink for UV filters and their chlorinated by-products, detected at the μg L^{-1} level.

In recent years, the occurrence of several PCPs, especially UV filters in coral species, is receiving attention, since they are suspected to be partly responsible of the bleaching and death of coral reefs. Several studies reported the presence of 5 UV filters in the coral tissues and surrounding waters [27, 28]. Significantly higher levels were found in the wet season, suggesting the contribution of sunscreen ingredients from swimmers.

3.2 Sample Preparation for PCPs in the Aquatic Environment

Sample preparation is crucial in environmental analysis. Choosing the appropriate sample preparation method represents a challenge, especially when a wide variety of compounds, from highly lipophilic to moderately polar or exhibiting basic, acidic, or neutral properties, must be analysed.

The main objective of this chapter is to provide an update on the sampling techniques, extraction methods, instrumental techniques, and transformation product screening methods of PCPs in the aquatic environment. In next pages, the most relevant analytical methods reported in the last decade for the extraction and determination of different PCPs families including UV filters, preservatives, plasticizers, siloxanes, and fragrance allergens will be presented. The novel developments and the latest technological trends will also be addressed. Water is the most studied environmental matrix and it is the preferred matrix for the development of new extraction methods. The most relevant sorbent and liquid-based methodologies for the determination of UV filters, preservatives, plasticizers, siloxanes, and fragrance allergens in water are described below and summarized in Tables 3.1–3.4, respectively, for the different PCPs families considered in this chapter.

3.2.1 Sorbent-based Methodologies

3.2.1.1 Solid-phase Extraction

Solid-phase extraction (SPE) is the most popular technique for the extraction of PCPs, including UV filters, preservatives, plasticizers, fragrances, or siloxanes from water matrices. In some cases, the SPE procedure is proposed for an individual group or family of PCPs, but in other cases multiresidue methodologies include several classes of PCPs and other contaminants such as pharmaceuticals, persistent organic pollutants (POPs), or endocrine disruptors (EDs).

The extraction is generally accomplished with commercial cartridges including classical normal phase (silica, alumina) or reversed-phase (C18, C8) and ion exchange sorbents [29–34].

Polymeric reversed-phase sorbents such as styrene-based polymers, polystyrene-divinylbencene (DVB) functionalized with acid groups, or ethynylbenzene-DVB (LiChrolut) have also been successfully employed for the determination of parabens [35], and other antimicrobial agents including triclosan and alkylphenols, and their chlorinated derivatives [36, 37]. However, one of the most employed is the commercially available N-vinylpyrrolidone-DVB sorbent, commonly named HLB. This reversed-phase water-wettable polymer with a hydrophilic-lipophilic balance can be use for acidic, basic, and neutral analytes. Probably for this reason, it has been employed as SPE sorbent to extract a wide variety of PCPs, including UV filters [38–41], preservatives [42–48], or plasticizers [49], simultaneously with phenols or other priority and emerging pollutants [50–52], from different water matrices such as river, tap, reservoir, or surface, and more complex matrices like wastewater.

The selection of the elution solvent depends on the analyte properties and on the determination technique. Since several PCPs such as the UV filters benzophenones or the preservative triclosan are relatively polar, the use of intermediate polarity solvents like dichloromethane (DCM) [27, 30], ethyl acetate (EA) [38], acetonitrile (ACN) [36], or a mixture of them [40] are preferred. However, more polar solvents such as methanol (MeOH) have been employed in multianalyte studies including polar compounds and medium polar ones [29, 41]. A further derivatization step is proposed for some polar compounds such as benzophenone derivatives [40] or preservatives like parabens or triclosan [36, 37], when the analysis is performed by gas chromatography (GC).

Table 3.1 Analytical methods for UV filters in environmental aqueous samples.

Analyte	Year	Sample matrix	Sample preparation (extraction method)	Determination	Recovery (%)	LOD (ng L^{-1})	RSD (%)	Concentration levels in real samples (ng L^{-1})	Ref.
Sorbent-based techniques									
OD-PABA, 4-MBC, BMDM, EHMC, IAMC, OC, BP3, ES, BP4, HMS, BP1, BP8	2014	River	SPE (C18, 350 mL sample, MeOH/EtAc)	LC-MS/MS	63–106	0.03–1.38		28–205	[39]
2-HBP, 4-HBP, THB, DHB, 4,4′- DHBP, 2,2′-DH-4-MBP	2020	Surface, wastewater	SPE (C18)	LC-MS/MS	79–81	0.5–2.9	<10		[31]
EHMC, OC, phthalates, phenols	2012	Run-off water	SPE (HLB, 820 mL sample)	GC-MS		<10		n.d.-3	[38]
BP3, BP4, 4-MBC, EHMC, UV-326, UV-329, OC	2011	Tap water	SPE (HLB, 1 L sample)	GC-MS/MS	71–150	0.3–5.6	<LOQ		[39]
BP1, BP3, BP8	2012	River water	SPE (HLB, 60 mg, 100 mL sample, EtAc, ACN, EA/DCM (1 : 1, 2 : 1) + derivatization)	GC-MS/MS	67–73	0.3–1.0		1.8–3.0	[40]
4-MBC, PBSA, BP3, EHMC, OC, OD-PABA, BP4, IAMC	2012	Tap water	SPE (HLB, 200 mg, 200–500 mL water, MeOH)	LC-MS/MS				10000–62000	[41]
BP3, BS, 4-MBC, OD-PABA, EHMC, OS, HMS, OC + 10 UV light stabilizers	2011	River, wastewater, lake	SPE (C18 + DSC-PH, 1 L water, DCM)	GC-MS	80–113	0.1–3.0		n.d.-1040	[30]

BP3, BS, 4-MBC, OD-PABA, EMC, EHS, HMS, OC	2013	Seawater	(C18 + DSC-PH, 1 L water, DCM)	GC-MS	80–113	0.1–3.0		2–1258	[27]
BP3, BP1, 4HB, 4DHB, BP8, BP2, BP4, 4-MBC, Et-PABA	2013	River water	On-line SPE (PLRP-s polymer, 5 mL sample, H_2O + ACN with 0.1% formic acid)	LC-MS/MS	84–113	0.5–3.5		7.5–862	[53]
BP1, BP2, BP3, BP4, 4DHB, 4HB, BP8, Et-PABA, 4-MBC	2014	Groundwater	On-line SPE (PLRP-s polymer, 5 mL sample, H_2O + ACN with 0.1% formic acid)	LC-MS/MS				0.6–7.9	[54]
MBP	2012	Seawater	On-line SPE (HLB, MeOH)	UHPLC-MS/MS	61–66	1.1			[56]
11 UV filters		Lake, river, sea, swimming pool	SPE (MIL-101)	GC-MS/MS	82–105	1–12		<4000	[60]
3 UV filters, parabens	2020	River, swimming pool	SPE (MCM-41)	UHPLC-MS/MS	62–90 (UV filters) 94–108 (parabens)	10–50	<11		[67]
6 UV filters	2018		GS-SPE	HPLC-UV	89–105	20–80	<10		[66]
10 UV filters	2018	River	MSPE Fe_3O_4-mGCB)	UHPLC-MS/MS	85–114	1–5	<14		[81]
10 UV filters	2018	Wastewater	MSPE (Fe_3O_4@PDA)	LC-MS	96–104	56–128			[82]
11 multiclass UV filters	2019	Lake, river, sea, swimming-pool water	FPSE (PDMS)	GC-MS/MS	80–110		<13	123 (ng mL^{-1})	[92]
6 benzotriazole UV filters	2016	Seawater	FPSE (PDMSPS)	UHPLC-MS/MS	41–51	1.0–8.9	<20	41–544	[93]

(Continued)

Table 3.1 (Continued)

Analyte	Year	Sample matrix	Sample preparation (extraction method)	Determination	Recovery (%)	LOD (ng L^{-1})	RSD (%)	Concentration levels in real samples (ng L^{-1})	Ref.
7 benzotriazole UV filters	2015	Wastewater	FPSE (PDMS)	UHPLC-MS/MS	56–105	6–61	<30	1.7–9900 (ng mL^{-1})	[94]
BP3, OC, OD-PABA, EHMC, EHS, HMS	2013	Wastewater	SBSE	LC-MS/MS				2–390	[97]
DHMB, BP3, OC, OD-PABA, 2 antimicrobials (TCS, TCC)	2010	River, wastewater	SBSE	UHPLC-MS/MS	25–87	2.5–10	<26	8–220	[100]
BP3, OC, BP10, 2-OHBP, 3-OHBP, 4-OHBP, OD-PABA, HMS, BS, EHS, EHMC, 4-MBC, BPA, TCS, MTCS,	2014	Sea, sewage, river, groundwater	SBSE (in situ derivatization)	APGC-TOF-MS	50–100	<1	<20	n.d.–56	[103]
BP3, OC, OMC, OD-PABA, EHS, HMS	2012	River, sea, wastewater	SBSE (PDMS)	LC-MS/MS	64–85	0.9–114	<20	n.d.–551	[105]
14 UV filters	2017	Sea, river, swimming-pool, aquapark	SPME (PDMS/DVB/CAR, DI acetylation)	GC-MS/MS	80–106	0.05–6	<11	n.d.–258000	[106]

14 UV filters	2016	Sea, river, swimming-pool, aquapark	SPME (PA, HS)	GC-MS/MS	70–123	0.07–12	<15	n.d.–692000	[107]
BP3, IAMC, 4-MBC, OC, OMC, OD-PABA, EHS, HMS	2017	River, sea, swimming-pool	SBDLME $[P^{+}_{6,6,6,14}]$ $[Ni(hfacac)^{-}_{3}]$	GC-MS	91–117	10–27	<15	n.d.–745	[108]
BP3, IAMC, 4-MBC, OCR, OMC, OD-PABA, EHS, HMS	2014	Seawater	SBSDME ($CoFe_2O_4$-oleic acid)	HPLC-UV	83–120	3000–30600	<20		[109]
BP3, IAMC, 4-MBC, OC, OMC, OD-PABA, EHS, HMS	2016	River, sea, swimming-pool	SBSDME ($CoFe_2O_4$-oleic acid)	GC-MS	88–112	13–148	<12	n.d.–914	[110]
BP4, DPDT, PMDSA, PDSA	2016	Sea, river, swimming-pool	SBSDME ($CoFe_2O_4$-SiO_2-nylon)	HPLC-UV	91–115	1600–2900	<13		[111]
HBP, DHB, BP, HMB	2013	Sea, wastewater	BAμE	LC-DAD	76–103	300–500	<9	2200–2500	[114]
PBSA, BP4	2019	Tap, sea, estuarine, wastewater	BAμE (polystyrene-DVB)	μLD-HPLC-DAD	61–70	40–200	<12		[113]
EHPABA, 4-MBC, PAHs	2020	River, lake	RDSE (cork)	GC-MS	80–120	140–510	<13		[120]
EHS, HMS, BS, BP3, MA, EHPABA, ETO, EHMC, OC	2020	Tap, lake	VASE	GC-MS	70–120	0.5–80	<14		[122]

(*Continued*)

Table 3.1 (Continued)

Analyte	Year	Sample matrix	Sample preparation (extraction method)	Determination	Recovery (%)	LOD (ng L^{-1})	RSD (%)	Concentration levels in real samples (ng L^{-1})	Ref.
BP1, BP2, BP3, BP4, BP6, BP8, IAMC, EHMC, ETO, BS, EHS, HMS, 4-MBC, DRT, MA, OC, DHHB, EHPABA, PBSA, BMDM, PABA	2019	River, sea, aquapark, swimming-pool	SPME (PA)	GC-MS/MS, LC-MS/MS, LC-QTOF-MS					[126]
10 UV filters, 2 antioxidants, 9 phthalates, 15 allergen fragrances, 7 synthetic musks	2019	River	SPME (PDMS/DVB, HS)	GC-MS/MS	90–106	0.003–6.6	<15	0.2–379	[135]
BP3, EHMC, OD-PABA, EHS	2014	River, wastewater	SPME (Ti-TiO_2-ZrO_2)	HPLC-UV	82–107	32–82	<12	n.d.–7000	[136]
BP3, OD-PABA, EHMC, phthalates	2015	River, wastewater	SPME (Zn-ZnO)	HPLC-UV	80–99	32–64	<11		[137]
UV filters, phthalates	2014	River, wastewater	SPME (TiO_2 nanosheets)	HPLC-UV	88–107	26–89			[138]
UV filters, phthalates, PAHs	2015	Environmental waters	SPME (Zn-ZnO)	HPLC-UV	86–105	52–84	<7		[139]
EHS, HMS, BS, BP3, MA, EHPABA, ETO, EHMC, OC	2018	Tap, lake, swimming-pool	PIL-SPME	GC-MS	79–120	2.8–26	<15		[148]
4-MBC, BP3, EDP, EHMC, HMS, OC	2018	Tap, dam, river, spa, wastewater	SPME (MWNTs)	GC-MS	71–114	10–60 (LOQ)	<15		[150]

BP3, 4-MBC, EHMC, OD-PABA, OS	2017	Lake, river, wastewater	ME-MB-IT/SPME	HPLC-DAD	44–100	40–260	<10		[153]
BP3, BS, EHMC, EPP, EHS, ETO, OCT, HMS, BMDM	2018	Lake, swimming-pool	PIL-SPME	HPLC-QTOF	66–123	100–5000	<11		[149]
OMC, BP1, BP2, BP3, BP4, PBSA, OC, BMDBM, TCS, TCC	2017	Wastewater	SPME (TWA-TFME)	LC-MS/MS			<19		[156]
BP3, 4-MBC, EHMC, BP4	2010	River	SPMDs (POCI)	GC-MS, LC-MS/MS	70–105	5–400 pg/POCI		n.d.–56	[157]
BP, 4-MBC, EHMC, TCC, BPA	2017	River	SPME (recycled diatomaceous earth)	GC-MS	70–117	1000–8000	<15		[158]
Liquid-based techniques									
MBC, OMC, OD-PABA, EHS	2015	Seawater	LLE (hexane)	GC-MS	89–120	0.08–0.15			[161]
BH, BP, EHS, HMS, 4-MBC, BP3	2018	Swimming-pool	mLLE	GC-MS	81–115	0.6–9.7	<7		[166]
BP3, BP8, MBC	2020	Seawater	DLLME	LC-MS/MS	67–115	3.3–5.7	<7	n.d.–515	[168]
Benzophenone-derived UV filters	2020	Environmental waters	DLLME (Fe_3O_4 nanoparticles)	HPLC-UV	80–103	700–800			[180]
6 UV filters, 13 musks	2019	Tap, river, sea, wastewater	DLLME	GC-MS/MS	80–120	1–20	<10	n.d.–6248	[174]
BP, BP1, BP3, BP8, 4-MBC, parabens, alkylphenols	2018	Surface water of coral reef	DLLME	LC-MS/MS	77–116	0.002–0.244	<20	n.d.–1233	[28]

(*Continued*)

Table 3.1 (Continued)

Analyte	Year	Sample matrix	Sample preparation (extraction method)	Determination	Recovery (%)	LOD (ng L^{-1})	RSD (%)	Concentration levels in real samples (ng L^{-1})	Ref.
EHS, BS, HMS, BP3, MA, EMC	2018	Lake, swimming-pool	IL-DLLME [$BMIM^+$] [Br^-], [$BeEOHIM^+$] [Br^-]	HS-GC-MS	101–107	500–1000	<14		[181]
BP, BP1, BP2, BP6, 4OH-BP	2018	Swimming-pool, river, wastewater	Air-assisted-DLLME (DES)	HPLC-DAD	89–106	50–200	<6	n.d.–4800	[185]
BP, BP1, BP3, PhS, BS	2020	Swimming-pool, well, river	DLLME (DES)	HPLC-DAD	88–106	45–540	<4	n.d.	[186]
EHS, HS, IAMC, 4-MBC, BP3, BP31, DHMB, EHPABA, THB, EHMC, OC, BMDM, DBENZO, BPA	2015	Wastewater	DLLME-silylation MAE	GC-MS	82	2–30	<22		[196]
BP1, BP3, BP8, BP12, UV-P, UV-327	2017	Tap, river, wastewater	MW-DLLME	LC-MS/MS	93–104	0.5–19	<3		[195]
BP3, IAMC, 4-MBC, OC, EDB, EHMC	2010	Swimming-pool, river, seawater	IL-SDME	HPLC-UV	92–115	60–3000	<9	n.d.–700	[198]
EHMC, 4-MBC, BS, MA, OC, EHPABA, EHS, HMS, IAMC, ETO		Sea, spa, river, swimming-pool, aquapark	USAEME	GC-MS, GC-MS/MS	<90	0.2–25 (GC-MS), 0.08–1.5 (GC-MS/MS)	<10	n.d.–3800000	[203]

BP, 4-MBC, BP3, 4-DHB	2012	Tap, river	IL-USAEME	HPLC-UV	96–108	500–1000	<7		[204]
BP, BP1, BP3, EHS, HMS	2020	River, swimming-pool	ET-USAEME	HPLC-DAD	89–98	800–1400	<5		[206]

4-MBC: 4-methylbenzylidencamphor; ACN: Acetonitrile; APGC: Atmospheric pressure gas chromatography; BAμE: Bar-assisted microextraction; [BeEOHIM$^+$] [Br$^-$]: 1-benzyl-3-(2-hydroxyethyl)imidazolium bromide; BMDM: Avobenzone; BS: Benzyl salicylate; BMIM: 1-butyl-3-methylimidazolium hexafluorophosphate; BP: Benzophenone; BPA: Bisphenol A; CAR: Carboxen; DAD: Diode array detection; DCM: Dichloromethane; DCS-PH: Monomerically bonded, phenyl-7% C; DES: Deep eutectic solvent; DHB; 2,4-dihydroxybenzophenone; DHMB; 2,2′-dihydroxy-4-methoxybenzophenone; DLLME: Dispersive liquid-liquid microextraction; DI: Direct immersion; DVB: Divinylbenzene; EDB: 2-ethylhexyl 4-dimethylaminobenzoate; EHMC: Ethylhexylmethoxycinnamate; EHS: Ethyl hexyl salicylate; EtAc: Ethyl acetate; EHPABA: EthylhexylPABA; ET: Elevated temperature; ETO: Etocrylene; FPSE: Fabric phase sorptive extraction; GC: Gas chromatography; GS: Graphene sponge; HBP: Hydroxybenzophenone; HLB: Hydrophilic-lipophilic balance; HMB: 2-hydroxy-4-methoxybenzophenone; HMS: Homosalate; HPLC: High-performance liquid chromatography; HS: Headspace; IAMC: Isoamyl methoxycinnamate; IL: Ionic liquid; IT: In-tube; LC: Liquid chromatography; LLE: Liquid-liquid extraction; LOD: Limit of detection; LOQ: Limit of quantification; MA: Menthylanthranilate; MAE: Microwave-assisted extraction; MBP: Bisoctrizole; mGCB: Graphitized carbon black; ME-MB: Magnetism-enhanced monolith; MeOH Methanol; MSPE: Magnetic solid-phase extraction; MS: Mass spectrometry; MTCS: Methyl triclosan; MW: Microwave-assisted; MWNTs: Magnetic multiwalled carbon nanotubes; OC: Octocrylene; OD: Octyldimethyl; PA: Polyacrylate; PABA: Para aminobenzoic acid; PDMS: Polydimethylsiloxane; PDMSPS: Polydimethyldiphenylsilicone; PDSA: Terephtalydene dicamphor sulfonic acid; PIL: Polymeric ionic liquid; PLRP: Polymeric reverse phase; PMDSA: Phenylbenzimidazole sulfonic acid; POCI: Polar organic chemical integrative; PhS: Phenyl salicylate; RDSE: Rotative disk sorptive extraction; RSD: Relative standard deviation; SBDLME: Stir bar dispersive liquid microextraction; SBSDME: Stir-bar sorptive-dispersive microextraction; SBSE: Stir bar sorptive extraction; SDME: Single-drop microextraction; SPE: Solid-phase extraction; SPMD: Semipermeable membrane device; SPME: Solid-phase microextraction; TCC: Triclocarban; TCS: Triclosan; TFME: Thin-film microextraction; THB: 2,3,4-trihydroxybenzophenone; TOF: Time of flight; TWA: Time-weighted average; USAEME: Ultrasound-assisted emulsification microextraction; UV: Ultraviolet; VASE: Vacuum-assisted sorbent extraction.

Table 3.2 Analytical methods for preservatives in environmental aqueous samples.

Analyte	Year	Sample matrix	Sample preparation (extraction method)	Determination	Recovery (%)	LOD (ng L^{-1})	RSD (%)	Concentration levels in real samples (ng L^{-1})	Ref.
Sorbent-based techniques									
MeP, EtP, PrP	2020	Wastewater	SPE	LC-MS	78–128	40–120	<11	<3000	[42]
MeP, EtP, PrP, BuP, TCS	2018	River	SPE (HLB)	GC-MS	61–95	4–8	<11	<3500	[43]
MeP, PrP, pharmaceuticals, and their metabolites	2019	Tap, wastewater	SPE (HLB)	LC-MS/MS	66–120	0.5–33	<17	<720 (MeP, PrP)	[44]
MeP, EtP, PrP, BuP, TCS	2018	Water	SPE (C18)	GC-MS	56–117	900–1500	<15		[32]
MeP, EtP, PrP, BuP, TCS, lipid regulators	2020	River	SPE (C18)	GC-MS/MS	93–100	0.9–14		n.d–487	[33]
MeP, EtP, BuP, UV filters, pharmaceuticals	2017	Reservoir water	SPE (HLB)	LC-MS/MS	70–120		<20	<0.122	[45]
MeP, EtP, PrP, BuP, BzP	2016	River, lake	SPE (polystyrene-DVB copolymer functionalized with acid groups)	LC-MS	57–96	0.3–30	<10	140–470	[35]

MeP, EtP, PrP, iPrP, iBuP, BzP, TCS, alkylphenols	2014	Drinking, river, pond, well, swimming pool, wastewater	SPE (LiChrolut sorbent, ethinylbenzene-DVB copolymer, ACN + derivatization)	GC-MS	90–101	<0.8	<7	30–390	[36]
MeP, EtP, PrP, fluoroalkyl compounds, BPA	2014	Surface, wastewater	SPE (HLB)	LC-MS/MS	98–113	<3	<5	10–90	[46]
MeP, EtP, PrP, iPrP, BuP, iBuP, BzP, and 14 chlorinated derivatives	2012	River, wastewater	SPE (styrene polymer sorbent + derivatization)	GC-MS	89–104	0.8–40	<11	<28 (river) <2600 (wastewater)	[37]
7 parabens and 3 halogenated MeP derivatives	2011	Wastewater	SPE (HLB)	LC-QTOF-MS	92–125	0.3–40	<6	300–1000	[47]
MeP, EtP, PrP, BuP, BzP, alkylphenols, BPA	2010	Estuary	SPE (HLB)	LC-MS/MS	107–118	0.2–20 (LOQ)		<64	[48]
MeP, PrP, 2 UV filters	2018	Swimming pool, wastewater	On-line-SPE (C18, 100 mg, 1.5 mL sample, ACN/H_2O)	HPLC-UV	65–107	1200–1700	<12	n.d.–93000	[55]
MeP, EtP, PrP, BuP, PeP, PhP, BzP	2019	Well, river, tap, wastewater	UA-SPE (γ-Fe_2O_3@HAP)	GC-MS	95–106	5000–10000	<5	n.d.–21000	[59]

(*Continued*)

Table 3.2 (Continued)

Analyte	Year	Sample matrix	Sample preparation (extraction method)	Determination	Recovery (%)	LOD (ng L^{-1})	RSD (%)	Concentration levels in real samples (ng L^{-1})	Ref.
MeP, EtP, iPrP, BuP	2013	River	μ-SPE (aminopropyl MNPs)	GC-PID	87–103	5–300	<8	n.d.	[58]
EtP, PrP, iPrP, BuP, iBuP, BzP	2015	Tap, spa, swimming pool	VA-d-μ-SPE	HPLC-DAD	78	1500–2600	<15	n.d.	[68]
MeP, EtP, PrP, BuP	2018	Tap, wastewater	In-syringe μSPE	HPLC-PDA	97–107	6–15	<5	n.d.	[69]
MeP, BuP, phthalates	2019	Tap, bottled water	DMμSPE (MWCNTs@(CR))	HPLC-DAD	38–70	30–150	<8		[84]
PrP	2018	Wastewater	MISPE	HPLC-DAD		0.2–0.8	<3	n.d.	[90]
EtP, BuP, TCS, plasticizers, pharmaceuticals	2019	Tap, groundwater, wastewater	FPSE (CW-20 M)	GC-MS	93–99	30–210	<6		[95]
MeP, EtP, PrP, BuP, iPrP, musks	2012	River, wastewater	SBSE (PDMS, in situ derivatization)	GC-MS	4.5–58	0.2–30	<14	n.d.–23593	[101]
MeP, iPrP, PrP, BuP, BzP, preservatives	2011	Tap, wastewater	SBSE (PDMS, in situ derivatization)	TD-GC-MS	22–79	500–4100	<10	n.d.–502	[98]
MeP, EtP, PrP, BuP	2020	River	BAμE (cork pellets)	HPLC-DAD	53–124	300–3000	<20	n.d.	[115]
MeP, EtP, BP, TCC	2015	Lake	BAμE (cork pellets)	HPLC-DAD	65–110	1000–20000 (LOQ)	<22		[116]

MeP, EtP, PrP, BuP	2020	Wastewater	RDSE (HLB)	UHPLC-QTOF/MS	60–67	8–18	<10	n.d.–14000	[118]
MeP, EtP, PrP, BuP	2018	Wastewater, tap water	RDSE (HLB)	GC-MS	79–91	20–50	<9	440 (MeP)	[119]
MeP, EtP, PrP, BuP, preservatives, anti-inflammatory drugs	2019	River, well, wastewater	RDSE	GC-MS	74–97		<16	n.d.–6000	[121]
MeP, EtP, PrP, BuP	2016	Urban river, stormwater	SPME (PA)	GC-MS	99	1000–2000	<12	n.d.–306000	[124]
PrP, iPrP, iBuP, BuP	2018	Surface water	ME-MB/IT-SPME	HPLC-DAD	87–109	10–26	<8	70	[154]
MeP, EtP, PrP, iPrP, iBuP, MeBz, EtBz, BuBz	2014	Chlorinated water, tap water	SPME (CAR/PDMS)	GC-MS					[125]
Liquid-based techniques									
MeP, EtP, PrP, BzP	2020	Water	CPE, VA, LLE	HPLC-DAD	41–148 (CPE), 27–134 (VE), 31–142 (LLE)		<18		[162]
MeP, EtP, PrP, BuP	2020	River, wastewater	LLE (EtAc)	LC-DAD	84–93	2000–3000	<9	300–1100	[163]
MeP, EtP, PrP, BuP, BzP, TCS	2010	River, wastewater	Non-porous membrane-assisted-LLE	LVI-GC-MS/MS	46–110	0.1–1.4	<8	n.d.–26000	[164]
MeP, EtP, PrP, BuP	2019	River, seawater	DLLME	HPLC-UV	93–105	2000–5000	<5	n.d.	[167]

(*Continued*)

Table 3.2 (Continued)

Analyte	Year	Sample matrix	Sample preparation (extraction method)	Determination	Recovery (%)	LOD (ng L^{-1})	RSD (%)	Concentration levels in real samples (ng L^{-1})	Ref.
MeP, EtP, PrP, BuP, alkylphenols, bisphenols	2014	Tap, river, wastewater	UA-DLLME-auto-IPS	GC-MS/MS	82–98	8–230	<7	n.d.–8000	[175]
MeP, EtP, PrP, iPrP, BuP, iBuP, BzP, TCS, estrogens, bisphenols	2017	Wastewater	UA-DLLME	UHPLC-MS/MS	86–116	0.02–0.05	<10	n.d.–73	[176]
MeP, EtP, PrP, BuP	2019	Tap, river, lake, wastewater	DES-LLME	HPLC-DAD		600–800	<7	n.d.–3500	[184]
MeP, EtP, PrP, SB, PHBA	2012	Tap, wastewater	USAEME	HPLC-UV	23–102	200–8300	<9.8	7000–79000	[200]
MeP, EtP, PrP, BuP, BA	2017	Surface, tap water	UASEME	HPLC-UV	70–138	30–10000	<7	n.d.	[201]

ACN: Acetonitrile; BA: Benzoic acid; BAμE: Bar-assisted microextraction; BPA: Bisphenol A; BuBz: Butyl benzoate; BzP: Benzyl paraben; CAR: Carboxen; CPE: Cloud-point extraction; CW-20: Carboxen-20; DAD: Diode array detection; DES: Deep eutectic solvent; DLLME: Dispersive liquid-liquid microextraction; DVB: Divinylbenzene;: EtAc: Ethyl acetate; EtBz: Ethyl benzoate; EtP: Ethyl paraben; FPSE: Fabric phase sorptive extraction; GC: Gas chromatography; HLB: Hydrophilic-lipophilic balance; HPLC: High-performance liquid chromatography; iBuP: Isobutyl paraben; iPrP: Isopropyl paraben; IPS: In-port silylation; IT: In-tube; LC: Liquid chromatography; LLE: Liquid-liquid extraction; LOD: Limit of detection; LOQ: Limit of quantification; LVI: Large volume injection; MeBz: Methyl benzoate; ME-MB: Magnetism-enhanced monolith; MeOH Methanol; MeP: Methyl paraben; MISPE: Molecularly imprinted solid-phase extraction; MS: Mass spectrometry; MWNTs: Magnetic multiwalled carbon nanotubes; PA: Polyacrylate; PDMS: Polydimethylsiloxane; PeP: Pentyl paraben; PID: Photoionization detector; PHBA: p-hydroxybenzoic acid; PrP: Propyl paraben; RDSE: Rotative disk sorptive extraction; RSD: Relative standard deviation; SB: Sodium benzoate; SBSE: Stir bar sorptive extraction; SPE: Solid-phase extraction; SPME: Solid-phase microextraction; TCS: Triclosan; TCC: Triclocarban; TFME: Thin-film microextraction; TOF: Time of flight; UA: Ultrasound-assisted; USAEME: Ultrasound-assisted emulsification microextraction; UASEME: Ultrasound-assisted surfactant-enhanced emulsification microextraction; UV: Ultraviolet; VA: Vortex-assisted.

Table 3.3 Analytical methods for plasticizers in environmental aqueous samples.

Analyte	Year	Sample matrix	Sample preparation (extraction method)	Determination	Recovery (%)	LOD (ng L^{-1})	RSD (%)	Concentration levels in real samples (ng L^{-1})	Ref.
Sorbent-based techniques									
21 phthalate metabolites	2021	Wastewater	SPE (Anion exchange, MAX, 60 mg)	UHPLC-MS/MS	80–140	0.08–4.40 (LOQ)	<17	14–3340	[34]
DPrP, DPP, DBP, DIPP, DNPP, BBP, DCHP, DEHP, DEHA	2020	Tap and wastewater	SPE (HLB, 200 mg, 50 mL water, EA)	GC-MS	75–112	0.8–71 (LOQ)		<LOQ	[49]
DEHP, BPA, TCPP pesticides, nitrosamines, pharmaceuticals	2014	Surface and wastewater	SPE (HLB)	LC-TOFMS	51–66	8–96 (LOQ)	<10		[50]
20 phthalates, phenols	2019	Seawater	SPE (HLB, 500 mg)	UHPLC-Q-orbitrap-MS	99–106	5–25	<20	n.d.–2645	[51]
DMP, DEP, DBP, BBP, DEHP, DnOP, BPA, PCBs, PAHs, estrogens	2012	Wastewater	SPE (HLB, 200 mg)	LVI-PTV-GC-MS	69–101	12–322	<30	n.d.–30755	[52]
DBP, DEHP, DOA	2016	Mineral water	SPE (Zn (II)-MOF)	GC-FID	88–110	200–700	<6	n.d.–1900	[63]

(*Continued*)

Table 3.3 (Continued)

Analyte	Year	Sample matrix	Sample preparation (extraction method)	Determination	Recovery (%)	LOD (ng L^{-1})	RSD (%)	Concentration levels in real samples (ng L^{-1})	Ref.
DMP, DEP, DPP, DIBP, DBP	2018	Well water	dMSPE (MIL-101(Cr)-Fe_3O_4 MNPs, 15 mg, hexane/acetone)	GC-MS	85–106	80–150	<8	n.d.	[61]
DMP, DEP, DBP, BBP, DEHP, DnOP	2018	Tap, drinking, mineral water	MSPE (15 mg Fe_3O_4@ MIL-100 + 15 mg Fe_3O_4@ SiO_2@polythiophene)	GC-MS	77–109	350–910	<9	<LOD	[62]
DMP, DEP, DIBP, DBP, DEHP, BBP, DnOP	2011	Tap water	SPE (C18-Fe_3O_4@mSiO_2 microspheres)	GC-MS		25000–77000	<14	n.d.-750000	[64]
DMP, DEP, BBP, DBP, DEHP, DnOP	2011	Tap, rain, river, pond water	SPE (electrospun nylon 6 nanofibers)	HPLC-UV	89–103	2–33	<7	n.d.-1700	[72]
DMP, DEP, DBP, DEHP, DnOP	2010	Tap, rain, river, pond water	SPE (nylon6 nanofibers)	HPLC-UV	84–102	3–33	<7	n.d.- 1500000	[71]
DBP, DEHP	2020	River, tap water	On-line-PET (nanofiber sheet)	HPLC-UV	95–102	100–300	<11	n.d.	[70]
DBzP, DPhP, DBP, DnPP, DCHP	2020	Lake water	MSPE (PAMAM dendrimers-grafted MNPs)	UHPLC-DAD	94–102	25–160	<4	<LOQ	[73]
BBP, DBEP, DBP, DIBP, DCHP, DEEP, DIDP, DINP, DIPP, DMEP, DMP, DnOP, DNPP, DPP	2018	Mineral, pond, wastewater	dMμSPE (Fe_3O_4 nanoparticles coated with reduced-graphene oxide)	UHPLC-MS	70–120	6–178	<20	n.d.-1947	[74]

DEP, DPP, DBP, DIPP, DNPP, BBP, DCHP, DEHP, DnOP, DINP, DIDP, DEHA	2017	Mineral, tap, wastewater	MSPE (core-shell poly(dopamine) MNPs)	GC-MS/MS	71–120	9–20	<19	1230–4200	[75]
DEP, DPP, DPrP	2017	Lake water	MSPE (magnetic spherical carbon)	HPLC-UV	80–113	50–80	<6	n.d.	[65]
DPP, DBP, BBP, DPIP, DCHP	2016	Lake, river, reservoir water	MSPE (@SiO_2@pPEI)	HPLC-UV	82–119	260–450	<5	n.d.-6500	[76]
DMP, DBP, DEHP	2015	Mineral water	MSPE (zeolite/ Fe_3O_4)	GC-FID	97–111	2800–3200	<14	n.d.-210000	[77]
DMP, DEP, DIBP, DBP, DEHP, BBP, DnOP	2014	River, pond water	MSPE (magnetic graphene composites)	GC-MS	88–110	10–56	<9	n.d.-151000	[78]
DPP, DPrP, BBP, DCHP, DEHP	2013	Environmental water	MSPE (magnetic microsphere-confined graphene, Fe_3O_4@SiO_2-G)	HPLC-UV	87–109	70–100	<6		[79]
DMP, DEP, DIBP, DBP, BBP, DEHP, DnOP	2011	Water	MSPE (PPy-coated Fe_3O_4 magnetic microspheres)	GC-MS	91–113	6–68			[83]
DEHP	2017	River, lake	MISPE (silica gel modified by aminopropyltriethoxysilane and acryloyl chloride)	HPLC-UV	93–102	12000	<5		[86]
DEHP	2012	River water	MISPE (methacrylamide and N-N'-methylene-bis-acrylamide)	GC-FID	95–99	11000	<5		[87]

(*Continued*)

Table 3.3 (Continued)

Analyte	Year	Sample matrix	Sample preparation (extraction method)	Determination	Recovery (%)	LOD (ng L^{-1})	RSD (%)	Concentration levels in real samples (ng L^{-1})	Ref.
5 phthalates	2016	Environmental water	MSPE (Fe_3O_4@ void@C-MIPs)	GC-MS	86–103	2–5	<7		[88]
DEHP	2016	Water	MMIP-SPE	GC-MS	93–103	20	<3		[89]
15 phthalates	2016	Seawater	SBSE (PDMS)	GC-MS		70–5710			[99]
2 plasticizers, 4 preservatives, 1 UV filter, perfluorinated, surfactants, flame retardants, pharmaceuticals, pesticides	2017	Surface, tap water	SBSE (ethylene glycol modified silicone)	LC-MS/MS	61–130	7–177 (LOQ)	<20		[104]
DBP, DMP, DEP	2014	Lake water	SBSE (neodymium magnet ($Nd_2Fe_{14}B$) powders)	HPLC-UV	89–97	90–210			[102]
DIBP, DBP, DEHP, DnOP	2013	Lake water	SPME (PDMS, DI, 50 mL sample)	GC-MS	90–94	30–400	<8	<20000	[127]
4 phthalates	2021	Mineral water	SPME ($CuFe_2O_4$ NPs)	GC-FID	81–103	120–400	<15		[141]
DBP, BBP, DEHP, DnOP	2020	River and mineral water	SPME (Mg/Al-LDH)	GC-FID	83–107	420–1290	<19	n.d.	[141]
BBP, DBP, DnOP, DEHP	2018	River	SPME (Mg/Al-LDH)	GC-FID		240–2310	<15		[142]
Phthalates	2019	Tap, Seawater	SPME (1-2(aminopropyl)-3-vinyl imidazolium bromide/tetrafluoroborate)	GC-MS	88–101	5–30	<5		[143]

Phthalates, PAHs, UV filters	2015	Water	SPME (AuNPs/SS)	HPLC-UV		8–37 1			[144]
Phthalates	2010	Mineral water	SPME (MWCNTs-PPy)	GC-MS	90–113	50–100	<11		[151]
DnOP, BBP, DBP, DIBP	2020	Wastewater	SPME (Polythionine)	GC-FID	83–120	340–1670			[145]
DMP, DEP, DBP, DEHP	2012	Water	SPME (Nano-structured self-doped polythiophene, SPT)	GC-FID	90–107	<120			[146]
DEP, DBP, BBP, BPA	2013	Seawater	EE-SPME	GC-MS		4–15	<5		[147]
DEHP, pesticides	2011	Water	IT-SPME	HPLC-UV					[152]
DIBP, DBP, BMPP, DPP, DHXP, BBP, DCHP, DEHP, DIPP, DnOP, DINP	2013	Tap, river water	SPME (bamboo charcoal)	GC-MS	62–87	4–23	<10	n.d.	[159]
5 phthalates	2020	Aqueous samples	SPME (CDs)	GC-FID	82–119	320–3130	<13		[160]
Liquid-based techniques									
DnOP, DBP, DEHP	2011	River	LLE	GC				40–1523	[165]
DMP, DEP, DIBP, DBP, DEHP, DnOP	2020	Water	DLLME	GC-FID	91–97	1000–1600	<10		[169]
	2015		DLLME	GC-FID	57–98	250–1000	<8		[170]
BBP, DEHP, DBP, DEP, DMP, DnOP	2014	River, tap water	DLLME	GC-MS	84–113	10–20	<6	n.d.	[171]

(Continued)

Table 3.3 (Continued)

Analyte	Year	Sample matrix	Sample preparation (extraction method)	Determination	Recovery (%)	LOD (ng L^{-1})	RSD (%)	Concentration levels in real samples (ng L^{-1})	Ref.
6 phthalates	2014	River, rain, wastewater	DLLME	LC-UV	98–106	10–30	<5		[172]
DBP, DMP, DnOP	2016	River	DLLME	GC-MS	106–141	1.5–3	<4	n.d.-5700	[173]
Phthalates, alkylphenols, hormones, pharmaceuticals	2019	Wastewater	DLLME	GC-MS	79–113	100–16000			[177]
DMP, DEP, DBP, DEHP	2015	Tap, river	In-syringe-DLLME	GC-FID	94–105	406–1330	<7	n.d.	[178]
DMP, DEP, DBP, DEHP, BBP, DnOP, BBP	2014	Ground, mineral, tap water	In-syringe-DLLME	GC-MS	88–102	35–110	<5	n.d.	[179]
BBP, DBP, DCHP, DEHP	2011	Tap, lake, wastewater	IL-DLLME	LC-UV	83–92	10600–28500 (LOQ)	<15		[182]
BBP, DnOP, DBP, insecticides	2011	Snow, Tap, lake, wastewater	IL-DLLME [C_8MIM] [PF_6]	HPLC-UV	85–103	230–470	<6	n.d	[183]
15 phthalates	2019	Tap, river	DLLME-SFO	HPLC-UV	87–119	13–160	<13		[187]
DEP, DPP, BBP, DCHP	2016	River, lake, reservoir	SFIL-LPME	GC-MS	73–95	270–1140	<8		[188]

DMP, DEP, BBP, DBP	2012	Drinking and environmental	MSA-DLLME	LC-UV		130000–380000	<4		[189]
DMP, DEP, BBP, DBP	2018	Sea, river	IL-UA-ISFME	HPLC-UV	90–107	222–333	<6	n.d	[190]
DMP, DEP, DBP, DEHP, BBP, DnOP	2017	Mineral water	IT-UAA-LLME	GC-MS	88–112	100–2000	<8	n.d-106 (ng mL^{-1})	[191]
DMP, DEP, DIBP, DBP, DEHP	2012		Air-assisted LLME	GC-FID	89–102	120–1150	<4		[192]
DMP, DEP, DnOP, BBP	2013	Wastewater	SDME	GC-FID	81–89	1000–9400	<4		[197]
DMP, DEP, DPP, DIBP, DBP	2011	Tap, river	USAEME-SFO	LC-UV		5–10	<13		[205]

BBP: Benzylbutyl phthalate; BMPP: bis-(4-methyl-2-pentyl) phthalate; BPA: Bisphenol A; DAD: Diode array detection; DBP: Dibutyl phthalate; DBzP: Dibenzyl phthalate; DBEP: Bis(2-butoxyethyl) phthalate; DCHP: Dicyclohexyl phthalate; DEEP: Bis(2-ethoxyethyl) phthalate DEHA; Diethylhexyl adipate; DEHP: Bis-2-(ethylhexyl)phthalate; DEP: Diethyl phthalate; DHXP: Dihexyl phthalate; DIBP: Diisobutyl phthalate; DIDP: Diisodecyl phthalate; DIPP: Diisopentyl phthalate; DLLME: Dispersive liquid-liquid microextraction; DMEP: Dimethoxyethyl phthalate; DMP: Dimethyl phthalate; DnOP: Di-n-octyl phthalate; DNPP: Di-n-pentyl phthalate; DOA: Dioctyl adipate; DPP: Dipentyl phthalate; DPrP: Dipropyl phthalate; EE: Electrochemically enhanced; Fe_3O_4@void@C: Yolk-shell magnetic mesoporous carbon; FID: Flame ionization detector; GC: Gas chromatography; HLB: Hydrophilic-lipophilic balance; HPLC: High-performance liquid chromatography; IL: Ionic liquid; IPS: In-port silylation; ISFME: In situ solvent formation microextraction; IT: In-tube; LC: Liquid chromatography; LDH: Layered double hydroxide; LLE: Liquid-liquid extraction; LLME: Liquid-liquid microextraction; LOD: Limit of detection; LOQ: Limit of quantification; LPME: Liquid phase microextraction; LVI: Large volume injection; MISPE: Molecularly imprinted solid-phase extraction; MMIP: Magnetic molecularly imprinted polymer; MOF: Metal-organic framework; MS: Mass spectrometry; MSA: Magnetic stirring assisted; MSPE: Magnetic solid-phase extraction; MWNTs: Magnetic multiwalled carbon nanotubes; NPs: Nanoparticles; PAH; Policyclic aromatic hydrocarbons; PAMAM: Poly amido amide; PCB: Polychlorinated biphenyls; PDMS: Polydimethylsiloxane; PTV: Programmable temperature vaporization; RSD: Relative standard deviation; SBSE: Stir bar sorptive extraction; SDME: Single-drop microextraction; Floating ionic liquid; SFO: Solidification of floating drop; SPE: Solid-phase extraction; SPME: Solid-phase microextraction; TCPP: (tris (1-chloro-2-propyl) phosphate); TOF: Time of flight; UA: Ultrasound-assisted; USAEME: Ultrasound-assisted emulsification microextraction; UUA: Ultrasonic and air assisted; UV: Ultraviolet; VA: Vortex-assisted.

Table 3.4 Analytical methods for fragrance allergens and siloxanes in environmental aqueous samples.

Analyte	Year	Sample matrix	Sample preparation (extraction method)	Determination	Recovery (%)	LOD (ng L^{-1})	RSD (%)	Concentration levels in real samples (ng L^{-1})	Ref.
Geraniol, coumarin, limonene, isoeugenol, amylcin namyl alcohol, bezyl cinnamate, ionone, HC, HCA, musks, phthalates	2015	Lake, wastewater	dSPE	GC-MS	83–105	0.3–3.2	<10	n.d.-2900	[80]
Fragrance allergens, fatty acids, phenols, phthalates	2016	Wastewater	SPME	GC-MS					[128]
26 fragrance allergens	2010	Bathwater, swimming-pool, wastewater	SPME (PDMS/DVB)	GC-MS	80–120	1–300	<12	3–845 (ng mL^{-1})	[130]
26 fragrance allergens	2009	Baby bathwater, swimming-pool	SPME (PDMS/DVB)	GC-MS	40–124	1–300	<10	n.d.-568 (ng mL^{-1})	[129]
16 fragrance allergens, 2 musks	2015	Wastewater	SPME	GC-MS		10–1700	<23		[131]
Fragrance allergens, pesticides, VOCs	2013	Surface, sea, wastewater	SPME	GC-MS		0.5–100	<21	n.d.-2500	[134]
26 fragrance allergens	2014	River, wastewater	UA-LLME	GC-FID, GC-MS		<100 (mg L^{-1})		170 (mg L^{-1})	[193]

Fragrance allergens	2015	Baby bathwater	USAEME	HPLC-DAD		1–154	<15		[202]
8 cyclic and linear siloxanes	2014	Wastewater	UA-DLLME	GC-MS	71–116	2–1400	<24		[194]
D3, D4, D5, D6, L3, L4	2013	Wastewater	SPME (PDMS/DVB)	GC-MS	78–96	2.6–7.8	<8	n.d.-7000	[132]

D3: Hexamethylcyclotrisiloxane; D4: Octamethylcyclo-tetrasiloxane; D5: Decamethylcyclo-pentasiloxane; D6: Dodecamethylcyclo-hexasiloxane; DAD: Diode array detector; DLLME: Dispersive liquid-liquid microextraction; DVB: Divinylbenzene; FID: Flame ionization detector; GC: Gas chromatography; HC: Hexylcinnamaldehyde; HCA: Hydroxycitronellal; HPLC: High-performance liquid chromatography; L3: Octamethyltrisiloxane; L4: Decamethyltetrasiloxane; LLME: Liquid-liquid microextraction; LOD: Limit of detection; MS: Mass spectrometry; PDMS: Polydimethylsiloxane; RSD: Relative standard deviation; SPME: Solid-phase microextraction; SPE: Solid-phase extraction; UA: Ultrasound-assisted; USAEME: Ultrasound-assisted emulsification microextraction; VOC: Volatile organic compound.

In general, SPE is considered a satisfactory method, easy to perform and generally achieving high recoveries between 60–100%. However, its main drawback is the fact that high sample volumes are frequently required, from 100 mL up to 1 L [2]. On-line SPE coupled with liquid chromatography (LC) is emerging as an effective technique since satisfactory recoveries have been reported employing it for UV filters in river and groundwater [53, 54], as well as for these compounds together with parabens in swimming-pool and wastewater [55]. The fully automated system offers better reproducibility, time reduction, and minimization of the sample contamination. In general, similar recoveries were obtained employing conventional SPE and on-line SPE. However, limits of detection (LODs) achieved using on-line SPE could be higher than those obtained with conventional SPE due to a lower preconcentration factor [56].

3.2.1.1.1 New Sorbent Materials and Approaches for SPE

In recent years, the development and application of new sorbent materials, as well as miniaturized procedures fulfilling green analytical chemistry (GAC) requirements, have received great attention. Metal-organic frameworks (MOFs) [2, 7, 21, 57] and magnetic nanoparticles (MNPs) [2, 7, 21, 22, 58, 59] have been used for the SPE of UV filters, parabens, phthalates, and adipates in the aquatic environment.

Recently, MIL-101 was successfully proposed for the effective extraction of 11 multiclass UV filters in lake, river, seawater and swimming pool waters [60], as well as for phthalates in different matrices including tap, drinking, or mineral water [61, 62]. The synthetized MOF could be reused for up to seven extraction cycles without losing efficiency. A Zn(II) based-MOF was also employed as sorbent to extract phthalates and adipates from water samples, allowing a preconcentration factor between 92–295 [63].

The use of Fe_3O_4 nanoparticles coating C18-microspheres was proposed prior to GC-MS for the determination of phthalates in environmental waters [64], whereas magnetic spherical carbon and nanoclusters Fe colloid were employed for the extraction and preconcentration of phthalates from lake water. The sorbent presented a high surface area and strong magnetism, with a high adsorption capacity [65]. Another interesting approach is based on the fabrication of graphene sponge (GS)-SPE to extract benzotriazole UV filters from water and cosmetic samples. The home-made GS-SPE cartridge could be successfully reused for more than 60 cycles [66]. Recently, a silica-based MCM-41 mesoporous material functionalized with cyanopropyl groups was synthesized, and applied to preconcentrate 6 parabens and 3 UV filters in river and swimming-pool waters [67].

Although on-column SPE is preferred, the on-batch mode is much used, especially in microextraction approaches (μ-SPE). On-batch μ-SPE employing HKUST-1 MOF has been employed to determine 7 parabens in tap and swimming-pool water [68]. Low LODs (<0.01 ng L^{-1}) were obtained applying an in-syringe μ-SPE-HPLC-PDA methodology for the analysis of 4 parabens in tap and wastewater matrices employing chitosan activated carbon (CAC) as sorbent [69]. A novel extraction medium packing PET nanofiber sheet into a stainless steel capillary of 0.8 mm inner diameter was also recently proposed for the determination of phthalates in water [70]. In this case, the capillary was directly connected to the HPLC instrument, and the extracted analytes were desorbed, and injected employing a small amount of organic solvent.

The use of green materials as SPE sorbents is increasing. Nylon6 nanofibers were proposed as green sorbent for the preconcentration of 6 phthalates from environmental waters [71, 72]. The efficiency of the sorbent was compared with classical C18 disks, cartridges, and with a national (China) standard methodology, providing better recoveries, LODs, and precision, allowing a miniaturized procedure, since only 2.5 mg of sorbent and 100 μL of solvent were used.

3.2.1.1.2 Magnetic Solid-phase Extraction

Magnetic solid-phase extraction (MSPE) is one of the new proposed methodologies based on dispersive (d)-SPE or d-μSPE [73–80]. Using an external magnet, the magnetic adsorbents and sample solution could achieve an ideal separation without facing time-consuming processes such as filtration or centrifugation. Besides, MSPE requires less organic solvent, which efficiently reduces the environmental pollution. The magnetic adsorbents play a crucial role for the effective extraction of the analytes. Several types of materials have been prepared for this purpose and they were successfully applied to the determination of UV filters, preservatives, and plasticizers in water, most of them employing Fe_3O_4 nanoparticles [74, 77, 79, 81–83]. The most relevant applications include the fabrication of a novel carbon-based magnetic material Fe_3O_4-graphitized carbon black (mGCB) composite, used in a dSPE-UPLC-MS/MS method to determine 50 pollutants from surface water. This method included 10 UV filters and reached LODs as low as 1–5 ng L^{-1}. Recently, the use of d-μSPE employing magnetic multiwalled carbon nanotubes (MMWCNTs@(CR)), followed by HPLC-DAD, has been proposed for the first time to simultaneously determine parabens and phthalates. Quantitative recoveries and LODs < 0.15 ng mL^{-1} were obtained [84].

3.2.1.1.3 Molecularly Imprinted Solid-phase Extraction

SPE based on molecularly imprinted polymers (MIPs), known as MISPE, is a novel approach for sample preparation and preconcentration, gaining interest in the field of environmental analysis. In MISPE, the high selectivity of the sorbent towards a given structure enables it to be selectively retained on the sorbent while the other compounds are not retained, resulting in the desired compound being extracted from most of the other compounds present in the sample [85].

The use of surface molecularly imprinted polymers (SMIP) for SPE has been proposed for the determination of plasticizers and preservatives in different environmental water matrices, including river, lake, or wastewater [86–90]. For phthalate determination, the selective adsorption of the target compounds was demonstrated by comparison with other compounds of this same family of plasticizers. Although the employed MIPs could be reused 6–8 times and enrichment factors up to 1400 were reported, the obtained LODs for DEHP were higher than those reported employing classical SPE or MSPE.

3.2.1.2 Fabric Phase Sorptive Extraction

Fabric phase sorptive extraction (FPSE) is a new type of solid-phase extraction (SPE) recently developed by Furton's and Kabir's group at Florida International University [91]. This technique uses small squares of cellulosic (or other) fabric coated with an ultra thin sol-gel, which are immersed directly in aqueous environmental water samples to sorb and extract analytes. The fabric piece is then submersed into a volume of solvent to elute the analytes, followed by injection onto a chromatography system. FPSE offers several advantages such as minimal solvent use, high surface area for the sorbent-analyte interaction, short extraction times, stability in a wide range of pH values and fast back extraction with a small solvent volume, achieving high preconcentration factors. FPSE usually uses less sample volume and organic solvent than stir bar sorptive extraction (SBSE) and, in addition, it is much faster.

Regarding PCPs, FPSE has been recently applied to the analysis of UV filters [92–94], and to the simultaneous determination of preservatives, plasticizers, and pharmaceuticals [95] in a broad range of water matrices, prior to GC-MS, GC-MS/MS [92, 93, 95], or LC-MS/MS [94]. In all cases, the most critical experimental parameters affecting FPSE, such as the type and size of sol-gel coated FPSE media, sample volume, extraction time, type and volume of desorption solvent, or ionic strength effect, were optimized to obtain the highest extraction efficiency. For the analysis of benzotriale UV filters, non-polar FPSE coatings such as PDMS and PDMSPD were employed [92–94], whereas for multiresidue analysis, a more polar coating, CW-20M, offered better efficiency [95]. In all cases, satisfactory recoveries and LODs at the low ng L^{-1}, allowing enrichment factors up to 40 times [92], were obtained. The validated methodologies were applied to different recreational and environmental waters including swimming-pool, lake, river, sea, tap, groundwater, or wastewater, demonstrating the ubiquity of these emerging contaminants. The highest concentration of PCPs was observed for the UV filter octocrylene (OCR) in a lake water sample (123 μg L^{-1}) [92].

3.2.1.3 Stir-bar Sorptive Extraction

Stir-bar sorptive extraction (SBSE) was developed one decade after SPME by Prof. Pat Sandra's group [96] and is based on the same principles. It consists of the extraction of the analytes from the liquid matrix onto a thick film (24–126 μL) coated on a magnetic stir bar incorporated into a glass jacket (10–20 mm in. length). This technique is usually followed by liquid (LD) or thermal desorption (TD). The applications are basically reduced to the use of PDMS as sorbent phase. In this way, SBSE was successfully applied for the extraction of individual PCPs families such as 6 UV filters [97], 5 parabens [98], and 15 phthalates [99] as well as several PCPs including UV filters and antimicrobial agents [100] or parabens and synthetic musks [101] in environmental waters. In all cases, the most critical experimental conditions affecting SBSE-LD or SBSE-TD sorption and desorption efficiency were optimized. PDMS was the preferable material massively employed as stir bar coating. A novel stir bar based on neodymium magnet ($Nd_2Fe_{14}B$) was developed for the extraction of 3 phthalates from lake water [102]. Good recoveries and precision were obtained, and LODs were similar to those reported employing classical SBSE for the determination of phthalates in seawater [99].

SBSE-LD followed by atmospheric pressure (AP)GC-QTOF was proposed as multiresidue methodology for the determination of 102 pollutants including UV filters, plasticizers, and fragrances in seawater, effluent wastewater, river water, and groundwater, obtaining LODs below 1 ng L^{-1}. The occurrence of 21 of 102 analytes, including 6 UV filters and 2 preservatives, was confirmed in several environmental aquatic matrices [103]. Another multiresidue methodology based on SBSE-LC-MS/MS was proposed for the determination of plasticizers, preservatives, UV filters, perfluorinated compounds, surfactants, flame retardants, pharmaceuticals, and pesticides. In this case, the extraction efficiency of PDMS coating was compared with EG-silicone, which offered better recoveries (60–130%) and LOQs [104].

3.2.1.3.1 Miniaturized SBSE Approaches: SBDLME, SBSDME, BAμE

Although SBSE presents many advantages, it requires a long extraction time reaching up to 5 hours [103, 105], which is its main disadvantage compared to other microextraction techniques such as SPME and in-vial acetylation SPME that only require 10 minutes to accomplish the analysis of UV filters [106, 107]. Different devices and miniaturized combined approaches have been developed to solve this problem, such as the use of magnetic ionic liquids with stir bar dispersive liquid microextraction (SBDLME). This approach has been successfully proposed for extracting UV filters requiring a short extraction time (10 minutes), with better recoveries (91–117%) [108] than classical SBSE [105]. Another recent approach is the stir-bar sorptive-dispersive microextraction (SBSDME), which uses a stir bar coated with a hydrophobic magnetic nanosorbent such as $CoFe_2O_4$-oleic acid [109, 110] and $CoFe_2O_4$-SiO_2-nylon [111]. Compared to SBSE, the extraction time is shorter (25–30 minutes) and the recoveries (83–120%) are favourable [109, 110], although the extraction time is longer than with SBDLME [108]. In addition, higher LODs are obtained compared with SBSE and SBDLME. BAμE uses a bar coated with an appropriate powdered sorbent subjected to an ultrasonic treatment [112]. Compared to traditional SBSE, it provides the possibility of selecting the most suitable sorbent for the target analytes. The use of disposable and tuneable devices makes BAμE much more user-friendly and suitable for the routine work, offering an alternative for trace analysis of PCPs in real matrices [113]. For the extraction of the UV filter BP3, better recoveries (77–98%) were obtained [114] than with SBSE [7]. In a recent study, the new generation of bar adsorptive microextraction devices combined with microliquid desorption were proposed for the determination of two polar UV filters in the aquatic media. Different sorbents were evaluated as BAμE coating phases, and polystyrene-DVB polymer showed the best selectivity for the analysis of the two UV filters, with mean extraction efficiency of 65% [113]. In recent years, methodologies based on the use of green sorbents, such as cork pellets as BAμE coatings, have been proposed for the determination of parabens [115], other antimicrobial agents such as triclocarban, and UV filters such as benzophenones [116]. However, recoveries were not completely satisfactory (53–123%) and higher LODs were obtained compared to SBSE.

3.2.1.3.2 Rotating-disk Sorptive Extraction

Rotating-disk sorptive extraction (RDSE) is an integrated extraction/stirring technique, proposed in 2009 by Richter and co-workers, which has been used for the extraction/preconcentration of various organic micro-pollutants from aqueous samples. Its general principles are based on SPME and SBSE and it also consists of two fundamental steps: extraction of the analytes from the sample to an extraction phase, followed by the desorption of the analytes to a small volume of a suitable solvent [117]. RDSE has been applied to determine parabens (MeP, EtP, PrP, BuP) employing HLB as sorbent phase, followed by LC-QTOF/MS [118] and GC-MS analysis [119]. In the last case, a derivatization step was necessary and, although the extraction time was shorter, LODs were higher than those reported employing SBSE-GC-MS [101]. Recently, the determination of multiclass organic micro-pollutants was also proposed, employing RDSE followed by GC-MS. Laminar cork was employed as extraction phase for the determination of 20 contaminants including UV filters, pesticides, and PAHs from aqueous samples [120], whereas HLB was employed as sorbent to determine parabens, triclosan, and the plasticizer bisphenol A together with anti-inflammatory drugs in river, well, and wastewater [121]. In both cases, the influencing parameters were carefully optimized using multivariate designs, and satisfactory analytical performance was achieved.

3.2.1.3.3 Vacuum-assisted Sorbent Extraction

Although their use is still not extended for PCPS analysis, vacuum-assisted sorbent extraction (VASE) has been recently applied in combination with GC-MS for the determination of UV filters in tap and lake water. This technique combines the advantages of SBSE and HS-SPME [122]. Reducing the pressure achieves a rapid extraction in comparison with analogous atmospheric pressure procedures, resulting in up to a 3-fold improvement under optimized conditions. Although good recoveries were obtained, its main drawback was the high incubation time required (14 hours).

3.2.1.4 Solid-phase Microextraction

Solid-phase microextraction (SPME) is nowadays the main microextraction technique to accomplish water analysis. Its popularity is increasing, and even routine laboratory protocols based on this technique are frequently used. Commercial fibres are preferentially used, although the development of new materials and formats is one of the most active research areas [123].

3.2.1.4.1 Classical SPME Coatings

The development of analytical methodology based on SPME employing classical coatings (PDMS/DVB, PDMS, PDMS/DVB/CAR, PA) to determine individual families of PCPs such as preservatives [124, 125], UV filters [106, 107, 126], plasticizers [127], fragrance allergens [128–131], or siloxanes [132], demonstrated the suitability of this technique, showing good performance in terms of accuracy, precision, and low LODs. Besides, SPME is a powerful tool to monitor PCPs' degradation (e.g., UV filters or preservatives) as well as to extract and preconcentrate photodegradation by-products [125, 126].

However, most attention is attracting the development of SPME-based methods to simultaneously analyse multiclass PCPs [133, 134]. Recently, SPME-GC-MS/MS employing PDMS/DVB coating fibre was proposed for the simultaneous determination of different PCPs families including UV filters, antioxidants, phthalates, fragrance allergens, and synthetic musks, up to 43 PCPs. The method was applied to water samples from the North Portuguese coast, and the target compounds were determined in a single chromatographic run within 23 minutes [135].

3.2.1.4.2 New Coating Materials

In recent years, the design and use of custom-made SPME fibres have increased. The use of metals nanoparticles such as TiO_2 or the combination of Zn-ZnO or Ti-TiO_2-ZrO_2 to functionalize SPME coatings, enhances the extraction efficiency and analyte selectivity [123]. These coating materials have been proposed for the determination of UV filters

[136], as well as for the extraction of UV filters and phthalates from different environmental waters including river and wastewater [137–139]. In all cases, the fibres showed high stability and long service time. Another possibility consists of the combination of organic/inorganic composite coated SPME fibres, which has been successfully employed for the simultaneous analysis of phthalates [140–143], plasticizers, UV filters, and PAHs in environmental waters [144]. The home-made fibres exhibited a high stability, performing up to 200 extractions without losing efficiency [143].

Recently, the use of polythionine as SPME coating, prepared by electrochemical procedures, was proposed to determine phthalates in wastewater water samples [145], whereas nano-structured self-dope polythiophene (STP) was applied as a thermally stable conductive polymer-based SPME to analyse these plasticizers in water [146], showing superior thermal stability. Electroenhanced SPME (EE-SPME)-GC-MS [147] was successfully applied to determine phthalates and BPA in seawater, showing LODs quite lower than those provided by other home-made coatings [143]. Crosslinked polymeric ionic liquids (PILs)-based materials have been also proposed as alternative SPME fibres, showing comparable performance to that of commercial PA and PDMS/DVB fibres for UV filter extraction [148, 149].

Other new materials involve the use of multiwalled carbon nanotubes (MWCNTs) as SPME coating for UV filters [150] and phthalates [151] before GC-MS analysis. The optimized methodologies were applied to real water samples with varied matrix complexity such as tap, river, spa, mineral, and wastewater.

In-tube SPME has been also proposed for the determination of phthalates and pesticides [152] and an original methodology combining magnetism-enhanced monolith-(ME-MB-IT/SPME) has been successfully applied for the extraction of UV filters [153] and parabens [154], in environmental waters. To perform the extraction, an IL-based monolithic capillary column doped with magnetic nanoparticles was prepared by in situ polymerization. In both cases, the key parameters affecting extraction efficiency were optimized. LODs at the low ng L^{-1} were obtained.

Thin film (TF)-SPME is another approach that has been applied to increase the volume of SPME extraction phase. In addition to the increased extraction phase, which is also offered by SBSE, TF-SPME presents a higher surface/volume ratio, providing not only an enhanced sensitivity, but also a faster sampling [155]. TF-SPME is a very suitable option to develop devices for the passive sampling of water, providing a drastic decrease of the LOQs in environmental applications. A time-weighted average (TWA) passive sampling approach using TF-SPME followed by LC-MS/MS analysis was employed for the determination of 8 UV filters of different polarities, and 3 preservatives in wastewater [156]. The proposed passive sampler contained an HLB coating, allowing long- and short-term monitoring of compounds with different polarities. The proposed method was validated, showing linearity for up to 70 days. The retracted device was successfully employed on-site for 90-day time periods in wastewaters. The obtained results were compared with a C18-coated open system, showing good agreement between the two methodologies.

Water may also be indirectly analysed (passive sampling) using semipermeable membrane devices (SPMDs). These devices consist of a thin lay-flat tube made from semipermeable polyethylene membranes. They are mounted on assemblies to give a spread configuration in perforated stainless steel containers, and exposed for a controlled time. Further extractions from these devices are usually performed by dialysis. SPMDs employing polar organic chemical integrative samplers (POCI) were applied to determine polar and mid-polar UV filters in different river locations, with a collection time of 4 weeks [157].

New trends in environmental SPME analysis involve the use of recycled materials as green sorbent phases. In this context, recycled diatomaceous earth from a beer purification process has been proposed as TF-SPME to extract several endocrine disruptors, including 3 UV filters, 1 preservative, and the plasticizer BPA, from environmental water samples [158]. The sample preparation time was only 2.8 minutes per sample since SPME was combined with a 96-well plate system, allowing for high-throughput analysis. Bamboo charcoal has also been employed as a green and inexpensive SPME coating for the determination of 11 phthalates in environmental waters, followed by GC-MS analysis [159]. The obtained LODs, 4–23 ng L^{-1}, were of the same order, or even lower, than those achieved employing Zn-ZnO nanoparticles [137] for the determination of phthalates, showing that the use of this inexpensive by-product may offer a promising environmentally friendly alternative for routine analysis. The use of carbon dots (CDs) immobilized on the surface of steel wire from candle soot (CS) was recently proposed to determine phthalates in aqueous samples [160] and, although this home-made fibre could be used up to 50 times, LODs were quite high in comparison with those obtained employing other home-made coatings [145, 146].

3.2.2 Liquid-based Extraction Techniques

Liquid-based extraction techniques are being employed for the analysis of PCPs in water samples. The most traditional been liquid-liquid extraction (LLE). However, it presents several disadvantages such as a long preparation time and consumption of large volumes of organic solvent and sample (200–500 mL). This technique is also difficult to automate, and it usually requires post-extraction treatment, such as filtration, purification, and solvent evaporation. Despite these limitations, classical LLE is still used and in recent years it has been applied to the extraction of UV filters [161], preservatives such as parabens and TCS [162–164], and plasticizers (phthalates) [165], offering good recoveries and LODs at the low ng L^{-1}.

The major trends in analytical chemistry, as previously mentioned, are the automatization and the development of miniaturized devices and procedures aiming at minimizing the organic solvent consumption and the sample volume, thereby reducing the environmental impact. In this regard, a fully automated analytical platform integrating water sampling-miniscale-LLE (mLLE) and a full evaporation dynamic headspace concentration has been recently proposed for the analysis of UV filters by GC-MS [166]. The optimized settings achieved good recoveries and LODs lower than 10 ng L^{-1}.

Regarding miniaturization, several techniques such as dispersive liquid-liquid microextraction (DLLME), single-drop microextraction (SDME), and ultrasound-assisted emulsification microextraction (USAEME) have been developed and applied to extract PCPs from water.

3.2.2.1 Microextraction Liquid Phase Approaches: DLLME, SDME, USAEME

3.2.2.1.1 Dispersive-liquid-liquid microextraction

In dispersive liquid-liquid microextraction (DLLME), a small volume of extractant solvent is dispersed by the action of a second solvent into the sample. The abundant contact surface of the fine droplets and the aquatic phase greatly enhances the extraction efficiency, thus reducing the solvent volume needed and the extraction time. DLLME has been applied to extraction of PCPs including parabens [167], UV filters [168], and phthalates [169–172] and analysis was mainly performed by LC followed by UV/DAD [167, 172]. GC-FID [169, 170] or GC-MS [171, 173] were sometimes employed. Low LODs were obtained allowing the measurement of trace levels of PCPs in natural waters including seawater, river, rain, or wastewater. In some cases, the use of DLLME allowed the simultaneous analysis of up to 21 samples per hour [173].

Multiresidue procedures to extract simultaneously PCPs from water have also been proposed. DLLME was applied to extract UV filters and synthetic musks from 5 types of water matrices [174], parabens and plasticizers [175] or estrogens [176, 177], as well as UV filters and preservatives in surface water of coral reef [28], where high levels of benzophenone-3 were detected ($\mu g\ L^{-1}$).

A variation of classical DLLME involves the use of in-syringe-DLLME, achieving dispersion by loading the aqueous sample at a high flow rate within a syringe already containing the extractant/dispersive mixture. This approach was applied to determine phthalates in different water samples [178, 179]. Although it offers a compact and versatile tool, the LODs were quite a bit higher than those reported employing classical DLLME [171, 173]. Very recently, a novel DLLME method employing a switchable hydrophilic solvent was developed for the extraction of 3 benzophenone-type UV filters from environmental waters [180]. After the extraction, unmodified Fe_3O_4 magnetic nanoparticles were used as a carrier to separate and retrieve the switchable hydrophilicity solvent from the sample solution. Similar performance in terms of recoveries and precision was obtained in comparison with classical DLLME.

Novel approaches include methodologies based on ionic liquids (ILs) that are considered environmentally friendly solvents, with unique and effective characteristics such as lower vapour pressure and high viscosity. An IL-DLLME method followed by HS-GC-MS was developed for the determination of 6 UV filters in pool and lake water. Up to 5 different ILs were specifically designed to favour the extraction of the target analytes. The optimized procedure required less than 10 minutes per extraction, allowing simultaneous multiple extractions [181]. This approach was also applied to UV filters, phthalates, and other contaminants such as insecticides in different water matrices including tap, lake, or wastewater [182, 183]. Good recoveries and precision were obtained, although LODs were higher than those obtained employing classical DLLME [172].

The combination of deep eutectic solvents (DES) with LLME is another of the novel trends. DES are considered as new and green solvents that can be used for the development of new liquid-based methods including the extraction of PCPs (preservatives, UV filters) in the aquatic environment [184–186]. DES-DLLME was recently proposed to determine benzophenone and salicylate derivatives UV filters in water [186], using different fatty acid-based hydrophobic DES as extraction solvents. Enrichment factors between 40 and 50 were obtained. Menthol hydrophobic DES was also proposed as an extractant for the preconcentration and extraction of benzophenone UV filters from water followed by LC-DAD [185]. Both methodologies were successfully validated and applied to different environmental water samples.

Other novel approach is the use of DLLME based on solidification of floating organic drop (SFO). It was successfully employed for phthalate extraction in different environmental water matrices [187, 188], showing quantitative recoveries and LODs at the low ng L^{-1} level.

To improve the dispersion of the solvent into the water matrix, LLME and DLLME have also been assisted by magnetic stirring (MSA-DLLME), ultrasound radiation (UA-LLME, UA-DLLME), or air (air-assisted-DLLME). These approaches were applied to extract preservatives [175, 176], phthalates [189–192], fragrances [193], or siloxanes [194] from water, showing in all cases good performance. Recently, a sensitive method using stable isotopic labelling (SIL) and microwave-assisted (MA)-DLLME was proposed for the determination of hydroxyl-UV filters in environmental waters including wastewater, river, and tap water. The whole sample treatment was carried out in only 9 minutes [195]. The use of DLLME followed by microwave (MW) energy has also proposed as an effective tool for PCPs in environmental aqueous matrices [196]. A multiresidue DLLME followed by rapid MW-assisted silylation was employed for the determination of 13 UV filters and the plasticizer BPA in wastewater samples. Under the optimal experimental conditions, good recoveries and low LODs were reached.

3.2.2.1.2 Single-drop Microextraction

After its introduction just over 20 years ago, single-drop microextraction (SDME) became a widely applied solvent-based microextraction for environmental analysis. In general, SDME is inexpensive, easy to operate, nearly solvent-free, and can be conducted using very basic apparatus. It can achieve high analyte enrichment in a short time, and it is easily combined with various instrumental techniques. SDME has been successfully applied to extract PCPs such as phthalates [197] and UV filters [198]. In the last case, an IL was employed as extractant (IL-SDME), and extraction was done with a drop volume of only 10 μL of 1-hexyl-3-methylimidazolium hexa-fluorophosphate ([C6MIM][PF6]) in 37 minutes. Enrichment factors up to 100-fold were obtained and recoveries were comparable to other liquid-phase extraction techniques.

3.2.2.1.3 Ultrasound-assisted Emulsification Microextraction

Ultrasound-assisted emulsification microextraction (USAEME) shows the advantage of using ultrasonic radiation to accelerate the mass-transfer process between two immiscible phases and therefore increases the extraction efficiency in a minimum amount of time. USAEME relies on the emulsification of the organic extractant in the sample by ultrasound without the need of a dispersive solvent, favouring efficiency, and achieving a complete extraction of the analytes into the organic phase. This technique was applied for the first time to extract musks and phthalates from water in only 10 minutes [199]. Later, it was successfully applied to extract other PCPs such as phenolic preservatives, fragrance allergens [200–202], and UV filters [203] from different water samples. In all cases, the extraction parameters were optimized, employing experimental designs to obtain the highest extraction efficiency. For UV filter analysis, only 5 minutes of extraction time were needed to achieve recoveries higher than 90%. This method was applied to a broad range of water matrices including sea and river water, as well as recreational waters such as spa, swimming-pool, or aquapark samples. The LODs were below 1 ng L^{-1} employing GC-MS/MS analysis.

The substitution of organic solvents by ILs has been proposed as an environmentally friendly alternative (IL-USAEME) prior to LC-UV to determine UV filters in water [204]. The combination of USAEME with solidification of floating drop (SFO) was applied to determine plasticizers [205] and UV filters [206] in water samples. With this technique, a mixture of extraction solvent and emulsifier was rapidly injected into the heated sample solution at elevated temperature. Several parameters influencing the extraction efficiency were optimized, and after a successful validation, the proposed method was applied to real water samples. LODs, about 1 μg L^{-1}, were similar to those reached employing IL-USAEME [204].

3.3 Determination of Personal Care Products

Gas chromatography (GC) and liquid chromatography (LC) are the techniques employed for the separation and determination of PCPs. The choice of the chromatography type depends on the chemical nature of the target analytes. As can be seen in Tables 3.1–3.4, LC is the preferred option for preservatives and UV filters, especially for the most polar ones such as hydroxyl preservatives or benzophenone derivatives. Separations are usually performed on reversed-phase columns, mainly C18 type and, to a lesser extent, C8. In reversed-phase chromatography, water, EtOH, ACN, MeOH, and tetrahydrofurane (THF), or combinations of them, are the solvents usually employed, either with isocratic or mostly with gradient elution modes, especially when multianalyte separation is required. Modifiers such as formic acid, acetic acid, phosphoric acid, perchloric acid, or formate and acetate buffers are commonly added to the mobile phase, to stabilize the compounds in their neutral forms or/and to enhance ionization if mass spectrometry is employed as detection technique.

On the other hand, GC is the preferred chromatographic techniques for the separation of the most volatile compounds such as fragrance allergens or siloxanes, although non-polar UV filters or preservatives are also successfully analysed by GC. However, in this case, for the determination of more polar compounds such as parabens or benzophenones, a derivatization step is usually included during the sample pre-treatment to modify several chemical functions of the original low-volatile compounds. Derivatization usually employs acetylation with acetic anhydride, or silyl derivatives to improve the chromatographic peak shape of the most polar compounds. The acetylation step is usually performed in situ, together with the extraction step to obtain a good method performance in a short time.

Traditionally, due to the capability of several PCPs such as UV filters or some plasticizers to absorb light in the UV-Vis spectrum, diode array detector (DAD) was the preferred detection method after LC separation. The main advantage of DAD detector is its capability of simultaneously recording absorbance data from 190 to 800 nm, matching the UV-Vis spectra obtained with spectral libraries of previously analysed standard compounds, to identify the eluted compounds. However, given the low specificity of UV detectors, which makes difficult to quantify multiple compounds in complex sample matrices, such as environmental ones, the combination of GC or LC with mass spectrometry (MS) or tandem mass spectrometry (MS/MS) detection became the most suitable option. In this way, the analytical selectivity and sensibility is improved, allowing the detection of PCPs at trace levels. In the case of GC analysis, MS or MS/MS detectors are the most employed for PCP analysis, whereas only few studies still use FID detection.

In recent years, the combination of chromatographic techniques with high-resolution mass spectrometry, mainly based on QTOF detection, has beginning to play a more important role in environmental analysis. This powerful technique is used to not only identify and sometimes quantify the target PCPs but also for the screening of non-target pollutants. However, the elucidation of chemical structures based on exact mass and other parameters such as isotope profile scores and high-resolution MS/MS spectra is a very challenging step requiring in-depth research and high skill. In addition, the development of analytical methods based on HRMS to identify unknown contaminants in many matrices is increasing. However, although great research efforts have been dedicated to the identification of transformation products (TPs), very few analytical methods including their quantification are reported, probably because few analytical standards are still available for these compounds.

3.4 Future Directions and Challenges

The frequent ocurrence of PCPs and other contaminants in different type of waters, has raised multiple concerns. For this reason, in recent years, great efforts have been made to develop robust and reliable analytical methodologies. However, there is still a long way to go before being able to elucidate the aquatic, environmental, and health issues of non-regulated compounds, monitor their presence, and study PCPs' degradation leading to the potential formation of other CECs.

In this context, the development of automated, integrated, high-throughput, and sensitive methodologies for the simultaneous extraction and determination of a high number of PCPs in a short time is necessary. The development of new sorbent materials with high adsorption capacity as well as the use of fast, cost-effective, and environmentally friendly sample pre-treatment techniques are needed. The improvements made in terms of miniaturization and portability of extraction devices greatly facilitate the implementation of on-site passive samplers. When coupled to portable analytical instruments, on-site analysis can also be implemented. Furthermore, the increasing performance of the modern analytical instrumentation would also facilitate the development of extraction technologies for applications involving direct coupling to this instrumentation. Further fields of research should also consist of developing analytical tools for conducting PCPs degradation studies in the environment, based on target and non-target analysis, to not only identify or quantify transformation products but also establish degradation patways to know their behaviour.

Acknowledgements

This work was supported by projects IN607B 2019/13 and ED431 2020/06 (Consolidated Research Groups Program, Xunta de Galicia). The authors belong to the National Network for the Innovation in miniaturized sample preparation techniques, RED2018-102522-T (Ministry of Science, Innovation and Universities, Spain). This study is based upon work from the Sample Preparation Study Group and Network, supported by the Division of Analytical Chemistry of the European Chemical Society. All these programmes are co-funded by FEDER (EU).

References

1 Brausch, J.M. and Rand, G.M. (2011). A review of personal care products in the aquatic environment: environmental concentrations and toxicity. *Chemosphere* 82(11): 1518–1532. doi: 10.1016/j.chemosphere.2010.11.018.

2 Ramos, S., Homem, V., Alves, A., and Santos, L. (2015). Advances in analytical methods and occurrence of organic UV-filters in the environment-a review. *Sci. Total Environ.* 526: 278–311. doi: 10.1016/j.scitotenv.2015.04.055.

3 Chisvert, A., Benedé, J.L., and Salvador, A. (2018). Current trends on the determination of organic UV filters in environmental water samples based on microextraction techniques. *A Review. Anal. Chim. Acta.* 1034: 22–38. doi: 10.1016/j.aca.2018.05.059.

4 Paxéus, N. (1996). Organic pollutants in the effluents of large wastewater treatment plants in Sweden. *Water Res.* 30(5): 1115–1122. doi: 10.1016/0043-1354(95)00278-2.

5 Haman, C., Dauchy, X., Rosin, C., and Muñoz, J.F. (2015). Occurrence, fate and behavior of parabens in aquatic environments: a review. *Water Res.* 68: 1–11. doi: 10.1016/j.watres.2014.09.030.

6 Lu, Y., Yuan, T., Wang, W., and Kannan, K. (2011). Concentrations and assessment of exposure to siloxanes and synthetic musks in personal care products from China. *Environ. Pollut.* 159: 3522–3528. doi: 10.1016/j.envpol.2011.08.015.

7 Cadena-Aizaga, M.I., Montesdeoca-Esponda, S., Torres-Padrón, M.A.E., Sosa-Ferrera, Z., and Santana-Rodríguez, J.J. (2020). Organic UV filters in marine environments: an update of analytical methodologies, occurrence and distribution. *Trend. Environ. Anal. Chem.* 25: 00079. doi: 10.1016/j.teac.2019.e00079.

8 Montes-Grajales, D., Fennix-Agudelo, M., and Miranda-Castro, W. (2017). Occurrence of personal care products as emerging chemicals of concern in water resources: a review. *Sci. Total. Environ.* 595: 601–614. doi: 10.1016/j.scitotenv.2017.03.286.

9 Ashfaq, M., Li, Y., Rehman, M.S.U., Zubair, M., Mustafa, G., Nazar, M.F., Yu, C.P., and Sun, Q. (2019). Occurrence, spatial variation and risk assessment of pharmaceuticals and personal care products in urban wastewater, canal surface water, and their sediments: a case study of Lahore, Pakistan. *Sci. Total Environ.* 688: 653–663. doi: 10.1016/j.scitotenv.2019.06.285.

10 Richardson, S.D. and Ternes, T.A. (2018). Water analysis: emerging contaminants and current issues. *Anal. Chem.* 90(1): 398–428. doi: 10.1021/acs.analchem.7b04577.

11 Regulation (EC) No 1223/2009 of the European Parliament and of the Council of 30 November 2009 on cosmetic products (recast) (text with EEA relevance), Off. J. Union, 2009, L342/59.

12 Dulio, V., van Bavel, B., Brorstrom-Lundén, E., Harmsen, J., Hollender, J., Schlabach, M., Slobodnik, J., Thomas, K., and Koschorreck, J. (2018). Emerging pollutants in the EU: 10 years of NORMAN in support of environmental policies and regulations. *Environ. Sci. Eur.* 30(1): 1–13. doi: 10.1186/s12302-018-0135-3.

13 Pena-Pereira, F., Bendicho, C., Pavlovic, D.M.I., Martín-Esteban, A., Díaz-Álvarez, M., Pan, Y., Cooper, J., Yang, Z., Safarik, I., Pospiskova, K., Segundo, M.A., and Psillakis, E.

(2021). Miniaturized analytical methods for determination of environmental contaminants of emerging concern-a review. *Anal. Chim. Acta.* 1158: 238108. doi: 10.1016/j.aca.2020.11.040.

14 Directive 2000/60/EC of the European Parliament and of the council of 23 October 2000 establishing a framework for Community action in the field of water policy. Off. J. L-327, 1–73.

15 Gomez Cortes, L., Marinov, D., Sanseverino, I., Navarro-Cuenca, A., Niegowska, M., Porcel Rodríguez, E., and Lettieri, T., (2020). Selection of substances for the 3rd Watch List under the Water Framework Directive, EUR 30297 EN, Off. Eur. Union, Luxembourg, 2020, ISBN 978-92-76-19425-5, doi:10.2760/240926,JRC121346.

16 Lores, M., Llompart, M., Sanchez-Prado, L., García-Jares, C., and Cela, R. (2005). Confirmation of the formation of dichlorodibenzo-p-dioxin in the photodegradation of triclosan by photo-SPME. *Anal. Bioanal. Chem.* 381(6): 1294–1298. doi: 10.1007/s00216-004-3047-6.

17 Sanchez-Prado, L., Llompart, M., Lores, M., García-Jares, C., Bayona, J.M., and Cela, R. (2006). Monitoring the photochemical degradation of triclosan in wastewater by UV light and sunlight using solid-phase microextraction. *Chemosphere* 65(8): 1338–1347. doi: 10.1016/j.chemosphere.2006.04.025.

18 Santos, A.J.M., Miranda, M.S., and Da Silva, J.C.G.E. (2012). The degradation products of UV filters in aqueous and chlorinated aqueous solutions. *Water Res.* 46(10): 3167–3176. doi: 10.1016/j.watres.2012.03.057.

19 Negreira, N., Canosa, P., Rodríguez, I., Ramil, M., Rubí, E., and Cela, R. (2008). Study of some UV filters stability in chlorinated water and identification of halogenated by-products by gas chromatography-mass spectrometry. *J. Chromatogr. A* 1178(1): 206–214. doi: 10.1016/j.chroma.2007.11.057.

20 Daughton, C.G. and Ternes, T.A. (1999). Pharmaceuticals and personal care products in the environment: Agents of subtle change? *Environ. Health. Persp.* 107(6): 907–938. doi: 10.1289/ehp.99107s6907.

21 Pedrouzo, M., Borrull, F., Marcé, R.M., and Pocurull, E. (2011). Analytical methods for personal-care products in environmental waters. *TrAC-Trends Anal. Chem.* 30(5): 749–760. doi: 10.1016/j.trac.2011.01.009.

22 Buchberger, W.W. (2011). Current approaches to trace analysis of pharmaceuticals and personal care products in the environment. *J. Chromatogr. A* 1218(4): 603–618. doi: 10.1016/j.chroma.2010.10.040.

23 Pérez-Lemus, N., López-Serna, R., Pérez-Elvira, S.I., and Barrado, E. (2020). Sample pre-treatment and analytical methodology for the simultaneous determination of pharmaceuticals and personal care products in sewage sludge. *Chemosphere* 258: 127273. doi: 10.1016/j.chemosphere.2020.127273.

24 Krzeminski, P., Schwermer, C., Wennberg, A., Langford, K., and Vogelsang, C. (2017). Occurrence of UV filters, fragrances and organophosphate flame retardants in municipal WWTP effluents and their removal during membrane post-treatment. *J. Hazard. Mater.* 323: 166–176. doi: 10.1016/j.jhazmat.2016.08.001.

25 Wang, D.G., Norwood, W., Alaee, M., Byer, J.D., and Brimble, S. (2013). Review of recent advances in research on the toxicity, detection, occurrence and fate of cyclic volatile methyl siloxanes in the environment. *Chemosphere* 93(5): 711–725. doi: 10.1016/j.chemosphere.2012.10.041.

26 Ramos, S., Homem, V., Alves, A., and Santos, L. (2016). A review of organic UV-filters in wastewater treatment plants. *Environ. Int.* 86: 24–44. doi: 10.1016/j.envint.2015.10.004.

27 Tashiro, Y. and Kameda, Y. (2013). Concentration of organic sun-blocking agents in seawater of beaches and coral reefs of Okinawa Island, Japan. *Mar. Pollut. Bull.* 77(1–2): 333–340. doi: 10.1016/j.marpolbul.2013.09.013.

28 Kung, T.A., Lee, S.H., Yang, T.C., and Wang, W.H. (2018). Survey of selected personal care products in surface water of coral reefs in Kenting National Park, Taiwan. *Sci. Total. Environ.* 635: 1302–1307. doi: 10.1016/j.scitotenv.2018.04.115.

29 Tsui, M.M.P., Leung, H.W., Wai, T.C., Yamashita, N., Taniyasu, S., Liu, W., Lam, P.K.S., and Murphy, M.B. (2014). Occurrence, distribution and ecological risk assessment of multiple classes of UV filters in surface waters from different countries. *Water Res.* 67: 55–65. doi: 10.1016/j.watres.2014.09.013.

30 Kameda, Y., Kimura, K., and Miyazaki, M. (2011). Occurrence and profiles of organic sun-blocking agents in surface waters and sediments in Japanese rivers and lakes. *Environ. Pollut.* 159(6): 1570–1576. doi: 10.1016/j.envpol.2011.02.055.

31 Chiriac, F.L., Paun, I., Pirvu, F., Pascu, L.F., Niculescu, M., and Galaon, T. (2020). Liquid chromatography tandem mass spectrometry method for ultra-trace analysis of organic UV filters in environmental water samples. *Rev. Chim.* 71(1): 92–99.

32 Filippe, T.C., Goulart, F.D.A.B., Mizukawa, A., and de Azevedo, J.L.C.S.R. (2018). Validation of analytical methodology for determination of Personal Care Products in environmental matrix by GC-MS/MS. *Eclet. Quim. J.* 43(3): 30–36. doi: 10.26850/1678-4618eqj.v43.3.2018.p30-36.

33 Reichert, G., Mizukawa, A., Antonelli, J., Goulart, F.D.A.B., Filippe, T.C., and de Azevedo, J.L.C.S.R. (2020). Determination of parabens, triclosan, and lipid regulators in a subtropical urban river: Effects of urban occupation. *Water Air Soil Pollut.* 231(3): 1–11. doi: 10.1016/j.scitotenv.2017.03.015.

34 Estévez-Danta, A., Rodil, R., Pérez-Castaño, B., Cela, R., Quintana, J.B., and González-Mariño, I. (2021). Comprehensive determination of phthalate, terephthalate and di-iso-nonyl cyclohexane-1, 2-dicarboxylate metabolites in wastewater by solid-phase extraction and ultra (high)-performance liquid chromatography-tandem mass spectrometry. *Talanta* 224: 121912. doi: 10.1016/j.talanta.2020.121912.

35 Zgola-Grzeskowiak, A., Jeszka-Skowron, M., Czarczynska-Goslinka, B., and Grzskowiak, T. (2016). Determination of parabens in Polish river and lake water as a function of season. *Anal. Lett.* 49(11): 1734–1747. doi: 10.1080/00032719.2015.1120739.

36 Azzouz, A. and Ballesteros, E. (2014). Trace analysis of endocrine disrupting compounds in environmental water samples by use of solid-phase extraction and gas chromatography with mass spectrometry detection. *J. Chromatogr. A* 1360: 248–257. doi: 10.1016/j.chroma.2014.07.059.

37 Terasaki, M., Takemura, Y., and Makino, M. (2021). Paraben-chlorinated derivatives in river waters. *Environ. Chem. Lett.* 10(4): 401–406. doi: 10.1007/s10311-012-0367-1.

38 Arukwe, A., Eggen, T., and Moder, M. (2012). Solid waste deposits as a significant source of contaminants of emerging concern to the aquatic and terrestrial environments – A developing country case study from Owerri, Nigeria. *Sci. Total. Environ.* 438: 94–102. doi: 10.1016/j.scitotenv.2012.08.039.

39 Liu, Y.S., Ying, G.G., Shareef, A., and Kookana, R.S. (2011). Simultaneous determination of benzotriazoles and ultraviolet filters in ground water, effluent and biosolid samples using gas chromatography-tandem mass spectrometry. *J. Chromatogr. A* 1218(31): 5328–5335. doi: 10.1016/j.chroma.2011.05.100.

40 Ho, Y.C. and Ding, W.H. (2012). Solid-phase extraction coupled simple on-line derivatization gas chromatography – Tandem mass spectrometry for the determination of benzophenone-type UV filters in aqueous samples. *J. Chin. Chem. Soc.* 59(1): 107–113. doi: 10.1002/jccs.201100317.

41 Rodil, R., Quintana, J.B., Concha-Graña, E., López-Mahía, P., Muniategui-Lorenzo, S., and Prada-Rodríguez, D. (2012). Emerging pollutants in sewage, surface and drinking water in Galicia (NW Spain). *Chemosphere* 86(10): 1040–1049. doi: 10.1016/j.chemosphere.2011.11.053.

42 Muckoya, V.A., Nomngongo, P.N., and Ngila, J.C. (2020). Factorial design optimisation of solid phase extraction for preconcentration of parabens in wastewater using ultra-high performance liquid chromatography triple quadrupole mass spectrometry. *Curr. Anal. Chem.* 16(4): 436–446. doi: 10.2174/1573411014666180627150854.

43 Mizukawa, A., Reichert, G., Filippe, T.C., Brehm, F.D.A., and Rodrigues de Azevedo, J.L.C.S. (2018). Occurrence and risk assessment of personal care products in subtropical urban rivers. *Environ. Eng. Sci.* 35(11): 1263–1272. doi: 10.1089/ees.2018.0066.

44 Malvar, J.L., Santos, J.L., Martín, J., Aparicio, I., and Alonso, E. (2019). Routine analytical method for monitoring the main metabolites for a recurrent group of parabens and pharmaceuticals in wastewater and tap water. *Anal. Bioanal. Chem.* 411(25): 6625–6635. doi: 10.1007/s00216-019-02035-2.

45 Aristizabal-Ciro, C., Botero-Coy, A.M.A., López, F.J., and Peñuela, G.A. (2017). Monitoring pharmaceuticals and personal care products in reservoir water used for drinking water supply. *Environ. Sci. Pollut. Res.* 24(8): 7335–7347. doi: 10.1007/s11356-016-8253-1.

46 Martín, J., Camacho-Muñoz, D., Santos, J.L., Aparicio, I., and Alonso, E. (2014). Determination of emerging and priority industrial pollutants in surface water and wastewater by liquid chromatography-negative electrospray ionization tandem mass spectrometry. *Anal. Bioanal. Chem.* 406(15): 3709–3716. doi: 10.1007/s00216-014-7689-8.

47 González-Mariño, I., Quintana, J.B., Rodríguez, I., and Cela, R. (2011). Evaluation of the occurrence and biodegradation of parabens and halogenated by-products in wastewater by accurate-mass liquid chromatography-quadrupole-time-of-flight-mass spectrometry (LC-QTOF-MS). *Water Res.* 45(20): 6770–6780. doi: 10.1016/j.watres.2011.10.027.

48 Jonkers, N., Sousa, A., Galante-Oliveira, S., Barroso, C.M., Kohler, H.-P.E., and Giger, W. (2010). Occurrence and sources of selected phenolic endocrine disruptors in Ria de Aveiro, Portugal. *Environ. Sci. Pollut. Res.* 17(4): 834–843. doi: 10.1007/s11356-009-0275-5.

49 Ortega-Zamora, C., González-Sálamo, J., Varela-Martínez, D.A.L., González-Curbelo, M., and Hernández-Borges, J. (2020). Extraction of phthalic acid esters and di (2-ethylhexyl) adipate from tap and waste water samples using Chromabond® HLB as sorbent prior to gas chromatography-mass spectrometry analysis. *Separations* 7(2): 21. doi: 10.3390/separations7020021.

50 Robles-Molina, J., Lara-Ortega, F.J., Gilbert-López, B., García-Reyes, J.F., and Molina-Díaz, A. (2014). Multi-residue method for the determination of over 400 priority and emerging pollutants in water and wastewater by solid-phase extraction and liquid chromatography-time-of-flight mass spectrometry. *J. Chromatogr. A* 1350: 30–43. doi: 10.1016/j.chroma.2014.05.003.

51 Huysman, S., Van Meulebroek, L., Janssens, O., Vanryckeghem, F., Van Langenhove, H., Demeestere, K., and Vanhaecke, L. (2019). Targeted quantification and untargeted screening of alkylphenols, bisphenol A and phthalates in aquatic matrices using ultra-high-performance liquid chromatography coupled to hybrid Q-Orbitrap mass spectrometry. *Anal. Chim. Acta.* 1049: 141–151. doi: 10.1016/j.aca.2018.10.045.

52 Bizkarguenaga, E., Ros, O., Iparraguirre, A., Navarro, P., Vallejo, A., Usobiaga, A., and Zuloaga, O. (2012). Solid-phase extraction combined with large volume injection-programmable temperature vaporization-gas chromatography-mass spectrometry for the multiresidue determination of priority and emerging organic pollutants in wastewater. *J. Chromatogr. A* 1247: 104–117. doi: 10.1016/j.chroma.2012.05.022.

53 Gago-Ferrero, P., Mastroianni, N., Díaz-Cruz, M.S., and Barceló, D. (2013). Fully automated determination of nine ultraviolet filters and transformation products in natural waters and wastewaters by on-line solid phase extraction-liquid chromatography-tandem mass spectrometry. *J. Chromatogr. A* 1294: 106–116. doi: 10.1016/j.chroma.2013.04.037.

54 Jurado, A., Gago-Ferrero, P., Vázquez-Suñé, E., Carrera, J., Pujades, E., Díaz-Cruz, M.S., and Barceló, D. (2014). Urban groundwater contamination by residues of UV filters. *J. Hazard. Mater.* 271: 141–149. doi: 10.1016/j.jhazmat.2014.01.036.

55 Chavez-Moreno, C.A., Hinojosaâ-Reyes, L., Ruiz-Ruiz, E.J., Hernández-Ramírez, A., and Guzmán-Mar, J.L. (2018). Optimization of solid-phase extraction of parabens and benzophenones in water samples using a combination of Plakett-Burman and Box-Behnken designs. *J. Sep. Sci.* 41(24): 4488–4497. doi: 10.1002/jssc.201800796.

56 Montesdeoca-Esponda, S., Sosa-Ferrera, Z., and Santana-Rodríuez, J.J. (2012). On-line solid-phase extraction coupled to ultra-performance liquid chromatography with tandem mass spectrometry detection for the determination of benzotriazole UV stabilizers in coastal marine and wastewater samples. *Anal. Bioanal. Chem.* 403(3): 867–876. doi: 10.1007/s00216-012-5906-x.

57 Manousi, N., Zachariadis, G.A., and Deliyanni, E.A. (2020). On the use of metal-organic frameworks for the extraction of organic compounds from environmental samples. *Environ. Sci. Pollut. Res.* Feb 19. doi: 10.1007/s11356-020-07911-4. Epub ahead of print. PMID: 32077018.

58 Abbasghorbani, M., Attaran, A., and Payehghadr, M. (2013). Solvent-assisted dispersive micro-SPE by using aminopropyl-functionalized magnetite nanoparticle followed by GC-PID for quantification of parabens in aqueous matrices. *J. Sep. Sci.* 36(2): 311–319. doi: 10.1002/jssc.201200556.

59 Ghasemi, E. and Sillanpaa, M. (2019). Ultrasound-assisted solid-phase extraction of parabens from environmental and biological samples using magnetic hydroxyapatite nanoparticles as an efficient and regenerable nanosorbent. *Microchim. Acta.* 186(9): 1–7. doi: 10.1007/s00604-019-3720-2.

60 Nurerk, P., Llompart, M., Donkhampa, P., Bunkoed, O., and Dagnac, T. (2020). Solid-phase extraction based on MIL-101 adsorbent followed by gas chromatography tandem mass spectrometry for the analysis of multiclass organic UV filters in water. *J. Chromatogr. A* 1610: 460564. doi: 10.1016/j.chroma.2019.460564.

61 Dargahi, R., Ebrahimzadeh, H., Asgharinezhad, A.A., Hashemzadeh, A., and Amini, M.M. (2018). Dispersive magnetic solid-phase extraction of phthalate esters from water samples and human plasma based on a nanosorbent composed of MIL-101 (Cr) metal-organic framework and magnetite nanoparticles before their determination by GC-MS. *J. Sep. Sci.* 41(4): 948–957. doi: 10.1002/jssc.201700700.

62 Li, H., Cao, Z., Cao, X., Jiang, Z., Abd El-Aty, A.M., Qi, Y., Shao, H., Jin, F., Zheng, L., and Wang, J. (2018). Magnetic solid-phase extraction using a mixture of two types of nanoparticles followed by gas chromatography-mass spectrometry for the determination of six phthalic acid esters in various water samples. *RSC Adv.* 8(69): 39641–39649. doi: 10.1039/C8RA08643E.

63 Tahmasebi, E., Masoomi, M.Y., Yamini, Y., and Morsali, A. (2016). Application of a Zn (II) based metal-organic framework as an efficient solid-phase extraction sorbent for preconcentration of plasticizer compounds. *RSC Adv.* 6(46): 40211–40218. doi: 10.1039/C6RA06560K.

64 Li, Z., Huang, D., Fu, C., Wei, B., Yu, W., Deng, C., and Zhang, X. (2011). Preparation of magnetic core mesoporous shell microspheres with C18-modified interior pore-walls for fast extraction and analysis of phthalates in water samples. *J. Chromatogr. A* 1218(37): 6232–6239. doi: 10.1016/j.chroma.2011.06.109.

65 Wu, J., Wang, C., Liang, X., Yang, X., Wang, C., Wu, Q., and Wang, Z. (2017). Magnetic spherical carbon as an efficient adsorbent for the magnetic extraction of phthalate esters from lake water and milk samples. *J. Sep. Sci.* 40(10): 2207–2213. doi: 10.1002/jssc.201601355.

66 Wang, X., Wang, J., Du, T., Kou, H., Du, X., and Lu, X. (2018). Determination of six benzotriazole ultraviolet filters in water and cosmetic samples by graphene sponge-based solid-phase extraction followed by high-performance liquid chromatography. *Anal. Bioanal. Chem.* 410(26): 6955–6962. doi: 10.1007/s00216-018-1301-6.

67 Kharbouche, L., Gil Garcia, M.D., Lozano, A., Hamaizi, H., and Martinez Galera, M. (2020). Determination of personal care products in water using UHPLC-MS after solid phase extraction with mesoporous silica-based MCM-41 functionalized with cyanopropyl groups. *J. Sep. Sci.* 43(11): 2142–2153. doi: 10.1002/jssc.201901148.

68 Rocio-Bautista, P., Martínez-Benito, C., Pino, V., Pasan, J., Ayala, J.H., Ruiz-Perez, C., and Afonso, A.M. (2015). The metal-organic framework HKUST-1 as efficient sorbent in a vortex-assisted dispersive micro solid-phase extraction of parabens from environmental waters, cosmetic creams, and human urine. *Talanta* 139: 13–20. doi: 10.1016/j.talanta.2015.02.032.

69 Mashile, G.P., Mpupa, A., and Nomngongo, P.N. (2018). In-syringe micro solid-phase extraction method for the separation and preconcentration of parabens in environmental water samples. *Molecules* 23(6): 1450. doi: 10.3390/molecules23061450.

70 Ueta, I., Sekiguchi, N., Suzuki, A., Kobayashi, Y., Kuwabara, T., and Saito, Y. (2019). Polyethylene terephthalate nanofiber sheet as the novel extraction medium for the determination of phthalates in water samples. *Anal. Sci.* 19: 317. doi: 10.2116/analsci.19P317.

71 Xu, Q., Yin, X., Wu, S., Wang, M., Wen, Z., and Gu, Z. (2010). Determination of phthalate esters in water samples using nylon6 nanofibers mat-based solid-phase extraction coupled to liquid chromatography. *Microchim. Acta.* 168(3–4): 267–275. doi: 10.1007/s00604-010-0290-8.

72 Xu, Q., Yin, X., Shen, Y., Zhang, N., Wang, M., and Gu, Z. (2011). Detection of phthalate esters in environmental water samples-comparison of nylon6 nanofibers mat-based solid phase extraction and other conventional extraction methods. *Chin. J. Chem.* 29(3): 567–574. doi: 10.1002/cjoc.201190124.

73 Wu, Y., Zhou, Q., Yuan, Y., Wang, H., Tong, Y., Zhan, Y., Sheng, X., Sun, Y., and Zhou, X. (2020). Enrichment and sensitive determination of phthalate esters in environmental water samples: A novel approach of MSPE-HPLC based on PAMAM dendrimers-functionalized magnetic-nanoparticles. *Talanta* 206: 120213. doi: 10.1016/j.talanta.2019.120213.

74 Santana-Mayor, A.L., Socas-Rodríguez, B., Del Mar Afonso, M., Palenzuela-López, J.A., and Rodríguez-Delgado, M. (2018). Reduced graphene oxide-coated magnetic-nanoparticles as sorbent for the determination of phthalates in environmental samples by micro-dispersive solid-phase extraction followed by ultra-high-performance liquid chromatography tandem mass spectrometry. *J. Chromatogr. A* 1565: 36–47. doi: 10.1016/j.chroma.2018.06.031.

75 González-Sálamo, J., Socas-Rodríguez, B., Hernández-Borges, J., and Rodriguez-Delgado, M.A. (2017). Determination of phthalic acid esters in water samples using core-shell poly (dopamine) magnetic nanoparticles and gas chromatography tandem mass spectrometry. *J. Chromatogr. A* 1530: 35–44. doi: 10.1016/j.chroma.2017.11.013.

76 Zhou, Q., Zheng, Z., Xiao, J., Fan, H., and Yan, X. (2016). Determination of phthalate esters at trace level from environmental water samples by magnetic solid-phase extraction with Fe@SiO2@ polyethyleneimine magnetic nanoparticles as adsorbent prior to high-performance liquid chromatography. *Anal. Bioanal. Chem.* 408(19): 5211–5220. doi: 10.1007/s00216-016-9613-x.

77 Mollahosseini, A., Toghroli, M., and Kamankesh, M. (2015). Zeolite/Fe_3O_4 as a new sorbent in magnetic solid-phase extraction followed by gas chromatography for determining phthalates in aqueous samples. *J. Sep. Sci.* 38(21): 3750–3757. doi: 10.1002/jssc.201500510.

78 Ye, Q., Liu, L., and Chen, Z. (2014). Analysis of phthalate acid esters in environmental water by magnetic graphene solid phase extraction coupled with gas chromatography-mass spectrometry. *J. Chromatogr. A* 1329: 24–29. doi: 10.1016/j.chroma.2013.12.086.

79 Wang, W., Ma, R., Wu, Q., Wang, C., and Wang, Z. (2013). Fabrication of magnetic microsphere-confined graphene for the preconcentration of some phthalate esters from environmental water and soybean milk samples followed by their determination by HPLC. *Talanta* 109: 133–140. doi: 10.1016/j.talanta.2013.02.008.

80 Maidatsi, K.V., Chatzimitakos, T.G., Sakkas, V.A., and Stalikas, C.D. (2015). Octyl-modified magnetic graphene as a sorbent for the extraction and simultaneous determination of fragrance allergens, musks, and phthalates in aqueous samples by gas chromatography with mass spectrometry. *J. Sep. Sci.* 38(21): 3758–3765. doi: 10.1002/jssc.201500578.

81 Piovesana, S., Capriotti, A.L., Cavaliere, C., La Barbera, G., Samperi, R., Chiozzi, R.Z., and Laganà, A. (2017). A new carbon-based magnetic material for the dispersive solid-phase extraction of UV filters from water samples before liquid chromatography-tandem mass spectrometry analysis. *Anal. Bioanal. Chem.* 409(17): 4181–4194. doi: 10.1007/s00216-017-0368-9.

82 Cheng, J., Kong, X., Liu, S., Che, D., Sun, Z., Li, G., Ping, M., Tang, J., and You, J. (2018). Determination of ultraviolet filters in domestic wastewater by LC-MS coupled with polydopamine-based magnetic solid-phase extraction and isotope-coded derivatization. *Chromatographia* 81(12): 1673–1684. doi: 10.1007/s10337-018-3650-x.

83 Meng, J., Bu, J., Deng, C., and Zhang, X. (2011). Preparation of polypyrrole-coated magnetic particles for micro solid-phase extraction of phthalates in water by gas chromatography-mass spectrometry analysis. *J. Chromatogr. A* 1218(12): 1585–1591. doi: 10.1016/j.chroma.2011.01.057.

84 Jalilian, N., Ebrahimzadeh, H., and Asgharinezhad, A.A. (2019). Preparation of magnetite/multiwalled carbon nanotubes/metal-organic framework composite for dispersive magnetic micro solid phase extraction of parabens and phthalate esters from water samples and various types of cream for their determination with liquid chromatography. *J. Chromatogr. A* 1608: 460426. doi: 10.1016/j.chroma.2019.460426.

85 Beltran, A., Borrull, F., Marcé, R.M., and Cormack, P.A.G. (2010). Molecularly-imprinted polymers: Useful sorbents for selective extractions. *TrAC-Trend Anal. Chem.* 29(11): 1363–1375. doi: 10.1016/j.trac.2010.07.020.

86 Chen, F., Yang, Z., Tang, Y., and Wang, X. (2017). Selective extraction and determination of di (2-ethylhexyl) phthalate in aqueous solution by HPLC coupled with molecularly imprinted solid-phase extraction. Iran. *J. Chem. Chem. Eng.* 36(3): 127–136. doi: 10.30492/IJCCE.2017.28070.

87 Shaikh, H., Memon, N., Khan, H., Bhanger, M.I., and Nizamani, S.M. (2012). Preparation and characterization of molecularly imprinted polymer for di (2-ethylhexyl) phthalate: Application to sample clean-up prior to gas chromatographic determination. *J. Chromatogr. A* 1247: 125–133. doi: 10.1016/j.chroma.2012.05.056.

88 Yang, R., Liu, Y., Yan, X., and Liu, S. (2016). Simultaneous extraction and determination of phthalate esters in aqueous solution by yolk-shell magnetic mesoporous carbon-molecularly imprinted composites based on solid-phase extraction coupled with gas chromatography-mass spectrometry. *Talanta* 161: 114–121. doi: 10.1016/j.talanta.2016.08.037.

89 Li, C., Ma, X., Zhang, X., Wang, R., Chen, Y., and Li, Z. (2016). Magnetic molecularly imprinted polymer nanoparticles-based solid-phase extraction coupled with gas chromatography-mass spectrometry for selective determination of trace di-(2-ethylhexyl) phthalate in water samples. *Anal. Bioanal. Chem.* 408(27): 7857–7864. doi: 10.1007/s00216-016-9889-x.

90 Vicario, A., Aragón, L., Wang, C.C., Bertolino, F., and Gomez, M.R. (2018). A simple and highly selective molecular imprinting polymer-based methodology for propylparaben monitoring in personal care products and industrial waste waters. *J. Pharmaceut. Biomed.* 149: 225–233. doi: 10.1016/j.jpba.2017.11.002.

91 Kabir, A. and Furton, K.G., Fabric phase sorptive extraction (FPSE): A new direction in sorptive-microextraction, Google Patents, 2017.

92 Celeiro, M., Acerbi, R., Kabir, A., Furton, K.G., and Llompart, M. (2020). Development of an analytical methodology based on fabric phase sorptive extraction followed by gas chromatography-tandem mass spectrometry to determine UV filters in environmental and recreational waters. *Anal. Chim. Acta* 4: 100038. doi: 10.1016/j.acax.2019.100038.

93 Garcia-Guerra, R.B., Montesdeoca-Esponda, S., Sosa-Ferrera, Z., Kabir, A., Furton, K.G., and Santana-Rodríguez, J.J. (2016). Rapid monitoring of residual UV-stabilizers in seawater samples from beaches using fabric phase sorptive extraction and UHPLC-MS/MS. *Chemosphere* 164: 201–207. doi: 10.1016/j.chemosphere.2016.08.102.

94 Montesdeoca-Esponda, S., Sosa-Ferrera, Z., Kabir, A., Furton, K.G., and Santana-Rodríguez, J.J. (2015). Fabric phase sorptive extraction followed by UHPLC-MS/MS for the analysis of benzotriazole UV stabilizers in sewage samples. *Anal. Bioanal. Chem.* 407(26): 8137–8150. doi: 10.1007/s00216-015-8990-x.

95 Kaur, R., Kaur, R., Rani, S., Malik, A.K., Kabir, A., and Furton, K.G. (2019). Application of fabric phase sorptive extraction with gas chromatography and mass spectrometry for the determination of organophosphorus pesticides in selected vegetable samples. *J. Sep. Sci.* 42(4): 862–870. doi: 10.1002/jssc.201800854.

96 Baltussen, E., Cramers, C., and Sandra, P. (2002). Sorptive sample preparation – a review. *Anal. Bioanal. Chem.* 373(1): 3–22. doi: 10.1007/s00216-002-1266-2.

97 Magi, E., Scapolla, C., Di Carro, M., Rivaro, P., and Ngoc Nguyen, K.T. (2013). Emerging pollutants in aquatic environments: Monitoring of UV filters in urban wastewater treatment plants. *Anal. Methods* 5(2): 428–433. doi: 10.1039/C2AY26163D.

98 Casas Ferreira, A.M., Moder, M., and Fernández Laespada, M.E. (2011). GC-MS determination of parabens, triclosan and methyl triclosan in water by in situ derivatisation and stir-bar sorptive extraction. *Anal. Bioanal. Chem.* 399(2): 945–953. doi: 10.1007/s00216-010-4339-7.

99 Si, Q., Li, F., Gao, C., Wang, C., Wang, Z., and Zhao, J. (2016). Detection of phthalate esters in seawater by stir bar sorptive extraction and gas chromatography-mass spectrometry. *Mar. Pollut. Bull.* 108(1): 163–170. doi: 10.1016/j.marpolbul.2016.04.034.

100 Pedrouzo, M., Borrull, F., Marcé, R.M., and Pocurull, E. (2010). Stir-bar-sorptive extraction and ultra-high-performance liquid chromatography-tandem mass spectrometry for simultaneous analysis of UV filters and antimicrobial agents in water samples. *Anal. Bioanal. Chem.* 397(7): 2833–2839. doi: 10.1007/s00216-010-3743-3.

101 Ramirez, N., Borrull, F., and Marcé, R.M. (2012). Simultaneous determination of parabens and synthetic musks in water by stir-bar sorptive extraction and thermal desorption-gas chromatography-mass spectrometry. *J. Sep. Sci.* 35(4): 580–588. doi: 10.1002/jssc.201100887.

102 Feng, R. (2014). Preparation of a novel magnetic stir bar for the pretreatment of the plasticizers in lake water. *Adv. Mat. Res.* 989–994: 164–167. doi: 10.4028/www.scientific.net/AMR.989-994.164.

103 Pintado-Herrera, M.G., González-Mazo, E., and Lara-Martín, P.A. (2014). Atmospheric pressure gas chromatography-time-of-flight-mass spectrometry (APGC-ToF-MS) for the determination of regulated and emerging contaminants in aqueous samples after stir bar sorptive extraction (SBSE). *Anal. Chim. Acta.* 851: 1–13. doi: 10.1016/j.aca.2014.05.030.

104 Aparicio, I., Martín, J., Santos, J.L., Malvar, J.L., and Alonso, E. (2017). Stir bar sorptive extraction and liquid chromatography-tandem mass spectrometry determination of polar and non-polar emerging and priority pollutants in environmental waters. *J. Chromatogr. A* 1500: 43–52. doi: 10.1016/j.chroma.2017.04.007.

105 Magi, E., Di Carro, M., Scapolla, C., and Nguyen, K.T.N. (2012). Stir bar sorptive extraction and LC—MS/MS for trace analysis of UV filters in different water matrices. *Chromatographia* 75: 973–982. doi: 10.1007/s10337-012-2202-z.

106 Vila, M., Celeiro, M., Lamas, J.P., García-Jares, C., Dagnac, T., and Llompart, M. (2017). Simultaneous in-vial acetylation solid-phase microextraction followed by gas chromatography tandem mass spectrometry for the analysis of multiclass organic UV filters in water. *J. Hazard. Mater.* 323: 45–55. doi: 10.1016/j.jhazmat.2016.06.056.

107 Vila, M., Celeiro, M., Lamas, J.P., Dagnac, T., Llompart, M., and García-Jares, C. (2016). Determination of fourteen UV filters in bathing water by headspace solid-phase microextraction and gas chromatography-tandem mass spectrometry. *Anal. Methods* 8(39): 7069–7079. doi: 10.1039/C6AY01787H.

108 Chisvert, A., Benedé, J.L., Anderson, J.L., Pierson, S.A., and Salvador, A. (2017). Introducing a new and rapid microextraction approach based on magnetic ionic liquids: Stir bar dispersive liquid microextraction. *Anal. Chim. Acta.* 983: 130–140. doi: 10.1016/j.aca.2017.06.024.

109 Benedé, J.L., Chisvert, A., Giokas, D.L., and Salvador, A. (2014). Development of stir bar sorptive-dispersive microextraction mediated by magnetic nanoparticles and its analytical application to the determination of hydrophobic organic compounds in aqueous media. *J. Chromatogr. A* 1362: 25–33. doi: 10.1016/j.chroma.2014.08.024.

110 Benedé, J.L., Chisvert, A., Giokas, D.L., and Salvador, A. (2016). Determination of ultraviolet filters in bathing waters by stir bar sorptive-dispersive microextraction coupled to thermal desorption-gas chromatography-mass spectrometry. *Talanta* 147: 246–252. doi: 10.1016/j.talanta.2015.09.054.

111 Benedé, J.L., Chisvert, A., Giokas, D.L., and Salvador, A. (2016). Stir bar sorptive-dispersive microextraction mediated by magnetic nanoparticles-nylon 6 composite for the extraction of hydrophilic organic compounds in aqueous media. *Anal. Chim. Acta.* 926: 63–71. doi: 10.1016/j.aca.2016.04.042.

112 Neng, N.R., Silva, A.R.M., and Nogueira, J.M.F. (2010). Adsorptive micro-extraction techniques-Novel analytical tools for trace levels of polar solutes in aqueous media. *J. Chromatogr. A* 1217(47): 7303–7310. doi: 10.1016/j.chroma.2010.09.048.

113 Ide, A.H. and Nogueira, J.M.F. (2019). Determination of hydrophilic UV filters in real matrices using new-generation bar adsorptive microextraction devices. *Separations* 6(4): 45. doi: 10.3390/separations6040045.

114 Almeida, C., Stepkowska, A., Alegre, A., and Nogueira, J.M.F. (2013). Determination of trace levels of benzophenone-type ultraviolet filters in real matrices by bar adsorptive micro-extraction using selective sorbent phases. *J. Chromatogr. A* 1311: 1–10. doi: 10.1016/j.chroma.2013.08.007.

115 Morelli, D.C., Mafra, G., Santos, A.V., Merib, J., and Carasek, E. (2020). Designing a green device to BAμE: Recycled cork pellet as extraction phase for the determination of parabens in river water samples. *Talanta* 219: 121369. doi: 10.1016/j.talanta.2020.121369.

116 Dias, A.N., Da Silva, A.C., Simao, V., Merib, J., and Carasek, E. (2015). A novel approach to bar adsorptive microextraction: Cork as extractor phase for determination of benzophenone, triclocarban and parabens in aqueous samples. *Anal. Chim. Acta.* 888: 59–66. doi: 10.1016/j.aca.2015.07.011.

117 Richter, P., Leiva, C., Choque, C., Giordano, A., and Sepúlveda, B. (2009). Rotating-disk sorptive extraction of nonylphenol from water samples. *J. Chromatogr. A* 1216(49): 8598–8602. doi: 10.1016/j.chroma.2009.10.044.

118 Becerra-Herrera, M., Miranda, V., and Richter, P. (2020). Rapid determination of parabens in water samples by ultra-high performance liquid chromatography coupled to time of flight mass spectrometry. *Anal. Sci.* 10(36): 675–679. doi: 10.2116/analsci.19P409.

119 Becerra-Herrera, M., Miranda, V., Arismendi, D., and Richter, P. (2018). Chemometric optimization of the extraction and derivatization of parabens for their determination in water samples by rotating-disk sorptive extraction and gas chromatography mass spectrometry. *Talanta* 176: 551–557. doi: 10.1016/j.talanta.2017.08.071.

120 Vieira, C.M.S., Mafra, G., Brognoli, R., Richter, P., Rosero-Moreano, M., and Carasek, E. (2020). A high throughput approach to rotating-disk sorptive extraction (RDSE) using laminar cork for the simultaneous determination of multiclass organic micro-pollutants in aqueous sample by GC-MS. *Talanta* 208: 120459. doi: 10.1016/j.talanta.2019.120459.

121 Arismendi, D., Becerra-Herrera, M., Cerrato, I., and Richter, P. (2019). Simultaneous determination of multiresidue and multiclass emerging contaminants in waters by rotating-disk sorptive extraction-derivatization-gas chromatography/mass spectrometry. *Talanta* 201: 480–489. doi: 10.1016/j.talanta.2019.03.120.

122 Trujillo-Rodríguez, M.J., Anderson, J.L., Dunham, S.J.B., Noad, V.L., and Cardin, D.B. (2020). Vacuum-assisted sorbent extraction: An analytical methodology for the determination of ultraviolet filters in environmental samples. *Talanta* 208: 120390. doi: 10.1016/j.talanta.2019.120390.

123 Llompart, M., Celeiro, M., García-Jares, C., and Dagnac, T. (2019). Environmental applications of solid-phase microextraction. *TrAC-Trends Anal. Chem.* 112: 1–12. doi: 10.1016/j.trac.2018.12.020.

124 Evans, W.A., Davies, P.J., and McRae, C. (2016). The occurrence of methyl, ethyl, propyl, and butyl parabens in the urban rivers and stormwaters of Sydney, Australia. *Environ. Sci. Water. Res. Tech.* 2(4): 733–742. doi: 10.1039/C5EW00240K.

125 Alvarez-Rivera, G., Llompart, M., García-Jares, C., and Lores, M. (2014). Identification of halogenated photoproducts generated after ultraviolet-irradiation of parabens and benzoates in water containing chlorine by solid-phase microextraction and gas chromatography-mass spectrometry. *J. Chromatogr. A* 1349: 105–115. doi: 10.1016/j.chroma.2014.05.009.

126 Celeiro, M., Facorro, R., Dagnac, T., Vilar, V.J.P., and Llompart, M. (2019). Photodegradation behaviour of multiclass ultraviolet filters in the aquatic environment: Removal strategies and photoproduct identification by liquid chromatography-high resolution mass spectrometry. *J. Chromatogr. A* 1596: 8–19. doi: 10.1016/j.chroma.2019.02.065.

127 Liu, P., Tian, T., Barreto, J., and Chou, J. (2013). Assessment and analysis of phthalate esters, in Lake Pontchartrain, by SPME combining with GC-MS. *Environ. Technol.* 34(4): 453–462. doi: 10.1080/09593330.2012.698653.

128 Martinez, C., Gomez, V., Borrull, F., and Pocurull, E. (2016). Headspace-solid phase microextraction: Useful technique to characterize volatile and semi-volatile organic compounds in water reuse applications. *Desalin. Water Treat.* 57(48–49): 23176–23184. doi: 10.1080/19443994.2015.1137145.

129 Lamas, J.P., Sanchez-Prado, L., García-Jares, C., and Llompart, M. (2009). Solid-phase microextraction gas chromatography-mass spectrometry determination of fragrance allergens in baby bathwater. *Anal. Bioanal. Chem.* 394(5): 1399–1411. doi: 10.1007/s00216-009-2829-2.

130 Becerril, E., Lamas, J.P., Sanchez-Prado, L., Llompart, M., Lores, M., and García-Jares, C. (2010). Analysis of regulated suspected allergens in waters. *Talanta* 83(2): 464–474. doi: 10.1016/j.talanta.2010.09.038.

131 Godayol, A., Besalú, E., Anticó, E., and Sanchez, J.M. (2015). Monitoring of sixteen fragrance allergens and two polycyclic musks in wastewater treatment plants by solid phase microextraction coupled to gas chromatography. *Chemosphere* 119: 363–370. doi: 10.1016/j.chemosphere.2014.06.072.

132 Xu, L., Shi, Y., and Cai, Y. (2013). Occurrence and fate of volatile siloxanes in a municipal wastewater treatment plant of Beijing, China. *Water Res.* 47(2): 715–724. doi: 10.1016/j.watres.2012.10.046.

133 Lima Gomes, P.C.F., Barnes, B.B., Santos-Neto, A.L.J., Lancas, F.M., and Snow, N.H. (2013). Determination of steroids, caffeine and methylparaben in water using solid phase microextraction-comprehensive two dimensional gas chromatography-time of flight mass spectrometry. *J. Chromatogr. A* 1299: 126–130. doi: 10.1016/j.chroma.2013.05.023.

134 Martinez, C., Ramirez, N., Gomez, V., Pocurull, E., and Borrull, F. (2013). Simultaneous determination of 76 micropollutants in water samples by headspace solid phase microextraction and gas chromatography-mass spectrometry. *Talanta* 116: 937–945. doi: 10.1016/j.talanta.2013.07.055.

135 Celeiro, M., Lamas, J.P., Vila, M., García-Jares, C., Homem, V., Ratola, N., Dagnac, T., and Llompart, M. (2019). Determination of multiclass personal care products in continental waters by solid-phase microextraction followed by gas chromatography-tandem mass spectrometry. *J. Chromatogr. A* 1607: 460398. doi: 10.1016/j.chroma.2019.460398.

136 Li, Y., Yang, Y., Liu, H., Wang, X., and Du, X. (2014). Fabrication of a novel Ti-TiO_2-ZrO_2 fiber for solid-phase microextraction followed by high-performance liquid chromatography for sensitive determination of UV filters in environmental water samples. *Anal. Methods* 6(21): 8519–8525. doi: 10.1039/C4AY01776E.

137 Song, W., Zhang, Y., Guo, M., Wang, T., Yang, Y., Wang, X., and Du, X. (2015). Rapid electrochemical preparation of porous sponge-like zinc-zinc oxide coating deposited on an etched stainless steel fiber for selective determination of UV filters in environmental water samples. *Anal. Methods* 7(16): 6619–6628. doi: 10.1039/C5AY01248A.

138 Li, Y., Zhang, M., Yang, Y., Wang, X., and Du, X. (2014). Electrochemical in situ fabrication of titanium dioxide-nanosheets on a titanium wire as a novel coating for selective solid-phase microextraction. *J. Chromatogr. A* 1358: 60–67. doi: 10.1016/j.chroma.2014.06.094.

139 Song, W., Guo, M., Zhang, Y., Zhang, M., Wang, X., and Du, X. (2015). Fabrication and application of zinc-zinc oxide nanosheets coating on an etched stainless steel wire as a selective solid-phase microextraction fiber. *J. Chromatogr. A* 1384: 28–36. doi: 10.1016/j.chroma.2015.01.059.

140 Wu, D., Liu, F., Tian, T., Wu, J.F., and Zhao, G.C. (2021). Copper ferrite nanoparticles as novel coating appropriated to solid-phase microextraction of phthalate esters from aqueous matrices. *Microchem. J.* 162: 105845. doi: 10.1016/j.microc.2020.105845.

141 Tian, T., Wang, F., and Zhao, G.C. (2020). Magnesium/aluminum-layered double hydroxide modified with hydrogen peroxide as a novel fiber coating for solid-phase microextraction of phthalate esters in aqueous samples. *Microchem. J.* 153: 104510. doi: 10.1016/j.microc.2019.104510.

142 Wang, F.Q., Li, J., Wu, J.F., and Zhao, G.C. (2018). Layered double hydroxides as a coating for the determination of phthalate esters in aqueous solution with solid-phase microextraction followed by gas chromatography. *Chromatographia* 81(5): 799–807. doi: 10.1007/s10337-018-3507-3.

143 Tashakkori, P., Erdem, P., Merdivan, M., and Bozkurt, S.S. (2019). Determination of phthalate esters in water and coffee by solid-phase microextraction using vinyl terminated imidazolium based ionic liquid grafted on graphene oxide coatings. *Chem. Select.* 4(8): 2307–2313. doi: 10.1002/slct.201803214.

144 Zhang, Y., Yang, Y., Li, Y., Zhang, M., Wang, X., and Du, X. (2015). Growth of cedar-like Au nanoparticles coating on an etched stainless steel wire and its application for selective solid-phase microextraction. *Anal. Chim. Acta.* 876: 55–62. doi: 10.1016/j.aca.2015.03.044.

145 Moahammadi, A.A., Davarani, S.S.H., Jafari, M., and Mehdinia, A. (2021). Preparation and evaluation of a new solid-phase microextraction fiber based on polythionine for analysis of phthalate esters in aqueous samples. *J. Iran Chem. Soc.* 18(2): 385–391. doi: 10.1007/s13738-020-02034-6.

146 Mehdinia, A., Bashour, F., Roohi, F., Jabbari, A., and Saleh, A. (2012). Preparation and evaluation of thermally stable nano-structured self-doped polythiophene coating for analysis of phthalate ester trace levels. *J. Sep. Sci.* 35(4): 563–570. doi: 10.1002/jssc.201100713.

147 Mousa, A., Basheer, C., and Al-Arfaj, A.R. (2013). Application of electro-enhanced solid-phase microextraction for determination of phthalate esters and bisphenol A in blood and seawater samples. *Talanta* 115: 308–313. doi: 10.1016/j.talanta.2013.05.011.

148 Trujillo-Rodríguez, M.J., Nan, H., and Anderson, J.L. (2018). Expanding the use of polymeric ionic liquids in headspace solid-phase microextraction: Determination of ultraviolet filters in water samples. *J. Chromatogr. A* 1540: 11–20. doi: 10.1016/j.chroma.2018.01.048.

149 An, J. and Anderson, J.L. (2018). Determination of UV filters in high ionic strength sample solutions using matrix-compatible coatings for solid-phase microextraction. *Talanta* 182: 74–82. doi: 10.1016/j.talanta.2018.01.052.

150 Vosough, M., Hassanbeigi, Z., and Salemi, A. (2018). Determination of ultraviolet filter compounds in environmental water samples using membrane-protected micro-solid-phase extraction. *J. Sep. Sci.* 41(11): 2401–2410. doi: 10.1002/jssc.201701082.

151 Asadollahzadeh, H., Noroozian, E., and Maghsoudi, S. (2010). Solid-phase microextraction of phthalate esters from aqueous media by electrochemically deposited carbon nanotube/polypyrrole composite on a stainless steel fiber. *Anal. Chim. Acta.* 669(1–2): 32–38. doi: 10.1016/j.aca.2010.04.029.

152 Moliner-Martínez, Y., Molins-Legua, C., Verdú-Andrés, J., Herráez-Hernández, R., and Campíns-Falcó, P. (2011). Advantages of monolithic over particulate columns for multiresidue analysis of organic pollutants by in-tube solid-phase microextraction coupled to capillary liquid chromatography. *J. Chromatogr. A* 1218(37): 6256–6262. doi: 10.1016/j.chroma.2011.07.026.

153 Mei, M. and Huang, X. (2017). Online analysis of five organic ultraviolet filters in environmental water samples using magnetism-enhanced monolith-based in-tube solid phase microextraction coupled with high-performance liquid chromatography. *J. Chromatogr. A* 1525: 1–9. doi: 10.1016/j.chroma.2017.09.065.

154 Mei, M., Pang, J., and Huang, X. (2018). Development of a sensitive method for the determination of parabens in complex samples by online coupling of magnetism-enhanced monolith-based in-tube solid phase microextraction with high performance liquid chromatography. *Anal. Methods* 10(17): 1977–1985. doi: 10.1039/C8AY00026C.

155 Barceló, D. and Petrovic, B. Eds. (2007). *Comprehensive Analytical Chemistry*. Boston: Elsevier.

156 Ahmadi, F., Sparham, C., Boyaci, E., and Pawliszyn, J. (2017). Time weighted average concentration monitoring based on thin film solid phase microextraction. *Environ. Sci. Technol.* 51(7): 3929–3937. doi: 10.1021/acs.est.6b06465.

157 Fent, K., Zenker, A., and Rapp, M. (2010). Widespread occurrence of estrogenic UV-filters in aquatic ecosystems in Switzerland. *Environ. Pollut.* 158(5): 1817–1824. doi: 10.1016/j.envpol.2009.11.005.

158 Kirschner, N., Dias, A.N., Budziak, D., Da Silveira, C.B., Merib, J., and Carasek, E. (2017). Novel approach to high-throughput determination of endocrine disruptors using recycled diatomaceous earth as a green sorbent phase for thin-film solid-phase microextraction combined with 96-well plate system. *Anal. Chim. Acta.* 996: 29–37. doi: 10.1016/j.aca.2017.09.047.

159 Zhao, R.S., Liu, Y.L., Zhou, J.B., Chen, X.F., and Wang, X. (2013). Bamboo charcoal as a novel solid-phase microextraction coating material for enrichment and determination of eleven phthalate esters in environmental water samples. *Anal. Bioanal. Chem.* 405(14): 4993–4996. doi: 10.1007/s00216-013-6865-6.

160 Wu, D., Chen, X., Liu, F., Wu, J.F., and Zhao, G.C. (2020). A carbon dots-based coating for the determination of phthalate esters by solid-phase microextraction coupled gas chromatography in water samples. *Microchem. J.* 159: 105563. doi: 10.1016/j.microc.2020.105563.

161 Sankoda, K., Murata, K., Tanihata, M., Suzuki, K., Nomiyama, K., and Shinohara, R. (2015). Seasonal and diurnal variation of organic ultraviolet filters from personal care products used along the Japanese coast. *Arch. Environ. Contam. Toxicol.* 68(2): 217–224. doi: 10.1007/s00244-014-0106-7.

162 Saleh, N.M., Hafiz, N.M., and NikWee, N.N.A. (2020). Determination of parabens from water samples using cloud point extraction, vortex extraction and liquid-liquid extraction method coupled with high performance liquid chromatography. *J. Comput. Theor. Nanos.* 17(2): 765–772. doi: 10.1166/jctn.2020.8717.

163 Derisso, C.R., Pompei, C.M.O.E., Spadoto, M.N., Da Silva Pinto, T., and Vieira, E.M. (2020). Occurrence of parabens in surface water, wastewater treatment plant in Southeast of Brazil and assessment of their environmental risk. *Water Air Soil Pollut.* 231(9): 1–13. doi: 10.1007/s11270-020-04835-0.

164 Villaverde de Saa, E., González-Mariño, I., Quintana, J.B., Rodil, R., Rodríguez, I., and Cela, R. (2010). In-sample acetylation-non-porous membrane-assisted liquid-liquid extraction for the determination of parabens and triclosan in water samples. *Anal. Bioanal. Chem.* 397(6): 2559–2568. doi: 10.1007/s00216-010-3789-2.

165 Zhu, X.J. and Qiu, Y.Y. (2011). Measuring the phthalates of xiangjiang river using liquid-liquid extraction gas chromatography. *Adv. Mater. Res.* 301: 752–755. doi: 10.4028/www.scientific.net/AMR.301-303.752.

166 Liew, C.S.M., Li, X., Zhang, H., and Lee, H.K. (2018). A fully automated analytical platform integrating water sampling-miniscale-liquid-liquid extraction-full evaporation dynamic headspace concentration-gas chromatography-mass spectrometry for the analysis of ultraviolet filters. *Anal. Chim. Acta.* 1006: 33–41. doi: 10.1016/j.aca.2017.12.035.

167 Farahani, H. (2019). Screening of parabens in natural water by salting-out based centrifugeless dispersive liquid-liquid microextraction combined with HPLC-UV. *Acta. Chim. Slov.* 66(3): 576–583. doi: 10.17344/acsi.2018.4889.

168 Ku, P.C., Liu, T.Y., Lee, S.H., Kung, T.A., and Wang, W.-H. (2020). An environmentally friendly strategy for determining organic ultraviolet filters in seawater using liquid-phase microextraction with liquid chromatography-tandem mass spectrometry. *Environ. Sci. Pollut. Res.* 27: 9818–9825. doi: 10.1007/s11356-020-07599-6.

169 Zakharkiv, I., Zui, M., and Zaitsev, V. (2020). Determination of phthalate esters in water and liquid pharmaceutical samples by dispersive liquid-liquid microextraction (DLLME) and gas chromatography with flame ionization detection (GC-FID). *Anal. Lett.* 53(10): 1536–1553. doi: 10.1080/00032719.2019.1711384.

170 Farajzadeh, M.A., Aghdam, S.R., Mogaddam, M.R.A., and Ghorbanpour, H. (2015). An elevated temperature-dispersive liquid-liquid microextraction method combined with GC-flame ionization detection as a sensitive method for determining phthalate esters. *Anal. Methods* 7(10): 4269–4277. doi: 10.1039/C4AY02998D.

171 Guo, L. and Lee, H.K. (2014). Automated dispersive liquid-liquid microextraction-gas chromatography-mass spectrometry. *Anal. Chem.* 86(8): 3743–3749. doi: 10.1021/ac404088c.

172 Xue, L., Zhang, D., Wang, T., Wang, X.-M., and Du, X. (2014). Dispersive liquid-lliquid microextraction followed by high performance liquid chromatography for determination of phthalic esters in environmental water samples. *Anal. Methods* 6(4): 1121–1127. doi: 10.1039/C3AY41996G.

173 Leng, G., Chen, W., and Wang, Y. (2016). Simple field-based automated dispersive liquid-liquid microextraction of trace level phthalate esters in natural waters with gas chromatography and mass spectrometric analysis. *J. Sep. Sci.* 39(17): 3392–3399. doi: 10.1002/jssc.201600383.

174 Ramos, S., Homem, V., and Santos, L. (2019). Simultaneous determination of synthetic musks and UV-filters in water matrices by dispersive liquid-liquid microextraction followed by gas chromatography tandem mass-spectrometry. *J. Chromatogr. A* 1590: 47–57. doi: 10.1016/j.chroma.2019.01.013.

175 Mudiam, M.K.R., Jain, R., and Singh, R. (2014). Application of ultrasound-assisted dispersive liquid-liquid microextraction and automated in-port silylation for the simultaneous determination of phenolic endocrine disruptor chemicals in water samples by gas chromatography-triple quadrupole mass spectrometry. *Anal. Methods* 6(6): 1802–1810. doi: 10.1039/C3AY41658E.

176 Wei, N., Zheng, Z., Wang, Y., Tao, Y., Shao, Y., Zhu, S., You, J., and Zhao, X.E. (2017). Rapid and sensitive determination of multiple endocrine-disrupting chemicals by ultrasound-assisted in situ derivatization dispersive liquid-liquid microextraction coupled with ultra-high-performance liquid chromatography/tandem mass spectrometry. *Rapid. Commun. Mass. Spectrom.* 31(11): 937–950. doi: 10.1002/rcm.7865.

177 Caglak, A., Chormey, D.S., Engin, G.O., and Bakirdere, S. (2019). Determination of micropollutants in wastewater matrix using gas chromatography-mass spectrometry after optimization of dispersive liquid-liquid microextraction. *Int. J. Environ. Sci. Technol.* 16(11): 7285–7292. doi: 10.1007/s13762-019-02319-1.

178 Sargazi, S., Mirzaei, R., Rahmani, M., and Sheikh, M. (2015). Application of in-syringe dispersive liquid-liquid microextraction coupled to GC/FID for determination of trace contamination of phthalate esters in water samples. *J. Chromatogr. Sci.* 53(8): 1420–1426. doi: 10.1093/chromsci/bmv007.

179 Clavijo, S., Del Rosario Brunetto, M., and Cerdá, V. (2014). In-syringe-assisted dispersive liquid-liquid microextraction coupled to gas chromatography with mass spectrometry for the determination of six phthalates in water samples. *J. Sep. Sci.* 37(8): 974–981. doi: 10.1002/jssc.201301176.

180 Çabuk, H. and Kavaraci, O. (2020). Magnetic retrieval of a switchable hydrophilicity solvent: Fast homogeneous liquid-liquid microextraction for the determination of benzophenone-type UV filters in environmental waters. *Int. J. Environ. Anal. Chem.* in press. doi: 10.1080/03067319.2020.1757088.

181 Pierson, S.A., Trujillo-Rodríguez, M.J., and Anderson, J.L. (2018). Rapid analysis of ultraviolet filters using dispersive liquid-liquid microextraction coupled to headspace gas chromatography and mass spectrometry. *J. Sep. Sci.* 41(15): 3081–3088. doi: 10.1002/jssc.201800415.

182 Sha, C., Yi-Sheng, Z., Shuia-Yuan, C., Tian, Q., and Hao, S. (2011). Development of an ionic liquid-based dispersive liquid-liquid micro-extraction method for the determination of phthalate esters in water samples. *J. Sep. Sci.* 34(13): 1503–1507. doi: 10.1002/jssc.201000855.

183 Zhou, Q., Zhang, X., and Xie, G. (2011). Simultaneous analysis of phthalate esters and pyrethroid insecticides in water samples by temperature-controlled ionic liquid dispersive liquid-phase microextraction combined with high-performance liquid chromatography. *Anal. Methods* 3(8): 1815–1820. doi: 10.1039/C1AY05137G.

184 Ge, D., Wang, Y., Jiang, Q., and Dai, E. (2019). A deep eutectic solvent as an extraction solvent to separate and preconcentrate parabens in water samples using in situ liquid-liquid microextraction. *J. Brazil. Chem. Soc.* 30(6): 1203–1210. doi: 10.21577/0103-5053.20190014.

185 Ge, D., Zhang, Y., Dai, Y., and Yang, S. (2018). Air-assisted dispersive liquid-liquid microextraction based on a new hydrophobic deep eutectic solvent for the preconcentration of benzophenone-type UV filters from aqueous samples. *J. Sep. Sci.* 41(7): 1635–1643. doi: 10.1002/jssc.201701282.

186 Zhang, K., Li, S., Wang, Y., Fan, J., and Zhu, G. (2020). Air-assisted liquid-liquid microextraction based on solidification of floating deep eutectic solvent for the analysis of ultraviolet filters in water samples by high performance liquid chromatography with the aid of response surface methodology. *J. Chromatogr. A* 1618: 460876. doi: 10.1016/j.chroma.2020.460876.

187 Yang, D., Yang, Y., Li, Y., Yin, S., Chen, Y., Wang, J., Xiao, J., and Sun, C. (2019). Dispersive liquid-liquid microextraction based on solidification of floating organic drop combined with high performance liquid chromatography for analysis of 15 phthalates in water. *J. AOAC Int.* 102(3): 942–951. doi: 10.5740/jaoacint.18-0160.

188 Hu, L., Shan, W., Zhang, Y., Li, S., Gao, H., Lu, R., Zhang, S., and Zhou, W. (2016). Liquid phase microextraction based on the solidification of a floating ionic liquid combined with high-performance liquid chromatography for the preconcentration of phthalate esters in environmental waters and in bottled beverages. *RSC Adv.* 6(43): 36223–36230. doi: 10.1039/C6RA00788K.

189 Ranjbari, E. and Hadjmohammadi, M.R. (2012). Magnetic stirring-assisted dispersive liquid-liquid microextraction followed by high performance liquid chromatography for determination of phthalate esters in drinking and environmental water samples. *Talanta* 100: 447–453. doi: 10.1016/j.talanta.2012.08.019.

190 Zeeb, M. and Farahani, H. (2018). Trace monitoring of phthalate esters in environmental water samples by ionic liquid-based ultrasound-assisted in-situ solvent formation microextraction combined with high-performance liquid chromatography. *J. Chem. Health Risks* 8(3): 245–254. doi: 10.22034/JCHR.2018.544220.

191 Farahani, A., Ramezani, M., Hassan, J., and Niazi, A. (2017). In tube ultrasonic and air assisted liquid-liquid microextraction-gas chromatography-mass spectrometry determination: A novel method for the determination of phthalate esters in aqueous samples. *J. Braz. Chem. Soc.* 28(6): 967–974. doi: 10.21577/0103-5053.20160247.

192 Farajzadeh, M.A. and Mogaddam, M.R.A. (2012). Air-assisted liquid-liquid microextraction method as a novel microextraction technique; Application in extraction and preconcentration of phthalate esters in aqueous sample followed by gas chromatography-flame ionization detection. *Anal. Chim. Acta* 728: 31–38. doi: 10.1016/j.aca.2012.03.031.

193 Szreniawa-Sztajnert, A. and Zabiegala, B.E. (2014). Direct-ultrasonic assisted microextraction coupled with RTL-GC-FID/GC-MS as a future standard procedure for monitoring 26 potentially allergenic fragrances in water samples. *Int. Proc. Chem. Biol. Environ. Eng.* 69: 60–64. doi: 10.7763/IPCBEE.

194 Cortada, C., Dos Reis, L.C., Vidal, L., Llorca, J., and Canals, A. (2014). Determination of cyclic and linear siloxanes in wastewater samples by ultrasound-assisted dispersive liquid-liquid microextraction followed by gas chromatography-mass spectrometry. *Talanta* 120: 191–197. doi: 10.1016/j.talanta.2013.11.042.

195 Li, X., Chen, G., Liu, J., Liu, Y., Zhao, X., Cao, Z., Xia, L., Li, G., Sun, Z., and Zhang, S. (2017). A rapid, accurate and sensitive method with the new stable isotopic tags based on microwave-assisted dispersive liquid-liquid microextraction and its application to the determination of hydroxyl UV filters in environmental water samples. *Talanta* 167: 242–252. doi: 10.1016/j.talanta.2017.02.028.

196 Cunha, S.C., Pena, A., and Fernandes, J.O. (2015). Dispersive liquid-liquid microextraction followed by microwave-assisted silylation and gas chromatography-mass spectrometry analysis for simultaneous trace quantification of bisphenol A and 13 ultraviolet filters in wastewaters. *J. Chromatogr. A* 1414: 10–21. doi: 10.1016/j.chroma.2015.07.099.

197 Sun, J.Q. (2013). Determination of phthalate esters in wastewater by single-drop microextraction coupled to capillary gas chromatography. *Appl. Mech. Mater.* Trans. Tech. Publ. 295–298: 483–486. doi: 10.4028/www.scientific.net/AMM.295-298.483.

198 Vidal, L., Chisvert, A., Canals, A., and Salvador, A. (2010). Ionic liquid-based single-drop microextraction followed by liquid chromatography-ultraviolet spectrophotometry detection to determine typical UV filters in surface water samples. *Talanta* 81(1–2): 549–555. doi: 10.1016/j.talanta.2009.12.042.

199 Regueiro, J., Llompart, M., García-Jares, C., Garcia-Monteagudo, J.C., and Cela, R. (2008). Ultrasound-assisted emulsification-microextraction of emergent contaminants and pesticides in environmental waters. *J. Chromatogr. A* 1190(1–2): 27–38. doi: 10.1016/j.chroma.2008.02.091.

200 Yamini, Y., Saleh, A., Rezaee, M., Ranjbar, L., and Moradi, M. (2012). Ultrasound-assisted emulsification microextraction of various preservatives from cosmetics, beverages, and water samples. *J. Liq. Chromatogr. Relat. Technol.* 35(18): 2623–2642. doi: 10.1080/10826076.2011.637272.

201 Jan-E, S., Santaladchaiyakit, Y., and Burakham, R. (2016). Ultrasound-assisted surfactant-enhanced emulsification micro-extraction followed by HPLC for determination of preservatives in water, beverages and personal care products. *J. Chromatogr. Sci.* 55(1): 90–98. doi: 10.1093/chromsci/bmw140.

202 Pérez-Outeiral, J., Millán, E., and Garcia-Arrona, R. (2015). Ultrasound-assisted emulsification microextraction coupled with high-performance liquid chromatography for the simultaneous determination of fragrance allergens in cosmetics and water. *J. Sep. Sci.* 38(9): 1561–1569. doi: 10.1002/jssc.201401330.

203 Vila, M., Lamas, J.P., García-Jares, C., Dagnac, T., and Llompart, M. (2016). Ultrasound-assisted emulsification microextraction followed by gas chromatography-mass spectrometry and gas chromatography-tandem mass spectrometry for the analysis of UV filters in water. *Microchem. J.* 124: 530–539. doi: 10.1016/j.microc.2015.09.023.

204 Ge, D. and Lee, H.K. (2012). A new 1-hexyl-3-methylimidazolium tris(pentafluoroethyl) trifluorophosphate ionic liquid based ultrasound-assisted emulsification microextraction for the determination of organic ultraviolet filters in environmental water samples. *J. Chromatogr. A* 1251: 27–32. doi: 10.1016/j.chroma.2012.06.048.

205 Kamarei, F., Ebrahimzadeh, H., and Yamini, Y. (2011). Optimization of ultrasound-assisted emulsification microextraction with solidification of floating organic droplet followed by high performance liquid chromatography for the analysis of phthalate esters in cosmetic and environmental water samples. *Microchem. J.* 99(1): 26–33. doi: 10.1016/j.microc.2011.03.006.

206 Ge, D., Zhong, G., Zhou, X., Dai, E., and Wang, Y. (2020). Elevated temperature-assisted surfactant-enhanced emulsification microextraction based on solidification of a floating organic drop for the determination of UV filters in water samples by high-performance liquid chromatography coupled to a diode array detector. *J. Iran Chem. Soc.* 17(2): 499–506. doi: 10.1007/s13738-019-01785-1.

4

New Psychoactive Substances

Noelia Salgueiro-González, Ettore Zuccato, and Sara Castiglioni

Department of Environmental Health Science, Istituto di Ricerche Farmacologiche Mario Negri – IRCCS, Milan, Italy

4.1 Overview of New Psychoactive Substances

The "new psychoactive substances (NPS)" have emerged in the mid-2000s as a legal alternative to established illicit drugs. These substances are synthetized adding to or modifying the chemical structure of traditional psychoactive substances (e.g. cannabis, cocaine, heroin, MDMA (ecstasy) and amphetamines) to have similar effects but circumventing legal restriction on purchasing [1]. The term "new" does not always refer to newly synthetized substances (some of them were synthesized or patented more than 40 years ago for pharmacological treatment) but to those that have recently emerged for recreational use [2].

Discriminating between NPS and more traditional drugs can be challenging because: (i) there is not a formal and universally accepted definition for this changeable and heterogeneous group of substances; the definition of NPS is not descriptive of inherent characteristics and therefore it varies across the organizations that highlight different issues related to drug surveillance and control [3]; (ii) several terms are used to define NPS – the most common ones are "new psychoactive substances", "novel psychoactive substances" or "emerging psychoactive substances", but others are used to define NPS by source, legal status or by market type: "legal highs", "bath salts", "plant food", "research chemicals" or "internet drug" [4]; and (iii) some definitions are time-dependent as they refer to the recently detected/used substance or to those which are not classified in the international drug controls [3].

4.1.1 Properties

In this chapter, we adopted the definition proposed by the United Nation Office on Drugs and Crime (UNODC), considering NPS as "substances of abuse, either in a pure form or a preparation, that are not controlled by the 1961 Single Convention on Narcotic Drugs

Analytical Methods for Environmental Contaminants of Emerging Concern, First Edition. Edited by Núria Fontanals and Rosa Maria Marcé.

or the 1971 Convention on Psychotropic Substances, but which may pose a public health threat" [2]. To date, a total of 1047 individual substances have been reported to the UNODC Early Warning Advisory (EWA) on NPS by governments, laboratories and partner organizations from 125 countries [5].

This wide number of substances can be grouped considering different criteria, such as the mode of action and/or effect (e.g. stimulants, synthetic cannabinoids receptor agonists, depressants, hallucinogens) or the chemical class/physico-chemical properties (e.g. synthetic cannabinoids, synthetic cathinones, phenetylamines, tryptamines) [1, 6]. It is important to mention that NPS categorization also varies across organizations and usually correspond to a simplification of the real situation, as not all the individual substances produce the expected effects or have the identical chemical properties of the corresponding established substance that they are intended to mimic [3].

Table 4.1 summarizes the main categories of NPS present in the market, considering their chemical class and their mode of action [6]. Characteristics, routes of administration and the most common NPS of each category were also included.

The individual and public harm associated with NPS is probably more serious than those associated with established illicit drugs [7]. The information on NPS toxicity, pharmacology and long-term risks, usually provided by clinical and forensic cases, is still

Table 4.1 NPS categories present in the market: mode of action and main characteristics. Based on [6].

NPS	Mode of action and characteristics	Availability and administration	Common NPS
Synthetic cannabinoids	Diverse group of synthetized compounds acting on the same receptors as THC found in cannabis. Effects are similar but more severe to those resulting from cannabis (e.g. euphoria, sedation and drowsiness).	Herbal mixtures (sprayed on) or liquids *Routes: oral and smoking*	JWH-210 AMB-CHMINACA CUMYL-PINACA
Synthetic cathinones	Synthetic analogues/ derivatives of the international controlled substance cathinone (found in khat plant). They produce stimulant effects and are often sold as similar to methamphetamine, ecstasy and cocaine.	Powder, pills or capsules *Routes: oral, intranasal, intravenous and smoking*	Methcathinone Mephedrone α-PVP
Phenethylamines	Synthetic substances related with amphetamine and methamphetamine. They generally produce stimulant effects, but a modification of these compounds can lead to potent hallucinogens.	Pills, powder or fixed on blotter paper *Routes: oral*	PMA 6-APB 25C-NBOMe

(continued)

Table 4.1 (Continued)

NPS	Mode of action and characteristics	Availability and administration	Common NPS
Arylcycloalkylamines *(ketamine analogues)*	This group includes ketamine and other phencyclidine-type substances, which produce dissociative, anesthetic and hallucinogenic effects.	Powder *Routes: intranasal, oral or intravenous*	Ketamine, Methoxetamine 2-FDCK
Tryptamines	They are derivatives of the naturally occurring tryptamine (found in plants, fungi and animals) and act as hallucinogens.	Capsule, tablet, powder or liquids *Routes: oral, intranasal, smoking or intravenous*	5-MeO-DiPT 4-AcO-DMT
Piperazines	They are often sold as similar to ecstasy because of their central nervous system stimulant properties.	Pills, capsules or powder *Routes: Oral, intranasal, smoking, snorting or intravenous*	BZP mCPP
Plant-based substances	This group includes plants with psychoactive properties, acting as stimulants, hallucinogens and sedatives. The most common are: kratom (stimulant or sedative at high doses), khat (stimulant) or salvia (hallucinogenic effects).	Plant leaves Commercial products like pills and capsules. *Routes: Oral (mainly chewing) or smoking*	DMT Mytragina (kratom)
Synthetic opioids	Centrally-acting analgesics that also result in drowsiness and euphoria similar to heroin, morphine and fentanyl.	Tablets, patches *Routes: oral, transdermal and intravenous*	Acetylfentanyl U-47700 Isotonitazene
Benzodiazepines	Synthetic substances which act as depressants.	Tablets, capsules and occasionally as injectable *Routes: oral and intravenous*	Flualprazolam Etizolam

Abbreviations: 6-APB, 6-(2-aminopropyl)benzofuran; 4-AcO-DMT, 4-acethoxy-dimethyltryptamine; BZP, Benzylpiperazine; DMT – N,N-Dimethyltryptamine; 2-FDCK, 2-fluorodeschloroketamine; mCPP, meta-Chlorophenylpiperazine; 5-MeO-DiPT, 5-Methoxy-N,N-dipropyltryptamine; PMA, para-methoxy-amphetamine; α-PVP, α-pyrrolidinovalerophenone; THC, Δ-9-tetrahydronnabinol

scant and therefore the real effects of NPS are often unknown. Moreover, several NPS are usually consumed in combination with other NPS/substances (poly-drug use), causing fatal intoxications [8]. An unconscious use of NPS should be also considered as consumers do not always know exactly which substances they are consuming: NPS are used as adulterants of established illicit drugs, and legally sold with different names or incorrectly/inadequately labelled, with variable contents both in type and quantity [3].

A correlation between NPS consumption and hospital emergencies and acute adverse health consequences has already been evidenced in the last years [7, 9]. In most of the cases, stimulants and synthetic cannabinoid receptor agonists (SCRAs) were involved, but most of the fatal intoxications were related to SCRAs, synthetic opioids (mainly fentanyl analogues) and benzodiazepine-type NPS [9]. The harm caused by depressants is expected, considering the potent effect of licit opioids and benzodiazepines; however, risks for use of SCRAs can be more serious than those caused by cannabis because of their higher potential, their withdrawal effects and their potency of addiction [10].

4.1.2 NPS Market, Dynamics and International Control

NPS and precursor products are usually produced in bulk quantities by pharmaceutical and chemical companies in China. They are then shipped by sea or air to other countries around the world, where NPS are processed into products and packaged [4]. These substances can be also produced in clandestine laboratories elsewhere, usually linked to organized crime, that use similar equipment and chemical expertise to those required for the manufacture of established illicit drugs [1]. NPS are sold legally on the surface web or physical smart shops, labelled as research chemicals, food supplements or legal highs, and commonly recommended "not for human consumption" to avoid drug control. Moreover they are sold on "darknet" markets and on the illicit market, sometimes falsely labeled as illicit drugs such as heroin, cocaine, ecstasy and benzodiazepines [4].

The NPS market is characterized by the high number and continuous change of substances, with the introduction of new NPS and the disuse of controlled ones. Figure 4.1 illustrates the trend of the total number of NPS reported to the UNODC-EWA from 2009 to 2019 [5]. This number has exponentially increased from 2009 (131 NPS) to 2015 (510) and maintained constant at this high level until 2019 (542). A similar trend can be observed considering the NPS reported for the first time since 2009 (Figure 4.1, line red), with a maximum peak of 163 new NPS identified in 2013 and 129 in 2015. After that, the

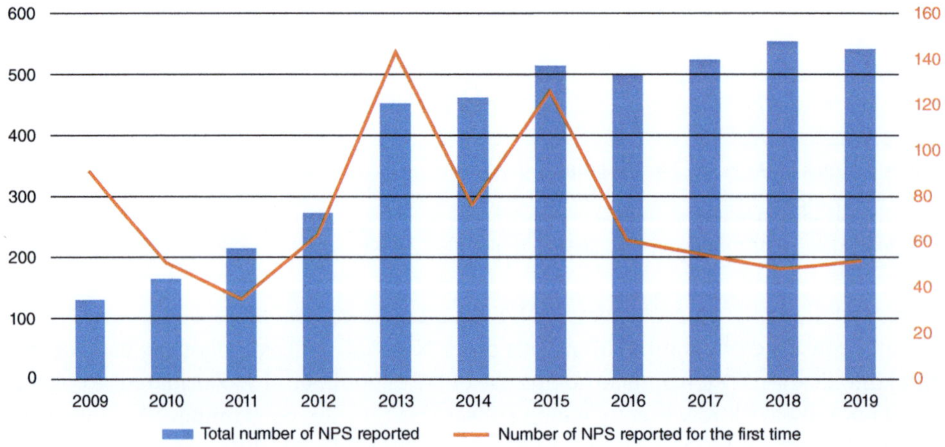

Figure 4.1 Number of NPS reported to UNODC-EWA each year from 2009 to 2019: total number of individual NPS (blue) and NPS identified for the first time (orange). Based on [5].

number of new NPS decreased to about 60 per year and remained constant until 2019, demonstrating the still ongoing innovation and adaptability of the NPS market. The categories of NPS mostly consumed are SCRAs and stimulants, with more than 200 substances reported per year. However, different trends can be observed since 2017, with a significant increase in the number of reported synthetic opioids and sedative/hypnotics, being three times higher than those reported in previous years [5].

A wide range of legal responses have been explored to address the dynamics of NPS market. The Commission on Narcotic Drugs placed under international control a total of 60 NPS and precursors between 2015 and 2020 [5]. Most of these NPS are SCRAs (30%), followed by stimulants (28%) and synthetic opioids (17%), mostly fentanyl analogues. These control legislative measures and/or regulations have to be implemented at regional or national level. Up to 2020, more than 50 countries have implemented legal responses to control NPS, amending the existing drug legislation (including individual substances or "generic" groups) or using innovative legal measures [5]. For example, consumer safety laws have been applied in some European countries, resulting in a mass closures of smart shops [11]. The consumption of the NPS included in the first group placed under control in 2015 (e.g. BZP, methylone, 25B-NBOMe) was reduced in the first year after the restriction [5], demonstrating the effectiveness of these measures. However, the number of new and total NPS remains high, and therefore new and common strategies are necessary to ensure public health.

4.1.3 Potential Metabolites and/or Transformation Products

It is crucial to investigate the occurrence of NPS metabolites and transformation products (TPs) that can be potentially found in the environment after chemical and biological degradation (from now on called *targets*). However, the selection of NPS *targets* is a challenging task because of the high number of substances and the continuous introduction of new ones, which creates a changeable scenario.

After human consumption, NPS undergo absorption, distribution, metabolism and excretion processes like other drugs. Knowing the metabolic pathways of NPS and their main urinary excretion products (parent compounds or free and/or conjugate metabolites) is fundamental for choosing suitable *targets*. The best option for the identification of metabolites is to analyze human urine samples; however, they are often not available, and clinical studies could not be performed due to ethical reasons [12]. *In vitro* and *in vivo* studies and *in silico* predictions are usually performed, elucidating a large number of possible metabolites to be monitored [13]. Up-to-date, very limited information concerning NPS pharmacokinetics is available, and therefore more studies are required for understanding their metabolism. However, some different metabolic behaviors have been already reported [14, 15]. For example, stimulants (phenetylamines and synthetic cathinones) are mainly excreted (partly) unchanged via urine because of their poor metabolizing capacity [15], while synthetic cannabinoids and compounds like NBOMes are highly metabolized in the human body and are excreted as complex mixture of metabolites [12, 14]. NPS *targets* from human metabolism may enter the urban sewage network and reach wastewater treatment plants (WWTP), where they can escape degradation and be discharged in the environment [16].

NPS TPs could be generated in the environment, as a result of a multitude of abiotic and biotic processes (such as hydrolysis, photolysis, oxidation and microbial metabolism), which can be found in WWTP and in the environment [17]. The most common strategies adopted to investigate NPS stability were wastewater incubations, activated sludge models or exposure to isolated microbes [18]. TPs of some synthetic cathinones and phenetylamines were identified and found in the environment [18, 19]; however, the information is scant in comparison with other chemicals of emerging concern, such as pharmaceuticals or pesticides. Thus, more biotransformation studies should be promoted, to select the most appropriate *targets* for environmental analysis.

The number of *targets* generated by each individual NPS could be very high for being monitored. Moreover, analytical standards are not always commercially available for quantitative analysis. Thus, the inclusion of these *targets* in mass spectral libraries is mandatory for their detection in screening approaches. New and combined analytical strategies should be developed to investigate NPS *targets* in the environment.

4.1.4 Occurrence in the Environment

The main sources of NPS are related to human use and production: (i) substances are consumed, metabolized and excreted via urine/feces; and (ii) substances are synthetized in clandestine drug laboratories and illegally discharged to avoid police controls (disposal of unused drugs). In both cases, NPS arrive in the sewage and if they are not totally degraded by traditional WWTP processes, they can enter the aquatic ecosystems at concentrations ranging from few ng/L to μg/L. Once there, NPS and metabolites/TPs could be adsorbed into sediments or bio-accumulate in biota, posing a risk to wildlife.

The presence of NPS and metabolites in raw wastewater has been widely demonstrated in several studies conducted for assessing NPS consumption by the general population [20–22]. Different categories of NPS were investigated in these studies, with stimulants being the most common (synthetic cathinones and phenetylamines). The number of NPS studies concerning other environmental compartments is lower; however, some NPS have been found in wastewater effluents [23, 24], surface water [25], tap water [26], sludge [27], sediments [28] and biota [29]. These NPS belong to different categories and have been studied in different seasons/years and in a unique environmental compartment. This heterogeneous information cannot be extrapolated to the whole NPS group due to the high number of substances with different chemical properties, so it is tough to create a current and global picture of NPS in the environment.

It is also difficult to determine the most dangerous NPS, as no individual data is available for almost all compounds. Only the most common and "older" NPS, such as ketamine (KET), have been studied in terms of distribution and ecotoxicological effects. Ketamine, used as a potent anesthetic and sedative since 1990s, has been found in wastewater effluents (together with its main metabolite norketamine) and surface water [27], sediments [28] and biota [29]. Behavioral effects and oxidative stress in some species have been reported after exposure to KET environmental levels [30, 31].

The environmental concentrations of NPS are expected to be lower than those of established drugs, according to their lower consumption; however, there is no evidence about the effects that these lower concentrations can cause on aquatic organisms. Moreover, the behavior of some substances such as benzodiazepine-type NPS or synthetic opioids will be probably similar to their licit analogues that pose potential risk for aquatic organisms [32, 33]. Thus, further research about their possible impact on the environment are required, including new strategies to speed up the prioritization of the most consumed NPS and their *targets*. Anyway, the occurrence and diffusion of NPS in the environment have been demonstrated and therefore their importance as environmental pollutants should not be neglected.

4.2 Sample Preparation and Collection

Selective and sensitive analytical methods are necessary to achieve the low concentrations expected for NPS in the environment. A well-designed protocol for sample collection and storage is necessary to ensure the representativeness of samples and the stability of *targets*. Moreover, an adequate sample (pre)treatment procedure is essential to isolate the selected *targets*, and reduce matrix interferences in complex environmental matrices.

As mentioned before, NPS have been mainly investigated in wastewater (raw wastewater, as an indicator of human consumption) and, for this reason, this section will be focused on this matrix. However, analytical methods developed for NPS analysis in other environmental media (i.e. surface water, sludge or biota) will also be discussed.

4.2.1 Urban Wastewater

4.2.1.1 Protocols for Collecting and Preparing Samples

Urban wastewater samples are usually collected by auto-samplers, which take multiple individual samples (aliquots) proportionally to specific periods of time or wastewater volume. These sampling modes are used to generate a "24-hour composite" or "daily" sample [34]. Another common technique is passive sampling, even if it is poorly explored in the case of NPS [35]. Passive samplers consist of polymeric-based sorbent materials (e.g. polar organic chemical integrative sampler (POCIS)) which can accumulate trace of *targets* for longer times (days or weeks), increasing the concentration of NPS and reducing the number of samples to be analyzed. However, their cost and the scant information available for NPS partitioning and stability, which makes calibration and data interpretation difficult, may limit their use for NPS analysis [35].

After collection, samples should be correctly transported and stored before analysis to preserve the stability of *targets*. This step is essential in the case of NPS, as little information is available on their (bio) transformation and stability. The most common procedure is to freeze samples at −20°C after collection and maintain this temperature until laboratory analysis [20, 22]. Stability of these NPS can be extended, adding preservatives to the samples, like acidification with hydrochloric acid (HCl). It has been demonstrated that sample acidification at pH = 2 improves the stability of synthetic cathinones, phenethylamines and opioid derivatives up to 14 days [36].

Filtration of samples is normally recommended to eliminate solid particles and particulate matter depending on the extraction technique selected. In the case of solid phase extraction (SPE), filtration is mandatory to avoid plugging of cartridges. Glass microfiber filters and nitrocellulose filters are commonly used for wastewater extraction [36]. While no sorption of phenetylamines and synthetic cathinones has been reported using mixed cellulose filters, losses of some synthetic cannabinoids and metabolites were observed after filtration on glass microfiber filters [37]. However, few information is available for the vast majority of NPS, and therefore further research about NPS stability during collection, transport, storage and filtration should be done to avoid losses of *targets*.

4.2.1.2 Extraction Procedures and Clean-up

Information about the available extraction protocols for NPS analysis in wastewater (i.e. extraction techniques, characteristics and sample volume) is reported in Table 4.2. In most of the cases, SPE has been selected as it allows the simultaneous extraction, concentration and clean-up of samples.

Table 4.2 Summary of different analytical methods used for NPS analysis in wastewater samples from 2015 to the present. Information about extraction, determination and validation steps is reported.

Extraction technique	Sample volume and characteristics	Number of NPS/ categories	Instrumental analysis and characteristics	Figures of merit	Ref
SPE	50 mL Oasis MCX® cartridges	7 NPS/3 categories (synthetic cathinones, phenetylamines and ketamine analogues)	Quantitative analysis. HPLC-MS/MS. Mass analyzer: QqQ	R: 94–105% RSD < 4% LOQ: 0.5–5 ng/L	[45]
SPE	5 mL Oasis HLB® cartridges	16 NPS/3 categories (synthetic cathinones, phenetylamines and ketamine analogues)	Quantitative analysis. UPLC-MS/MS. Mass analyzer: QqQ	R: 76–112% RSD < 17% LOQ: 3–70 ng/L	[46]
SPE	100 mL UCT XtracT Clean Screen® cartridges (C8 + ion exchange)	21 NPS/4 categories (synthetic cathinones, phenetylamines, synthetic opioids and ketamine analogues)	Quantitative analysis. UPLC-MS/MS. Mass analyzer: hybrid QTrap	R: 80–110% RSD < 17% LOQ: 0.05–1 ng/L	[36]
SPE	50 mL Oasis MCX® cartridges	30 NPS/6 categories (mainly synthetic cathinones and phenetylamines)	Quantitative analysis. HPLC-MS/MS. Mass analyzer: QqQ	R: 75–116% RSD < 14% LOQ: 0.06–15 ng/L	[20]

(continued)

Table 4.2 (Continued)

Extraction technique	Sample volume and characteristics	Number of NPS/ categories	Instrumental analysis and characteristics	Figures of merit	Ref
SPE	100 mL UCT XtracT Clean Screen® cartridges (C8 + ion exchange)	187 NPS/ 9 categories (all)	Qualitative analysis. Target screening. HPLC-MS. Mass analyzer: hybrid QTOF	Similar NPS considered for validation (R: 20–250%)	[51]
SPE	50–100 mL Oasis MCX® and Oasis HLB® cartridges	198 NPS/ 9 categories (all)	Qualitative analysis. Target screening. HPLC-HRMS. Mass analyzer: Q-Exactive™ Hybrid Quadrupole-Orbitrap™	Illicit drugs considered for validation	[38]
SPE	100 mL In-house multilayer cartridges (polymeric RP + weak cationic exchange)	278 NPS/ 9 categories (all)	Qualitative analysis. Target screening. HPLC-MS. Mass analyzer: hybrid QTOF	Similar NPS considered for validation (R: 55–118%; RSD < 24%)	[39]
On-line SPE	5 mL (10-fold diluted sample) Extraction with polymeric sorbent (PLRP)	7 NPS/3 categories (synthetic cathinones, benzodiazepines and ketamine analogues)	Quantitative analysis. HPLC-MS/MS. Mass analyzer: hybrid QTrap	R: 96–119% RSD < 30% LOQ: 1.9–13 ng/L	[40]
LLE	20 mL Extraction with hexane:ethylacetate (10 mL). Agitation for 30 seconds	4 NPS/1 category (synthetic cannabinoids)	Quantitative analysis. UHPSFC-MS/ MS. Mass analyzer: QqQ	R: 60–75% RSD < 13% LOQ: 1–6.5 ng/L	[42]
LLE	20 mL Extraction with ethylacetate (10 mL). Agitation for 30 seconds	26 NPS/1 category (synthetic cannabinoids)	Quantitative analysis. LC-MS/MS. Mass analyzer: hybrid QTrap	R: 73–112% RSD: n.a. LOQ: 1–100 ng/L	[37]

Abbreviations: HPLC, high-performance liquid chromatography; LLE, liquid-liquid extraction; LOQ, limit of quantitation; MS, mass spectrometry; n.a. not available; QqQ, triple quadrupole; QTOF, quadrupole time of flight; QTrap, triple quadrupole linear ion trap; R, recovery; RSD, relative standard deviation; SPE, solid phase extraction; UHPSFC, ultra-high-performance supercritical fluid chromatography; UPLC, ultra performance liquid chromatography

Aliquots of filtered wastewater samples (from 5 to 100 mL) are usually extracted using different SPE sorbents in order to cover the highest number of NPS with broad chemical properties. *Targets* are retained and then eluted with an adequate solvent, avoiding matrix interferences. The most common SPE sorbents used are reversed phase (RP) polymers with properties built of generic hydrophilic and lipophilic balanced monomers (e.g. Oasis HLB®, Strata-X®) or mixed mode RP-cationic exchange materials (e.g. Oasis MCX®) [34]. In the second case, samples should be acidified to pH = 2–3 to ensure that *targets* are positively charged during extraction. This especially aids the recovery of synthetic cathinones, synthetic opioids and phenethylamines [20, 36]. Multiple SPE cartridges can be used to perform several separate extractions [38]; however, using cartridges consisting of several layers with different sorbents could reduce time consumption [39].

The application of SPE requires a considerable lab-effort and is time-consuming, thus coupling SPE to the analytical system (i.e. mass spectrometer) can be helpful to speed up the procedure [40]. On-line SPE contributes to make sample preparation faster (increasing the sample throughput), minimize sample manipulation (improving methods robustness), increase the sensitivity and reduce sample volume (generally less than 10 mL). Accurate and precise results have been obtained for synthetic cathinones, benzodiazepine-type NPS and ketamine analogues (recoveries 70–120%) using 5 mL of sample [41].

Liquid-liquid extraction (LLE) has been also applied, but limited to synthetic cannabinoids analysis [37, 42]. Because of their lower polarity, these substances can be adsorbed to suspended particulate matter and glass surfaces [42], and therefore filtration step and SPE may result in significant losses. Alternative extraction procedures like LLE are recommended to avoid poor method performance of synthetic cannabinoids, improving quantitation limits and also reducing sample volume, solvent consumption and extraction time [37].

The high pre-concentration factors normally applied during sample treatment (e.g. 100–1000 in SPE) to deal with the low NPS concentrations could result in strong matrix effects, affecting the precision and robustness of the method. Internal standards (IS), i.e. isotopically labelled analogues which are expected to behave in the same manner of *targets* during the whole procedure, are normally added before the extraction to compensate matrix effects and correct for losses during extraction. However, it is not always possible to select the correct IS, especially in the case of NPS, because of their cost and availability. If an appropriate IS is not available, an additional clean-up step is recommended to reduce matrix effects and improve the sensitivity of the method [43].

4.2.2 Other Environmental Matrices

Table 4.3 summarizes different analytical methods for the determination of NPS in various additional environmental matrices, and includes information related to the extraction and clean-up steps. The number of methods available for surface and drinking water or soils/sludge is lower compared to those for wastewater. Moreover, these methods normally include, with NPS, other psychoactive substances (i.e. illicit drugs, prescription opioids or amphetamine-type substances). However, the available studies conducted in

Table 4.3 Analytical methods used for NPS analysis in different environmental samples from 2015 to the present. Information about extraction, clean-up determination and validation steps is reported.

Environ-mental matrix	Extraction technique and characteristics	Additional clean-up step	Instrumental analysis and characteristics	Figures of merit	Number of NPS/ categories	Ref
River water	SPE 250 mL Oasis MCX® cartridges.	no	Quantitative analysis. HPLC-MS/MS. Mass analyzer: QqQ	R: 64–110% RSD < 26% LOQ: 0.1–16 ng/L	20 NPS/2 categories (synthetic cathinones and phenteylamines)	[23]
River water	SPE 250 mL Strata-X® cartridges (polymeric sorbent).	no	Quantitative analysis. HPLC-MS/MS. Mass analyzer: hybrid QTOF	R: 65–94% RSD < 15% LOQ: 1–50 ng/L	21 NPS/6 categories (synthetic cathinones, phenetylamines, synthetic cannabinoids, ketamine analogues, piperazines and tryptamines)	[25]
River water	SPE 100 mL Oasis HLB® cartridges.	no	Quantitative analysis. UPLC-MS/MS. Mass analyzer: QqQ	R: 97–115% RSD: n.a. LOQ: 0.37–4.8 ng/L	5 NPS/4 categories (synthetic cathinones, phenetylamines, ketamine analogues and piperazines)	[27]
Tap water	SPE 400 mL Oasis HLB® cartridges.	no	Quantitative analysis. UPLC-MS/MS. Mass analyzer: QqQ	R: 60–139% RSD < 14 LOD: 0.01–0.04 ng/L	7 NPS/5 categories (synthetic cathinones, phenetylamines, synthetic opioids, ketamine analogues and piperazines)	[26]
Sludge and sediments	UAE 1 g. Extraction with mixture citrate/ phosphate buffer:methanol (1 : 1 v/v); t, 10 minutes.	SPE Strata-X polymeric RP	Quantitative analysis. UPLC-MS/MS. Mass analyzer: QqQ	R: 25–109% RSD < 21% LOQ: 0.3–3.9 ng/g	17 NPS/5 categories (synthetic cathinones, phenetylamines, ketamine analogues, piperazines and tryptamines)	[28]

(continued)

Table 4.3 (Continued)

Environ-mental matrix	Extraction technique and characteristics	Additional clean-up step	Instrumental analysis and characteristics	Figures of merit	Number of NPS/ categories	Ref
Sludge	MAE 500 mg. Extraction with methanol:water (50 : 50); T, 110°C; t, 40 minutes; 800 W.	SPE Oasis MCX® cartridges.	Quantitative analysis. UPLC-MS/MS. Mass analyzer: QqQ	R: 99–105% RSD: n.a. LOQ: 0.1–0.3 ng/g	2 NPS/2 categories (synthetic cathinones and ketamine analogues)	[27]
Fish tissues	UAE 1 g. Extraction with mixture 0.1%acetic acid in methanol; t, 10 minutes.	SPE Oasis MCX® cartridges.	Quantitative analysis. UPLC-MS/MS. Mass analyzer: QqQ	R: 60–127% RSD < 12% LOQ: 0.005–0.05 ng/g	3 NPS/2 categories (synthetic cathinones and ketamine analogues)	[29]
Mussels	QuEChERS 10 g. Extraction with 10 mL acetonitrile and clean-up with different sorbents.	no	Quantitative analysis. UPLC-MS/MS. Mass analyzer: hybrid QTrap	R: 81–95% RSD < 14% LOQ: 1 ng/g	4 NPS/3 categories (synthetic cathinones, synthetic opioids and ketamine analogues)	[41]

Abbreviations: HPLC, high-performance liquid chromatography; LLE, liquid-liquid extraction; LOD, limit of detection; LOQ, limit of quantitation; MAE, microwave-assisted extraction; MS, mass spectrometry; QqQ, triple quadrupole; QTOF, quadrupole time of flight; QTrap, triple quadrupole linear ion trap; QuEChERS, "Quick, Easy, Cheap, Effective, Rugged, and Safe"; R, recovery; RSD, relative standard deviation; SPE, solid phase extraction; UAE, ultrasonic-assisted extraction; UPLC, ultra performance liquid chromatography

aqueous and solid environmental matrices may be a starting point for future research to investigate the occurrence of NPS in the environment.

The (pre)treatment techniques adopted for aqueous matrices as surface water or tap water are similar to those described for wastewater [23, 27]. Grab samples are collected and stored at −20°C until analysis in order to avoid losses of analytes [25], as scant information about the stability of NPS in these matrices has been reported so far. Before the analysis, water samples are filtered on glass fiber filters [23, 25] and then extracted by SPE using the aforementioned sorbents. The main difference is the volume of sample, which is higher for surface water and tap water (between 100 and 500 mL) compared to wastewater in order to consider the lower expected concentration for NPS.

Solid samples (i.e. sediment, sludge and biota) are usually collected as grab samples and transported refrigerated (4°C) to the laboratory. Once there, samples can be immediately analyzed [27, 41] or pretreated and stored at −20°C until analysis [28, 29]. The stability of some stimulants, ketamine analogues and piperazines was demonstrated in sediments and sludge over three months at −20°C [28]. Sample pre-treatment varies for sediment and biota samples: biota samples are usually homogenized, frozen and lyophilized [29], while an additional step is used for sediment and sludge samples (i.e. sieved across less than 2 mm to obtain a homogeneous mixture) [28].

The most common extraction technique for NPS is the ultrasound-assisted extraction (UAE), probably because of the easier handling procedure, the shorter extraction times (around 10 minutes) and the reasonable cost of the equipment [28, 29]. Microwave-assisted extraction (MAE) has been also applied, despite its moderate cost, improving the extraction yields and the capability of analyzing a larger quantity of sample in a single run [27].

An additional clean-up step is frequently applied to avoid matrix effects (e.g. humic substances in sediments or lipids in organisms) as well as possible damages to the chromatographic system. The most common clean-up technique is SPE, by using polymeric reversed phase and mixed cationic exchange sorbents [27–29]. This additional step may be avoided applying a user-friendly alternative technique, called Quick, Easy, Cheap, Effective, Rugged, and Safe (QuEChERS). Extraction and clean-up steps are combined in a unique established protocol [44], reducing time and solvent consumption. Briefly, an acetonitrile salting-out extraction of a solid sample is followed by liquid–solid extraction (called dispersive SPE, abbreviated as d-SPE) to remove the remaining matrix interferences. Finally, extracts are analyzed by different instrumental techniques. Satisfactory results (recoveries between 81 and 95%) were obtained for synthetic cathinones, synthetic opioids and ketamine analogues in mussels samples [41].

4.3 Determination of New Psychoactive Substances

NPS analysis is challenging because of the high number of different substances available on the market, their similar chemical structure that may impair the identification and the dynamic drug market where new substances appear continuously. An efficient chromatographic separation is needed to allow an unquestionable identification of substances, especially in the case of isomers. Furthermore, highly selective and sensitive instrumental methods are required for achieving the low levels expected in environmental matrices.

Different instrumental methods can be applied depending on the scope of the analysis, the number of NPS considered and the availability of reference standards. Qualitative analysis allow us to screen for the presence of NPS in the environment, but a following quantitative analysis may be required to know their concentrations. The most common technique for NPS quantitative analysis is high-performance liquid chromatography coupled to (tandem) mass spectrometry (HPLC-MS/MS), whereas new strategies based on high-resolution mass spectrometry (HPLC-HRMS) have been developed for qualitative NPS analysis.

In this section, different analytical strategies applied for NPS identification in environmental samples are discussed, including aspects related to method validation. Moreover, the most relevant issues ("hits and tips") in NPS analysis are highlighted in order to help the readers for future research.

4.3.1 Development of the Instrumental Method

4.3.1.1 Chromatographic Separation

The high polarity and low volatility of most NPS and metabolites make liquid chromatography (LC) the most appropriate technique for their analysis in environmental samples, avoiding the derivatization step required in gas chromatography (GC) applications. Most of the studies apply multi-residue methods based on reverse-phase LC (RPLC) that allows the separation of a broad range of NPS from low-polar to non-polar [43]. Nevertheless, smaller and more polar molecules may be not retained with RPLC columns, and therefore more specific stationary phases should be employed. For example, HILIC (hydrophilic interaction liquid chromatography) columns have been satisfactorily used for phenethylamines and synthetic cathinones analysis, ensuring reproducibility and repeatability between injections [45].

Recent innovations in RPLC systems and columns have improved the efficiency, robustness and sensitivity achieved with HPLC systems. Ultra-high-performance LC (UHPLC) is based on LC columns containing stationary-phase packing with small particle sizes (<2 μm; normally 1.7 μm), which improves sensitivity with shorter chromatographic run times. The reduction of the internal diameter of LC column and the use of lower flow rates (<100 μL/min) improve the ionization efficiency in micro-liquid chromatography (μLC), enhancing the sensitivity [46]. A μLC-MS/MS methodology has been proposed for the analysis of illicit drugs and several synthetic cathinones and phenetylamines; however, the method was not enough robust for NPS analysis, probably because of the poor retention time stability and the extended analysis time [46].

Another chromatographic technique employed for NPS analysis is ultra-high-performance supercritical fluid chromatography (UHPSFC), which uses (sub) supercritical carbon dioxide (CO_2) with diverse organic modifiers as the mobile phase and allows the direct injection of final organic extracts. This technique has been satisfactorily applied to synthetic cannabinoids analysis, increasing chromatography efficiency and resolution [22].

Enantiomeric analysis is another important issue in NPS determination. Racemic mixtures of NPS (R and S enantiomers) are usually distributed in the drug market and are differently metabolized in the human body after consumption [34]. Thus, the presence of R-(+) and S-(-) enantiomers and the enantiomeric fraction can identify the source of NPS, i.e. distinguish between human consumption and direct disposal of unused substances [47]. This analysis has been demonstrated to be helpful for analysis of stimulants such as mephedrone [48].

4.3.1.2 Detection

LC-MS/MS methods have been used for reliable identification and quantitation of NPS in environmental samples at low levels (<10 ng/L) [34]. Low-resolution (LRMS) instruments have commonly been employed for this scope, using triple quadrupole (QqQ) [20, 23] and ion trap [36, 41] mass analyzers because of their robustness, selectivity and sensitivity [43]. Multiple Reaction Monitoring (MRM), which consists of selecting precursor and product ions (MRM transitions) for each target analyte, is used as preferred acquisition method in tandem MS analysis. In the case of NPS, two or three MRM transitions are usually selected to avoid false positives: the most sensitive is used for quantitation and the less sensitive for confirmation purposes. According to different guidelines [49, 50], the chromatographic retention time (RT) and the MRM ratio between transitions obtained in positive environmental samples should match with those observed in reference standards.

Synthetic cathinones and phenetylamines are the categories most studied in environmental samples [43]. However, the number of NPS considered in quantitative analysis is limited, and depends on the availability of NPS reference standards, as they are required for optimization of LC and MS/MS parameters and for the identification, confirmation and quantitation of NPS. This point is particularly critical for NPS because reference standards are available only for a restricted number of substances. Thus, complementary approaches are needed to keep up with the rapid innovation of NPS and changes in the drug market, as this may also influence their presence in the environment. New strategies based on qualitative analysis (screening approaches) using high-resolution mass spectrometry (HRMS) instruments have been developed to deal with NPS dynamics, and covering the limitations of LRMS.

Different screening approaches (i.e. targeted, suspect or untargeted screening) may be applied for NPS qualitative analysis using hybrid mass analyzers such as quadrupole time of flight (QTOF) and quadrupole-Orbitrap (Q-Orbitrap) [38, 51]. Moreover, a retrospective analysis of samples can be done in HRMS, avoiding the re-extraction and re-analysis of samples and minimizing time consumption [52].

The vast majority of the approaches are based on "targeted" or "suspect" screening, in which a known group of substances is investigated. In both cases, a database containing a list of selected NPS and metabolites should be created before analysis, including the accurate mass ion (normally the protonated molecule) and RT of each substance [38]. This information can be obtained using reference standards ("targeted screening") or theoretically estimated ("suspect screening"). Nowadays, different HRMS databases containing more than 500 NPS and metabolites (i.e. HighResNPS [53] and NPSHub [54]) are available for users, simplifying data interpretation and allowing the analysis of a wide range of substances. Non-targeted (or untargeted) screening allows the investigation of other NPS not included in these databases, increasing the number of NPS considered. However, this analysis is laborious and time-consuming, as each chromatographic peak with the corresponding accurate mass spectrum should be individually examined [43]. This strategy is normally applied for the elucidation of metabolic pathways and biotransformation processes [13, 19].

Hybrid mass analyzers may provide relevant structural information by obtaining accurate mass product-ion spectra after MS/MS experiments, using different acquisition modes [52]. However, the later identification and confirmation of substances merit a particular attention. Although there is not an established consensus, different confidence levels of identification were proposed for identification/confirmation purposes, depending on the availability of reference standards [52, 55]. When available, compounds can be detected/identified or confirmed using reference standards: detection/identification occurs when the most abundant ion (normally the protonated molecule) is found at an expected RT (0.1 minutes; ±2%) and with a mass error <5 ppm; two most abundant ions (protonated molecule and products) should be found for confirmation, under the aforementioned conditions [50, 56]. When reference standards are not available, a "tentative identification" can be performed considering diagnostic evidence or spectra data from libraries/databases or literature. In this case, the accurate mass error should be <5 ppm and different product ions should be identified. Moreover, a reference standard is mandatory to confirm the substance.

HRMS approaches are a complementary tool for quantitative analysis, so as to allow the identification of a selected number of NPS and reduce the number of substances for MS/MS determination. Thus, a combination of both techniques is strongly recommended for NPS analysis.

4.3.2 Figures of Merit

As non-standardized methods are available for NPS analysis, the establishment of quality assurance/quality control (QA/QC) protocols is needed to ensure the reliability of results. Instrumental and procedural blanks and quality controls are normally included in each batch to check for eventual contamination and verify the acceptability of results.

Method validation in quantitative analysis is normally performed in terms of linearity, sensitivity, accuracy, precision and matrix effects, following the recommendation of different guidelines [49, 50]. Accuracy and precision are calculated by recovery assays, spiking the environmental samples at one or two different known concentrations within the linear range (see Tables 4.2 and 4.3). In general, satisfactory results (recovery 80–120% and relative standard deviation <20%) were achieved for almost all substances investigated; however, recoveries lower than 50% were observed for some *targets*, which can be explained taking into account the complexity of environmental matrices and the high number and different characteristics of the studied NPS [23, 42]. Low limits of quantitation (LOQ), estimated as the concentration of *targets* in samples giving a signal/noise of 10, were usually achieved (e.g. <50 ng/L in water samples). Matrix effects (signal enhancement and suppression caused by matrix interferences) is usually evaluated by comparing the peak area obtained for each *target* in the final extract and in a standard solution. When necessary, it can be corrected using IS [20] or matrix-calibration curves [28] for quantitation purposes.

For validating screening approaches, and taking into account the scant availability of standards, a group of representative compounds with wide physico-chemical properties is normally chosen. Recovery and precision for these selected NPS is assumed to be similar to their analogues. NPS belonging at different categories or illicit drugs were selected for the validation step, with satisfactory results [39, 51].

4.3.3 Hits and Tips

The wide number of substances and different categories, the similar chemical properties of each NPS category and the continuous variations in NPS market make their analysis challenging. Flexible and sensitive methods are required to investigate a broad number of substances at low concentration levels. However, more specific methods should be applied for special categories/groups of NPS. For example, alternative extraction procedures are recommended for synthetic cannabinoids which are usually adsorbed to particulate matter. Moreover, LC-MS/MS analysis using more selective LC columns may be applied for small or polar compounds or for enantiomeric separation. To date, HRMS has been mainly applied for NPS analysis in wastewater; thus, further research is needed to apply this technique to other environmental matrices, as has been done for other emerging pollutants such as pharmaceutical compounds [17] or illicit drugs [52]. The combination of qualitative analysis (to select the most interesting NPS for analysis) and quantitative analysis (to know exactly the concentration) is thus required. In general, new strategies are needed for the determination of NPS in environmental matrices, as there is a lack of information related to the occurrence, behavior and impact of NPS in the aquatic environment.

4.4 Future Direction and Challenges

This chapter summarized the analytical methods used for investigating NPS in environmental samples. To date, the vast majority of the studies have been focused on urban wastewater analysis and only few works investigated the occurrence of NPS in other environmental matrices. The behavior of these substances in the environment is still unexplored and further research is therefore required to assess the potential risks that NPS can pose to aquatic organisms.

The following actions can be a starting point to achieve this aim:

- *Selection of NPS and metabolites.* The high number of reported NPS and the constant introduction of new substances on the drug market makes their monitoring difficult, expensive and time-consuming. A selection of NPS is required to focus the analysis on a limited number of substances, and consider those more interesting, for instance those consumed most frequently or with a high dose and those excreted partially unchanged. Unfortunately, scant information on NPS metabolism is currently available, and further pharmacological studies are needed to identify their excretion pattern and to identify the main urinary metabolites to be searched in environmental samples.
- *NPS stability and distribution in the environment.* Another important issue is the stability of NPS in environmental samples. The occurrence of NPS and their metabolites in raw wastewater has been demonstrated, but little information about their (bio)transformation and elimination in WWTP can be found in literature. Consequently, the distribution of NPS in the environment is still unexplored. Although low concentrations of NPS are expected to enter the environment, they may cause potential effects to the aquatic organisms due to their high biological activity. Further research is needed in this direction to understand the occurrence and distribution of these substances in the environmental compartments and to identify the most dangerous NPS.
- *Impact of NPS in the environment.* The impact of NPS in aquatic organisms is also unknown. The effect of established illicit drugs such as cocaine, methamphetamine and ecstasy on organisms at environmental concentrations has been published so far [57]. However, no ecotoxicological studies have been performed to date for NPS, and therefore it is not possible to evaluate their potential risks in the environment. Further research is mandatory to evaluate NPS effects in the aquatic and terrestrial organisms at the measured concentrations.

- *Development of new analytical strategies for NPS analysis.* An improvement of the current analytical methods for NPS determination is required to keep up with the continuous changes in drug market and trends. A combination of qualitative and quantitative strategies is recommended in order to investigate a broad number of substances and select the most important ones. Flexible, fast and sensitive quantitative methods are needed to achieve the NPS low concentrations found in the environment. To date, almost all the studies are focused on wastewater analysis, and therefore new methods for the analysis of other aqueous and solid environmental matrices should be considered in future research to preserve the environment.

Acknowledgments

The authors acknowledge EuSeME (project number 861602) funded by the European Union's Justice Programme – Drugs Policy.

References

1 EMCDDA. European Monitoring Centre for Drugs and Drug Addiction (2015). New psychoactive substances in Europe. An update from the EU Early Warning System. Publications Office of the European Union, Luxembourg.

2 UNODC. United Nations Office on Drugs and Crime (2013). The challenge of new psychoactive substances. A Report from the Global SMART Programme. UNODC Laboratory and Scientific Section. Vienna.

3 Peacock, A., Bruno, R., Gisev, N., Degenhardt, L., Hall, W., Sedefov, R., White, J., Thomas, K.V., Farrell, M., and Griffiths, P. (2019). New psychoactive substances: challenges for drug surveillance, control, and public health responses. *The Lancet* 394: 1668–1684. doi: 10.1016/S0140-6736(19)32231-7.

4 EMCDDA. European Monitoring Centre for Drugs and Drug Addiction (2018). *European Drug Report 2018: Trends and Developments*. Luxembourg: Publications Office of the European Union.

5 UNODC. United Nations Office on Drugs and Crime (2021). Regional diversity and the impact of scheduling on NPS trends. Global SMART Update, vol. 25. UNODC Laboratory and Scientific Section. Vienna.

6 UNODC. United Nations Office on Drugs and Crime (2017). Early warning advisory on new psychoactive substances. https://www.unodc.org/LSS/Home/NPS. Accessed 14 December 2021.

7 EMCDDA. European Monitoring Centre for Drugs and Drug Addiction (2017). *High-Risk Drug Use and New Psychoactive Substances. Results from an EMCDDA Trendspotter Study*. Luxembourg: Publications Office of the European Union.

8 EMCDDA. European Monitoring Centre for Drugs and Drug Addiction (2016). *Health Responses to New Psychoactive Substances*. Luxembourg: Publications Office of the European Union.

9 UNODC. United Nations Office on Drugs and Crime (2019). *Current NPS Threats. Global SMART Programme, Vol. I-III*. Vienna: UNODC Laboratory and Scientific Section.

10 Tracy, D.K., Wood, D.M., and Baumeister, D. (2017). Novel psychoactive substances: types, mechanisms of action, and effects. *BMJ (Clinical Research Ed.)*. 356: ji6848. doi: 10.1136/bmj.i6848.

11 EMCDDA. European Monitoring Centre for Drugs and Drug Addiction (2016). *New Psychoactive Substances in Europe. Legislation and Prosecution – Current Challenges and Solutions*. Luxembourg: Publications Office of the European Union.

12 Wagmann, L., Gampfer, T.M., and Meyer, M.R. (2021). Recent trends in drugs of abuse metabolism studies for mass spectrometry-based analytical screening procedures. *Anal. Bioanal. Chem.* 413: 5551–5559. doi: 10.1007/s00216-021-03311-w.

13 Diao, X. and Huestis, M.A. (2019). New synthetic cannabinoids metabolism and strategies to best identify optimal marker metabolites. *Front. Chem.* 7: 109. doi: 10.3389/fchem.2019.00109.

14 Meyer, M.R. (2016). New psychoactive substances: an overview on recent publications on their toxicodynamics and toxicokinetics. *Arch. Toxicol.* 90: 2421–2444. doi: 10.1007/s00204-016-1812-x.

15 Tyrkkö, E., Andersson, M., and Kronstrand, R. (2016). The toxicology of new psychoactive substances: synthetic cathinones and phenylethylamines. *Ther. Drug Monit.* 38: 190–216. doi: 10.1097/FTD.0000000000000263.

16 Zuccato, E., Castiglioni, S., Bagnati, R., Chiabrando, C., Grassi, P., and Fanelli, R. (2008). Illicit drugs, a novel group of environmental contaminants. *Water Res* 42: 961–968. doi: 10.1016/j.watres.2007.09.010.

17 Picó, Y. and Barceló, D. (2015). Transformation products of emerging contaminants in the environment and high-resolution mass spectrometry: a new horizon. *Anal. Bioanal. Chem.* 407: 6257–6273. doi: 10.1007/s00216-015-8739-6.

18 Mardal, M., Bischoff, M., Ibáñez, M., Ruffing, U., Hernández, F., and Meyer, M.R. (2017). Microbial biotransformation of five pyrrolidinophenone-type psychoactive substances in wastewater and a wastewater isolated *Pseudomonas putida* strain: microbial biotransformation of psychoactive substances. *Drug Test. Anal.* 9: 1522–1536. doi: 10.1002/dta.2165.

19 Kinyua, J., Negreira, N., McCall, A.K., Boogaerts, T., Ort, C., Covaci, A., and van Nuijs, A.L.N. (2018). Investigating in-sewer transformation products formed from synthetic cathinones and phenethylamines using liquid chromatography coupled to quadrupole time-of-flight mass spectrometry. *Sci. Total Environ.* 634: 331–340. doi: 10.1016/j.scitotenv.2018.03.253.

20 Castiglioni, S., Salgueiro-González, N., Bijlsma, L., Celma, A., Gracia-Lor, E., Beldean-Galea, M.S., Mackulak, T., Emke, E., Heath, E., Kasprzyk-Hordern, B., Petkovic, A., Poretti, F., Rangelov, J., Santos, M.M., Sremacki, M., Styszko, K., Hernández, F., and Zuccato, E. (2021). New psychoactive substances in several European populations assessed by wastewater-based epidemiology. *Water Res.* 195: 116983. doi: 10.1016/j.watres.2021.116983.

21 Bade, R., Bijlsma, L., Sancho, J.V., Baz-Lomba, J.A., Castiglioni, S., Castrignanò, E., Causanilles, A., Gracia-Lor, E., Kasprzyk-Hordern, B., Kinyua, J., McCall, A.K., van

Nuijs, A.L.N., Ort, C., Plósz, B.G., Ramin, P., Rousis, N.I., Ryu, Y., Thomas, K.V., de Voogt, P., Zuccato, E., and Hernández, F. (2017). Liquid chromatography-tandem mass spectrometry determination of synthetic cathinones and phenethylamines in influent wastewater of eight European cities. *Chemosphere* 168: 1032–1041. doi: 10.1016/j.chemosphere.2016.10.107.

22 González-Mariño, I., Gracia-Lor, E., Rousis, N.I., Castrignanò, E., Thomas, K.V., Quintana, J.B., Kasprzyk-Hordern, B., Zuccato, E., and Castiglioni, S. (2016). Wastewater-based epidemiology to monitor synthetic cathinones use in different European countries. *Environ. Sci. Technol.* 50: 10089–10096. doi: 10.1021/acs.est.6b02644.

23 Senta, I., Krizman, I., Ahel, M., and Terzic, S. (2015). Multiresidual analysis of emerging amphetamine-like psychoactive substances in wastewater and river water. *J. Chromatogr. A.* 1425: 204–212. doi: 10.1016/j.chroma.2015.11.043.

24 Borova, V.L., Gago-Ferrero, P., Pistos, C., and Thomaidis, N.S. (2015). Multi-residue determination of 10 selected new psychoactive substances in wastewater samples by liquid chromatography–tandem mass spectrometry. *Talanta* 144: 592–603. doi: 10.1016/j.talanta.2015.06.080.

25 Andrés-Costa, M.J., Andreu, V., and Picó, Y. (2016). Analysis of psychoactive substances in water by information dependent acquisition on a hybrid quadrupole time-of-flight mass spectrometer. *J. Chromatogr. A.* 1461: 98–106. doi: 10.1016/j.chroma.2016.07.062.

26 Wang, Z., Gao, S., Dai, Q., Zhao, M., and Yang, F. (2020). Occurrence and risk assessment of psychoactive substances in tap water from China. *Environ. Pollut.* 261: 114163. doi: 10.1016/j.envpol.2020.114163.

27 Styszko, K., Proctor, K., Castrignanò, E., and Kasprzyk-Hordern, B. (2021). Occurrence of pharmaceutical residues, personal care products, lifestyle chemicals, illicit drugs and metabolites in wastewater and receiving surface waters of Krakow agglomeration in South Poland. *Sci. Total Environ.* 768: 144360. doi: 10.1016/j.scitotenv.2020.144360.

28 Álvarez-Ruiz, R., Andrés-Costa, M.J., Andreu, V., and Picó, Y. (2015). Simultaneous determination of traditional and emerging illicit drugs in sediments, sludges and particulate matter. *J. Chromatogr. A* 1405: 103–115. doi: 10.1016/j.chroma.2015.05.062.

29 Yin, X., Guo, C., Teng, Y., and Xu, J. (2019). Development and application of the analytical method for illicit drugs and metabolites in fish tissues. *Chemosphere* 233: 532–541. doi: 10.1016/j.chemosphere.2019.06.018.

30 Liao, P.H., Yang, W.K., Yang, C.H., Lin, C.H., Hwang, C.C., and Chen, P.J. (2018). Illicit drug ketamine induces adverse effects from behavioral alterations and oxidative stress to p53-regulated apoptosis in medaka fish under environmentally relevant exposures. *Environ. Pollut.* 237: 1062–1071. doi: 10.1016/j.envpol.2017.11.026.

31 Wang, Z., Xu, Z., Wu, Y., Zhang, Z., and Li, X. (2020). Impact of ketamine on the behavior and immune system of adult medaka (Oryzias latipes) at environmentally relevant concentrations and eco-risk assessment in surface water. *J. Hazard. Mater.* 393: 121577. doi: 10.1016/j.jhazmat.2019.121577.

32 Tanoue, R., Margiotta-Casaluci, L., Huerta, B., Runnalls, T.J., Eguchi, A., Nomiyama, K., Kunisue, T., Tanabe, S., and Sumpter, J.P. (2019). Protecting the environment from psychoactive drugs: Problems for regulators illustrated by the possible effects of tramadol on fish behaviour. *Sci. Total Environ.* 664: 915–926. doi: 10.1016/j.scitotenv.2019.02.090.

33 Cerveny, D., Brodin, T., Cisar, P., McCallum, E., and Fick, J. (2020). Bioconcentration and behavioral effects of four benzodiazepines and their environmentally relevant mixture in wild fish. *Sci. Total Environ.* 702: 134780. doi: 10.1016/j.scitotenv.2019.134780.

34 Gent, L. and Paul, R. (2021). The detection of new psychoactive substances in wastewater. A comprehensive review of analytical approaches and global trends. *Sci. Total Environ.* 776: 146028. doi: 10.1016/j.scitotenv.2021.146028.

35 Baz-Lomba, J.A., Reid, M.J., and Thomas, K.V. (2016). Target and suspect screening of psychoactive substances in sewage-based samples by UHPLC-QTOF. *Anal. Chim. Acta* 914: 81–90. doi: 10.1016/j.aca.2016.01.056.

36 Bade, R., Abdelaziz, A., Nguyen, L., Pandopulos, A.J., White, J.M., and Gerber, C. (2020). Determination of 21 synthetic cathinones, phenethylamines, amphetamines and opioids in influent wastewater using liquid chromatography coupled to tandem mass spectrometry. *Talanta* 208: 120479. doi: 10.1016/j.talanta.2019.120479.

37 Pandopulos, A.J., Bade, R., O'Brien, J.W., Tscharke, B.J., Mueller, J.F., Thomas, K.V., White, J.M., and Gerber, C. (2020). Towards an efficient method for the extraction and analysis of cannabinoids in wastewater. *Talanta* 217: 121034. doi: 10.1016/j.talanta.2020.121034.

38 Salgueiro-González, N., Castiglioni, S., Gracia-Lor, E., Bijlsma, L., Celma, A., Bagnati, R., Hernández, F., and Zuccato, E. (2019). Flexible high resolution-mass spectrometry approach for screening new psychoactive substances in urban wastewater. *Sci. Total Environ.* 689: 679–690. doi: 10.1016/j.scitotenv.2019.06.336.

39 Diamanti, K., Aalizadeh, R., Alygizakis, N., Galani, A., Mardal, M., and Thomaidis, N.S. (2019). Wide-scope target and suspect screening methodologies to investigate the occurrence of new psychoactive substances in influent wastewater from Athens. *Sci. Total Environ.* 685: 1058–1065. doi: 10.1016/j.scitotenv.2019.06.173.

40 López-García, E., Mastroianni, N., Postigo, C., Barceló, D., and López de Alda, M. (2018). A fully automated approach for the analysis of 37 psychoactive substances in raw wastewater based on on-line solid phase extraction-liquid chromatography-tandem mass spectrometry. *J. Chromatogr. A* 1576: 80–89. doi: 10.1016/j.chroma.2018.09.038.

41 López-García, E., Postigo, C., and López de Alda, M. (2019). Psychoactive substances in mussels: analysis and occurrence assessment. *Mar. Pollut. Bull.* 146: 985–992. doi: 10.1016/j.marpolbul.2019.07.042.

42 González-Mariño, I., Thomas, K.V., and Reid, M.J. (2018). Determination of cannabinoid and synthetic cannabinoid metabolites in wastewater by liquid-liquid extraction and ultra-high performance supercritical fluid chromatography-tandem mass spectrometry: Cannabinoid derivatives in wastewater by UHPSFC-MS/MS. *Drug Test. Anal.* 10: 222–228. doi: 10.1002/dta.2199.

43 Bijlsma, L., Bade, R., Been, F., Celma, A., and Castiglioni, S. (2021). Perspectives and challenges associated with the determination of new psychoactive substances in urine and wastewater – A tutorial. *Anal. Chim. Acta* 1145: 132–147. doi: 10.1016/j.aca.2020.08.058.

44 Anastassiades, M., Lehotay, S., Stajnbaher, D., and Schenck, F. (2003). Fast and easy multiresidue method employing acetonitrile extraction/partitioning and "dispersive solid-phase extraction" for the determination of pesticide residues in produce. *J. AOAC Int.* 86: 412–431. doi: 10.1093/jaoac/86.2.412.

45 Kinyua, J., Covaci, A., Maho, W., McCall, A.K., Neels, H., and van Nuijs, A.L.N. (2015). Sewage-based epidemiology in monitoring the use of new psychoactive substances: validation and application of an analytical method using LC-MS/MS: NPS in wastewater. *Drug Test. Anal.* 7: 812–818. doi: 10.1002/dta.1777.

46 Celma, A., Sancho, J.V., Salgueiro-González, N., Castiglioni, S., Zuccato, E., Hernández, F., and Bijlsma, L. (2019). Simultaneous determination of new psychoactive substances and illicit drugs in sewage: potential of micro-liquid chromatography tandem mass spectrometry in wastewater-based epidemiology. *J. Chromatogr. A* 1602: 300–309. doi: 10.1016/j.chroma.2019.05.051.

47 Castrignanò, E., Lubben, A., and Kasprzyk-Hordern, B. (2016). Enantiomeric profiling of chiral drug biomarkers in wastewater with the usage of chiral liquid chromatography coupled with tandem mass spectrometry. *J. Chromatogr. A* 1438: 84–99. doi: 10.1016/j.chroma.2016.02.015.

48 Castrignanò, E., Mardal, M., Rydevik, A., Miserez, B., Ramsey, J., Shine, T., Pantoş, G.D., Meyer, M.R., and Kasprzyk-Hordern, B. (2017). A new approach towards biomarker selection in estimation of human exposure to chiral chemicals: a case study of mephedrone. *Sci. Rep.* 7: 13009. doi: 10.1038/s41598-017-12581-3.

49 European Commission (2002). Commission Decision 2002/657/EC of 12 August 2002 implementing Council Directive 96/23/EC concerning the performance of analytical methods and the interpretation of results. *Off. J. Eur. Commun.* L221/8.

50 European Commission (2017). Guidance document on analytical quality control and method validation procedures for pesticide residue and analysis in food and feed. SANTE/12682/2019. Directorate General for Health and Food Safety.

51 Bade, R., Tscharke, B.J., White, J.M., Grant, S., Mueller, J.F., O'Brien, J., Thomas, K.V., and Gerber, C. (2019). LC-HRMS suspect screening to show spatial patterns of new psychoactive substances use in Australia. *Sci. Total Environ.* 650: 2181–2187. doi: 10.1016/j.scitotenv.2018.09.348.

52 Hernández, F., Castiglioni, S., Covaci, A., de Voogt, P., Emke, E., Kasprzyk-Hordern, B., Ort, C., Reid, M., Sancho, J.V., Thomas, K.V., van Nuijs, A.L.N., Zuccato, E., and Bijlsma, L. (2018). Mass spectrometric strategies for the investigation of biomarkers of illicit drug use in wastewater. *Mass Spectrom. Rev.* 37: 258–280. doi: 10.1002/mas.21525.

53 Mardal, M., Andreasen, M.F., Mollerup, C.B., Stockham, P., Telving, R., Thomaidis, N.S., Diamanti, K.S., Linnet, K., and Dalsgaard, P.W. (2019). HighResNPS.com: an online crowd-sourced HR-MS database for suspect and non-targeted screening of new psychoactive substances. *J. Anal. Toxicol.* 43: 520–527. doi: 10.1093/jat/bkz030.

54 Urbas, A., Schoenberger, T., Corbett, C., Lippa, K., Rudolphi, F., and Robien, W. (2018). NPS Data Hub: a web-based community driven analytical data repository for new psychoactive substances. *Forensic Chem.* 9: 76–81. doi: 10.1016/j.forc.2018.05.003.

55 Schymanski, E.L., Jeon, J., Gulde, R., Fenner, K., Ruff, M., Singer, H.P., and Hollender, J. (2014). Identifying small molecules via high resolution mass spectrometry: communicating confidence. *Environ. Sci. Technol.* 48: 2097–2098. doi: 10.1021/es5002105.

56 WADA. World Anti-doping Agency (2010). *WADA Technical Document – TD2010IDCR. Identification Criteria for Qualitative Assays Incorporating Column Chromatography and Mass Spectrometry*. WADA Laboratory Committee.

57 Fontes, M.K., Maranho, L.A., and Pereira, C.D.S. (2020). Review on the occurrence and biological effects of illicit drugs in aquatic ecosystems. *Environ. Sci. Pollut. Res.* 27: 30998–31034. doi: 10.1007/s11356-020-08375-2.

5

Artificial Sweeteners

Konstatinos Vasilatos, Maria-Christina Nika, Georgios Gkotsis, and Nikolaos S. Thomaidis

Laboratory of Analytical Chemistry, Department of Chemistry, National and Kapodistrian University of Athens, Athens, Greece

5.1 Overview of Artificial Sweeteners

Artificial sweeteners (ASs) have been recently included in the group of contaminants of emerging concern, since monitoring studies revealed their widespread occurrence in the environment, as a result of their high production and their environmental persistence. ASs are not as well-investigated as other emerging contaminants, like pharmaceuticals, and, therefore, the development and application of accurate and reliable analytical methods for their determination in different environmental matrices remains critical [1].

5.1.1 Properties

Sweeteners consist of two main groups: caloric or nutritive, and noncaloric or non-nutritive compounds. Nutritive sweeteners are carbohydrates or their derivatives such as glucose, fructose, and maltose. Artificial high-intensity sweeteners are non-nutritive and provide no or trivial energy and do not belong to any particular chemical group of compounds. ASs are high production volume substances due to their global use as tabletop sweeteners and food additives in diet beverages, pharmaceuticals, and some personal care products, including toothpastes. When consumed, they cause an artificial, metallic aftertaste with no glycemic effect/insulin response or calorie intake, since they are not metabolized as carbohydrates. ASs are used alone or as a mixture to provide sweet flavor and as sugar substitutes to help reduce insulin resistance and obesity. They are considered to be safe for human consumption with acceptable daily intake (ADI), although accumulating evidence suggests that frequent consumers of ASs may also be prone to excessive weight gain, metabolic syndrome, type 2 diabetes, and cardiovascular diseases [2].

The most frequently used ASs are saccharine (SAC), cyclamate (CYCL), aspartame (ASP), sucralose (SUC), acesulfame (ACE), neotam (NEO), neohesperidin dihydrochalcone (NHDC), and alitame (ALI). Their structures and physicochemical properties are summarized in Table 5.1. SAC, the oldest-used AS, is produced by an initial reaction

Analytical Methods for Environmental Contaminants of Emerging Concern, First Edition. Edited by Núria Fontanals and Rosa Maria Marcé.

Table 5.1 Structure and physicochemical properties of artificial sweeteners [42, 53, 54].

Artificial sweetener	Structure	CAS RN	Molecular formula	Molecular weight (g/mol)	$logK_{ow}$	pK_a	Water solubility (g/L)
SAC	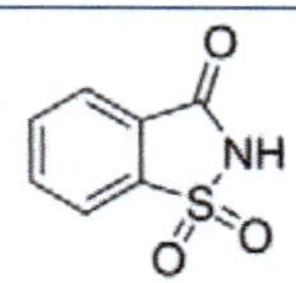	81-07-2	$C_7H_5NO_3S$	183.2	0.91	2.2	4 (25°C)
CYCL (sodium)	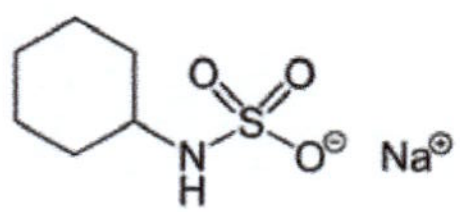	139-05-9	$C_6H_{13}NaNO_3S$	201.2	−1.61	1.9	1000 (25°C)
ASP	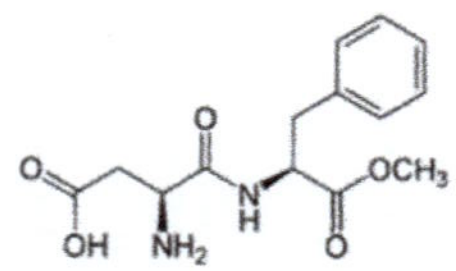	22839-47-0	$C_{14}H_{18}N_2O_5$	294.3	0.07	3.19 acid 7.87 basic	10.2 (pH = 7.00, (25°C))
SUC	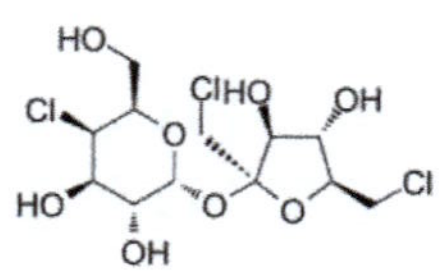	56038-13-2	$C_{12}H_{19}Cl_3O_8$	397.6	−1.00	11.8	20.3 (25°C)

ACE-K	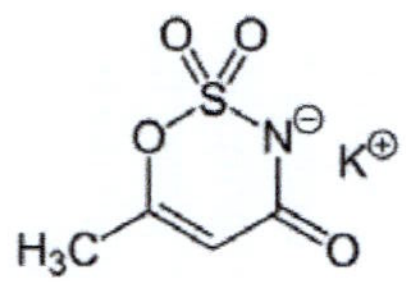	55589-62-3	$C_4H_4KNO_4S$	201.2	−1.33	2.0	270 (20°C)
NEO		165450-17-9	$C_{20}H_{30}N_2O_5$	378.5	2.39	3.01 $^{acid}_{basic}$ 8.02	12.6 (25°C)
NHDC	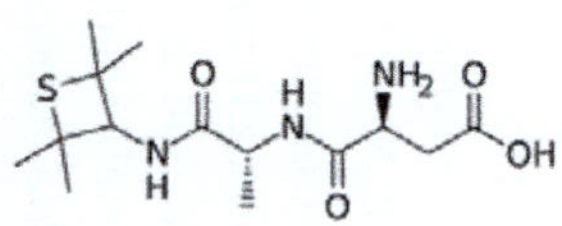	20702-77-6	$C_{28}H_{36}O_{15}$	612.6	0.75	9.7	0.4–0.5 (20°C)
ALI	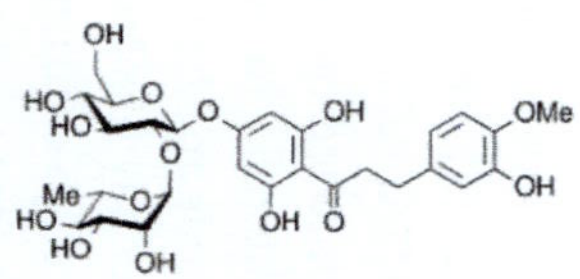	80863-62-3	$C_{14}H_{25}N_3O_4S$	331.2	−0.37	3.71	130 (pH = 5.00, (25°C))

between toluene and chloro-sulfonic acid. It is a white crystal that melts at 229°C. SAC is stable at a pH range of 2–7 and at temperatures up to 150°C, and its sodium and calcium salts are very soluble in water. CYCL commonly exists in sodium or calcium salt forms, which are both colorless and odorless, while in its acidic form (cyclamic acid) it is a strong acid with pK_a of 1.71. ASP, a dipeptide which is composed only of the L-isomers, is non-polar, odorless, and has high water solubility (10.2 g/L). SUC is a polar, chlorinated sugar which is formed synthetically from SAC by selectively replacing three hydroxyl groups with chlorine atoms. SUC presents high solubility in water (20.3 g/L) and stability over a wide range of pH and temperature (melting point: 130°C) and therefore is used in cooking and baking. Its sweetness action remains much longer than ASP. ACE-K occurs as a colorless to white-colored, odorless, crystalline powder with an intensely sweet taste and stability under high temperatures (melting point: 123°C) and thus it is often used as an ingredient in baked products. The solid state of ACE-K presents high stability and, for this reason, it can be stored at ambient temperature for 10 years without decomposition [3]. NEO, a derivate of ASP, is an odorless white to gray-white powder with a strong sweetness and is readily soluble in alcohols and water (12.6 g/L). The consumption of NEO by people who suffer from phenylketonuria is considered safe due to the presence of the 3,3-dimethylbutyl moiety, which blocks peptidases and prevents the peptide bond between the aspartic acid and phenylalanine moieties from breaking. NHDC is produced by alkaline hydrogenation of neohesperidin, and has a molecular weight at 612.6 g/mol. It can be isolated from the peel and pulp of orange, grapefruit, and other citrus fruit. It is a flavonoid and in pure form is found as an odorless white to pale yellow substance not unlike powdered sugar. ALI is a dipeptide analog composed of L-aspartic acid and D-alanine and contains a thioalkyl amide group at the carboxyl end. It is a crystalline, odorless, and nonhygroscopic powder, with a good solubility in most polar solvents such as water (130 g/L at pH 5.6) and alcohol.

5.1.2 Legislation and Environmental Risk Assessment

Before being approved for consumption, ASs have been extensively investigated and they have been Generally Recognized As Safe (GRAS) for humans, although scientific evidence for possible risks and other safety issues are reported in the literature. However, considering their recent release in the market, it is too early to evaluate any potential risk to induce cancer [3]. Five ASs are approved by the US Food and Drug Administration (FDA) in the USA: ACE-K, ASP, NEO, SAC, and SUC, whereas NHDC and CYCL are additionally approved by the European Food Safety Authority (EFSA). In order to prevent conceivable risks to humans, EU regulations set an upper limit on the concentration of AS in foods and beverages and an acceptable daily intake from 0.5 to 50 mg/kg of body weight per day has been established [4, 5].

Due to their high consumption, aqueous solubility, and environmental persistence, ASs have been frequently detected in the aquatic environment, mainly through the releases of effluent wastewater into receiving water bodies. Their omnipresence in the environment give rise to concerns regarding their ecotoxicological impacts on both the aquatic and terrestrial ecosystem. However, data concerning their ecotoxicity are scarce and so far indicate low risk. SUC has been reported as non-toxic to plant growth; however, its high persistence might lead to chronic low-dose exposure with unknown effects. Moreover, the physiology and locomotion behavior of crustaceans exposed to SUC may be affected. Another recent study reports that ASP affects the swimming behavior of zebrafish and can be lethal at high concentrations [4]. So far, ASs have not been included in the European Watch List of substances for Union-wide monitoring in the field of water policy pursuant to Directive 2008/105/EC [6–9] and no Environmental Quality Standards have been set. Therefore, for assessing the potential environmental risk of ASs, Predicted No Effect Concentration (PNEC) values can be used. PNEC values are usually determined on the basis of results from single species laboratory tests or, in a few cases, established effect and/or no-effect concentrations from model ecosystem tests, taking into account adequate assessment factors. The PNEC can be derived using an assessment factor approach. Lowest PNECs can be primarily used for prioritization. The NORMAN Ecotoxicology Database [10] contains PNECs for thousands of contaminants of emerging concern, including ASs, derived from Quantitative Structure Activity Relationships (QSAR) prediction models or obtained experimentally. The NORMAN lowest PNECs for ASs in freshwater and marine water and respective biota are presented in Table 5.2. In-depth ecotoxicological studies are required to fully assess the potential risk of AS occurrence in the environment and to understand the effects of chronic exposure to ASs and their transformation products, in order to evaluate whether regulatory measures for their maximum allowable detected environmental concentration should be established.

5.1.3 Reported or Potential Metabolites and/or Transformation Products

All ASs are highly water-soluble and present high stability. Two representative compounds, ACE and SUC, are excreted unchanged from the human body, remain stable during the water treatment processes, and therefore could be detected in source waters of waterworks and in tap waters in concentrations up to mg/L levels. Many pathways result in the formation of ASs transformation products (TPs), including their metabolism in mammals, abiotic transformation procedures (hydrolysis, and oxidative processes such as ozonation and photo-oxidation), and bio-transformation. The identification of TPs and the investigation of their ecotoxicological behavior are crucial to evaluate the

Table 5.2 NORMAN lowest PNECs for ASs (water in μg/L, biota in μg/kg w.w.).

Artificial sweetener	Species	Fresh water	Marine water	Biota-fresh	Biota-marine
ACE	*Daphnia magna*	72.4	7.24	162	16.2
CYCL (acid)		69	6.9	266	26.6
NEO	*Pimephales promelas*	1.37	0.14	9.43	0.94
ASP	*Selenastrum capricornutum*	39.3	3.93	93.6	9.36
NHDC		0.43	0.043	1.34	0.13
SAC		35.9	3.59	66.3	6.63
SUC		29.7	2.97	24.6	2.46

overall impact of ASs in the environment. Concerning ASs' metabolism, SAC is reported not to be metabolized in mammals, including humans, whereas according to the World Health Organization (WHO), ACE is excreted unchanged, even after repeated administration. ASP is metabolized in the gastrointestinal tract into three metabolites: methanol, phenylalanine, and aspartic acid, which is further metabolized by transamination into oxaloacetate. Although CYCL was initially considered to be excreted unchanged after consumption, thorough investigation revealed the detection of cyclohexylamine metabolite, which can be further metabolized in humans to cyclohexanol and trans-cyclohexane-1,2-diol. Significant differences in the conversion rates and in the time needed to develop the ability to metabolize CYCL have already been reported. Although SUC is mainly excreted unchanged, in the recent literature, two more polar metabolites, glucuronide conjugates of SUC, have been tentatively identified in lower concentrations in contrast with the parent compound [11]. In human bodies, NEO is metabolized into its de-esterified derivative and methanol by esterase, and eliminated in the urine and feces within 72 hours [12]. NHDC is primarily metabolized through deglycosylation to hesperetin dihydrochalcone 4′-β-d-glucoside and subsequently to the aglycon hesperetin dihydrochalcone. The latter is further hydrolyzed to the corresponding 3-(3-hydroxy-4-methoxyphenyl)propionic acid and probably phloroglucinol [13].

Despite the fact that ASs are considered to be resistant to hydrolysis, some intermediate compounds and hydrolysis TPs have been reported in the literature and their formation was proven to be highly affected by pH, time, and temperature. SAC is degraded to 2-sulfobenzoic acid, under acidic conditions, whereas at an alkaline pH, 2-sulfamoylbenzoic acid is formed. Both compounds are present at trace levels in commercially available SAC. Even if negligible degradation was noticed for CYCL during hydrolysis in alkaline and neutral conditions, cyclohexylamine was identified as a TP of CYCL in acidic pH values. The stability of ASP in aqueous solution is generally reported to be limited. Diketopiperazine (DKP) formed by intramolecular cyclization, methanol, L-phenylalanine methyl ester, and aspartylphenylalanine, which is further hydrolyzed to aspartic acid and phenylalanine, have been identified as ASP hydrolysis products. 4-chloro-4-deoxygalactose (4-CG) and 1,6-dichloro-1,6-dideoxyfructose (1,6-DCF) are generated though the hydrolysis of SUC and are stable to further degradation. Acetone, CO_2, ammonium hydrogen sulfate, or amidosulfate are produced as ACE hydrolysis intermediates, after long storage periods under extreme pH conditions [11].

To date, little information is available concerning the transformation of ASs in reaction with oxidants/disinfectants. Seven TPs of SAC, including mono-, di-, and tri-hydroxylated products, as well as TPs that have been formed after cleavage of the C–N bond of sulfonamide group and further oxidation, are formed through the photocatalytic transformation by UV-LED and UV-BL. These compounds were intermediates, as they were proven unstable to long exposure time to UV irradiation, whereas short-chain carboxylic acids such as oxalic, formic, and maleic acid have been reported as end-products of SAC and its TPs during oxidation [14, 15].

Cyclohexanone and amidosulfonic acid have been reported as the main TPs of CYCL during ozonation experiments, while three isomeric TPs of oxocylohexyl sulfamate, having a keto moiety most likely at the cyclohexyl ring of CYC, have been identified [16].

SUC was efficiently removed in reaction with ozone and hydroxyl radicals in alkaline aqueous solutions (pH 7–10). Formaldehyde, glyoxal, and oxalic acid were detected in ozonated samples of SUC. Moreover, several aldehydes, carboxylic acids, and chloride-containing TPs were identified and tentative structures were proposed, among which the unsaturated aldehyde of SUC, sucralose uronic acid and 4-CG [17].

Eleven TPs were structurally elucidated based on their full scan mass spectrometry (MS) and tandem mass spectrometry (MS/MS) spectra information, in aqueous solutions of ACE oxidized by permanganate (Mn(VII)), while ecotoxicity tests revealed higher toxicity of some TPs than ACE [18]. ACE was reported to be highly reactive with ozone, leading to its complete removal, and the parallel formation of acetic acid and ACE dihydroxyacetyl sulfamate TP as main reaction products [16]. Hydroxylation, hydrolysis, oxidation, and (de)hydration appeared to be the main reaction mechanisms during ACE peroxone treatment, generating 15 tentatively identified TPs [19]. Several chlorinated derivatives of ACE, as well as several known and regulated disinfection by-products were reported when investigating the fate of ACE during chlorination experiments, whereas toxicity assessment results revealed increased toxicity for the treated (chlorinated) solution [20]. Additionally, 12 TPs of ACE have been generated and tentatively identified by UV-LED process, while in another study amidosulfonic acid, sulfate, hydroxy acesulfame and iso-acesulfame were identified as ACE TPs by UV irradiation [21].

For the suspect screening of potential ASs metabolites or TPs in environmental samples, a list of candidates may be compiled. Apart from the available information in the scientific literature, additional suspect TPs, produced during human metabolism or any abiotic process, can be generated following *in-silico* strategies. The available software or programs typically require a starting parent compound and employ pattern recognition techniques while using a known list of rules including possible reactions [22].

5.1.4 Occurrence in the Environment

Due to their water solubility and persistence, ASs are present in water bodies and have been reported in various environmental samples. Table 5.3 summarizes ASs occurrence data in different environmental matrices worldwide.

In most cases AS are not metabolized in human bodies and are mainly discharged through urine and feces, reaching wastewater treatment plants (WWTPs) at moderately high concentrations. WWTPs are considered to be the dominant source of ASs in receiving bodies, like surface and/or ground water, since many ASs are not totally eliminated by the applied treatment processes. SUC and ACE appear to be recalcitrant to conventional treatment processes and, therefore, are being discharged unchanged at high concentration levels. On the other hand, CYCL and SAC are typically degraded more than 90% during biological wastewater treatment, hence they are hardly detected or only at low levels in effluents wastewaters. The fate of treated wastewater in the urban environment is an emerging area of research, due to the demand for a sustainable water management. Consequently, the identification of chemical tracers or indicators of wastewater discharges is of imperative research need, in order to evaluate quantitatively to what extent the wastewater has mixed with surrounding groundwater, and to estimate the migration of wastewater to the embedded groundwater. ACE and SUC have been

Table 5.3 Occurrence data of ASs in different environmental matrices worldwide.

Determined sweeteners	Matrix of analysis	Sampling location	Sampling year	Concentration range	Reference
SUC	water – DWTP ($n = 19$)	USA	2010	100–2900 ng/L	[55]
	water ($n = 56$), DW ($n = 43$)	Florida, USA	2012	Reclaimed water: 4.1–18 μg/L, DW: 100–465 ng/L	[26]
	GW ($n = 37$)	California, USA	2018	21.8–285 ng/L	[56]
	WW ($n = 5$), SW ($n = 5$), Reservoir ($n = 5$)	Colorado, USA	2012	WW: 1317–29096, SW: 70.6–484, Reservoir: 32–84 ng/L	[51]
	WW ($n = 28$)	Uppsala, Sweden	2016	200–3000 ng/L	[56]
	WW ($n = 67$)	Narragansett Bay, USA	2016	18–3180 ng/L	[57]
	Leachates ($n = 11$)	Atlantic region, USA	2014	0.296–621 μg/L	[58]
	SW ($n = 17$)	Broward County, USA	2014	8–148 ng/L	[59]
ACE	SW ($n = 10$)	Ontario, Canada	2013	4000–6500 ng/L	[60]
ACE, SUC	GW ($n = 100$)	Alberta, Canada	2013	ACE: 1.2–1534, SUC: 65–541 ng/L	[61]
ACE, CYCL, SAC	WW ($n = 12$)	Giessen, Germany	2018	ACE: 500–1800, CYCL: 16.2–32.4, SAC: 8.0–30.7 ng/L	[62]
	GW ($n = 143$) Leachates ($n = 7$)	Ontario, Canada	2013	Leachates: ACE: 0.5–85, SAC: 0.5–18, GW: ACE: 0.05–32, CYCL: 0.5–14, SAC: 0.5–250 μg/L	[63]
	GW ($n = 24$)	Chungnam, Korea	2017	ACE: 8.1–2695, CYCL: 13.5–860, SAC: 12.7–1156 ng/L	[64]
	Soils, Dust ($n_{total} = 98$)	China	2013	ACE: 0.08–569, CYCL: 0.01–1280, SAC: 0.11–34.7 ng/g	[65]
	SW ($n = 16$), GW ($n = 8$), DW ($n = 18$)	Bassel, Germany	2013	SW: ACE: 619–916, CYCL: 25–51, SAC: 37–76, GW: ACE: 180–524, DW: ACE: 92–144 ng/L	[25]

ACE, SUC, SAC	WW($n = 40$), GW ($n = 48$)	Kumamoto, Japan	2016	INF WW: ACE: 800–4900, SUC: 1960–13000, SAC: 3190–11000, EFF WW: ACE: 4.6–2120, SUC: 2380–11000, SAC: 3190–11000, GW: ACE: 1.3–42, SUC: 12–160, SAC: 0.29–29 ng/L	[66]
	Leachates ($n = 25$)	Northern Greece	2018	Raw: ACE: < LOD–56, SUC: < LOD–36.8, SAC: < LOD–77, Biologically treated: ACE: < LOD–26.1, SUC: < LOD–28.6, SAC: < LOD–14.2, after reversed osmosis: SUC: < LOD, SAC: < LOD, ACE: < LOD–4.5 μg/L	[41]
ACE, CYCL, SAC, SUC	SW ($n = 57$)	South Ontario, Canada	2012	ACE: < 0.008–3.6, CYCL: < 0.003–0.88, SAC: < 0.021–7.2, SUC: < 5–21 μg/L	[67]
	SW ($n = 48$)	6 rivers & 3 lakes, Finland	2011–2012	ACE: 34–9600, CYCL: < 13–210, SAC: < 25–490, SUC: < 200–1000 ng/L	[36]
	SW ($n = 188$)	Southern Ontario, Canada	2008–2015	ACE: < 2.0–225, CYCL: < 3.0–204, SAC: < 2.0–380, SUC: < 20–291 ng/L	[68]
	SW ($n = 3$), GW ($n = 4$)	North China Plain	2015	SW: ACE: 32.2–955, CYCL: < LOD–85.7, SAC: 38.2–212.6, SUC:36.1–315.2, GW: ACE: 5.9–92.5, CYCL: < LOD, SAC: < LOD, SUC: < LOD–811 ng/L	[38]
	SW ($n = 13$), GW ($n = 11$)	Dongjiang River basin, South China	2017	SW: ACE: 80–5300, CYCL: 27.9–22900, SAC: 11.3–12000, SUC:53.9–3640, GW: ACE: 20–4580, CYCL: 4.0–110, SAC: 25–700, SUC: 70–2440 ng/L	[40]
	SW ($n = 5$), GW ($n = 37$)	Aosta Plain, Italy	2017	SW: ACE: 4.93–7.05, CYCL: 0.12–0.32, SAC: 0.53, SUC: 3.21–3.43, GW: ACE: 0.68–9.69, CYCL: 0.14–29.6, SAC: 0.5–5.44, SUC: 3.0 μg/L	[69]
	SW ($n = 16$), S ($n = 16$)	Wuhan, China	2017–2018	SW: ACE: 0.9–19.5, CYCL: 1.5–12, SAC: 0.89–20.6, SUC: 0.86–18, S: ACE: 0.02–7.0, CYCL: 0.8–6.5, SAC: 0.01–8.9, SUC: 3.0–6.2 μg/Kg	[39]
	SW ($n = 13$)	Hong Kong, China	2011–2012	Winter conc.: ACE: 0.12–0.31, CYCL: 0.01–0.27, SAC: 0.06–0.23, SUC: 0.01–0.17, Summer conc.: ACE: 0.01–0.38, CYCL: 0.01–0.19, SAC: 0.02–0.27, SUC: 0.02–0.23 μg/L	[70]

(*Continued*)

Table 5.3 (Contiuned)

Determined sweeteners	Matrix of analysis	Sampling location	Sampling year	Concentration range	Reference
ACE, CYCL, SAC, SUC	WW, GW ($n = 250$)	Canada (Whitehorse, Jasper, Barrie, Hamilton)	2009–2010	WW: ACE: 7–19, SUC: 20–43, SAC: < LOD–14, CYCL: < LOD–22, GW: ACE: 2–9, SUC: 7–24, SAC: < LOD–2, CYCL: < LOD–1 μg/L	[63]
	WW ($n = 317$)	Ontario, Canada	2011–2016	WW: ACE: 0.01–2500, CYCL: 0.03–250, SAC: 1.1–140, SUC: 0.4–360 μg/L	[71]
	WW($n = 20$), SW	Switzerland	2008	INF: ACE: 12–36, CYCL: 10–65, SAC: 4.8–18, SUC:2.0–9.1, EFF: ACE:14–46, CYCL:< LOD-0.82, SAC: < LOD-3.2, SUC:2.0–8.8, DW: ACE:0.02–0.07 μg/L SW: SUC: > LOD only in rivers, ACE: C(rivers) > C(lakes)	[31]
	INF (n = 16) & EFF (n = 32) WW, Storm water runoff (urban (n = 51) & agriculture (n = 19)), Reservoirs (n = 83)	Republic of Singapore	2018	INF WW: ACE: 4647–29552, CYCL: 3114–50362, SAC: 4408–52962, SUC: 2265–12320, EFF WW: ACE: 899–19495, CYCL: 164–2711, SAC: < LOQ-5663, SUC: 2575–9533, Urban stormwater: ACE: 69–2013, CYCL: 26–2098, SAC: 45–9125, SUC: < LOQ-1616, Agricultural stormwater: ACE: 7.8–392.5, CYCL: < LOQ-565, SAC: 17–1028, SUC: < LOQ-223, Reservoirs: ACE: < LOQ-312, CYCL: < LOQ-148, SAC: < LOQ-339, SUC: 1.25–251 ng/L	[33]
	Soils, Feces, Manure, SS, Leachates	Switzerland	2009	Digested SS:ACE: 23–43, CYCL: 0.6–5.5, SAC: 10–16, SUC: 5.4–8.6 μg/L, Piglet feces: SAC: 240–240 mg/Kg, Liquid manure: SAC: 0.3–6.3, ACE: < 0.001–0.048 mg/L	[45]
	Manure ($n = 5$), WW ($n = 5$)	Tianjin, China	2016	Pig manure: ACE: 0.32–12.3, CYCL: 0.2–97.8, SAC: 14–151 mg/kg, INF WW: ACE: 33, CYCL: 52, SAC: 780, SUC: < LOD, EFF WW: ACE: 3.26, CYCL: 0.98, SUC: < LOD, SAC: 5.64 μg/L	[43]

ACE, ASP, CYCL, SAC, SUC	SW ($n = 14$), GW ($n = 14$)	Ganges River Basin, India	2014	SW: ACE: < 1.6–2.92, CYCL: 0.17–0.46, SAC: 1.0–85.4, ASP: < 2.0, SUC: 0.53–24.6, GW: ACE: < 1.6–2.31, CYCL: 0.14–0.50, SAC: < 1.0, ASP: < 2.0, SUC: 0.53–24.64 ng/L	[37]
	WW ($n = 14$), SW ($n = 31$), GW ($n = 12$)	Vietnam, Myanmar, Philippines	2013–2015	INF WW: ACE: 55–580, ASP: 570–3100, CYCL: 87–810, SAC: 7600–13,000, SUC: 750–1500, EFF WW: ACE: 150–210, SAC: 23–36, SUC: 990–1100, SW: ACE: 0.69–7800, ASP: 1700–2600, CYCL: 1.9–1900, SAC: 0.092–17,000, SUC: 340–8300, GW: ACE: 0.30–130, ASP: < LOD, CYCL: 1.8–65, SAC: 0.44–13, SUC: 8.1–47 ng/L	[34]
	WW, SW ($n = 4$)	Cadiz Bay, Spain	2015	INF WW: ACE: 12.1–93.2, ASP: 0.01–0.1, CYCL: 7.24–78.3, SAC: 9.24–50.8, SUC: 2.96–16.6, EFF WW: ACE: 0.48–32.9, ASP: < 0.01–0.09, CYCL: 0.03–0.19, SAC: 0.14–2.37, SUC: 5.68–39.5, SW: ACE: < 0.01–0.05, ASP: < 0.01–0.02, CYCL: < 0.01–0.08, SAC: < 0.01–24, SUC: 0.43–2.95 μg/L	[29]
	WW($n = 12$), SW ($n = 6$)	Tarragona, Spain	2014	INF WW: ACE: 49–149, ASP: 0.05–0.07, CYCL: 0.8–3.7, SAC: 7.4–155, SUC: 12–60, EFF WW: ACE: 11–88, CYCL: 0.05–0.5, SAC: 0.3–0.5, SUC: 12–60, SW: ACE: 0.12–1.62, ASP: 0.01–0.03, CYCL: 0.02–0.08, SAC: < LOD, SUC: 0.04–3.6 μg/L	[30]
	WW, SW, GW ($n = 8$)	Rastatt, Germany	2001–2008	Max conc. in WW: ACE: 54, SAC: 55, ASP: 20,SUC: 50, CYCL: 17, SW: ACE: 2010, SAC: 10, ASP: 20, SUC: 50, CYCL: 245, GW: ACE:2870, SAC: 10, ASP: 20, SUC: 50, CYCL: 1200 ng/L	[35]
	SS ($n = 40$)	South Korea	2013	ACE: 14–201, CYCL: 5.0–49.2, SAC: 7.08–3240, SUC: 21.1–122, ASP: 28.4–5220 ng/g	[72]

(Contiuned)

Table 5.3 (Contiuned)

Determined sweeteners	Matrix of analysis	Sampling location	Sampling year	Concentration range	Reference
ACE, ASP, CYCL, SAC, SUC, NHDC	SW ($n = 18$)	Ontario, Canada	2019	ACE: 0.22–1.41, ASP: 0.04, CYCL: 0.03–0.05, SAC: 0.28–0.31, SUC: 1.6–20.3, NHDC: LOQ µg/L	[46]
	WW ($n = 30$), SW ($n = 42$), GW ($n = 20$)	Republic of Singapore	2012	GW: ACE: < LOQ-95, CYCL: < LOQ-87, SAC: < LOQ-210, NHDC: < LOQ-95, SW: ACE: < 5.0–350, CYCL: < 28–1406, SAC: < 40–810, NHDC: < LOQ, WW: ACE: < 187–75093, ASP: < LOQ-2262, CYCL: < 300–250348, SAC: < 500–135759, SUC: < 100–4719 ng/L	[45]
	WW, SW ($n_{total} = 14$)	Spain	2011	SW: ACE: 1.2–53.7, CYCL: 0,5–15.7, SAC: 0.5–19.7, SUC: 1.1–5.3, INF: ACE: 24.8–26.7, CYCL: 25.9–36.4, SAC: 18.4–21.3, SUC: 3.0–5.3, EFF: ACE: 30.7–32.7, CYCL: 17.2–20.0, SAC: 6.4–9.1, SUC: 16.1–18.1 µg/L	[28]
ACE, ASP, CYCL, SAC, SUC, NHDC	SS ($n = 8$)	Spain		SS: ACE: 36–151, CYCL: 17–326, SAC: 124–255, SUC: 38–628 ng/g	[42]

ACE, ASP, CYCL, SAC, SUC, NHDC, NEO	WW, SW, GW, TW ($n_{total} = 70$)	Tianjin, China	2012	SW: ACE: 9.9, CYCL: 1.5, SAC: 1.3 μg/L, GW: SUC: 9.6, SAC: 52–63, TW: ACE: 680, SUC: 120, SAC: 100, CYCL: 36 ng/L	[27]
	WW ($n = 5$), SW ($n = 5$), GW ($n = 5$), S ($n = 5$)	Belgrade, Serbia	2019	SW: ACE: 19–74, CYCL: 66, SAC: < LOQ-37, SUC: 20–80, NEO: 27, GW: ACE: < LOQ-20, SUC: < LOQ-17, WW: ACE: < 670–7194, ASP: 48–99., CYCL: 1664–4942, SAC: 2857–13460, SUC: 1163–4057, NHDC: < LOD, NEO: 66–129 ng/L, S: ASP: 117–568, NEO: 2.0–56 μg/Kg	[73]
	WW, SW, TW, GW, Sea water ($n_{total} = 13$)	Tianjin, China	2012	SW: ACE: 2730–4650, ASP: 62.5–76.8, CYCL: 205–313, SAC: 193–746, SUC: 287–311, NHDC: 18.9, NEO: 25.4, Sea water: ACE: 875–3550, CYCL: 62.7–252, SAC: 99.7–249, SUC: 53.9–97.4, TW: ACE: 786–824, CYCL: 29–35.4, SAC: 47.5–65.1, SUC: 113–171, GW: ACE: 24.8 & 45.7, ASP: < LOD, CYCL: 6.10 & 50.6, SAC: 11.3 & 80.0, SUC: 12.9, WW: ACE: 11600 (INF) & 10300 (EFF), ASP: 64.7 (INF), CYCL: 2760 (INF) & 133 (EFF), SAC: 0.000103 (INF) & 242 (EFF), SUC: 2060 (INF) & 1850 (EFF), NHDC: 144 (INF), NEO: 36.6 (INF) ng/L	[4]
	Manure ($n = 48$), WW($n = 32$), Compost ($n = 11$), Soils ($n = 11$)	Tianjin, China	2018	Pig manure: ACE: 0.11–7.92, CYCL: 0.03–72.6, SAC:1.26–401, SUC: < LOD, ASP: < LOD–0.11, NEO: 0.02–66.1, NHDC: 1.06–2.85, INF WW: ACE: 0.7–8.68, CYCL: 0.75–59.5, SAC: 43.2–1440, ASP: < LOD, NEO: 0.08–1.21, NHDC: < LOD–4.45, EFF WW: ACE: 0.5–1.25, CYCL: 0.06–4.65, SAC: 1.82–103, ASP: < LOD, NEO: 0.02–0.34, NHDC: < LOD–1.25, Compost: ACE: 0.15–1.84, CYCL: 0.12–9.14, SAC: 3.44–94.2, NEO: 0.04–10.3, Soil: ACE: 1.02–3.12, CYCL: 1.6–8.32, SAC: 8.23–127, NEO: 1.92–9.11 mg/kg	[66]

(*Contiuned*)

Table 5.3 (Contiuned)

Determined sweeteners	Matrix of analysis	Sampling location	Sampling year	Concentration range	Reference
ACE, ASP, CYCL, SAC, SUC, NHDC, ALI, NEO	WW ($n = 12$)	Athens, Greece	2012	INF: ACE: 11.9–25.3, CYCL: 6.04–57.8, SAC: 15.0–46.0, SUC: 6.25–25.4 EFF: ACE: 12.1–27.2, CYCL: 1.3–4.48, SAC: < LOD–0.27, SUC: 6.35–26.7 μg/L	[74]
	Fish ($n = 6$)	Tarragona, Spain	2017	SAC < LOQ (2.5 ng/g) in *Scomber scombrus*	[44]
ACE, ASP, CYCL, DKP, NEO, NHDC, SAC, SUC	WW ($n = 28$), SW ($n = 48$), GW ($n = 23$), DW ($n = 7$)	Switzerland	2011	Tap water: ACE: < LOQ-69, GW: ACE: < LOQ-50, SAC: < LOQ-15, ASP: < LOQ-13, CYCL: < LOQ-21, Lakes: ACE: 36–573, SAC: < LOQ-23, CYCL: < LOQ-36, SUC: < LOQ-81, SW: ACE: < LOQ-7346, SAC: < LOQ-310, CYCL: < LOQ-1959, SUC: < LOQ-949, WW: ACE: 9385–61249, SAC: 86–16201, ASP: < LOQ-11.5, DKP: < LOQ-219, CYCL: < LOQ-19226, SUC: 681–7667 ng/L	[24]

WW: Wastewater; SW: Surface water; DW: Drinking water; TW: Tap water; GW: Groundwater; SS: Sewage sludge; INF WW: Influent wastewater; EFF: Effluent wastewater; DWTP: Drinking water treatment plant identified as ideal tracers in groundwater affected by domestic wastewater due to their hydrophilicity, anthropogenic sources, and ubiquity in groundwater. Moreover, both compounds can be used to separate human and animal waste contributions, due to their exclusive consumption by humans, and indicate wastewater existence, dilution, and transformation in surface and ground waters [23].

Several studies worldwide have so far reported the occurrence of ASs in different environmental compartments, including drinking water, ground water, surface water, and wastewater, leachates as well as sewage sludge, manure, compost, and biota. Berset and Ochsenbein [24] investigated the presence of AS levels in several environmental matrices, including effluent wastewater, surface water (rivers, lakes), groundwater, and tap water. ACE was the most abundant and most frequently detected AS and, along with SAC, CYCL, and SUC, with concentration levels up to 61.2, 16.2, 19.2, and 7.6 μg/L, respectively. ACE was also detected in tap water at <69 ng/L levels [24]. Similar results for ACE (92–144 ng/L) were found in drinking water from Bassel (Germany) by Ens et al. [25] and in Florida (USA) at concentration levels ranging from 100 to 465 ng/L [26]. Several AS were also detected in drinking water in Tianjin (China) at the following concentrations; ACE: 0.68, SUC: 0.12, SAC: 0.10, and CYCL: 36 μg/L [27]. ACE levels up to 70 ng/L were also reported in groundwater and tap water samples from Switzerland [24]. Moreover, ACE and SUC were also determined at concentrations up to 7 and 2.4 μg/L, respectively, in effluents of drinking water treatment plants [28].

Due to their high human consumption, ASs are typically detected in wastewater and affected surface water worldwide. Ordóñez et al. [28]. reported the detection of ACE, CYCL, and SUC in wastewater and surface water samples from Spain at concentration levels ranging from 0.4 to 53.7 μg/L. According to the literature, concentrations at the same order of magnitude were also found in wastewater (influents) from Spain and especially in Cadiz Bay (ACE: 12.1–93.2, CYCL: 7.24–78.3, SAC: 9.24–50.8, SUC: 2.96–16.6 μg/L) [29] and Tarragona (ACE: 49–149, CYCL: 0.8–3.7, SAC: 7.4–155, SUC: 12–60 μg/L) [30]. Comparable concentration levels were also detected in wastewater from Switzerland (ACE: 12–36, CYCL: 10–65, SAC: 4.8–18, SUC: 2.0–9.1 μg/L) [31], Canada (ACE:7–19, SUC: 20–43, SAC: 0–14, CYCL: 0–22 μg/L) [32], and Singapore (ACE: 0.899–194, CYC: 0.164–2.71, SAC: < LOQ–5.66, SUC: 2.57–9.53 μg/L) [33], whereas rather lower concentrations for most ASs were detected in wastewater from Vietnam, Myanmar, and Philippines (ACE: 0.055–0.580, ASP: 0.570–3.10, CYCL: 0.087–0.810, SAC: 7.60–13.0, SUR: 0.750–1.50 μg/L) [34]. Diketopiperazine (DKP), the metabolite of ASP, was also reported in wastewater at concentration levels up to 200 ng/L and is proven to be unstable in environmental water samples, through its further transformation to unknown compounds or total mineralization. For preventing the formation of DKP, the pH of the samples should be adjusted to 4.3. As far as surface waters are concerned, at a European level, ASs were detected in the rivers Sars and Sarela in Spain at significantly high concentration levels (ACE: 1.2–53.7, CYCL: 0,5–15.7, SAC: 0.5–19.7, SUC:1.1–5.3 μg/L) [28], in Rastatt, Germany (ACE: 2.01, SAC: 0.010, ASP: 0.020, SUC: 0.050, CYCL: 0.245 μg/L) [35], in Finnish surface water, including rivers (Aura, Kokemäki, Kymi, Porvoo, Vantaa, and Vuoksi) and lakes (Tuusula, Vanajavesi, and Päijänne) (ACE: 0.034–9.60, CYCL: < 0.013–0.210, SAC: < 0.025–0.490, SUC: < 0.200–1.00 μg//L) [36], and in Belgrade, Serbia (ACE: 0.019–0.074, CYCL: 0.066, SAC: < LOQ–0.037, SUC: 0.020–0.080 μg/L) [37]. ASs were also reported in North China Plain (ACE: 32.2–955, CYCL: below the limit of detection (LOD)–85.7, SAC: 38.2–213, SUC:36.1–315 μg/L) [38], in Wuhan, China (ACE: 0.9–19.5, CYCL: 1.5–12, SAC: 0.89–20.6, SUC: 0.86–18 μg/L) [39], and in Dongjiang River basin, South China (ACE: 0.080–5.30, CYCL: 0.028–22.9, SAC: 0.011–12.0, SUC: 0.053–3.64 μg/L) [40].

Nika et al. [41] evaluated the occurrence of ASs among other compounds in raw and treated landfill leachates after biological treatment and/or application of reverse using HRMS analysis. The concentration ranges in raw leachate samples were ACE: <LOD–56 μg/L, SUC: <LOD–36.8 μg/L, SAC: <LOD–77 μg/L. After biological treatment, maximum values of the AS concentration in the leachate samples were decreased (<LOD–26.1 μg/L, SUC: <LOD–28.6 μg/L, SAC: <LOD–14.2 μg/L). Finally, after the application of the reverse osmosis process, the concentration ranges were as follows; ACE: <LOD–4.5 μg/L, SUC: <LOD, SAC: <LOD. The results showed that biological treatment appears to be effective for reducing the concentration of the three ASs; however, reverse osmosis is the ideal process for accomplishing removal efficiency higher than 98% [41].

SUC has been mostly detected in surface waters from Western Europe and Scandinavia up to 1 μg/L, whereas lower concentrations have been reported in Germany and Eastern Europe. In another study, Fu et al. [39] reported the variation in concentration of the ASs in surface water and sediments under different weather conditions in China. SUC, ACE, SAC, and CYCL were the dominant compounds during summer with mean concentrations at 3.11 ± 3.26, 1.10 ± 0.62, 0.38 ± 0.58, and 0.21 ± 0.50 μg/L in surface water, respectively. During the winter, the mean concentrations were at 3.78 ± 4.26, 1.36 ± 0.62, 0.48 ± 0.69, and 0.29 ± 0.52 μg/L for the same sample sites, respectively, indicating rather higher levels for SUC and ACE. The same trend of increasing concentrations was also observed in sediments. A possible explanation for this variation maybe the heavy rain during the summer, diluting the concentrations of ASs in surface water. Another potential factor might be the high summer temperatures that may have accelerated the degradation or transformation of ASs [39].

Ordóñez et al. [42] investigated the accumulation of AS in sewage sludge samples taken from different WWTPs in the northwestern region of Spain. ASP and NHDC were not found in any of the analyzed samples, whereas ACE, CYCL, and SUC were omnipresent in sewage sludges. ACE concentration ranged from 50 to 63 and from 32 to 151 ng/g in primary and secondary sludge, respectively. CYCL was determined at considerably higher concentration in primary (175–326 ng/g) than in secondary sludge (17–76 ng/g), whereas SUC followed the opposite detection pattern, as it was found at higher concentrations in secondary (68–628 ng/g) than in primary sludge (38–59 ng/g). SAC was detected in all primary sludge samples at concentrations levels ranging from 141 to 255 ng/g, but only in one sample of the secondary sludge at 124 ng/g. Typically, there is a 75–95% water content in the original sludge, before the freeze-drying. Considering that the concentration of the same analytes from the same geographical area range from 5 to 35 μg/L levels in wastewater, it can be supported that absorption of the analytes to sludge is not significant. The most possible scenario is that most of ASs residues in the sludge, have remained from the evaporation of water, taking into account their strong polar character. Nevertheless, the re-use of sludge as a fertilizer in compost formulations should not be neglected as a potential source of these chemicals in soils and aquifers [28, 42]. Pig farms have also been proven as a potential source of ASs distribution in the environment as a result of their extensive use as feed additives for taste improving of pig feed and the use of manure as fertilizers. Ling Ma et al. [43] reported the occurrence of SAC, CYCL, and ACE in several matrices (pig feed, manure, wastewater effluent, soil, and plants) from a pig farm and the neighboring farmland that received fermented manure fertilizer and irrigated with wastewater. The average levels of ASs in pig feed were 66, 13, and 4 mg/kg for SAC, CYCL, and ACE, respectively, whereas SUC was below the limit of detection. AS, and especially SAC, were detected in considerable quantities in pig manure and soils, supporting the idea that pig manure used as a fertilizer can be a significant source of soil contamination and accumulation for ASs. Moreover, SAC, CYCL, and ACE were substantially removed from the on-farm WWTP with removal ratios higher than 90%, supporting that the contribution of wastewater in soils irrigated with wastewater is considerably lower than that of pig manure [42].

Núñez et al. [43] studied the occurrence of selected sweeteners in different fish species with different lipid content, collected from Tarragona market and Ebro River in Spain. None of the tested sweeteners was detected apart from SAC, which was detected at trace levels below the quantification limit (12.5 ng/g) [43].

Apart from ACE, SAC, SUC, CYCL, and ASP, other AS such as NHDC, NEO, and ALI are less frequently been investigated and their detected environmental levels are significantly lower than those of the most popular ASs.

5.2 Sample Preparation and Collection

5.2.1 Protocols for Collecting and Preparing Samples

Sample stability during transport and storage under laboratory conditions may significantly affect the determination of organic contaminants in environmental matrices, such as water. Although the ISO 5667-3 guide comprehensively describes different processes of sample transport, treatment, preservation processes, and storage conditions of water samples for several classes of pollutants (such as hydrocarbons, pesticides, etc.) (International Organization for Standardization, 2018), there are no established recommendations for water samples intended for ASs analysis, so far [24]. As reported in the literature, the commonly used conditions of collecting and storing environmental samples are followed for the determination of ASs. In most studies, water samples are collected either in pre-cleaned high-density plastic bottles or amber flasks and transferred to the laboratory at 4°C. If not analyzed immediately, samples are usually acidified to pH 3 with HCl and stored at −20°C until analysis. The addition of preservative agent is not reported in the literature [4]. After sampling or prior to the extraction step, water samples (especially wastewater) are filtered through glass fiber or cellulose nitrate membrane filters (0.22–0.45 μm) [4, 23]. Solid samples with high % water content, such as sewage sludge and fish, are being transferred frozen to the laboratory, where they are lyophilized, ground to obtain a homogeneous powder, sieved, and stored in freezers or fridge until analysis [42, 44].

5.2.2 Sample Extraction and Clean-up

Considering the low concentrations of ASs in environmental samples (at ppt to ppb levels) and the complex matrix of analysis (especially for wastewater and sewage sludge), the pre-concentration of analytes, as well as a clean-up step are recommended prior to instrumental analysis. The technique of choice for the determination of ASs should be taken into consideration during the sample extraction and clean-up step for selecting compatible solvents and pre-concentration factors to be in line with the sensitivity of the method. The most prevalent extraction technique for water samples is solid-phase extraction (SPE), both off- and on-line, whereas solid-phase microextraction (SPME) was also tested as an alternative. In the case of solid samples, liquid extraction and pressurized liquid extraction (PLE) are mainly applied in the literature.

Currently, SPE is the most frequently used technique for sample pre-treatment due to its unique advantages in enriching the final extract with the analytes of interest and removing interferences that may vigorously restrain the chromatographic separation and ionization efficiency in mass spectrometry, due to high matrix effect. Table 5.4 summarizes the most frequently used sorbent materials in SPE sample clean-up and extraction solvents. The predominant sorbent used in the literature for the determination of ASs is the polymeric reversed-phase material Oasis HLB, and methanolic solutions are most frequently used during elution.

For reaching the required method LODs, high volumes of water samples are being loaded into the SPE cartridges and a pre-concentration of 100–1000 times is usually applied. For off-line SPE the water samples' volumes range from 50 to 200 mL for wastewater and from 200 to 1000 mL for surface or underground samples. However, cartridge breakthrough phenomena or increased matrix effect should be thoroughly evaluated when large volumes are used.

The efficiency of the SPE and, consequently, the recoveries of the determined contaminants, are mainly affected by the absorbent material, the pH of the sample and the elution solvents [3]. For multi-residue and multi-analyte methods, the optimization of SPE conditions can be really challenging, considering the wide spectrum of analytes with different physico-chemical characteristics (such as their pK_a values varying from 0.9 to 4 and the polarity ranging from $\log K_{ow}$ 0 to 4.9). ASs contain nitrogen and/or oxygen in their molecular structure, belong to three different chemical classes (sulfamates, peptides, and carbohydrate derivatives), and cover a wide range of polarities. Polymer and silica-based sorbents are the most frequently used for ASs analysis. Modified sorbents with hydrophobic and/or hydrophilic chemical moieties are also used, in order to give either reversed-phase (e.g. Strata X-RP, Oasis HLB) or both reversed-phase and ion-exchange functionality (mixed-mode sorbents, Oasis MAX, and Strata X-AW). For sorbents with hydrophilic chemical fractions, the retention of the analytes is typically favored and, therefore, better recoveries are obtained. Strata X-RP and Oasis HLB sorbents consist of poly divinylbenzene-N-vinylpyrrolidone and are preferred during the analysis of ASs, due to their ability to apply hydrophobic, p–p, and even hydrophilic intermolecular forces to these compounds. On the contrary, sorbents with poly divinylbenzene-ethylvinylbenzene groups such as styrene-divinylbenzene (SDB) copolymers are not preferred in the clean-up step, because hydrophobic interactions with ASs are not possible. For ionizable target compounds, an alternative would be the use of ion-exchange sorbents having opposite charge from the determined analytes. A thorough investigation and evaluation of SPE sorbents performance in ASs determination is crucial to achieve low LODs, high precision, low matrix effects, and high repeatability. Conventional off-line

Table 5.4 Most frequently used SPE cartridges and elution solvents for the extraction of ASs in environmental samples.

Artificial sweetener	Matrix of analysis	SPE cartridge	Elution solvent	Ref.
ACE, ASP, CYCL, DKP, NEO, NHDC, SAC, SUC	TW, GW, SW, WW	Oasis HLB (*500 mg, 6 cc*)	MeOH & water	[24]
ACE, CYCL, SAC, SUC	GW, SW	Ethyl vinyl benzene divinyl benzene polymer (*200 mg, 3 cc*)	MeOH	[38]
	GW, SW	HR-X (*500 mg, 6 cc*)	MeOH	[71]
	SW	Oasis WAX (*150 mg, 6 cc*)	1% NH_3/ACN (*ACE, CYCL, SAC*), MeOH (*SUC*)	[36]
ACE, ASP, CYCL, SAC, SUC	SW, WW	Oasis HLB (*500 mg, 6 cc*)	MeOH	[29]
			MeOH & MeOH/ NH_4OH (*95/5*)	[30]
	SS	C_{18} endcapped (*1 g, 6 cc*)	MeOH & MeOH/ Ace/ AcOEt (*2/2/1*)	[72]
ACE, ASP, CYCL, SAC, SUC, NHDC	SW, WW	Oasis HLB (*500 mg, 6 cc*)	MeOH	[28]
	SS			[42]
	GW, SW, WW	HR-X (*500 mg, 6 cc*)		[45]
ACE, ASP, CYCL, SAC, SUC, NHDC, NEO	Soil, Compost, Manure, WW	Poly-Sery PWAX (*60 mg, 3 cc*)	MeOH/TRIS & 5% NH_3 (*pH 11*)	[43, 75]
	Fish tissue	Oasis HLB (*500 mg, 6 cc*)	MeOH & MeOH/ NH_4OH (*95/5*)	[44]
	TW, GW, SW, WW	Poly-Sery PWAX (*60 mg, 3 cc*)	MeOH with TRIS	[4]
	TW, WW	Strata X RP (*200 mg, 6 cc*)	MeOH	[74]

WW: wastewater; SW: Surface water; TW: Tap water; GW: Groundwater; SS: Sewage sludge.

SPE demands large sample volumes and is too laborious as it includes several steps, such as pH adjustment, buffering of water samples, solvent evaporation, and reconstitution. Consequently, for large-scale environmental studies, sample pre-treatment may be extremely time-consuming.

Ordóñez et al. [28] evaluated different SPE cartridges for the extraction of ASs: polymeric reversed-phase (Oasis HLB, Env +, Plexa, and Strata X), and mixed-mode with either weak (Oasis WAX) or strong anionic-exchange (Oasis MAX and Plexa PAX) sorbents. Among them, reversed-phase sorbents, especially Oasis HLB and Strata X, presented the best performance. Oasis HLB provided good accuracy (recoveries: 73–112%), precision (RSD $<$ 10%), and limits of quantification (LOQ: 0.01–0.5 μg/L), whereas mixed-mode sorbents provided low recoveries (below 20%) for some compounds, such as ASP with Oasis WAX and Plexa PAX, ACE with Oasis MAX and Plexa PAX, and CYCL and SAC with Oasis MAX. Strong anionic-exchange mixed-mode cartridges like Oasis MAX (for ACE, CYCL, and SAC) and Plexa PAX (for ACE), provided low recoveries which are mainly attributed to the incomplete elution due to the strong interaction of the most acidic analytes with the anionic-exchange moieties. Thus, the elution of the analytes from the Oasis MAX and Plexa PAX cartridges was re-evaluated using a higher formic acid content in the eluent. At 5% formic acid, only the recovery of CYCL was slightly improved, but was still lower than 60% [28]. Common conclusions were drawn by Tran et al. [45], who investigated five different SPE cartridges, including HR-X, HLB Plus, HR-WAX, C18 endcapped, and ElutPlexa. Plexa and HR-XAW showed low recoveries for ACE, ASP, and CYCL. HR-X cartridge was found to be the most effective for most of the investigated analytes. This study also revealed that ACE, CYCL, ASP, SUC, and SAC presented very low recoveries either at basic or neutral pH values. A potential cause might be the poor cartridge retention and a possible explanation could be the pH-induced molecular conformation changes, which would increase or decrease the formation of ions in the electrospray ionization (ESI) source [45]. The adjustment of pH in SPE is critical for its overall performance. Therefore, based on the physicochemical parameters of the determined analytes and the used sorbent material, the pH of the sample and the elution solvents should be investigated in depth. Gan et al. [4] investigated the influence of sample pH on ASs recoveries in SPE procedure using CNW Poly-Sery PWAX cartridges in tap water samples spiked at a concentration of 1 ng/mL and at pH range of 3–10. Although the sample pH had little influence on the recoveries of the seven ASs within the pH range of 3–9, at pH 10 the recoveries of ASP, CYCL, and ACE decreased significantly, which can be explained by the pK_a of PWAX (approximate 9.8). Relatively lower recoveries for all the investigated ASs were observed with Waters Oasis WAX SPE cartridges compared to WAX materials by other vendors, such as CNW Poly-Sery PWAX, which are packed with amino functionalized styrene-divinylbenzene copolymer. In the same study, the authors also investigated ten disposable SPE cartridges and five wash buffers i.e. acetic acid-sodium acetate buffer solutions at pH 3.6, 4.0, 4.5, and 6.0, and 1 mmol/L TRIS water solution, while Milli-Q water was used as a reference. Methanol, methanol containing 2% ammonia hydroxide (v/v), and methanol containing 1 mm TRIS were compared as elution solvents. For all the investigated ASs, satisfactory recoveries were obtained only from Waters Oasis WAX, CNW Poly-Sery PWAX, and CNW Poly-SeryPSD, among the tested cartridges under appropriate conditions. CNW Poly-Sery PSD was able to completely retain all tested ASs when using an acetic acid-sodium acetate buffer at pH levels of 3.6, 4.0, and 4.5. However, the recovery of ACE using the CNW Poly-Sery PSD cartridge was relatively low (approximately 50%). The most satisfactory recoveries were obtained with CNW Poly-Sery PWAX when using 25 mm acetic acid-sodium acetate buffer at pH of 4 as wash buffer and methanol containing 1 mm TRIS as elution solvent. Under these conditions, the recoveries for all analyzed ASs were higher than 95% (RSD < 5%), except for ASP, which was recovered up to 77%. This may be supported by its strong adsorption on the sorbent and the low efficiency of the elution solvent and wash buffer [4]. In order to remove organic matter and other interferences, and thus reduce the matrix effect, a washing step is applied in SPE procedure. However, potential losses of the analytes of interest during this step should be taken into account for selecting the suitable washing solution. When using organic wash solvents in non-selective sorbents (either silica or polymeric-based), a significant decrease in the recovery rate is observed as a result of the disturbance of the non-specific interactions within the target analytes and the sorbent. On the other hand, through aqueous clean-up, salts and other matrix ions are removed, but the high organic matter content normally found in complex samples is not. Addition of organic modifiers to the aqueous solution up to 10%, may result to the partial elution (losses) of the polar compounds and it has limited removal effectiveness. Actually, a washing solution based on 10% MeOH in water appears to be the best compromising between washing of interferences and elution for the determination of ASs.

Although SPE is mainly performed off-line, it has the potential to operate in on-line mode coupling to liquid chromatography (LC). Sorbents used for on-line SPE included both traditional (alkyl-bonded silicas and polymers) and novel (restricted access materials (RAMs), molecularly imprinted synthetic polymers (MIPs), and immobilized receptors or antibody (immune sorbents) materials. This technique has many advantages such as high sensitivity and selectivity, minimum sample manipulation, and automation. Moreover, on-line SPE presents significantly good performance criteria in terms of accuracy, reproducibility, robustness, reliability of results, and capacity for multi-analyte determination. Batchu et al. [26] developed and validated an automated on-line SPE LC-APCI-MS/MS method for the determination of SUC at trace levels in drinking and reclaimed water. The SPE pre-concentration column was a Hyper Sep Retain PEP (20 mm × 3.0 mm I.D)). An HTC-PAL autosampler (Thermo Scientific, San Jose, CA, USA) was set to perform up to 10 mL injections. The automated on-line SPE clean-up and pre-concentration step was performed using only 10 mL of untreated water samples. The method detection limits were 8.5 ng/L and 2.7 μg/L in drinking and reclaimed waters, respectively, whereas the recoveries in the tested matrices ranged from 85 to 113% [26].

An alternative to SPE, is the performance of the fully automated sample preparation technique of SPME, which has already been applied in environmental analysis, as well as for the determination of ASs. In the SPME workflow, an extraction phase (coating) is exposed to the sample for the passive extraction of analytes. Although pre-equilibrium extraction can be performed, maximum sensitivity is achieved when analyte equilibrium is reached between the sample matrix and the extraction phase. Recently, thin-film solid-phase microextraction (TF-SPME) was introduced as another geometry of SPME, in which the thin-film configuration provides a higher surface-area to volume ratio of the extraction phase in comparison to the conventional SPME fiber, which results in higher extraction efficiency and faster equilibration time due to rapid diffusion through the thin layer of coating. A TF-SPME method coupled with LC can be fully automated by using a robotic autosampler designed to perform analysis of up to hundreds of samples. In this set-up, a solid extraction phase is immobilized by polyacrylonitrile (PAN) deposited onto metal blades. The coated blades are used to extract small organic compounds from samples; after that, desorption is carried out in an appropriate solvent. TF-SPME methods have been successfully applied for the determination of ASs. Cárdenas-Soracá et al. [46] tested three different coatings (Oasis HLB, Oasis HLB:PSDVB-WAX(50 : 50), and HLB-WAX) and three ratios of water/methanol (9/1, 7/3, and 6/4). The solid extraction phase was dispersed in PAN glue for immobilization of particles onto the stainless steel blade. The HLB-WAX showed the best performance. For WAX coatings, an anion exchanger was needed to desorb the analytes; hence, desorption solvents contained 0.5% ammonium formate (NH_4HCO_2). The composition of the tested desorption solvents was chosen according to the polarity of the compounds and the compatibility of the solvents with the LC-MS/MS conditions (to prevent tailing and bad peak shape). The developed method was simple, reproducible, environmentally friendly, and capable of quantifying ASs at an accuracy comparable with those obtained *via* traditional SPE methods. The remarkable low method LOQs (0.02–0.15 ng/mL), enabled the detection of ASs in surface water samples, at ultra-trace levels [46].

Regarding the pre-treatment of solid samples, PLE is the most prevalent technique, following in-cell clean-up or followed by SPE extraction. During the in-cell clean-up, a sorbent (most frequently alumina, silica, or florisil) instead of inactive material is placed into the extraction cell with the sample, for the retention of the interfering substances. Alternatively, the PLE extract is collected and loaded to a SPE cartridge, with appropriate sorbent already mentioned for the application of SPE for ASs. Following this approach, complete or partial evaporation of the PLE extract is required, which is much more laborious and time-consuming. However, the combination of PLE and SPE steps is currently the most frequently followed approach, considering the complexity of solid environmental samples. An analytical method for the determination of six ASs in sewage sludge has been developed by Ordóñez et al. [42], based on PLE followed by SPE. The PLE extraction was implemented in a single cycle of 21 minutes static time at 80°C using an aqueous 500 mm formate buffer, adjusted to pH 3.5, as extraction solvent. The acquired water extract was further concentrated by SPE, using Oasis HLB (200 mg) cartridges. The PLE extract (adjusted to pH 2) was loaded into the cartridge and the analytes were eluted with 2 mL of MeOH. Then, the SPE extract was evaporated to dryness under a N_2 stream and reconstituted with MeOH/water (25/75) to 0.5 mL. Satisfactory recoveries were obtained for all the compounds ranging from 94% to 116%. This confirms that the SPE protocol performs well also with sludge extracts [42].

5.3 Determination of Artificial Sweeteners

5.3.1 Development of the Instrumental Method

The development of highly sensitive and accurate analytical methods is crucial for the simultaneous determination of ASs, at trace levels, in complex environmental samples. To achieve the highest analytical sensitivity, liquid chromatography coupled to mass spectrometric techniques is mainly performed. The instrumental LOD and signal-to-noise value significantly affect the sample preparation method that should be followed (pre-concentration factor) in order to enable the determination of ASs in various environmental samples. The optimization and validation of parameters (such as the mobile phase composition and elution program, chromatographic columns, and analytic operational parameters of ESI and MS conditions (such as the declustering potential, collision energy, and cell exit potential)) are fundamental for the achievement of sufficient separation and lower detection limits of ASs.

5.3.1.1 Chromatography

The efficiency of a chromatographic separation is a measure of the dispersion of the analyte band as it travels through the chromatographic system. Several parameters contribute to column efficiency: flow rate, column dimensions, substrate particle size, and extra-column effects (including instrumental and plumbing effects). As regards the chromatographic techniques, mainly liquid chromatography (LC) and to a lesser extent gas chromatography (GC) have been used to determine ASs in several environmental matrices. Reversed-phase liquid chromatography (RP-LC), using C_{18} or C_8 stationary phases, is the predominant technique for the determination of ASs. Typically, the mobile phase combines water and an organic solvent (methanol or acetonitrile), using buffers such as ammonium acetate, acetic acid, or formic acid, considering an efficient chromatographic separation and ESI sensitivity. Through solvent composition, pH adjustment, column temperature, and flow rate optimization, RP can provide good peak shape and enable the separation of co-eluted compounds in multi-analyte methods and the separation of the compounds of interest from matrix interferences responsible for ion suppression. In cases where there is insufficient retention of polar or ionic compounds, an alternative analytical approach may be the use of hydrophilic interaction chromatography (HILIC). Chromatographic separations may be faster by increasing the flow rate of the mobile phase, decreasing the length of the column, reducing the stationary phase particle diameter, or increasing the temperature of analysis. The organic solvent and the mobile phase additives can play a crucial role in chromatographic separation and detection sensitivity of RP-LC-MS/MS methods. The presence of additives might either enhance or suppress the signal of the determined analytes, which is highly affected by the applied concentration level. Selected liquid chromatographic conditions reported in the literature for the determination of ASs are listed in Table 5.5.

RP-LC is the most frequently applied chromatographic technique for the determination of ASs. Several studies investigated the optimization of chromatographic parameters for the efficient separation of ASs and the achievement of low method limits of detection. Tran et al. [45] evaluated the effects of different organic mobile phases and additives on the separation efficiency and detection sensitivity of ASs. Three organic solvents were examined, including methanol (MeOH), acetonitrile (ACN), and a mixture of MeOH/ACN (50/50). MeOH presented notably higher signal intensity for all tested ASs compared to the rest solvents and was therefore selected for further investigation. In the next step, the addition of ammonium acetate (NH_4Ac) and formic acid (FA), at different concentration levels, into both water and organic phase was evaluated. Based on the results, the addition of both additives resulted in better peak shapes of the analytes. However, the presence of FA in the mobile phase seemed to suppress the signal of most compounds of interest. The addition of NH_4Ac, at a concentration of 2 mM, proved to enhance the signal intensity of most ASs and particularly to improve the chromatographic separation efficiency and peak shapes of all determined analytes [45]. For the separation of seven AS, Gan et al. [4] tested two chromatographic columns, a Zorbax Eclipse XDB-C_8 (4.6 mm × 150 mm, 5 μm) and an Athena C_{18}-WP (4.6 mm × 150 mm, 3 μm), which is compatible with 100% aqueous solutions, and the addition of TRIS in the elution solvents. The mobile phase consisted of water and ACN, both containing 5 mM ammonium acetate and 1 mM TRIS. Gradient elution was carried out at a flow rate of 0.4 mL/min, and the system was equilibrated for 8 minutes before the next injection. Both columns provided excellent retentions at 30°C, however, Athena C_{18}-WP column presented better separation for all tested analytes. As an ion-pair reagent, TRIS can hold

Table 5.5 Stationary and mobile phases commonly reported in the literature for the determination of ASs by LC.

Artificial sweeteners	Stationary phase	Mobile phases	Ref.
ACE, SUC	Dikma C_{18} (*3 mm × 100, 3 μm*)	(A) water, (B) MeOH (*both cont. 0.1% formic acid*)	[61]
ACE, CYCL, SAC, SUC	Athena C_{18}-WP (*4.6 × 150 mm, 3 μm*)	(A) water, (B) ACN (*both 5 mM Amm. Ac. and 1 mM TRIS*)	[43]
	Kinetex C_{18} (*4.6 × 100, 2.6 μm*)	(A) water (*5 mM Amm. Ac.*), (B) MeOH	[40]
	Acquity UPLC HSS T3 (*1.8 × 50 mm, 2.1 μm*)	(A) water, (B) MeOH (*both cont. 2 mM Amm. Ac.*)	[36]
ACE, ASP, CYCL, SAC, SUC	ZORBAX Eclipse XDB-C_{18} (*2.1 × 150 mm, 3.5 μm*)	(A) water (*5 mM of dibutylammonium acetate*), (B) ACN/water (*4/1*)	[34]
	Ascentis Express RP-amide (*2.1 × 100 mm, 2.7 μm*)	(A) acidified water (*formic acid, pH 2.5*), (B) ACN	[30]
	Hypersil™ GOLD (*2.1 × 150 mm, 3 μm*)	(A) acidified water (*0.1% formic acid*), (B) MeOH	[72]
ACE, ASP, CYCL, NHDC, SAC, SUC	Agilent ZORBAX SB-C_{18} (*2.1 × 150 mm, 3.5 μm*)	(A) water, (B) MeOH (*both cont. 2 mM Amm. Ac.*)	[45]
	Luna C_{18} (*2.0 × 100 mm, 3 μm*)	(A) water, (B) MeOH (*both cont. 5 mM Amm. Ac.*)	[42]
ACE, ASP, CYCL, NEO, NHDC, SAC, SUC	Athena C_{18}-WP (*4.6 × 150 mm, 3 μm*)	(A) water, (B) ACN (*both cont. 5 mM Amm. Ac. and 1 mM TRIS*)	[4]
ACE, ALI, ASP, CYCL, NEO, NHDC, SAC, SUC	XBridge HILIC (*2.1 × 150 mm, 3.5 mm*)	(A) water/ACN (*1/4*), (B) water/ MeOH/ACN (*15/5/80*), (*both cont. 5 mM Amm. Formate*)	[74]
	Zorbax Eclipse XDB-C_8 (*4.6 × 150 mm, 5 μm*)	(A) acidified water (*formic acid, pH 2.5*), (B) ACN	[44]
ACE, ASP, CYCL, DKP, NEO, NHDC, SAC, SUC	Atlantis dC_{18} (*2.1 × 150 mm, 3 μm*)	(A) water, (B) MeOH (*both cont. 1 mM Amm. Ac.*)	[24]

anionic compounds in a RP column. The signal was enhanced by 30–290% with the post-column injection of TRIS buffer, since it can facilitate deprotonation of weakly acidic analytes and enhance the response of some ASs. Moreover, better separation was achieved using 1 mM TRIS as an ion-pair reagent in mobile phase; while without TRIS, peaks were broader and were poorly resolved [4].

Although most of published methods used RP columns for the separation of ASs, other LC retention mechanisms, such as hydrophilic interaction (HILIC), have also been investigated considering the fact that ASs are polar contaminants. Ordóñez et al. [28] investigated the efficiency of two chromatographic columns for the separation of ACE, ASP, CYCL, SAC, SUC, and NHDC in wastewater and surface water samples. The first column was a 100 mm × 2.0 mm Luna C_{18} (3 μm, 100 Å) and the second one a 100 mm × 2.0 mm (3 μm, 200 Å) Luna HILIC, both protected with compatible pre-columns. In both cases, the separation was performed at 40°C, using a flow rate of 0.2 mL/min and an injection volume of 10 μL. For the C_{18} column, a dual eluent system of water (A) and methanol (B), both with 5 mM ammonium acetate, was used, while for the separation in HILIC, different modifiers (ammonium acetate and formate) and pH values (between 3.5 and 6.5) of the mobile phase were considered. Based on the method development results, water (A) and acetonitrile/water (95/5) (B), both containing 5 mM ammonium acetate, adjusted to pH 3.5, provided the best separation efficiency. Overall, RP-LC separation provided a significantly better performance than HILIC, with matrix effect values in the 77–103% range, while in HILIC high signal suppression for SAC and SUC (matrix effect ≤ 10%) were observed [28].

SUC has lower sensitivity in LC, compared to other ASs, due the hydroxyl group which increases the difficulties during its detection in trace levels in environmental samples. An alternative analytical option for SUC determination is the application of GC-MS/MS analysis through derivatization and especially silylation (the replacement of active hydrogens (e.g. –OH, =NH, $-NH_2$, –SH, –COOH) with silyl groups). The silyl derivatives have, relative to their parent molecules, decreased polarity, increased volatility, and increased thermal stability, making this derivatization procedure favorable for GC-MS analysis. In addition to the primary purpose of increasing suitability for separation by GC, silylation can also improve mass spectrometric identification by producing distinctive fragmentation patterns. However, the derivatization step prior to GC-MS analysis is a time-consuming sample preparation step [47]. Qiu et al. [48] determined SUC in Splenda® and performed silylation (trimethylsilyl) prior to analysis, while Voss et al. [49] recently performed silylation followed by GC-analysis for the determination of SUC in groundwater samples. The results showed good accuracy and recovery and yielded a method detection limit of 21.8 ng/L. Moreover, the use of deuterated internal standard ensured method accuracy and recovery through internal standard calibration and significantly contributed to fragments annotation and to calculating the degree of derivatization.

5.3.1.2 Detection

HPLC-UV is used for the determination of the most popular ASs, such as ACE, ASP, SAC, and SUC, at selected wavelengths with high absorbance yield (at 226, 193, 202, and 314 nm, respectively). SUC cannot be detected directly due to lack of chromophores, therefore a derivatization step before the analysis is required. Nevertheless, UV detectors are mainly used for determinations of ASs in foodstuff and soft drinks. Given that environmental matrices are complex, and thus interference components may absorb in the same wavelength as selected ASs, and that AS are present in low concentration levels, more accurate, selective, and sensitive detectors, such as mass spectrometers, are the key choice for AS reliable detection in environmental samples [50].

Mass spectrometry (MS) is a highly sensitive technique for the successful separation and detection of ASs in environmental samples. GC systems typically use electron ionization (EI) for mass spectrometry, whereas HPLC systems typically use ESI to ionize the separated compounds by applying high voltage to create a charged aerosol. Coupled with GC or HPLC, mass spectrometry allows the identification of the compound fragments (or intact molecule), from which parent compound structure can be deduced. The most common type of mass analyzer is a quadrupole mass analyzer. For reducing the background signal and increasing the sensitivity and selectivity of the determined compounds, more sophisticated configurations, like tandem mass spectrometry (MS/MS) is used, providing more reliable identification. HPLC technique is mostly hyphenated to MS/MS, since the formation of fragments is essential for the identification of compounds, since poor fragmentation occurs using ESI. Recently, high-resolution mass spectrometry (HRMS) has also been used for the detection of ASs. Although hybrid mass analyzers like Quadrupole Time of Flight (Q-ToF) provide high-resolution mass spectra, the sensitivity of the methodologies has to be evaluated based on the expected levels of detection.

Ferrer et al. [51] evaluated the differences of SUC sensitivity in water samples using both low- (triple quadrupole, QqQ) and high-resolution (Q-ToF) mass analyzers. Based on their findings, in positive ionization mode, triple quadrupole detection using multiple reaction monitoring (MRM) was about 30 times more sensitive than Q-ToF full scan MS measurement. However, in negative ion mode, both analyzers and scan modes provided comparable results in terms of sensitivity. Moreover, the effect of mobile phase additives on analytes sensitivity was tested, using formic acid and acetic acid as a modifier in the aqueous phase. In positive ion mode, no significant difference was observed, whereas in negative mode, the sensitivity obtained with acetic acid was slightly better (20% enhancement) compared with that of formic acid. Signal suppression is one of the most common matrix effects in ESI due to the competition between the analyte and the matrix components to form ions in solution at the surface of the droplet. For this reason, the quantification of ASs by LC-MS detection can be affected when co-extracted matrix components are present. Ferrer et al. [51] investigated matrix effects in several types of water samples (deionized water, drinking water, groundwater, surface water, and wastewater) during the determination of SUC. The results showed that the different extracted water matrices may change the ion intensity of a SUC deuterated standard, spiked after pre-concentration of the water samples. As expected, extracted de-ionized water and drinking water did not show considerable ion suppression for SUC, compared with a standard solution at the same concentration level. However, surface water and groundwater showed a considerable matrix effect (more than 60%), and even more severe suppression occurred with wastewater (up to 81%). This observation highlights the importance of quantifying with isotopically labeled standards to account for signal suppression and differences in sodium adduct formation and to achieve better quantitation in water samples [51]. Through the analysis by MS/MS, multiple transitions can be monitored simultaneously, and enable the selection of the transition which gives the highest sensitivity and selectivity in the tested matrix, for quantification of the determined analyte, during method development. Multi-analyte LC-MS/MS methodologies are mainly applied for the simultaneous determination of ASs in environmental samples. Negative ionization mode is mainly used; however, ASP, SAC and SUC can be ionized in positive mode as well and detected either as pseudo-molecular ions or as sodium adducts. Table 5.6 summarizes commonly used MRM transition of ASs acquired by ESI-MS/MS analysis, both in negative and positive ionization mode.

For quantification purposes, the use of internal standards, using isotopically labeled compounds, is mostly used in environmental studies, due to the need to sufficiently counterbalance for matrix effects. According to the literature, it is clear that the deuterated internal standard (IS) of SUC (SUC-d6) has been used in most cases. Furthermore, other lately released IS, such as aspartame-d5 (ASP-d5), neotame-d3 (NEO-d3), neotame-d5 (NEO-d5), and neohesperidin dihydrocalcone-d3 (NHDC-d3), have been applied. Apart from the use of IS, quantification by external calibration has also been reported. ^{1}H and ^{2}H (D) have considerable differences in their physical properties compared to ^{12}C and ^{13}C; however, deuterated IS are mainly used due to the high cost of ^{13}C labeled compounds.

5.3.2 Figures of Merit

HPLC-MS/MS is the technique of choice for the determination of AS in complex environmental matrices, due to its high sensitivity and selectivity. The combination of the separative efficiency of HPLC with the power of MS to detect and identify molecular structures enable their determination even at ppt to ppb concentration levels at different environmental matrices, including ground water, surface water, wastewater, sewage

Table 5.6 MRM transitions of AS applied in LC-MS/MS methodologies [24, 76, 77].

Artificial sweetener	Precursor ion type	Precursor ion (m/z)	Product ions (m/z)
ACE	$[M]^-$	162	82, 78
SUC	$[C_{12}H_{19}{}^{35}Cl_3O_8\text{-}H]^-$	395	359, 35
	$[C_{12}H_{19}{}^{37}Cl_3O_8\text{-}H]^-$	397	361, 359, 35
	$[(M + HCOOH)\text{-}H]^-$	441	395
	$[M + Na]^+$	419	239, 221, 199, 180, 163, 145
CYCL	$[M]^-$	178	80, 96
ASP	$[M\text{-}H]^-$	293	261, 200
	$[M + H]^+$	295	120, 235, 260, 180
SACH	$[M]^-$	182	106, 62, 42
	$[M + H]^+$	184	166
NHDC	$[M\text{-}H]^-$	611	303
NEO	$[M\text{-}H]^-$	377	195, 345
DKP	$[M\text{-}H]^-$	261	146, 217

sludge, and biota. Moreover, the extraction method, the application of a clean-up step and the pre-concentration of the analytes play an important role on method detection limits. The figures of merit for validated protocols available in current literature are provided in Table 5.7. Method LODs, the linear working range, and recoveries are provided for every developed methodology. For multi-analyte methods the linear working range has to be considered, to avoid any saturation in the ion source or the detector. All methods presented satisfactory correlation coefficient (>0.99) and low LODs at ng/L or ng/g levels.

5.3.3 Hints and Tips

Based on the literature review and the conclusions of previous studies, the following hits and tips could be proposed for the successful determination of ASs in environmental samples.

- Polymer- or silica-based SPE sorbents are recommended for the extraction of ASs and the clean-up of complex matrices extracts. Strata X-RP and Oasis HLB sorbent materials consisting of poly(divinylbenzene-N-vinylpyrrolidone) are mainly used for the extraction of ASs, presenting good recoveries.
- The sensitivity of ASs detection in RPLC-ESI(-)-MS/MS can be improved when extra basic buffers are added in mobile phases, facilitating the deprotonation of weakly acidic analytes and enhancing their response. For instance, TRIS (hydroxyl-methyl)-amino methane) was proven to improve the ionization efficiency and sensitivity of selected ASs in a reverse chromatographic column, acting like an ion-pair reagent in mobile phase.
- Concerning the organic mobile phase in RPLC-ESI(-)-MS/MS, MeOH presented notably higher signal intensity and more efficient separation for all tested ASs compared to ACN or ACN/MeOH (50/50, v/v) solution.
- Mainly reversed-phase chromatographic columns are being used for the separation of ASs; however, HILIC has also been applied as an alternative analytical approach.
- Although negative electrospray ionization is mainly used for the determination of ASs in LC-ESI-MS/MS analysis, ASP, SAC, and SUC can also be determined also in positive ionization mode as pseudo-molecular ions or as sodium adducts.
- In RPLC-MS/MS in positive ionization mode, no significant difference was observed when testing formic acid and acetic acid as modifiers in the aqueous phase, while in negative ionization mode the sensitivity obtained with acetic acid (at 2 mM) was reported to be higher compared with that obtained with formic acid. Moreover, both chromatographic separation efficiency and peak shapes of all determined ASs are improved with acetic acid.
- The use of isotopically labeled internal standards contributes significantly to the accuracy of the developed methodology. The most commonly used internal standards for the determination of AS are SUC-d6, ASP-d3 or ASP-d5, SAC-d4, ACE-d4, and CYCL-d11.
- Regarding the use of low- and high-resolution mass spectrometers and the application of different acquisition modes, it was noticed that triple quadrupole using MRM was about 30 times more sensitive than full scan measurements of time-of-flight in positive

Table 5.7 Figures of merit of validated protocols. For the determination of ASs in different environmental matrices.

Artificial sweeteners	Matrix	Extraction	Analytical technique	Linear working range	Correlation coefficient (R_2)	(%)Recoveries range (analyte, matrix)	%RSD range (analyte, matrix)	LODs (analyte, matrix)	Ref.
ACE, CYCL, SAC	TW, GW, SW	(Direct injection)	LC-MS/MS	10–1,600 ng/L	≥0.999	100 (*CYCL, TW*)-102 (*ACE, SW*)	11 (*ACE, SW*)-25 (*ACE, GW*)	0.3 (*CYCL, TW*)-2.3 ng/L (*SAC, DW*)	[14]
ACE, CYCL, SAC, SUC	GW, SW	SPE		0.500–500 µg/L	≥0.998	77.8 (*SUC, GW*)-92.1 (*SAC, GW*)	2.9 (*CYCL, GW*)-6.1 (*SUC, GW*)	1.7 (*CYCL, GW*)-12.5 ng/L (*SUC, GW*)	[38]
ACE, ASP, CYCL, SAC, SUC	SW, WW			0.030–160 µg/L	≥0.998	24 (*NHDC*)-90 (*SAC*)	>8	1.0 (*ACE, SW*)-500 ng/L (*ACE, WW*)	[30]
	SS			0.1–2,000 ng/g	≥0.998	78 (*SUC*)-126 (*CYCL*)	14 (*CYCL*)-19 (*SUC*)	1.7 ng/g	[72]
ACE, ASP, CYCL, SAC, SUC, NHDC	GW, SW, WW			0.100–1,000 µg/L	≥0.997	62.5 (*ACE, WW*)-102 (*ASP, WW*)	3.3 (*SUC, GW*)-12.2 (*SAC, WW*)	1.0 (*SAC, SW & ACE, GW*)-15 ng/L (*SUC, SW/W*)	[45]
ACE, ASP, CYCL, SAC, SUC, NHDC, NEO	TP, GW, SW, WW			0.100–1,000 µg/L	≥0. 999	41 (*ASP, WW*)-107 (*CYCL, GW*)	2 (*NHDC, SW*)-10 (*SUC, SW*)	0.1 (*ACE, TW & ASP, SW*)-2.3 ng/L (*SUC, SW*)	[4]
ACE, CYCL, SAC, SUC, ASP, NHDC, NEO, DKP	TW, GW, SW, WW			10–1,000 ng/L	≥0.997	55 (*NHDC, SW*)-115 (*SUC, GW*)	1.8 (*NEO, SW*)-9.1 (*SAC, GW*)	3.0 ng/L	[24]
ACE, ASP, CYCL, SAC, SUC, NHDC, ALI, NEO	TW, WW			5–200 µg/L	≥0.993	40 (*CYCL, TW*)-117 (*NEO, TW*)	0.7 (*SUC, TW*)-21 (*NEO, WW*)	4.8 (*CYCL*)–88 ng/L (*SUC*)	[61]
	Fish samples	PLE-SPE	LC-HRMS	12.5–1,000 ng/g	≥0.991	11 (*ASP*)-122 (*ACE*)	2.0 (*SACC*)-15 (*NHDC*)	2.5 (*ACE*)–125 ng/g (*NHDC*)	[45]

WW: wastewater; SW: Surface water; TW: Tap water; GW: Groundwater; SS: Sewage sludge.

ionization mode, whereas in negative mode both low- and high-resolution mass analyzers presented comparable sensitivities for SUC.

5.4 Future Directions and Challenges

The increasing use of ASs in food and other industries such as pharmaceuticals has led to the detection of ASs in environmental samples, which may pose significant environmental risk since their bioaccumulation and ecological effects still remain unknown. Based on their increasing use and detection in environmental samples, and due to their persistence, ASs should be integrated in future water monitoring and risk assessment programs. Long-term monitoring data worldwide may provide valuable information on their occurrence and the potential need to include some ASs in the European Watch Lists or future legislation. Therefore, much is still needed to be done to accurately assess the ecotoxicological implications that ASs may pose to the ecosystems worldwide. The long-term exposure of the aquatic organisms to significantly high AS concentrations downstream of urban centers that discharge WWTP effluents has not been evaluated so far. Therefore, the effects of ASs in biota and especially in association with the biochemical response in plants and soil invertebrates, could be thoroughly investigated. Moreover, the formation, occurrence, and ecotoxicity of AS transformation products has not been studied in depth, therefore more harmful chemicals than precursor AS may be produced and could be further examined. Since reference standards for most TPs are not commercially available, suspect and not target screening approaches should be employed in data acquired by high-resolution mass spectrometric instruments. For suspect screening purposes, computational (*in-silico*) prediction tools can be used for the prediction of possible TPs structures and the compilation of the suspect list to be used for screening the environmental samples. For the prediction of TPs, commercially available or freely accessible programs are available [52].

Given that WWTPs are the main source of ASs in the environment and given the persistence of some ASs, like SUC and ACE, during the conventional treatment processing methods that are being applied in WWTPs, the effectiveness of advanced oxidation treatment processes on their elimination could be evaluated and optimized to reinforce their adequate removal.

From an analytical point of view, most published studies provide limited data on the method development and validation, which are necessary to fully evaluate the method performance. Detailed information on preliminary test and optimization of the separation and determination of AS would be valuable to save future efforts and select suitable conditions for the determination of AS in multi-class monitoring studies, when experimental conditions and instrumental parameters should be compromised. Moreover, new sampling approaches, like passive sampling, could be developed and optimized to reduce the time and efforts during the sampling and improve the representativeness of the obtained water samples. Regarding the instrumentation, the technological advances and the use of state-of-the art instruments, like ion mobility spectrometry, will contribute to the improvement of ASs and their TPs detection.

References

1 Luo, J., Wu, L., Zhang, Q., Wu, Y., Fang, F., Feng, Q., Li, C., Xue, Z., and Cao, J. (2019). Review on the determination and distribution patterns of a widespread contaminant artificial sweetener in the environment. *Environ. Sci. Pollut. Res.* 26: 19078–19096. doi: 10.1007/s11356-019-05261-4.

2 Swithers, S.E. (2013). Artificial sweeteners produce the counterintuitive effect of inducing metabolic derangements. *Trends Endocrinol. Metab.* 24: 431–441. doi: 10.1016/j.tem.2013.05.005.

3 Scheurer, M., Brauch, H.J., and Lange, F.T. (2009). Analysis and occurrence of seven artificial sweeteners in German waste water and surface water and in soil aquifer treatment (SAT). *Anal. Bioanal. Chem.* 394: 1585–1594. doi: 10.1007/s00216-009-2881-y.

4 Gan, Z., Sun, H., Wang, R., and Feng, B. (2013). A novel solid-phase extraction for the concentration of sweeteners in water and analysis by ion-pair liquid chromatography-triple quadrupole mass spectrometry. *J. Chromatogr. A* 1274: 87–96. doi: 10.1016/j.chroma.2012.11.081.

5 Schiano, C., Grimaldi, V., Scognamiglio, M., Costa, D., Soricelli, A., Nicoletti, G.F., and Napoli, C. (2021). Soft drinks and sweeteners intake: Possible contribution to the development of metabolic syndrome and cardiovascular diseases. Beneficial or detrimental action of alternative sweeteners? *Food Res. Int.* 142: 110220. doi: 10.1016/j.foodres.2021.110220.

6 European Commission (2015). Commission Implementing Decision (EU) 2015/495 of 20 March 2015 establishing a watch list of substances for Union-wide monitoring in the field of water policy pursuant to Directive 2008/105/EC of the European Parliament and of the Council (notified under). *Off. J. Eur. Union. L* 78: 40–42. https://eur-lex.europa.eu/legal-content/EN/TXT/?uri=uriserv:OJ.L_.2015.078.01.0040.01.ENG.

7 European Commission (2018). Commission Implementing Decision (EU) 2018/840 of 5 June 2018 establishing a watch list of substances for Union-wide monitoring in the field of water policy pursuant to Directive 2008/105/EC of the European Parliament and of the Council and repealing. *Comm. Off. J. Eur. Union. L* 141: 9–12.

8 European Commission (2020). Commission Implementing Decision (EU) 2020/1161 of 4 August 2020 establishing a watch list of substances for Union-wide monitoring in the field of water policy pursuant to Directive 2008/105/EC of the European Parliament and of the Council (notified under. *Off. J. Eur. Union. L*257: 32–35. https://eur-lex.europa.eu/legal-content/EN/TXT/?uri=uriserv:OJ.L_.2020.257.01.0032.01.ENG&toc=OJ:L:2020:257:TOC. Accessed 14 December 2021.

9 European Commission. (2008). Directive 2008/105/EC of the European Parliament and of the Council of 16 December 2008 on environmental quality standards in the field of water policy, amending and subsequently repealing Council Directives 82/176/EEC, 83/513/EEC, 84/156/EEC, 84/491/EEC. *Off. J. Eur. Union.* 16: 84–97. https://eur-lex.europa.eu/eli/dir/2008/105/oj.

10 The NORMAN network, NORMAN Ecotoxicology database, https://www.norman-network.com/nds/ecotox (accesssed September 2021).

11 Scheurer, M., Brauch, H.J., and Lange, F.T. (2014). Transformation products of artificial sweeteners. *Transform. Prod. Emerg. Contam. Environ. Anal. Process. Occur. Eff. Risks.* 525–544. doi: 10.1002/9781118339558.ch17.

12 Chi, L., Bian, X., Gao, B., Tu, P., Lai, Y., Ru, H., and Lu, K. (2018). Effects of the artificial sweetener neotame on the gut microbiome and fecal metabolites in mice. *Molecules* 23: 367. doi: 10.3390/molecules23020367.

13 Braune, A., Engst, W., and Blaut, M. (2005). Degradation of neohesperidin dihydrochalcone by human intestinal bacteria. *J. Agric. Food Chem.* 53: 1782–1790. doi: 10.1021/jf0484982.

14 Lin, H., Wu, J., Oturan, N., Zhang, H., and Oturan, M.A. (2016). Degradation of artificial sweetener saccharin in aqueous medium by electrochemically generated hydroxyl radicals. *Environ. Sci. Pollut. Res.* 23: 4442–4453. doi: 10.1007/s11356-015-5633-x.

15 Davididou, K., McRitchie, C., Antonopoulou, M., Konstantinou, I., and Chatzisymeon, E. (2018). Photocatalytic degradation of saccharin under UV-LED and blacklight irradiation. *J. Chem. Technol. Biotechnol.* 93: 269–276. doi: 10.1002/jctb.5349.

16 Scheurer, M., Godejohann, M., Wick, A., Happel, O., Ternes, T.A., Brauch, H.J., Ruck, W.K.L., and Lange, F.T. (2012). Structural elucidation of main ozonation products of the artificial sweeteners cyclamate and acesulfame. *Environ. Sci. Pollut. Res.* 19: 1107–1118. doi: 10.1007/s11356-011-0618-x.

17 Hu, R., Zhang, L., and Hu, J. (2017). Investigation of ozonation kinetics and transformation products of sucralose. *Sci. Total Environ.* 603–604: 8–17. doi: 10.1016/j.scitotenv.2017.06.033.

18 Yin, K., Li, F., Wang, Y., He, Q., Deng, Y., Chen, S., and Liu, C. (2017). Oxidative transformation of artificial sweetener acesulfame by permanganate: reaction kinetics, transformation products and pathways, and ecotoxicity. *J. Hazard. Mater.* 330: 52–60. doi: 10.1016/j.jhazmat.2017.02.012.

19 Chow, C.H. and Leung, K.S.Y. (2019). Removing acesulfame with the peroxone process: transformation products, pathways and toxicity. *Chemosphere* 221: 647–655. doi: 10.1016/j.chemosphere.2019.01.082.

20 Li, A.J., Wu, P., Law, J.C.F., Chow, C.H., Postigo, C., Guo, Y., and Leung, K.S.Y. (2017). Transformation of acesulfame in chlorination: kinetics study, identification of byproducts, and toxicity assessment. *Water Res.* 117: 157–166. doi: 10.1016/j.watres.2017.03.053.

21 Scheurer, M., Schmutz, B., Happel, O., Brauch, H.J., Wülser, R., and Storck, F.R. (2014). Transformation of the artificial sweetener acesulfame by UV light. *Sci. Total Environ.* 481: 425–432. doi: 10.1016/j.scitotenv.2014.02.047.

22 Djoumbou-Feunang, Y., Fiamoncini, J., Gil-de-la-fuente, A., Greiner, R., Manach, C., and Wishart, D.S. (2019). BioTransformer: a comprehensive computational tool for small molecule metabolism prediction and metabolite identification. *J. Cheminform.* 11: 1–25. doi: 10.1186/s13321-018-0324-5.

23 Van Stempvoort, D.R., Roy, J.W., Grabuski, J., Brown, S.J., Bickerton, G., and Sverko, E. (2013). An artificial sweetener and pharmaceutical compounds as co-tracers of urban wastewater in groundwater. *Sci. Total Environ.* 461–462: 348–3359. doi: 10.1016/j.scitotenv.2013.05.001.

24 Berset, J.D. and Ochsenbein, N. (2012). Stability considerations of aspartame in the direct analysis of artificial sweeteners in water samples using high-performance liquid chromatography-tandem mass spectrometry (HPLC-MS/MS). *Chemosphere* 88: 563–569. doi: 10.1016/j.chemosphere.2012.03.030.

25 Ens, W., Senner, F., Gygax, B., and Schlotterbeck, G. (2014). Development, validation, and application of a novel LC-MS/MS trace analysis method for the simultaneous quantification of seven iodinated X-ray contrast media and three artificial sweeteners in surface, ground, and drinking water. *Anal. Bioanal. Chem.* 406: 2789–2798. doi: 10.1007/s00216-014-7712-0.

26 Batchu, S.R., Quinete, N., Panditi, V.R., and Gardinali, P.R. (2013). Online solid phase extraction liquid chromatography tandem mass spectrometry (SPE-LC-MS/MS) method for the determination of sucralose in reclaimed and drinking waters and its photo degradation in natural waters from South Florida. *Chem. Cent. J.* 7: 1–16. doi: 10.1186/1752-153X-7-141.

27 Gan, Z., Sun, H., Feng, B., Wang, R., and Zhang, Y. (2013). Occurrence of seven artificial sweeteners in the aquatic environment and precipitation of Tianjin, China. *Water Res.* 47: 4928–4937. doi: 10.1016/j.watres.2013.05.038.

28 Ordóñez, E.Y., Quintana, J.B., Rodil, R., and Cela, R. (2012). Determination of artificial sweeteners in water samples by solid-phase extraction and liquid chromatography-tandem mass spectrometry. *J. Chromatogr. A* 1256: 197–205. doi: 10.1016/j.chroma.2012.07.073.

29 Baena-Nogueras, R.M., Traverso-Soto, J.M., Biel-Maeso, M., Villar-Navarro, E., and Lara-Martín, P.A. (2018). Sources and trends of artificial sweeteners in coastal waters in the bay of Cadiz (NE Atlantic). *Mar. Pollut. Bull.* 135: 607–616. doi: 10.1016/j.marpolbul.2018.07.069.

30 Arbeláez, P., Borrull, F., Pocurull, E., and Marcé, R.M. (2015). Determination of high-intensity sweeteners in river water and wastewater by solid-phase extraction and liquid chromatography-tandem mass spectrometry. *J. Chromatogr. A* 1393: 106–114. doi: 10.1016/j.chroma.2015.03.035.

31 Buerge, I.J., Buser, H.R., Kahle, M., Müller, M.D., and Poiger, T. (2009). Ubiquitous occurrence of the artificial sweetener acesulfame in the aquatic environment: an ideal chemical marker of domestic wastewater in groundwater. *Environ. Sci. Technol.* 43: 4381–4385. doi: 10.1021/es900126x.

32 Van Stempvoort, D.R., Roy, J.W., Brown, G., and Bickerton, G. (2011). Artificial sweeteners as potential tracers in groundwater in urban environments. *J. Hydrol.* 401: 126–133. doi: 10.1016/j.jhydrol.2011.02.013.

33 Tran, N.H., Reinhard, M., Khan, E., Chen, H., Nguyen, V.T., Li, Y., Goh, S.G., Nguyen, Q.B., Saeidi, N., and Gin, K.Y.H. (2019). Emerging contaminants in wastewater, stormwater runoff, and surface water: application as chemical markers for diffuse sources. *Sci. Total Environ.* 676: 252–267. doi: 10.1016/j.scitotenv.2019.04.160.

34 Watanabe, Y., Bach, L.T., Van Dinh, P., Prudente, M., Aguja, S., Phay, N., and Nakata, H. (2016). Ubiquitous detection of artificial sweeteners and iodinated X-ray contrast media in aquatic environmental and wastewater treatment plant samples from Vietnam, the Philippines, and Myanmar. *Arch. Environ. Contam. Toxicol.* 70: 671–681. doi: 10.1007/s00244-015-0220-1.

35 Wolf, L., Zwiener, C., and Zemann, M. (2012). Tracking artificial sweeteners and pharmaceuticals introduced into urban groundwater by leaking sewer networks. *Sci. Total Environ.* 430: 8–19. doi: 10.1016/j.scitotenv.2012.04.059.

36 Perkola, N. and Sainio, P. (2014). Quantification of four artificial sweeteners in Finnish surface waters with isotope-dilution mass spectrometry. *Environ. Pollut.* 184: 391–396. doi: 10.1016/j.envpol.2013.09.017.

37 Sharma, B.M., Bečanová, J., Scheringer, M., Sharma, A., Bharat, G.K., Whitehead, P.G., Klánová, J., and Nizzetto, L. (2019). Health and ecological risk assessment of emerging contaminants (pharmaceuticals, personal care products, and artificial sweeteners) in surface and groundwater (drinking water) in the Ganges River Basin, India. *Sci. Total Environ.* 646: 1459–1467. doi: 10.1016/j.scitotenv.2018.07.235.

38 Gui, J.Y., Sun, W., Zhang, C.L., Zhang, Y.T., Zhang, L., and Liu, F. (2016). An innovative approach to sensitive artificial sweeteners analysis by ion chromatography-triple quadrupole mass spectrometry. *Chin. J. Anal. Chem.* 44: 361–366. doi: 10.1016/S1872-2040(16)60914-3.

39 Fu, K., Wang, L., Wei, C., Li, J., Zhang, J., Zhou, Z., and Liang, Y. (2020). Sucralose and acesulfame as an indicator of domestic wastewater contamination in Wuhan surface water. *Ecotoxicol. Environ. Saf.* 189. doi: 10.1016/j.ecoenv.2019.109980.

40 Yang, Y.Y., Zhao, J.L., Liu, Y.S., Liu, W.R., Zhang, Q.Q., Yao, L., Hu, L.X., Zhang, J.N., Jiang, Y.X., and Ying, G.G. (2018). Pharmaceuticals and personal care products (PPCPs) and artificial sweeteners (ASs) in surface and ground waters and their application as indication of wastewater contamination. *Sci. Total Environ.* 616–617: 816–823. doi: 10.1016/j.scitotenv.2017.10.241.

41 Nika, M.C., Ntaiou, K., Elytis, K., Thomaidi, V.S., Gatidou, G., Kalantzi, O.I., Thomaidis, N.S., and Stasinakis, A.S. (2020). Wide-scope target analysis of emerging contaminants in landfill leachates and risk assessment using Risk Quotient methodology. *J. Hazard. Mater.* 394: 122493. doi: 10.1016/j.jhazmat.2020.122493.

42 Ordóñez, E.Y., Quintana, J.B., Rodil, R., and Cela, R. (2013). Determination of artificial sweeteners in sewage sludge samples using pressurised liquid extraction and liquid chromatography-tandem mass spectrometry. *J. Chromatogr. A* 1320: 10–16. doi: 10.1016/j.chroma.2013.10.049.

43 Ma, L., Liu, Y., Xu, J., Sun, H., Chen, H., Yao, Y., Zhang, P., Shen, F., and Alder, A.C. (2017). Mass loading of typical artificial sweeteners in a pig farm and their dissipation and uptake by plants in neighboring farmland. *Sci. Total Environ.* 605–606: 735–744. doi: 10.1016/j.scitotenv.2017.06.027.

44 Núñez, M., Borrull, F., Pocurull, E., and Fontanals, N. (2017). Pressurised liquid extraction and liquid chromatography-high resolution mass spectrometry to determine high-intensity sweeteners in fish samples. *J. Chromatogr. A* 1479: 32–39. doi: 10.1016/j.chroma.2016.12.006.

45 Tran, N.H., Hu, J., and Ong, S.L. (2013). Simultaneous determination of PPCPs, EDCs, and artificial sweeteners in environmental water samples using a single-step SPE coupled with HPLC-MS/MS and isotope dilution. *Talanta* 113: 82–92. doi: 10.1016/j.talanta.2013.03.072.

46 Cárdenas-Soracá, D.M., Singh, V., Nazdrajić, E., Vasiljević, T., Grandy, J.J., and Pawliszyn, J. (2020). Development of thin-film solid-phase microextraction coating and method for determination of artificial sweeteners in surface waters. *Talanta* 211: 120714. doi: 10.1016/j.talanta.2020.120714.

47 Orata, F. (2012). Derivatization reactions and reagents for gas chromatography analysis. *Adv. Gas Chromatogr. – Prog. Agric. Biomed. Ind. Appl.* doi: 10.5772/33098.

48 Qiu, W., Wang, Z., Nie, W., Guo, Y., and Huang, L. (2007). GC-MS determination of sucralose in Splenda. *Chromatographia* 66: 935–939. doi: 10.1365/s10337-007-0422-4.

49 Voss, S., Newman, E., and Miller-Schulze, J.P. (2019). Quantification of sucralose in groundwater well drinking water by silylation derivatization and gas chromatography-mass spectrometry. *Anal. Methods* 11: 2790–2799. doi: 10.1039/c9ay00442d.

50 Serdar, M. and Knežević, Z. (2011). Determination of artificial sweeteners in beverages and special nutritional products using high performance liquid chromatography. *Arh. Hig. Rada Toksikol.* 62: 169–173. doi: 10.2478/10004-1254-62-2011-2084.

51 Ferrer, I., Zweigenbaum, J.A., and Thurman, E.M. (2013). Analytical methodologies for the detection of sucralose in water. *Anal. Chem.* 85: 9581–9587. doi: 10.1021/ac4016984.

52 Bletsou, A.A., Jeon, J., Hollender, J., Archontaki, E., and Thomaidis, N.S. (2015). Targeted and non-targeted liquid chromatography-mass spectrometric workflows for identification of transformation products of emerging pollutants in the aquatic environment. *TrAC – Trends Anal. Chem.* 66: 32–44. doi: 10.1016/j.trac.2014.11.009.

53 Sardarodiyan, M. and Hakimzadeh, V. (2016). Artificial sweeteners. *Int. J. PharmTech Res.* 8: 4304. https://sphinxsai.com/2016/ph_vol9_no4/2/(357-363)V9N4PT.pdf.

54 Kim, S., Thiessen, P.A., Bolton, E.E., Chen, J., Fu, G., Gindulyte, A., Han, L., He, J., He, S., Shoemaker, B.A., Wang, J., Yu, B., Zhang, J., and Bryant, S.H. (2016). PubChem substance and compound databases. *Nucleic Acids Res.* 44: D1202–D1213. doi: 10.1093/nar/gkv951.

55 Mawhinney, D.B., Young, R.B., Vanderford, B.J., Borch, T., and Snyder, S.A. (2011). Artificial sweetener sucralose in U.S. drinking water systems. *Environ. Sci. Technol.* 45: 8716–8722. doi: 10.1021/es202404c.

56 Gago-Ferrero, P., Gros, M., Ahrens, L., and Wiberg, K. (2017). Impact of on-site, small and large scale wastewater treatment facilities on levels and fate of pharmaceuticals, personal care products, artificial sweeteners, pesticides, and perfluoroalkyl substances in recipient waters. *Sci. Total Environ.* 601–602: 1289–1297. doi: 10.1016/j.scitotenv.2017.05.258.

57 Cantwell, M.G., Katz, D.R., Sullivan, J., and Kuhn, A. (2019). Evaluation of the artificial sweetener sucralose as a sanitary wastewater tracer in Narragansett Bay, Rhode Island, USA. *Mar. Pollut. Bull.* 146: 711–717. doi: 10.1016/j.marpolbul.2019.07.036.

58 Clarke, B.O., Anumol, T., Barlaz, M., and Snyder, S.A. (2015). Investigating landfill leachate as a source of trace organic pollutants. *Chemosphere* 127: 269–275. doi: 10.1016/j.chemosphere.2015.02.030.

59 Batchu, S.R., Ramirez, C.E., and Gardinali, P.R. (2015). Rapid ultra-trace analysis of sucralose in multiple-origin aqueous samples by online solid-phase extraction coupled to high-resolution mass spectrometry. *Anal. Bioanal. Chem.* 407: 3717–3725. doi: 10.1007/s00216-015-8593-6.

60 Liu, Y.Y., Blowes, D.W., Groza, L., Sabourin, M.J., and Ptacek, C.J. (2014). Acesulfame-K and pharmaceuticals as co-tracers of municipal wastewater in a receiving river. *Environ. Sci. Process. Impacts* 16: 2789–2795. doi: 10.1039/c4em00237g.

61 Wu, M., Qian, Y., Boyd, J.M., Hrudey, S.E., Le, X.C., and Li, X.F. (2014). Direct large volume injection ultra-high performance liquid chromatography-tandem mass spectrometry determination of artificial sweeteners sucralose and acesulfame in well water. *J. Chromatogr. A* 1359: 156–161. doi: 10.1016/j.chroma.2014.07.035.

62 Ribbers, K., Breuer, L., and Düring, R.A. (2019). Detection of artificial sweeteners and iodinated X-ray contrast media in wastewater via LC-MS/MS and their potential use as anthropogenic tracers in flowing waters. *Chemosphere* 218: 189–196. doi: 10.1016/j.chemosphere.2018.10.193.

63 Roy, J.W., Van Stempvoort, D.R., and Bickerton, G. (2014). Artificial sweeteners as potential tracers of municipal landfill leachate. *Environ. Pollut.* 184: 89–93. doi: 10.1016/j.envpol.2013.08.021.

64 Lee, C.M., Hamm, S.Y., Yun, S.M., Oh, J.E., Kim, M.S., and Kim, H.K. (2018). Indication of groundwater contamination using acesulfame and other pollutants in a rural area of Korea. *Water (Switzerland)* 10: 9–11. doi: 10.3390/w10121731.

65 Gan, Z., Sun, H., Yao, Y., Zhao, Y., Li, Y., Zhang, Y., Hu, H., and Wang, R. (2014). Distribution of artificial sweeteners in dust and soil in China and their seasonal variations in the environment of Tianjin. *Sci. Total Environ.* 488–489: 168–175. doi: 10.1016/j.scitotenv.2014.04.084.

66 Ishii, E., Watanabe, Y., Agusa, T., Hosono, T., and Nakata, H. (2021). Acesulfame as a suitable sewer tracer on groundwater pollution: A case study before and after the 2016 M_W 7.0 Kumamoto earthquakes. *Sci. Total Environ.* 754: 142409. doi: 10.1016/j.scitotenv.2020.142409.

67 Spoelstra, J., Schiff, S.L., and Brown, S.J. (2013). Artificial sweeteners in a large Canadian river reflect human consumption in the watershed. *PLoS One* 8. doi: 10.1371/journal.pone.0082706.

68 Spoelstra, J., Schiff, S.L., and Brown, S.J. (2020). Septic systems contribute artificial sweeteners to streams through groundwater. *J. Hydrol. X* 7: 1–8. doi: 10.1016/j.hydroa.2020.100050.

69 Stefania, G.A., Rotiroti, M., Buerge, I.J., Zanotti, C., Nava, V., Leoni, B., Fumagalli, L., and Bonomi, T. (2019). Identification of groundwater pollution sources in a landfill site using artificial sweeteners, multivariate analysis and transport modeling. *Waste Manag.* 95: 116–128. doi: 10.1016/j.wasman.2019.06.010.

70 Sang, Z., Jiang, Y., Tsoi, Y.K., and Leung, K.S.Y. (2014). Evaluating the environmental impact of artificial sweeteners: A study of their distributions, photodegradation and toxicities. *Water Res.* 52: 260–274. doi: 10.1016/j.watres.2013.11.002.

71 Snider, D.M., Roy, J.W., Robertson, W.D., Garda, D.I., and Spoelstra, J. (2017). Concentrations of artificial sweeteners and their ratios with nutrients in septic system wastewater. *Groundwater Monit. Remediat.* 37: 94–102. doi: 10.1111/gwmr.12229.

72 Subedi, B., Lee, S., Moon, H.B., and Kannan, K. (2014). Emission of artificial sweeteners, select pharmaceuticals, and personal care products through sewage sludge from wastewater treatment plants in Korea. *Environ. Int.* 68: 33–40. doi: 10.1016/j.envint.2014.03.006.

73 Gvozdić, E., Matić Bujagić, I., Đurkić, T., and Grujić, S. (2020). Trace analysis of artificial sweeteners in environmental waters, wastewater and river sediments by liquid chromatography-tandem mass spectrometry. *Microchem. J.* 157. doi: 10.1016/j.microc.2020.105071.

74 Kokotou, M.G. and Thomaidis, N.S. (2013). Determination of eight artificial sweeteners in wastewater by hydrophilic interaction liquid chromatography-tandem mass spectrometry. *Anal. Methods* 5: 3825–3833. doi: 10.1039/c3ay40599k.

75 Li, D., Yao, Y., Sun, H., Wang, Y., Pu, J., Calderón, R., Alder, A.C., and Kannan, K. (2020). Artificial sweeteners in pig feed: a worldwide survey and case study in pig farms in Tianjin, China. *Environ. Sci. Technol.* 54: 4059–4067. doi: 10.1021/acs.est.9b06853.

76 Kokotou, M.G., Asimakopoulos, A.G., and Thomaidis, N.S. (2012). Artificial sweeteners as emerging pollutants in the environment: analytical methodologies and environmental impact. *Anal. Methods* 4: 3057–3070. doi: 10.1039/c2ay05950a.

77 Oppenheimer, J., Eaton, A., Badruzzaman, M., Haghani, A.W., and Jacangelo, J.G. (2011). Occurrence and suitability of sucralose as an indicator compound of wastewater loading to surface waters in urbanized regions. *Water Res.* 45: 4019–4027. doi: 10.1016/j.watres.2011.05.014.

6

Perfluorinated Substances

Julian Campo and Yolanda Picó

Desertification Research Center (CIDE), Environmental and Food Safety Research Group of the University of Valencia (SAMA-UV), Montcada, Valencia, Spain

6.1 Overview of Perfluoroalkyl Substances

According to Buck et al. [1], the "perfluoroalkyl and polyfluoroalkyl substances" (PFASs) are a subset of highly fluorinated aliphatic substances that contain one or more C atoms, on which all the H substituents (present in the non-fluorinated analogues from which they are notionally derived) have been replaced by F atoms, in such a manner that they contain the perfluoroalkyl moiety C_nF_{2n+1}–.

There are numerous families of PFASs, consisting of a series of homologues and branched isomers (Table 6.1). One of the most discussed in literature is the family of perfluoroalkyl acids (PFAAs), which includes carboxylic (PFCAs, $C_nF_{2n+1}COOH$) and sulfonic (PFSAs, $C_nF_{2n+1}SO_3H$), such as, perfluorooctanoic acid, PFOA ($C_7F_{15}COOH$) and perfluorooctane sulfonic acid, PFOS ($C_8F_{17}SO_3H$). Other perfluoroalkyl acids families are sulfinic (PFSIAs, $C_nF_{2n+1}SO_2H$), phosphonic (PFPAs, $O = P(OH)_2C_nF_{2n+1}$) and phosphinic (PFPIAs $O = P(OH)(C_nF_{2n+1})(C_mF_{2m+1})$). PFAAs and their anions are frequently referred to as "long-chain" or "short-chain". Based on the definition provided by the Organisation for Economic Co-operation and Development [2], "long-chain" refers to PFCAs with eight carbons and greater, and PFSAs with six carbons and greater. Other families as those based on polyfluoroalkyl "fluorotelomer-based" products are perfluoroalkyl iodides (PFAIs, $CnF_{2n+1}I$), n:2 fluorotelomer iodides (n:2 FTIs, $C_nF_{2n+1}CH_2CH_2I$), fluorotelomer olefins (FTOs, $C_nF_{2n+1}CH = CH_2$), n:2 fluorotelomer alcohols (n:2 FTOHs, $C_nF_{2n+1}CH_2CH_2OH$) and their acrylic, methacrylic and phosphoric esters, semifluorinated alkanes and alkenes, fluorotelomer alcohols and their esters, fluorotelomer sulfonic acids, perfluoroalkane sulfonamide derivatives, and others explained extensively by Buck et al. [1].

Restrictions on the use related to human health risks have led, in the U.S. and in Europe, to voluntarily phase out initiatives on the production of legacy PFASs at the beginning of twenty-first century [3]. However, their replacements are still other poly- or perfluorinated compounds with shorter chain or the insertion of ester functional groups. Some examples of these replacement compounds (mainly of PFOS/PFOA) are

Analytical Methods for Environmental Contaminants of Emerging Concern, First Edition. Edited by Núria Fontanals and Rosa Maria Marcé.

Table 6.1 List of poly- and perfluoroalkyl substances (PFASs).

Group	Structure
Perfluoroalkyl sulphonic acids (PFSAs)	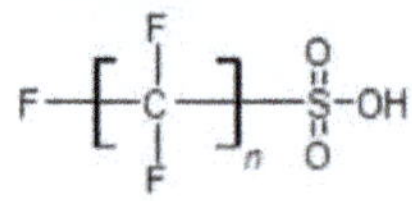
Perfluoroalkyl carboxylic acids (PFCAs)	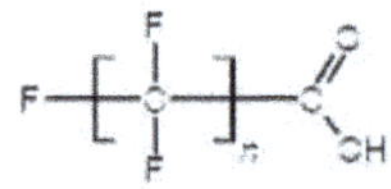
Perfluoroalkyl phosphonic acids (PFPAs)	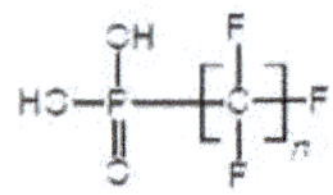
Perfluoroalkyl phosphinic acids (PFPiAs)	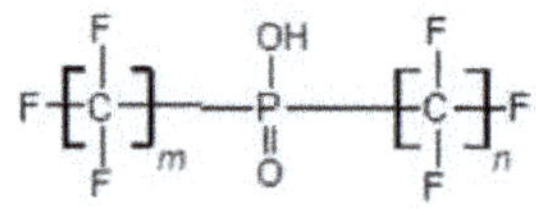
Perfluoroalkane sulphonamides (FASAs)	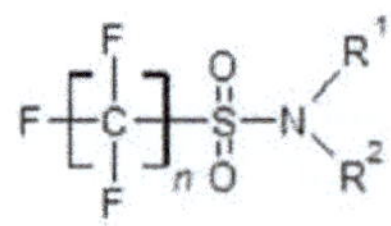
N-alkyl perfluoroalkane sulphonamido ethanols (FASEs)	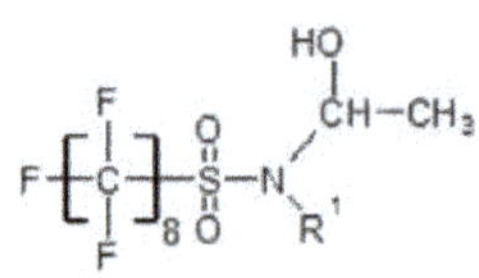
Perfluoroalkyl iodides (PFAIs)	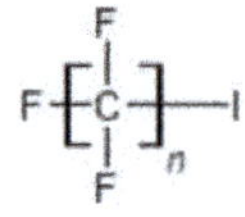
Perfluoroether sulphonic acids (PFESAs)	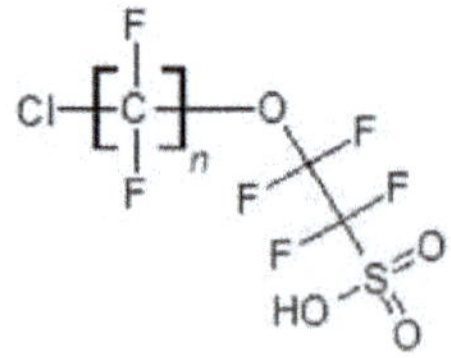
Perfluoroether carboxylic acids (PFECAs) (i.e. HFPO-DA, trade name: GenX)	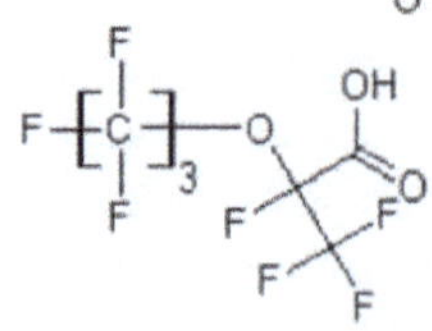
Perfluorooctane sulphonamido ethanol-based phosphate esters (SAmPAPs)	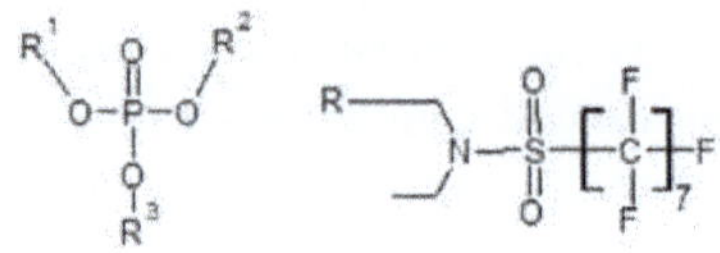

(*Continued*)

Table 6.1 (Continued)

Group	Structure
Cyclic perfluoroalkyl sulphonic acids (cyclic PFSAs)	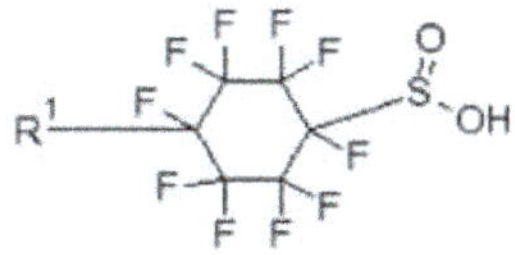
Fluorotelomer sulphonic acids (FTSAs)	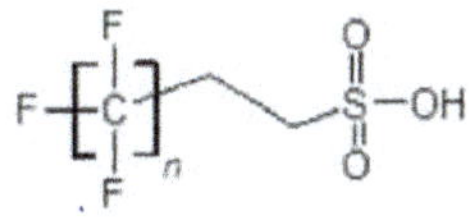
Fluorotelomer carboxylic acids (FTCAs)	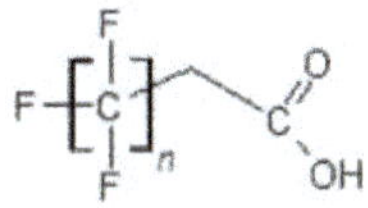
Fluorotelomer unsaturated carboxylic acids (FTUCAs)	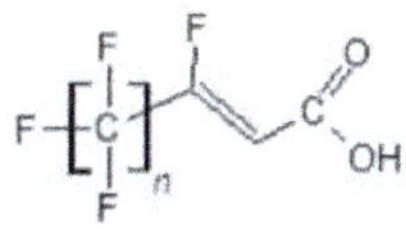
Fluorotelomer olefins (FTOs)	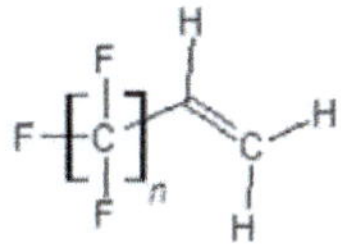
Fluorotelomer alcohols (FTOHs)	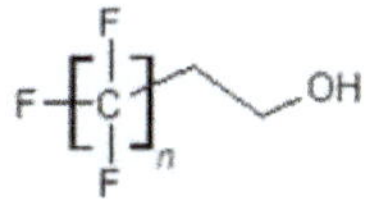
Fluorotelomer iodides (FTIs)	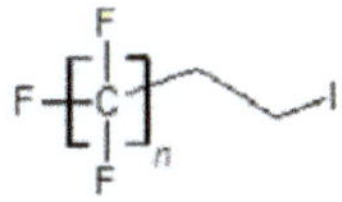
Fluorotelomer acrylates (FTAs)	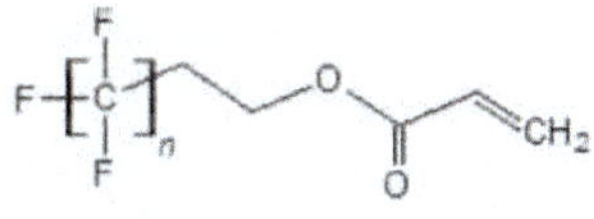
Fluorotelomer methacrylates (FTMAs)	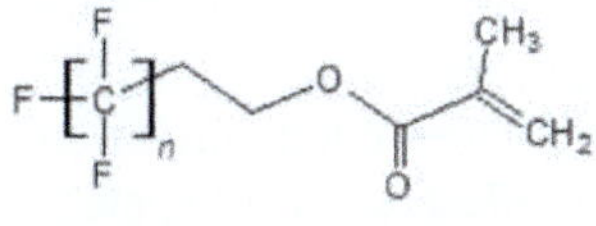
Polyfluoroalkyl phosphate monoesters (monoPAPs)	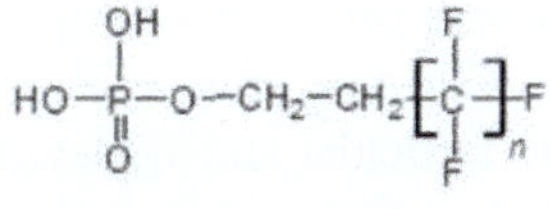
Polyfluoroalkyl phosphate diesters (diPAPs)	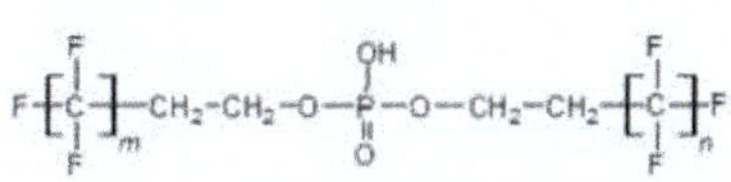

hexafluoropropylene oxide-dimer acid (HFPO-DA: GenX), 4,8-dioxa-3 H-perfluorononanoic acid (ADONA), 9-chlorohexadecafluoro-3-oxanonane-1-sulfonate (F53B major) and 11-chlororeicosafluoro-3-oxaundecane-1-sulfonic acid (F53B minor) [4], together with a number of emerging PFASs predicted by different software (e.g. SPARC, Marvin 15.10.26, ACD/Labs 12.0, EPISuite 4.1) and reported in different studies [5]. These emerging fully or partially fluorinated compounds include cyclic PFASs, perfluoroalkyl phosphonic acids (PFPAs) and other PFASs in aqueous film-forming foams (AFFFs) and commercial surfactant products as well as cationic and zwitterionic PFASs. A recent survey identified over 4700 PFASs currently in use on the global market [2].

6.1.1 Properties

PFAS molecules consist of hydrophobic tails and hydrophilic heads, such as carboxylic or sulfonic groups. Due to the high electronegativity and the small size of the F atom, the C-F bond is the strongest covalent bond in organic chemistry (~485 kJ/mol). The presence of three pairs of non-bonding electrons around each F, as well as the effective shielding of C atom by F atoms, lead to incredibly high thermal and chemical stabilities [6].

PFASs are non-flammable, thermodynamically stable and highly resistant to hydrolysis, metabolism, photolysis and other forms of degradation (acids, bases, oxidants, microbes) [7, 8]. Because of these properties and their lipophobic nature, PFASs and their polymers have been widely used since the 1950s in numerous industrial applications including surfactants, water repellents sprays, lubricants, adhesives, additives, coatings and paints [9, 10]. Surfactant applications that take advantage of the unparalleled aqueous surface tension-lowering properties include processing aids for fluoropolymer manufacture, and AFFFs used to extinguish fires involving highly flammable liquids [11]. PFASs have also been used in a number of commercial applications as non-stick cookware, food packaging, insecticides and waterproof and fire-resistant fabrics [4, 12]. According to Glüge et al. [13], more than 200 use categories and subcategories have been identified for more than 1400 individual PFASs. In their extensive review, besides well-known categories such as textile impregnation, firefighting foam and electroplating, other categories as ammunition, climbing ropes, guitar strings, artificial turf and soil remediation are also described.

On the other hand, because of their amphiphilic properties and preferential binding to proteins, PFASs have also been detected in wildlife, posing clear, potential hazards to human health [14]. In general, PFASs are considered extremely persistent in the environment, highly toxic and with the potential to bio-accumulate and bio-magnify through the food chain. These compounds may act as endocrine disruptors, and may have multiple other toxic effects such as hepatotoxicity, immunotoxicity, reproductive and developmental toxicity and tumorigenic effects [11]. Some of them have also been linked to neurodevelopmental toxicants at various levels [15]. There is some evidence of PFASs association with testicular and kidney cancer, and some studies even suggest an association with prostate cancer, although the data are inconsistent [16].

6.1.2 Legislation

Due to the persistent, bio-accumulative and toxic properties, PFOS, its salts and perfluorooctane sulfonyl fluoride (PFOSF) have been included in Annex B of the

Stockholm Convention on Persistent Organic Pollutants (POPs) in May 2009 [17]. European Commission also prohibited the general use of PFOS after June 2008 [18], and in 2012 proposed Maximum Allowable Concentrations for this sulfonate and its salts in inland waters [19]. In addition, C8–C14 PFCAs, perfluorohexane sulfonate (PFHxS), and their sodium and ammonium salts were included in the candidate list of Substances of Very High Concern (SVHC) in the EU and are candidates to be listed under the Convention [11].

PFASs were also prime candidates for chemicals that will need authorisation within the REACH regulation [20]. The production, placing on the market and use of PFOS is regulated by EU legislation on persistent organic pollutants [21]. Restrictions related to the manufacture and marketing of PFOA entered into force on 4 July 2020, following scientific assessments by the European Chemicals Agency (ECHA).

Furthermore, already in 2008, the European Food Safety Authority (EFSA) established tolerable daily intake (TDI) for PFOS of 150 ng/kg body weight bw and for PFOA 1500 ng/kg bw. Recently, EFSA's Contaminants Panel has established a new safety threshold for the main PFASs, which accumulate in the body. The threshold is a tolerable weekly intake (TWI) of 4.4 ng/kg bw applicable to the sum of the four PFASs (PFOA, PFNA, PFHxS and PFOS) with the longest half-life in the body [22].

The environmental quality safety standard (EQS) for PFOS established the annual average (AA) EQS for inland surface waters in 6.5×10^{-4} μg/L and the AA EQS for other surface waters in 1.3×10^{-4} μg/L. The biota EQS is 9.1 μg/kg wet weight (ww) [23]. PFASs have also been recently included in the new directive for drinking water where limit values for PFAS are 0.1 μg/L for the sum of specific PFASs considered to be of concern for drinking water (including 20 substances) as well as 0.5 μg/L for "PFASs total". Other countries such as Canada developed similar restricting regulations [24], and the United States Environmental Protection Agency (US EPA) established a short-term Health Advisory Level for a combined concentration of 70 pg/mL of PFOA and PFOS in 2016 [25]. Additionally, safety values for PFOA, PFOS, PFHxA, PFPeA, PFHxS, PFNA and PFBS were also established recently by other US local Agencies including those of Minnesota, New Hampshire, New Jersey, Vermont and California [6].

6.1.3 Reported or Potential Metabolites and/or Transformation Products

The production of isomers varies by manufacturing process [26]. The telomerisation process produces exclusively linear PFASs, whereas the electrochemical fluorination results in a mixture of branched and linear isomers [1]. There have been a handful of studies investigating the varying degradation kinetics of branched POSF-based chemicals that are metabolically transformed to PFOS [27]. Different detection techniques have been used to identify the isomers, but a major challenge is the lack of standards for the different branched variations, so isomers are generally named by their order of retention [28].

Some polyfluoroalkyl chemicals contain labile functional groups that enable abiotic and biotic transformations in the environment, generally resulting in the formation of terminal PFCA and PFSA degradation products [28] (Figure 6.1). Such precursors, already mentioned in the introduction, include FTOHs, fluorotelomer acrylates (FTAs), FTIs, FTOs, N-alkyl perfluoroalkane sulfonamides, N-alkyl perfluoroalkane sulfonamidoethanols and polyfluoroalkyl phosphates [1].

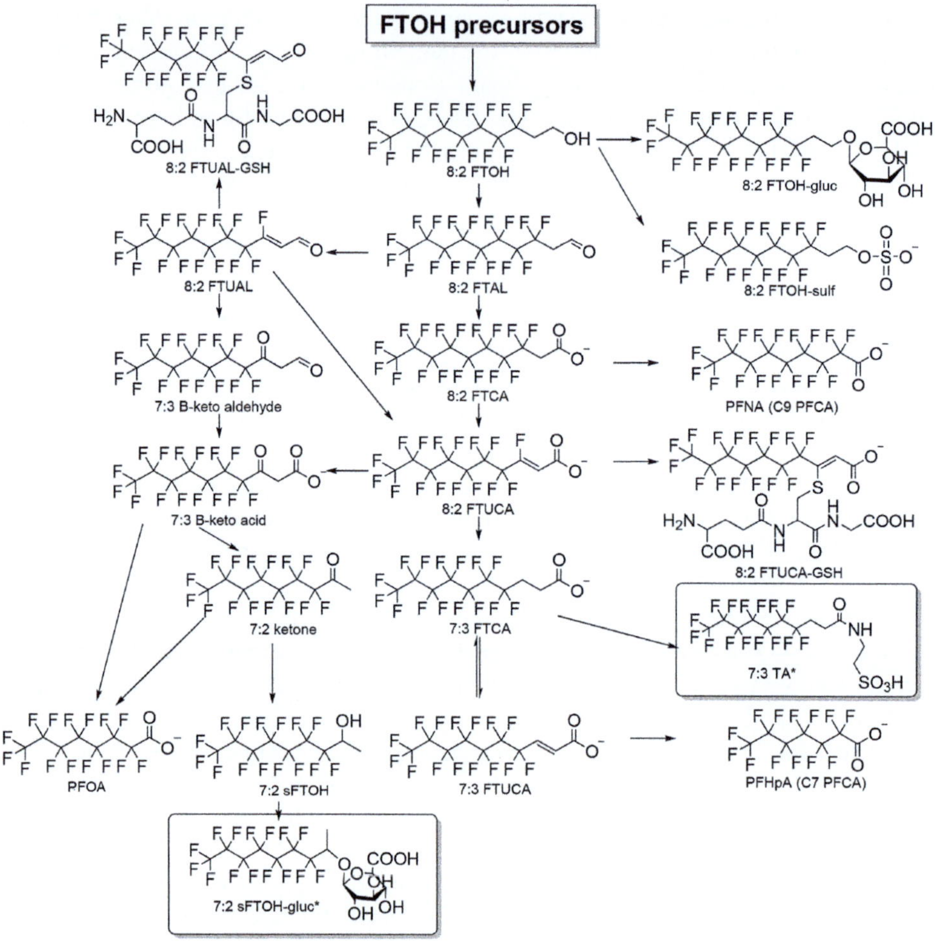

Figure 6.1 Biotransformation pathway of FTOH-based chemicals in juvenile rainbow trout [28]. Reproduced with the permission of Elsevier.

Other polyfluoroalkyl precursors include commercial surfactants such as the fluorotelomer disubstituted phosphate esters (diPAPs) and N-ethyl perfluorooctane sulfonamidoethanol-based phosphate diester (SAmPAPs), components of AFFFs such as 6:2 fluorotelomer alkyl betaine (6:2 FTAB), and polymers made from acrylate-linked fluorotelomers [28]. During this degradation, reactive intermediates can be generated, which can be more toxic and persistent in the environment than the PFCA and PFSA terminal degradation products [29].

Recent studies used pure chemicals to conclude that PFPiAs can form PFPAs in fish [30] and rats [31]. Perfluoroalkyl amides have been also identified as impurities in some perfluoroalkyl sulfonamide containing industrial products, especially the 3M N-Methyl perfluorooctane sulfonamidoethanol (MeFOSE) and N-Ethyl perfluorooctane sulfonamidoethanol (EtFOSE) [32]. Perfluoroalkyl amides hydrolyse to PFCAs. This explains partly the PFOA observed in degradation studies of POSF-based chemicals [33].

6.1.4 Occurrence in the Environment

PFASs have been ubiquitously reported throughout the environment, even in remote regions as the Arctic [34], because of the high volumes used and their ability to undergo long-range atmospheric [35] and/or oceanic [36] transport. In general, the more volatile and neutral PFASs (nPFASs) such as FTOHs, perfluorooctane sulfonamides (FOSAs), perfluorooctane sulfonamidoethanols (FOSEs), perfluoroalkane sulfonamidoethyl acrylates (FASACs), FTOs and FTAs are transported via the atmosphere. On the other hand, PFCAs and PFSAs are ionisable, and consequently more water-soluble, than nPFASs [1]. In fact, the first global extent report of such contamination was published in 2001 for PFOS in wildlife [8]. Despite long-chain PFASs remain in the environment for long periods, due to their high chemical, microbial and photolytic resistance, the new emerging short-chain PFASs have currently been detected more frequently because of their scarce degradability [37].

PFASs contamination across the world can be explained by local sources and/or long-range transport processes of PFASs or their precursors [6]. The main sources of PFASs identified are those directly related to industrial production [38, 39], use of AFFFs [11, 40], landfill leachates [41] and wastewater treatment plant (WWTP) effluents [42, 43]. Atmospheric deposition has also been reported as an important source [44].

PFASs have been detected in aqueous matrices including surface water such as rivers [5, 45], lakes [46] and sea [47, 48] as well as in drinking water [49, 50], wastewaters [42, 51], groundwater [52, 53] and even rainwater [54]. As PFASs manufacturing has moved from Western to Asian countries, an increasing number of studies have been conducted in Asia [55]. In a wide study in 2016, water samples were collected from 65 rivers and 34 coastal drain outlets around the Bohai Sea, China, where high levels of PFOS (0.13–881 ng/L) and PFOA (1.28–24,700 ng/L) were detected [47]. In the Albufera Natural Park (Valencia, Spain), a characteristic coastal wetland, samples of surface waters (including a Mediterranean lake) were analysed. Of the 21 PFASs included in this study, 10 were detected in surface waters with the most frequently detected being PFOA (100%), PFOS (92%) and perfluoropentanoic acid (PFPeA, 83%). The highest concentrations were found for PFOS (47.8 ng/L) and 2H-Perfluoro-2-decenoic acid (FOUEA, 46.1 ng/L) [46]. Recently, Zeng et al. [56] determined 58 PFASs (including chlorosubstituted perfluoroalkyl phosphonic acids (Cl-PFPAs), 9-chlorohexadeca fluoro-3-oxanonane-sulfonate (9Cl-PF3ONS) (trade name: Fe53B)) in surface water around Zhangjiang High-Technology Park in Shanghai. The concentration of ΣPFASs was 115–600 ng/L and the main pollutant (41.6% of ΣPFASs) was perfluoro-3-methylheptanoic acid (P3MHpA), which is isopropyl isomer of PFOA. It was detected in all samples and its concentration was far higher than that of PFOA with the maximum of 432 ng/L determined in surrounding integrated circuits company.

Concentrations of novel PFASs in aqueous samples, including cyclic PFSAs, PFPAs, PFESAs and PFECAs, have also been reported in a few studies, indicating that there is a paucity of information on many probable PFASs impacting the world's ecosystems (Figure 6.2). Wang et al. [39] presented results from a survey of PFASs in water samples collected at the mouths of Chinese rivers. F-53B was frequently detected at a maximum level of 78.5 ng/L in samples collected near metal-plating facilities. Similarly, information about HFPO-DA (GenX) and 6:2 FTAB occurrence is related to their local production and

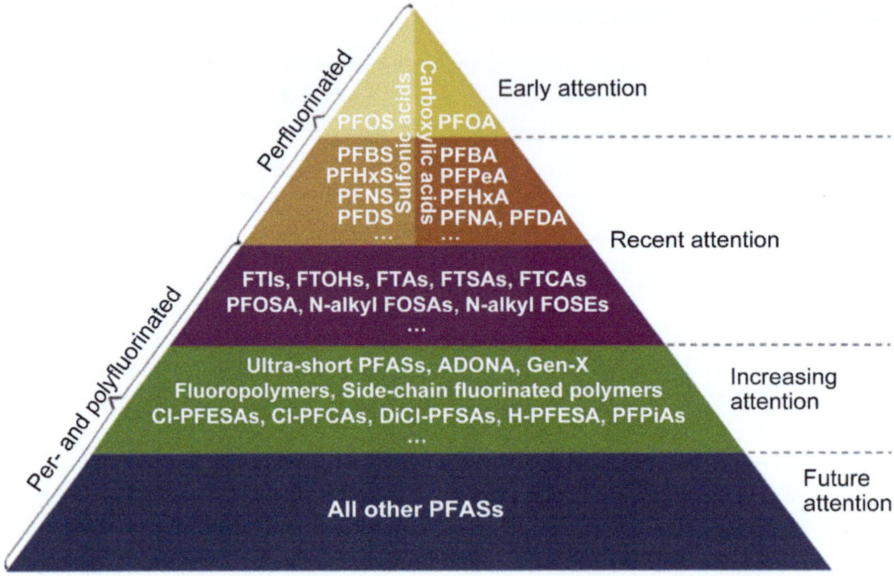

Figure 6.2 Occurrence of PFASs in the environment [48]. Reproduced with the permission of Elsevier.

application. In a large occurrence study, HFPO-DA was found to occur at a maximum concentration of 3825 ng/L (detection rate of 76%) in the Xiaoqing River System in China [57].

Concentrations of 23 PFASs, including new classes, were determined in samples from the South China Sea [58]. Total PFAS concentrations from 195 to 4925 pg/L were detected at 51 sampling stations. PFOS concentrations were found to be two times higher than those measured nine years ago at the same locations. In a recent study, PFOA, PFOS and its precursors were assessed in seawater samples from Ría de Vigo estuary (Galicia, Spain) [59]. Surface and depth samples were analysed from the inner to the outside of the estuary measuring levels of 0.64 ng/L and <12.6 ng/L for PFOS and PFOA, respectively. N-methyl perfluorooctanesulfonamide (n-MeFOSA) and N-ethyl perfluorooctanesulfonamide (n-EtFOSA) were detected at low concentrations in some samples.

Tap water samples (n = 14) were screened for 14 PFASs, including 4 PFSAs (C4-C10) and 10 PFCAs (C5-C14), in South Korea in 2017, reporting concentrations from nd to 189.6 ng/L [60]. In a national screening program carried out in two consecutive years (2010 and 2011) in Spain, 21 PFASs (PFCAs: C4–C14, C16, C18; PFSAs: C4, C6-C8 and C10 sulfonates and C8 sulfonamide) were determined in influent, effluent, and sludge from 16 different WWTPs located in the Ebro (6), Guadalquivir (5), Jucar (2) and Llobregat (3) Rivers [42]. All wastewater samples were contaminated with at least one PFAS, and PFBA, PFPeA and PFOS were the most frequently detected. The highest concentration in 2010 was in a Guadalquivir River WWTP (PFHxA: 5.60 μg/L) and, in 2011, in an Ebro River WWTP (PFBS: 0.31 μg/L). The total concentrations of 14 PFASs, including original PFASs and two novel PFASs such as 6:2 chlorinated polyfluorinated ether sulphonic acid (6:2 Cl-PFESA), with the trade name of F-53B, and 6:2 FTSA in ground water samples (n = 102) collected from 13 non-industrialised cities in Jiangsu Province, China, were 2.69–556 ng/L [52].

Boiteux et al. [38], reported the detection of 6:2 FTAB in a river near an industrial WWTP receiving wastewater from a fluorochemical manufacturing site at northern France. In 2013, 6:2 FTAB was measured at levels much higher than those of the other target PFASs. The concentration decreased from 968 ng/L near the WWTP to 248 ng/L at ~62 km away from the WWTP discharging point. The worldwide distribution of PFESAs and PFECAs in surface water was investigated using 160 samples collected between September and December 2016 [61]. HFPO-DA, hexafluoropropylene oxide trimer acid (HFPO-TA) and 6:2 Cl-PFESA were widely detected in all countries including China ($n = 106$), the United States ($n = 12$), the United Kingdom ($n = 6$), Sweden ($n = 10$), Germany ($n = 14$), the Netherlands ($n = 6$) and South Korea ($n = 6$), with median values of 0.95, 0.21 and 0.31 ng/L, respectively. The hydrogen-substituted analogue of 6:2 Cl-PFESA (6:2 H-PFESA) was only detected in China with a high detection rate (>95%).

In relation to PFASs' occurrence in air, the Global Atmospheric Passive Sampling (GAPS) survey researched the global occurrence and long-range atmospheric transport of PFASs [62]. In this survey, various PFASs were monitored with solvent-impregnated polyurethane foam (SIP) samplers including PFCAs, PFSAs, FASAs, FASEs and FTOHs at 21 locations around the world. In the 2009–2015 GAPS survey, FTOHs were detected at ranges <0.4–21 pg/m^3 in the polar region, and 40–238 pg/m^3 in urban sites [63]. Regarding indoor air monitoring of PFASs, Yao et al. [64] concluded that FTOHs were mostly detected in both hotels and houses ranging from 246 to 62,100 pg/m^3. These authors also reported concentrations of PFCAs and PFSAs in the range 90.9–1,970 pg/m^3 and 86.8–587 pg/m^3, respectively, and were higher in houses, while total FASA/FASE levels were higher in hotels with values from no detectable (ND) to 2,460 pg/m^3 [64].

In general, PFASs tend to exhibit strong adsorption to solid matrices [65, 66]. A number of studies have investigated PFASs and their alternative substances in soil [40, 67], sediment [68, 69], sewage sludge [42, 51] and dust samples [64, 70] collected from different environments. Additionally, soil core samples have been studied to assess temporal trends of both original and novel PFASs [71]. A recent study analysed four sets of soil samples collected days after the derailment of an unmanned train carrying crude oil and subsequent fire in the town of Lac-Mégantic, Quebec, which led to the use of 33,000 L of AFFF concentrate. A total of 33 PFASs were quantified in the soils. The highest concentrations corresponded to those of 6:2 fluorotelomer sulfonamidoalkyl betaine, 6:2 and 8:2 fluorotelomer sulfonates and short-chain PFCAs. Qualitative analysis revealed the presence in soil of 55 additional PFASs that had been previously identified in AFFF formulations [67].

Sediment samples from two important Spanish River basins (Ebro and Guadalquivir) were analysed for the occurrence of 21 PFASs [68]. Eight PFASs were detected in the samples from Ebro and nine in those from Guadalquivir River. The PFASs most frequently detected were PFBA, PFPeA, PFOA and PFOS, with maximum concentration in Ebro samples of PFOA (32.3 ng/g dry weight dw) and in Guadalquivir samples of PFBA (63.8 ng/g dw). Perfluorooctane sulphonamide ethanol-based phosphate (SAm-PAP) diester and triester were also detected in freshwater sediments from Taihu Lake in the range of <0.03–4.3 ng/g dw and <0.024–1.13 ng/g dw, respectively [72]. Novel PFASs such as 6:2 Cl-PFESA and its analogues and 6:2 fluorotelomer sulphamide alkylbetaines (FTABs) have been detected in abiotic solid matrices [5]. Seventeen target PFAAs were analysed in marine sediment samples from the Yangtze Estuary and East China Sea.

PFAAs were widely present in the samples, with PFOA and PFOS identified as dominant, and ΣPFAAs ranging from 0.67 ng/g dw to 36.75 ng/g dw. Results indicated that terrigenous input strongly influenced PFAA distribution in sediments from the study areas [69].

Temporal studies and model simulations conclude that sediment, soil and sludge are an important environmental sink for PFASs [55, 73]. Regarding PFCAs and PFASs, a number of novel PFASs, including cationic and zwitterionic species, have been detected in recent studies [40, 67]. PFAS levels were measured in core sediment samples collected from the Great Lakes region of North America in 2006 and 2009, with higher concentrations of PFASs detected in surface layers than in deeper layers [74]. Marine sediment core samples collected in China and Korea presented a similar trend [71]. In a systematic review about temporal tendencies, increasing trends for PFOS and some PFCAs were found in global sediment core samples in the period 1850–2013 [75]. Moreover, diPAPs and PFPiAs were detected in the most recent layer.

Sewage sludge from 16 Spanish WWTPs were analysed, in 2010 and 2011, and all samples, except two of Guadalquivir River basin, were contaminated with at least one PFAS. PFBA, PFPeA and PFOS were the most frequently detected, with maximum concentration in 2010 of 1.79 µg/g dw (PFOS, in Llobregat River), and in 2011, 1.88 µg/g dw (PFBA, in Guadalquivir River). According to authors, high PFAS values in sludge could be related to poor removal efficiencies and can be attributed to their adsorption [42]. Recently, Semerád et al. [51] determined 32 PFASs including hexafluoropropylene oxide-dimer acid, HFPO-DA (GenX), sodium dodecafluoro-3H-4, 8-dioxanonanoate (NaDONA) and other new PFASs in sludge of 43 WWTPs from Czech Republic. Values of ΣPFASs were from 5.6 to 963.2 ng/g. PFOS was the most abundant, reaching 932.9 ng/g. GenX and a high percentage of short-chain PFASs found in samples suggest that they are used as a replacement for long-chain PFASs (especially restricted PFOA and PFOS). Research on PFASs in dust samples from rooms of homes and hotels, textile shops and cinemas in Tianjin, China has shown that concentrations and composition are different depending on collection sites; samples collected from hotels presented higher levels of FTOHs (24.8–678 ng/g), and those from houses showed more short-chain PFCAs (41.6–226 ng/g) and much fewer FTOHs [64].

Since the first global extent report of PFOS contamination in wildlife published in 2001 [8], there has been an increasing trend in PFASs research, including their isomers, in all kinds of biota samples, especially in that living close to sites where PFASs are currently manufactured [75, 76]. Plankton samples from the tropical and subtropical Pacific, Atlantic and Indian Oceans were collected during the Malaspina 2010 circumnavigation expedition. Fourteen ionisable PFASs, including PFOA, PFOS and their respective linear and branched isomers, were screened. PFOA and PFOS levels in plankton were 0.1–43 ng/g dw and 0.5–6.7 ng/g dw, respectively. Authors concluded that plankton samples showing the highest PFOS concentrations also presented the largest relative abundances of branched PFOS, suggesting a selective cycling/fractionation in the surface ocean mediated by plankton [77]. A comprehensive survey carried out in a temperate macrotidal estuary (Gironde, SW France) examined the trophodynamics of a suite of 19 PFASs. In the 147 biota samples (18 taxa) analysed, PFOS, FOSA and C8–C14 PFCAs were the most-recurrent analytes. ΣPFASs ranged between 0.66 and 45 ng/g wet weight (ww) of the whole body. Benthic organisms showed higher ΣPFASs than demersal organisms and displayed specific composition profiles with higher relative abundances of C8 and C9 PFCAs [78].

Significant accumulation of PFASs has been reported in fish. A national campaign sampling fishes in main Spanish River basins (Ebro, Guadalquivir, Jucar and Llobregat) showed that PFOS was the most frequently detected compound. In all basins the high occurrence of short-chain PFASs intended to replace the long-chain ones was remarkable. Mean concentrations detected in the different species considered were 0.63–431 μg/kg ww. Maximum concentration in the Ebro samples was 1280.2 ng/g ww (PFHxA), and in Guadalquivir was 79.8 ng/g ww (PFOS) [45, 68, 79]. Perfluoroethylcyclohexanesulfonate (PFECHS) was reported in the top predator fish (<method detection limit MDL – 3.7 ng/g ww in whole body homogenate) from the Great Lakes, North America, in 2011. PFOS was found as a predominant compound in samples of fish, highlighting that concentrations of the detected PFASs were not statistically different from previously reported 2004 trout data in Lake Ontario [80].

Concentrations of individual PFASs in mussels and oysters from south China and Japan were 113.6–586.0 pg/g ww for PFOS, 63.1–511.6 pg/g ww for PFHxS, 9.3–30.1 pg/g ww for PFBS and 37.8–2957.0 pg/g ww for PFOSA. Some other PFASs such as PFDoDA (Tokyo Bay, Japan, 195.9 pg/g ww) and PFDA (Fuzhou, China, 131.7 pg/g ww, Tokyo Bay, 118.6 pg/g ww) were identified [81]. Grey mullet liver (*Chelon labrosus*) and mussel (*Mytilus galloprovincialis*) samples from the Basque Coast (North of Spain) and Yellowfin tuna muscle tissue (*Thunnus albacares*) samples from the Indian Ocean were screened for the determination of 14 PFASs and 10 potential precursors. Furthermore, 8:2 monosubstituted polyfluorodecyl phosphate (8:2 monoPAP) in mussels and 8:2 disubstituted polyfluorodecyl phosphate (8:2 diPAP) in tuna were detected [82]. Other novel PFAS (6:2 Cl-PFESA and HFPO-TA) were detected in various tissues from black spotted frogs collected in a rice paddy field near a large-scale fluorochemical production site in China [83]. Liver samples ($n = 56$) collected from all sampling sites contained 6:2 Cl-PFESA in the range of 0.13–119 ng/g ww, while HFPO-TA was found in samples ($n = 4$) from only one site (Huantai) in the range of 6.51–27.30 ng/g ww. In another study also developed in China (Bohai Sea), 152 biota samples including invertebrates, fishes, seabirds and mammals were analysed to assess the residual level, spatial distribution, bioaccumulation and bio-magnification of Cl-PFESAs. 6:2 Cl-PFESA was found in concentrations <MDL up to 3.84 ng/g ww and it is the dominant congener when compared with concentrations of 8:2 Cl-PFESA. Compared with other bays and regions, levels of 6:2 Cl-PFESA were considered relatively high in bivalves and fishes from Liaodong Bay. These levels appeared to increase in relation to previous measurements [76]. PFASs, especially those that have been detected more recently, are frequently determined in biota because this matrix also provides information on PFASs' toxicity.

6.2 Sample Preparation and Collection

6.2.1 Protocols for Collecting and Preparing Samples

Special attention is required for collecting and preparing samples for analysis since PFASs are ubiquitous, and can be present in any plastic or waterproof material used during sampling. For surface water samples (i.e. rivers, lakes, sea), the procedure is to collect a certain amount of water using a stainless steel bucket and transfer it into polypropylene (PP) bottles. Glass bottles must be avoided since PFASs can bind to glass in aqueous solutions. Teflon materials may be a source of PFASs contamination [84] and are not recommended. To reduce possible contamination, the bucket can be rinsed with methanol at the laboratory and with the water to be sampled at least twice prior to use [58, 85, 86]. The samples should be analysed within 24 h, otherwise they should be stored at 4°C (if the analysis is in the following days) or −20°C (for longer periods of storage). When vessels are used for transportation, attention should be paid to the paint of the boat and life jackets used during sampling because they may contain PFASs [86]. For obtaining the samples of water at a specific depth, Niskin bottles attached to a rosette sampler are recommended [87]. Passive samplers have also gained popularity within PFASs analyses, and the use of polar organic chemical integrative samplers (POCISs) is becoming a common technique for measuring concentrations of PFASs in water, obtaining concentrations in ng/L in water, or in ng/POCIS [88]. Wang et al. [89] used POCIS-immobilised ionic liquid (IIL) for 7 days to evaluate levels of five PFAAs in the influent and effluent of a WWTP in Tianjin (China). According to the authors, all five PFASs could be accumulated in the samplers, with concentrations ranging from 6.5×10^{-3} to 3.6×10^{-1} nmol/L in the former and from 1.3×10^{-2} to 2.2×10^{-1} nmol/L in the latter. Modified POCISs with anion exchange sorbents, POCIS-WAX and POCIS-HLB, have also been applied to study 26 PFAS levels in a drinking water treatment plant [90]. Both type of POCISs showed good agreement with composite water sampling.

PFASs in air might be volatile or non-volatile. The latter are commonly adsorbed onto the dust, which should be kept in mind for extraction and sample preparation. Air sampling techniques include glass-fibre filters (GFF), quartz-fibre filters (QFF), XAD resin sandwiched by polyurethane foam (PUF) and solid-phase extraction (SPE) cartridges, alone or in combination for the analysis of volatile and neutral PFASs [55]. Both outdoor and indoor air samples are collected by PUF/XAD/PUF using a high-volume air sampler [91], or by an SPE cartridge using a low-volume air sampler [34]. Air sampling volume of 300–2,000 m^3 for outdoor air and 20–200 m^3 for indoor air are recommended [92], although for indoor air this volume can be decreased to 0.2–8 m^3 [64]. Passive air samplers consisting of solvent-impregnated polyurethane foam (SIP) for the monitoring of PFASs in air have also been developed [93]. Background contamination must be eliminated during washing/preconditioning, storage and transport of samplers. XAD and PUF sorbents are commonly Soxhlet extracted (24–30 h) before being placed in the sampler to remove contaminants [94]. SPE cartridges are generally washed with methanol and dried with high-purity nitrogen before use, while samplers are wrapped with aluminium foil or stored in PP containers at −20°C until analysis [64, 92, 94].

In general, when solid samples are collected, they need to be dried, sieved (usually for soil, sediment and sludge) and homogenised before extraction. For sediment, the sampler should be selected based on the research objectives and sampling location [48]. A van Veen or portable Ekman grab sampler has been used [86, 87], as well as a stainless steel scoops, acrylic tubes and a hand piston sediment sampler [65, 66, 85]. Soil/sediment core samples can be collected using a Model MC-400 Multi-corer [73]. Dust is often collected from houses, offices and shops to estimate human exposure to PFASs [64, 95]. Dust collection is generally carried out using a vacuum cleaner [64]. Stainless steel tools should be pre-cleaned with methanol at the lab before manipulating the sediment.

Samples can be stored in PP bottles (pre-cleaned with methanol) or in any other non-plastic container. The samples should be stored at −20°C until further analysis [79].

For biota samples, a vast array of sampling procedures should be considered, depending on the type of organism to be collected. These procedures go from collecting plankton samples with a 50-mm mesh net using vertical trawls [77], to fish, crustaceans, bivalves, etc. that can be sampled by electric [45, 68], bottom trawl [76] and gill nets [80] fishing or purchased from markets [46, 78]. Wildlife samples (fish, frog, marine organisms, etc.) should be transported immediately to the laboratory, dissected with stainless steel tools pre-washed with methanol [76, 83], homogenised and stored at −20°C until analysis are carried out [45, 80].

6.2.2 Sample Extraction and Clean-up

Sample pre-treatment procedures are continuously improving to minimise PFASs background levels and optimise their recoveries. However, contamination and adsorption of PFASs into laboratory utensils can still happen. For example, materials as syringe nylon membrane PP filters [85], cellulose acetate membrane filters [86] and polyethylene filtration units [78] are recommended for filtering since they do not release PFASs. In any case, it is highly advised to include laboratory blanks to check possible contamination in each batch.

Aqueous samples usually contain lower levels of PFASs (ng/mL or pg/mL) compared to solid samples, with high levels of PFASs (ng/g or higher). Therefore, the aqueous sample preparation should pay more attention on the isolation and pre-concentration of PFASs [7]. Although liquid–liquid extraction (LLE) with ion-pair extraction (IPE) can be used, SPE is a commonly recommended technique for PFASs extraction from aqueous and, in general, for clean-up of liquid extract. A polymer-based SPE cartridge is commonly employed for PFASs analyses [55].

Different types of SPE cartridges have been employed as Sep-Pak tC18 (trifunctional octadecyl/silica) [96], Oasis Hydrophilic-Lipophilic Balance (HLB) [84], OasisWeak Anion Exchange (WAX) [85, 97], Strata-X or ENVI-Carb cartridges [66] among others. Elution with methanol allows the collection of the neutral fraction of PFASs, while methanol with 0.1% NH_4OH is recommended to collect the PFASs' anionic fraction. Possible PFASs contamination due to the cartridges needs to be considered. After SPE, samples are concentrated under a gentle stream of high-purity nitrogen gas and transferred to PP vials, with PP insert if possible, and PP septa for instrumental analysis [58, 86].

In addition to SPE, other extraction procedures proposed for tap, river water and urine samples are more in-line with green chemistry principles, which utilise less extraction solvents and thereby decrease the environmental contamination. These "greener" methods involved solid-phase micro extraction (SPME) [98], dispersive liquid–liquid micro extraction (DLLME) [99], ultra-performance convergence chromatography (UPC2) [100] and stir bar sorptive extraction (SBSE) [101].

The PFASs collected from air samples into XAD resin or PUF are extracted by Soxhlet [93, 94] or pressurised liquid extraction (PLE) [34]. PFASs collected on a SPE cartridge are eluted using organic solvents selected depending on target PFAS properties. Neutral PFASs are retained in HLB or ISOLUTE ENV + cartridges and usually eluted with methanol [63], while ionic PFASs such as PFCAs, PFSAs and diPAPs are isolated in WAX cartridges and generally eluted with methanol containing an ammonium solution to get a basic pH [64]. For solvent-impregnated polyurethane foam (SIP), collected PFASs are subjected to Soxhlet extraction with appropriate organic solvents such as acetone/petroleum ether (1:1), methanol or ethyl acetate, or cold column extraction with ethyl acetate [102]. Additional clean-up by ENVI-Carb is employed in some cases [102]. Airborne particulate matter collected on filters is generally subjected to Soxhlet extraction with dichloromethane or ultrasonic extraction with methanol [103].

PFASs extraction from sediment, soil or suspended particulate matter samples can be achieved by Soxhlet, PLE, or solid-liquid extraction (SLE), followed by additional clean-up procedures using graphite carbon materials such as ENVI-Carb, SLE, or ion-pair extraction (IPE) [92]. Organic solvents such as methanol and methyl tertbutyl ether (MTBE) [66], or mixtures of these with strong acids and bases such as methanol with acetic acid (acid digestion) [87], or with sodium hydroxide (alkaline digestion) [104] have been used. After vortexing and ultra-sonication, the extract is centrifuged and the supernatant is transferred. Two or three extraction cycles should be completed before clean-up to increase the extraction efficiency [73, 79]. A common combination is SLE followed by SPE clean-up using ENVI-Carb or an SPE cartridge (OASIS WAX, OASIS HLB or C18) under neutral or basic conditions. Wang et al. [69] compared SPE (WAX, HLB and C18) with solvent extraction and filtration as extraction procedure and concluded that solvent extraction achieves better recoveries.

In the case of dust, sample extraction procedures need to extract PFASs with diverse properties, especially highly hydrophobic compounds, cations and zwitterions [40] since all of them can be present. The analysis of both volatile and non-volatile PFASs implies that methods increase complexity involving most of the times fractionating with different extracting solvents and/or repeated extraction steps [64, 95]. Other methods such as dispersive SPE (dSPE) have emerged as a green clean-up possibility that can avoid potential contamination from polytetrafluoroethylene lines and SPE vacuum manifolds [105].

In general, biota samples are complex matrices. Then, the development of efficient extraction and clean-up methods has always been challenging [92, 106]. In these samples, special care must be taken with the ability of PFASs to form bonds with proteins, and therefore extraction methods are generally more thorough. Various extraction methods such as SLE [77], acidic/alkaline digestion coupled with SLE [107], PLE [78, 108], ion-pair extraction [108], alkaline digestion followed by LLE, and alkaline digestion followed by SPE [81] have been widely employed in the last decade. Matrix effect related to high lipid and protein contents can result in suppression or increase of detection. Eluates are typically cleaned-up with an SPE cartridge containing HLB, WAX [109] or graphite carbon (e.g., ENVI-Carb) [82]. PFSAs, PFPAs and FOSAs were extracted from fish samples by QuEChERS (Quick, Easy, Cheap, Effective, Rugged and Safe) with d-SPE followed by C18 sorbents, and ENVI-Carb to reduce matrix effect and remove the co-extracts [110]. Last developments in these clean-up techniques include specific sorbent modifications like the Captiva ND and Phree^{TM} to separate phospholipids, Ostro^{TM} to remove phospholipids and proteins and enhanced matrix removal (EMR-Lipid) to eliminate mostly triglycerides [111]. The most promising approach for minimising matrix effects is reducing the sample amount and the extract volume injected [55].

6.3 Determination of PFASs

6.3.1 Development of the Instrumental Method

Two types of method are used nowadays to determine PFASs: chromatographic methods and sensors.

6.3.1.1 Chromatography-Mass Spectrometry

PFASs have been determined using mostly liquid (LC) and gas (GC) chromatography. Table 6.2 summarises the chromatographic determination technique most used for each PFAS. LC techniques applied are high-performance liquid chromatography (HPLC), ultra-high-performance liquid chromatography or ultra-performance liquid chromatography (UHPLC/UPLC) and capillary liquid chromatography (CLC). LC is generally used for the screening of ionic PFASs (i.e., PFCAs, PFSAs), while GC is utilised for volatile, semi-volatile and neutral PFASs (i.e., FTOHs, FTSHs, FTACs). In LC, PFASs are well separated by conventional reverse phase C18 columns and mobile phases containing salts to improve the peak shapes. In this sense, gradient mixtures of water and methanol/acetonitrile containing 5–50 mm ammonium acetate/fluoride are the most common mobile phases [45, 46]. UHPLC – the advanced LC platform that allows working with <1.7 μm particle size – has shown outstanding performance in terms of column efficiency, and analytical resolution with less mobile phase consumption [69, 106]. For GC separation, it is generally employed a WAX column such as DB-WAX (column size of 0.25 mm diameter, 30 or 60 m length, film thickness of 0.25 μm) [94, 102].

Determination of PFASs is generally carried out by tandem mass spectrometry (MS) using LC (LC-MS/MS) or GC (GC-MS/MS). Polar perfluoroalkyl acids (PFAAs) are generally determined by LC-MS/MS using electrospray ionisation (ESI) through their deprotonated ionic forms $[M-H]^-$. PFAAs are characterised by strong surfactant properties, which increase partitioning to the droplet's surface, and their volatility helps in their transfer to the gaseous phase [28]. The extremely low pK_as are also advantageous as the analytes already contain a negative charge [28]. Semi-volatile and volatile analytes such as FTOHs, FTSHs, FTACs, FTIs, EtFOSA, 1H-perfluoroalkanes and other degradation products are generally separated and detected by GC-MS with chemical ionisation (CI), looking for protonated ions in positive mode, or radical anions in negative ionisation mode, although the recently introduced soft interfaces at atmospheric pressure chemical ionisation (APCI) [112] and atmospheric pressure photoionisation (APPI) [113] have also been tested. Electron ionisation (EI) in positive mode has been used for neutral, cationic or zwitterionic related chemicals in AFFF-impacted zones [5, 40]. However, the EI leads to a too much extensive fragmentation of the PFAS molecules and then reduces the sensitivity.

It is important to highlight within the multiclass analysis of PFASs that different chain lengths, functional groups and the extent of fluorination can make it difficult using the same type of instruments [6]. However, many studies are able to determine several PFAS

Table 6.2 Characteristics of poly- and perfluoroalkyl substances (PFASs) and determination techniques used as the last step of their determination.

Group	Characteristic	Determination	Reference
Perfluoroalkyl sulphonic acids (PFSAs)	Ionisable, water-soluble	LC and GC (after derivatisation)	[1, 42, 48, 60]
Perfluoroalkyl carboxylic acids (PFCAs)	Ionisable, water-soluble	LC and GC (after derivatisation)	[1, 42, 48, 60, 98]
Perfluoroalkyl phosphonic acids (PFPAs)	Ionisable	LC	[30, 56, 110, 114]
Perfluoroalkyl phosphinic acids (PFPiAs)	Ionisable	LC	[1, 30, 31, 75]
Perfluoroalkane sulphonamides (FASAs)	Volatile	GC	[1, 62, 110, 114]
N-alkyl perfluoroalkane sulphonamido ethanols (FASEs)	Volatile	GC	[1, 62, 64]
Perfluoroether sulphonic acids (PFESAs)	Ionisable	LC	[5, 76, 83]
Perfluoroether carboxylic acids (PFECAs) (i.e. HFPO-DA, trade name: GenX)	Ionisable	LC	[39, 51, 57, 110]
Perfluorooctane sulphonamido ethanol-based phosphate esters (SAmPAPs)	Ionisable	LC	[28]
Cyclic perfluoroalkyl sulphonic acids (cyclic PFSAs)	Ionisable	LC	[80]
Fluorotelomer sulphonic acids (FTSAs)	Ionisable	LC	[52, 110, 114]
Fluorotelomer carboxylic acids (FTCAs)	Ionisable	LC	[114]
Fluorotelomer olefins (FTOs)	Volatile	GC	[1]
Fluorotelomer alcohols (FTOHs)	Volatile	GC	[1, 62, 64]
Fluorotelomer iodides (FTIs)	Ionisable	LC	[1]
Fluorotelomer acrylates (FTAs)	Volatile	GC	[1, 110, 114]
Polyfluoroalkyl phosphate monoesters (monoPAPs)	Ionisable	LC	[82]
Polyfluoroalkyl phosphate diesters (diPAPs)	Ionisable	LC	[82, 110, 114]

families with different physicochemical properties mostly using any form of LC. As an exhaustive example, Gremmel et al. [114] analysed 52 PFASs in aqueous samples, including fluorinated telomeric carboxylic acids (FTCAs) (n:2/n:3), fluorinated telomeric sulfonic acids (FTSAs), PFPAs, fluorotelomer phosphonate diesters (diPAPs), PFCAs,

PFSAs, PFECAs, perfluoroalkyl phosphonic acids (PFPAs) and perfluoroalkyl sulfonamides (FASAs). In the same way, Zeng et al. [56] developed a method to determine 17 classes of 58 PFASs (including chlorosubstituted perfluoroalkyl phosphonic acids (Cl-PFPAs) and 9-chlorohexadeca fluoro-3-oxanonane-sulfonate (9Cl-PF3ONS or Fe53B)) in surface water by UPLC-MS/MS.

In a review on emerging persistent organic pollutants, Lorenzo et al. [106] noted that MS/MS sometimes shows inadequate selectivity. The use of high-resolution MS (HRMS) techniques, such as time-of-flight (TOF)-MS or Orbitrap-MS, as alternative tools for (ultra-)trace analyses of PFASs has been proposed. HRMS provides accurate mass and then information on the most probable empirical formula. Despite HRMS instruments not being able to provide MS/MS, their hybridation with quadrupoles or linear ion-traps can achieve the MS, pseudo MS/MS (two parallel MS at low and high energy) and real MS/MS fragmentation with high mass accuracy as well as acquisition in data dependent or independent analyses. The HRMS instruments provide higher LODs but they have the advantage of being able to detect and identify non-target PFASs, as well as their metabolites and transformation products [106].

Forty new classes of PFASs were detected by high-resolution mass spectrometry with quadrupole time-of-flight mass spectrometry (QTOF-MS) in AFFF-impacted groundwater [115]. A large number of PFAS precursors with different chain lengths and functional groups (i.e., anionic, cationic, zwitterionic, short- and ultra-short-chain) were identified in drinking water (tap and bottled water) [50]. Ultra-short-chain PFASs (i.e., C2–C3) were also detected by QTOF-MS [116] and ultra-performance convergence chromatography (UPC^2) coupled to a tandem mass spectrometer [100] in AFFF-impacted groundwater, river and rain water. A large number of PFASs (PFCAs (i.e., C4–C9, C11–C14, C16), PFSAs (i.e., C4, C7, C8, C10) and FOSA) have also been detected in snow [117]. Liu et al. [118] described HPLC coupled to ultra-high-resolution Orbitrap mass spectrometry (HPLC-uHRMS) for PFASs (C5–C17) monitoring. Samples were extracted from wastewater via an on-line SPE using C_{18} sorbent and obtained MDLs 0.005–0.2 ng/mL. Concha-Graña [59] reported a novel method for the determination of PFOS, PFOA, and PFOS precursors in seawater, in which liquid–liquid microextraction was combined with HPLC and LTQ-Orbitrap HRMS for PFASs determination.

HPLC-MS/MS has also been used to analyse the current and historical concentrations of 22 PFASs in sediment collected from Lake Superior and northern Lake Michigan in 2011 and Lake Huron in 2012 [73]. In this study, identified PFASs were not a significant fraction of fluorine-containing compounds detected in the sediment samples (<0.01% in extractable organic fluorine, EOF). Also using HPLC-MS/MS, Wang et al. [87] described a method for PFOS and PFOA identification in sewage, achieving concentrations in the range 0.62 ng/g–2.65 ng/g. Recently, Semerád et al. [51] developed a methodology to determine 32 PFASs, including hexafluoropropylene oxide-dimer acid, HFPO-DA (GenX), sodium dodecafluoro-3H-4, 8-dioxanonanoate (NaDONA) and other new replacements of older and restricted PFASs, in sludge by means of LC-MS/MS. They analysed 43 WWTPs from Czech Republic and found values of ΣPFASs from 5.6 to 963.2 ng/g. PFOS was the most abundant, reaching 932.9 ng/g. GenX and a high percentage of short-chain PFASs found in samples suggest they are used as a replacement for long-chain PFAS (especially restricted PFOA and PFOS). Munoz et al. [40] analysed

AFFF-amended soils with a sequential extraction method (i.e. methanol vs. HCl-methanol vs. CH_3COONH_4 -methanol vs. NaOH) and ultra-high-performance liquid chromatography coupled to high-resolution accurate mass and Orbitrap mass spectrometry (UHPLC-HRMS), and Karásková et al. [95] identified 20 PFASs in household dust using an HPLC-(ESI)MS/MS system.

Ultra-high-performance liquid chromatography tandem triple quadrupole mass spectrometry (UPLC-QqQ-MS/MS), operating in negative-ion mode with MRM, was used in the identification and quantification of PFASs (14 ionisable PFASs, including PFOA, PFOS and their respective linear and branched isomers) in plankton samples [77]. Based on a previous study [119], linear isomers were confirmed using authentic linear native and/or linear surrogate standards for PFOS, PFOA, PFNA, PFDA, PFDoA, NEtFOSA, and FOSA and could be inferred for all other perfluoalkyl acids (PFAs) since the latest eluting isomer was also the largest peak. The linear isomer for each PFA, or PFA-precursor, was always the latest eluting on perfluorooctyl (PFO). The elution order on PFO, from latest to earliest, was n-PFOS, iso-PFOS, 5m-PFOS, 4m-PFOS, B4-PFOS, 3m-PFOS and on perfluorophenyl was n-PFOS, iso-PFOS, 5m-PFOS, 1m-PFOS, 4m-PFOS, 3m-PFOS. Isomer structures were elucidated by product ion spectra using the missing "0-series" (m/z 130, 180, 230, etc.) and "9-series" techniques. Briefly, the absence of the "0-series" product ion identifies the location of the tertiary carbon. The absence of this fragment occurs due to the stability of the tertiary carbanion, which prevents simultaneous breaking of the CF_3 – and CF_3CxF_2x – groups, thus leading to an absence of this "0-series" ion. Correspondingly, for each branched isomer, a specific "9-series" fragment (e.g., m/z 119, 169, 219, etc.) cannot be formed, thus its absence should confirm the location of the branching point. For example, for *iso*-PFOS (i.e., $(CF_3)_2(CF_2)_6SO_3^-$) an ion having m/z 119 ($[CF_3CF_2]^-$) cannot form.

Other novel PFASs as Cl-PFESAs were screened in 152 biological samples including invertebrates, fishes, seabirds and mammals collected from Bohai Sea of China using HPLC-MS/MS [76]. 6:2 Cl-PFESA was detected in concentrations from < MDL to 3.84 ng/g ww, being the dominant congener. Fourteen PFASs and 10 potential precursors (e.g., 8:2 monoPAP, 8:2 diPAP) were detected by LC-MS/MS in grey mullet liver (*Chelon labrosus*) and mussel (*Mytilus galloprovincialis*) samples from the Basque Coast (North of Spain) and yellowfin tuna muscle tissue (*Thunnus albacares*) samples from the Indian Ocean [82]. Similarly, HFPO-TA and 6:2 Cl-PFESA were detected in various tissues from black spotted frogs collected in a rice paddy field near a large-scale fluorochemical production site in China by means of an UPLC coupled to a triple quadrupole mass spectrometer [83]. This information pointed out the possibilities of LC methods to detect these new types of PFASs.

One of the challenges in PFOA and PFOS analysis is the separation of linear and branched isomers that have the same mass-to-charge ratio but can have different toxicological properties. Current methods that are used for separating isomeric PFASs require relatively long analysis times (minutes to hours) that can limit throughput. A recent addition within the promising techniques able to help in this separation is high-resolution differential mobility spectrometry (DMS)-MS, which has demonstrated it is able to separate and detect isomeric PFAS compounds in milliseconds. In the study of Ahmed et al. [120], for multistage MS/MS experiments, a hybrid LTQ-Orbitrap equipped with a nano-EI ionisation source was used to obtain high-resolution collision-induced dissociation (CID) mass spectra (Figure 6.3). Good results obtained in this research indicate that high-resolution MS-MS could be a useful approach for the rapid analysis of isomeric PFASs.

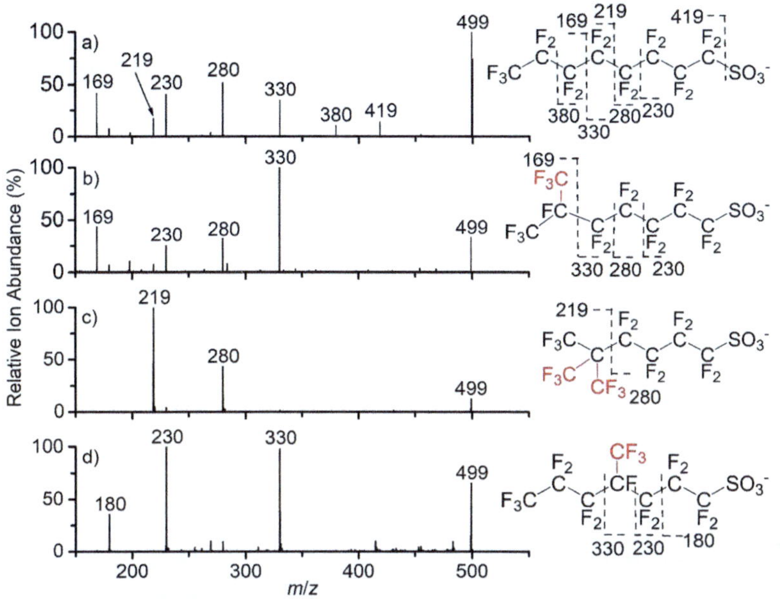

Figure 6.3 Collision-induced dissociation (CID) mass spectra of sulfonate isomers obtained for PFASs (a) 5, (b) 6, (c) 7, and (d) 8. A normalized collision energy of either (a) 40% or (b–d) 35% was used [120]. Reproduced with the permission of Elsevier.

Determination of PFASs by GC may require derivatisation because even though they are volatile enough, their ionic nature provides wide and tailing peaks difficult to quantify. This is a time-consuming procedure and a potential source of uncertainty [48]. The most used mass analyser in GC is the single quadrupole, although GC-MS/MS(QqQ) [98, 112] and GC-QTOF [112] have also been reported. Yao et al. [91] analysed a suite of neutral and ionic PFASs, around the Bohai Sea (China), from a high sampling volume of air using a positive (CI) GC-MS. Rapid and sensitive determination of PFCAs in aqueous matrices by GC-MS/MS(QqQ) is achieved by transforming the carboxylic groups in their esters [106]. This technique can be combined with a selective headspace solid-phase micro extraction (HS-SPME), as in the research of Monteleone et al. [98]. These authors, after a fast derivatisation of the carboxylate function by a propylchloroformate/propanol mixture, used 85 μm carboxen/polydimethylsiloxane fibre to extract PFCAs from aqueous matrices. By means of a GC-MS/MS(QqQ) system in negative chemical ionisation mode with ammonia as reagent gas, these authors achieved LODs and LOQs of 0.08–6.6 ng/L and 0.17–14.3 ng/L, respectively.

Pórtoles et al. [112] compared the ionisation and in-source-fragmentation behaviour of four FTOHs and four N-alkyl FOSAs/-FOSEs by APCI in relation to EI and CI. They also performed comparison of GC-QTOF to GC-MS/MS(QqQ) operating in APCI mode for the analysis of PFASs precursors. No matrix effects were observed and concentrations found in samples were in the range of 1–100 pg/L, far below the LODs achieved with other methods. Unknown related PFASs, such as methyl-sulfone and methyl-sulfoxide analogues for FTOHs, were also discovered and tentatively identified, showing the potential of HRMS in combination with GC. A new GC – APPI – HRMS method combined with HS – SPME was developed for the determination of FTOs, FTOHs, FOSAs and FOSEs in water samples (Figure 6.4). This procedure showed a good analytical

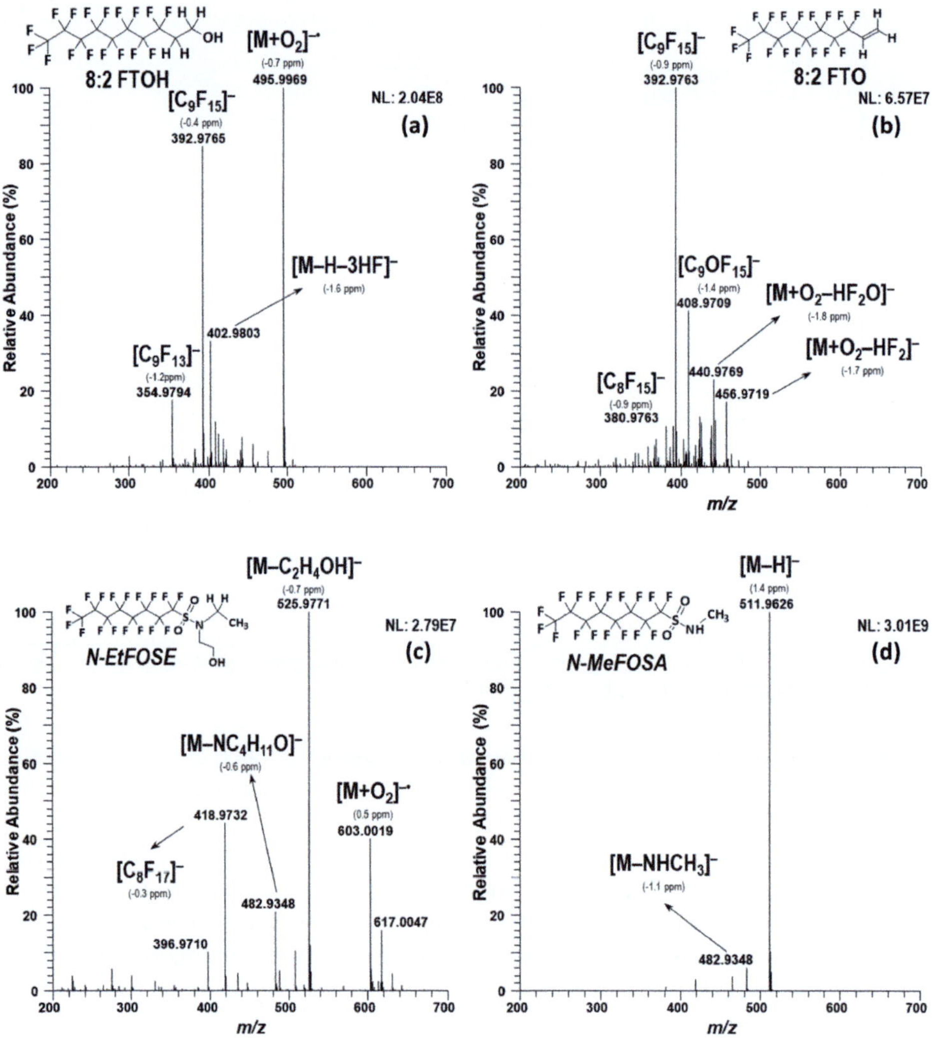

Figure 6.4 Negative-ion APPI mass spectra of (a) 8:2 FTOH, (b) 8:2 FTO, (c) N-EtFOSE and (d) N-MeFOSA using acetone as APPI-dopant [121]. Reproduced with the permission of Elsevier.

performance for the analysis of river water, providing very low limits of detection (0.02–15 ng/L), good repeatability (RSD < 11%) and trueness (relative error < 12%) [121].

It is noteworthy that recent advances in GC separation, such as comprehensive two-dimensional gas chromatography (GCxGC) or fast GC, have not been tested yet for PFASs, and many of the previously described techniques are only in the earlier stages of application. This is due to the low utilisation of these techniques.

6.3.1.2 Biosensors

Recently, nanoscale materials have emerged as an efficient and promising approach to be incorporated into biosensors for the monitoring of contaminants [7]. Because of their excellent physical, chemical and optical properties; unique behaviours (aggregation, disaggregation, adsorption and desorption); small size (1–100 nm) and outstanding sensitivity, different types of nano sensors, particularly nanoparticle (NP)-based sensors, have been developed for PFASs identification.

An early NP-based sensor was developed to trace PFASs in aquatic samples [122]. Nanoparticles were modified and functionalised with an octadecyl group. Subsequently, chitosan-coated magnetite nanoparticles (Fe_3O_4-C_{18}-chitosan-MNPs) were used to extract PFOS, PFNA, PFDA, PFUnDA, PFDoDA and PFTA from rain and tap water. In another example, a QD-based sensor was described by Liu et al. [123], who used 3-mercaptopropionic acid (MPA) for the modification of cadmium sulphide (CdS)-QDs to determine PFOA in raw textile ingredients. Gold nanoparticles (AuNP) are widely used as nano sensors and they have been also prevalent in research on the PFASs sensing. Xia et al. [124] developed a PPARα responsive element (PPRE) modified AuNPs sensor for the identification and quantification of PFASs in river water. Wang et al. [125] reported a very sophisticated and quick detection method of PFOS in aquatic samples by the aggregation of a multifunctional AuNP-polydopamine-polyvinylidene fluoro-membrane.

Molecular imprinted polymers (MIP) have also been used for PFASs identification, due to the target molecule-imprinted nano-cavity created during the polymerisation process. PFOA, PFOS and PFHA were analysed in tap, river and mountain waters combining TiO_2 nanotube arrays with MIP, obtaining a LOD of 86 ng/L [126]. Cennamo et al. [127] developed a method based on MIP-SPR (surface Plasmon resonance)-POF (plastic optical fibre), to detect PFOA, PFOS and PFAs in water samples with LODs in the range of 0.13–0.15 ng/mL.

Aggregation-induced emission (AIE) is another technique that is also receiving increasing attention. In contrast to the traditional aggregation-quenching, these synthesised AIE molecules emit fluorescence once aggregated [7]. Fang et al. [128] developed a glass-chip-AIE-based sensor in which AIE luminogens (AIEgen) were added into a solution (water: acetone) containing PFASs. After different vapour pressures of water and acetone prompted PFASs to reach a critical micelle concentration (CMC), the AIEgen molecules were trapped and aggregated by the hydrophobic core of PFOA micelles, which could be turned on by the AIEgen-fluorescence, allowing the concentration of analytes to be measured under UV illumination. The method is simple, fast (~1 min) and green (amount of sample 1–2 μL).

Recently, an app-based sensor has been developed by Fang et al. [129] to assess PFOA in tap and ground water. Extraction should be carried out by adding astkCARETM reagents (ethyl acetate and ethyl violet), to remove the background, and to react with anionic surfactants (AS) like PFOA/PFOS. As a result, an immiscible ion-pair is formed, which is hydrophobic, and their hydrophilic terminates are shielded by each other due to an ion-pairing interaction. This ion-pair is extracted from the aqueous phase to the organic phase, and after several steps of extraction, the smartphone's camera reads the colour of this organic phase (Figure 6.5). The app relates the colouration (Red, Green, and Blue) to the concentration of the AS, which can be in the range 10–1000 ng/mL with a standard deviation of <10%. The smartphone sensor device is sensitive and portable. On the other hand, it is not highly selective for particular anionic surfactant detection, which will be improved in future research.

6.3.2 Figures of Merit

Quantification is generally carried out by the precursor product ion and its multiple reaction monitoring transitions (MRM), or by single reaction monitoring transitions (SIM),

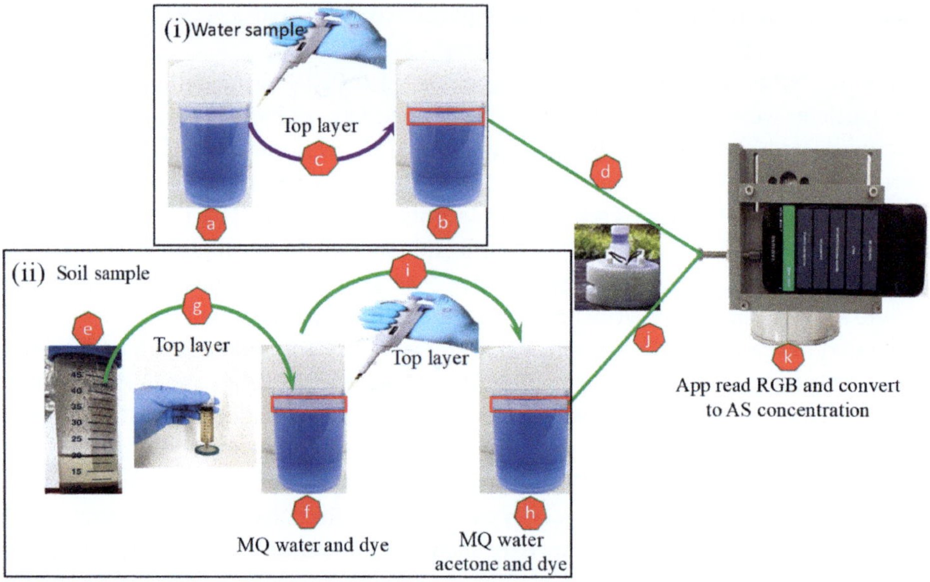

Figure 6.5 Smartphone app-based sensor for PFASs determination. Briefly, during the extraction process, anionic surfactant, AS, (i.e. PFOA/PFOS) reacts with cationic dye to form an ion-pair (a, e, f), which is extracted into the organic phase and transferred (c, g, i) to the next container (b, f, h) to read the concentration of AS with the smartphone app-based sensor (d, j, k) [7] M. Al Amin, et.al, (2020), Fig. 5, p. 16/with permission of Elsevier.

which provides a wide linear dynamic range of at least three orders of magnitude with excellent sensitivity at μg/L (even ng/L) levels [69, 103, 109]. A significant issue in quantitation of PFASs is that they occur at very low concentration levels (sub-ng/L range) in environmental samples, which means that a highly sensitive method is needed to their accurate quantitation. MS techniques have turned into indispensable tools for PFASs determination, due to their high accurate mass determination for compositional analysis, the ability of MS/MS fragmentation techniques for structural identification and the possibility to couple with separation techniques like LC or GC [106].

Excellent sensitivities are reported in recent studies achieving MDLs at ng/L and even pg/L levels. In this way, the method developed by Zeng et al. [56], who determined 17 classes of PFASs in surface water by UPLC-MS/MS, showed recoveries of 60–130%, relative standard deviation (RSD) less than 30%, and MDLs of 0.004 ng/L–2.0 ng/L. In a study carried out at the coastal area of Bangladesh, limits of quantification (LOQs) for target PFAAs ranged from 0.016 to 0.4 ng/L in seawater and from 0.006 to 0.102 ng/g d.w. in sediment [86]. Wang et al. [69] improved the recovery, time efficiency and detection limit relative to traditional methods with LODs varying from 0.003 to 0.045 ng/g, and spike recoveries to sediment ranging from 90% to 110% with suitable precisions (1.7%–14.6%). Gao et al. [109] determined simultaneously 21 PFASs in fish by isotope dilution UPLC-MS/MS attaining MDLs between 2 pg/g and 10 pg/g except for PFBA (120 pg/g). The method quantitation limits (MQLs) ranged from 5 pg/g to 30 pg/g except for PFBA (300 pg/g). The matrix spiked recoveries of three spiked levels were in the range of 79.6%–109%, while the intra-day RSD and inter-day RSD were from 0.94% to 13.9% and

0.36% to 11.2%, respectively. The quantification is commonly performed using isotopically labelled internal standards, which is enough to avoid the matrix effects in most matrices.

However, the analysis of biota also entails the problem of matrix effect, and it is not easy to develop and perform a pre-treatment/clean-up method that can eliminate it. To overcome this difficulty, many studies have improved quantitation results using isotope dilution [82, 109] and matrix-matched calibration curve approaches [59, 105]. However, problems remain since not all isotope labelled reference standards are currently available, and matrices representative of all samples are also not available.

6.3.3 Hints and Tips

Since PFASs are ubiquitous, special attention must be paid when collecting and preparing samples for analysis. Important rules for sampling include avoiding polytetrafluoroethylene, low-density polyethylene, glass materials, adhesive notes, waterproof material, Tyvek, some sunscreens, aluminium foil and so on. Accordingly, field quality control (QC) samples are extremely important when sampling for PFASs, probably more critical for this family of contaminants than others are, and they are needed to evaluate the suitability of any data set [4]. There are different regulatory programmes that require the collection of trip blanks, field blanks and equipment blanks.

PFASs can partition in the sampling bottle similar to the way they partition in the environment, to the surface. Consequently, it is important to utilise the whole sample volume in the extraction procedure. If subsampling is done, there can be a potential concentration bias for the PFSAs > C8 and the PFCAs > C10 [4]. It is also essential to rinse the bottles with a water-miscible solvent such as methanol, which should be also considered in the extraction to ensure a complete quantitative transfer, because long-chain PFAAs can still adhere to high-density polyethylene (HDPE) bottles.

In addition to possible PFASs contamination by the cartridges, there is a wide range of solubility from C4 to C14 PFASs that can make SPE challenging. If the pH of the sample is low (5–6), some PFASs may be lost on the cartridge, depending on the sorbent used. It is important to evaluate which SPE cartridge the laboratory is using and potential limitations, if any [4].

Background contamination issues have also been reported when using LC [130]. Consequently, particular attention should be paid to vial caps made of Teflon or Viton fluoropolymers, fluoropolymer tubing, solvent inlet filters and a variety of other laboratory products containing Teflon and perfluoroalkoxy compounds, as they are sources of PFASs [48]. It is advised to replace these parts and to use vials with PP septa, and/or to install a guard or trapping column before the injector to prevent contamination from the LC system [56, 68].

PFASs have great quantity in numbers and complexity in molecular structure, which bring more analytical issues on chromatograph separation, structure elucidation, accurate quantification and biological effect assessment. The current detection techniques are only capable of detecting a certain amount of them. This can be due either to commercial reasons, to the partial degradation as intermediates, to the unavailability of their standards for quantification or because the structure of the compound is unknown [131].

Different chain lengths, functional groups and the extent of fluorination make it difficult to analyse all the PFASs using the same analytical method [6]. Short-chain PFASs (C6 or shorter) have been increasingly manufactured and used worldwide as alternatives for long-chain PFASs, and analytical methods have always been a challenge when it comes to the detection of short or ultrashort-chain PFASs [100]. Additionally, the presence of branched isomers increases the challenges for the accurate quantification of many PFASs in environmental matrices [1]. There can be several peaks that are less defined for branched isomers relative to the single pronounced peak for the linear isomer.

Among the conclusions of the last PFASs Experts Symposium held in Arlington, Virginia on May 20 and 21, 2019 [4], assistants highlighted the following information gaps for the fundamental mechanisms controlling PFASs transport and fate: (1) the release of major PFASs from source zones and the separation of individual fractions; (2) the influence of physical processes and parameters for PFAS contaminants (dispersion, diffusion, interfacial flow, etc.); (3) the effect of co-contaminants on transport; (4) the differences in adsorption between various isomers; (5) the transport behaviour of PFASs other than PFOA, PFOS and, perhaps, a few other compounds studied to date such as PFNA and PFBS; (6) whether transport mechanisms are different than generally accepted adsorption isotherm theories at trace concentrations; and (7) the transport and potential abiotic reactions (e.g., fluorotelomers) and possible long-term biological reactions in groundwater plumes.

6.4 Future Directions and Challenges

This chapter highlights several challenging aspects related to the determination of PFASs. The first is that the determination of PFASs is no longer restricted to the determination of PFCAs and PFSAs, as it was before. It is now possible to determine some precursors and new compounds synthesised to replace the old ones. This has broadened the range of polarity of these compounds.

Concern about the possible toxic effects of PFASs has grown, as can be deduced from the recent update of the legislation that has led to more restrictive limits, which have been set in all areas of the environment. However, according to recent studies, known PFASs account for only a small fraction of extractable organic fluorine (EOF), with the majority remaining unidentified. Consequently, there is increasing scientific concern about the presence of novel and precursor PFASs in the environment. Most of these emerging pollutants have not been regulated and their ecological effects remain unknown. It is important to highlight that their production and use could raise serious human and environmental health issues, as well as presenting challenges in terms of environmental analysis and monitoring.

The analytical methods available to date are capable of monitoring the levels required by legislation. Nevertheless, the analyses have many complications because PFASs are ubiquitous compounds. It is very important to take extreme precautions to avoid contamination. Current extraction methods fulfil their function of isolating and pre-concentrating PFASs, but most of them are still time-consuming, tedious and have little emphasis on green chemistry. It is to be expected that in the near future new extraction methods will be developed that will reduce the amount of sample and especially the quantity of solvent and toxic reagents.

Liquid chromatography-mass spectrometry allows the identification of the largest possible number of PFASs. The possibilities of high-resolution masses to identify new structures are especially crucial to detect the related replacement products. In this sense, short-chain PFASs (C6 or shorter) have been increasingly manufactured and used worldwide as alternatives for long-chain PFASs, and analytical methods have always been a challenge, when it comes to the detection of short or ultrashort-chain PFASs. Therefore, developing robust analytical methods that can capture both original and novel PFASs possessing broad chemical properties in environmental matrices is a priority. One field to be developed in the future is biosensors. These devices will have a major advantageous application in routine analysis. These systems are the most incipient and therefore great development is to be expected in the next few years.

Another source of uncertainty is the unavailability of standards for quantification, particularly for those PFASs whose structure is still unknown. To fill this gap, the total oxidisable precursors assay (TOPA) has been introduced. In TOPA, an excessive amount of oxidant is used to oxidise PFAS precursors under an alkaline environment, to transform or convert the "unknown" PFAS precursors to the detectable PFASs, such as PFCAs. TOPA is an easy and simple way to quantify the mass balance of known and unknown PFAS precursors, but more work is needed to generalise this process. In the near future, TOPA could be applied for the assessing of fate and transport of PFASs from different segment of environmental matrices, which is still a research gap at this moment. Other methods have been introduced to detect fluorinated substances, such as through the total organic fluorine (TOF) analysis. The ratio of EOF to TOF can be used as a measure to assess the efficiency of the sample treatment, while the ratio of known PFASs to EOF can be considered an indication of the progress of current analytical efforts to the amount of work remaining in this field.

Another challenge about PFASs monitoring is related to air sampling issues. There is no standardised methodology, which hampers comparison of global studies being urgent the development of a globally applicable sampling method. Furthermore, simultaneous analytical methods for anionic and neutral PFASs in air samples are scarce. Dust is one of the most important routes of human exposure; hence, a method suitable for both volatile and non-volatile PFASs is warranted. Furthermore, passive sampling is an ideal monitoring tool for spatial-temporal trends and transport in air and water, and consequently to predict human exposure. It is low-cost and simple to deploy without electrical requirements, but inter-laboratory studies for a wide range of PFASs, besides PFOA and PFOS, are needed to standardise protocols in order to improve comparison of calibration results.

Remediation strategies for PFASs are extremely complex and challenging. The lack of proven destructive in-situ remediation technologies, and limited information on natural complete degradation processes, may make active in-situ remediation exceptionally costly and impracticable at many PFASs' sites. Emerging new trends of PFASs remediation using multiple synergetic technologies simultaneously (known as treatment train processes) to effectively achieve in-situ remediation of PFASs are promising. However, there is still a lot of work to do because of the broad diversity of PFASs' chemical structures and corresponding physicochemical properties. Besides this, transformation pathways for destructive remedial technologies are not well characterised, and incomplete mineralisation may result in the formation of intermediate compounds that would present new risks.

References

1 Buck, R.C., Franklin, J., Berger, U., Conder, J.M., Cousins, I.T., de Voogt, P., Jensen, A.A., Kannan, K., Mabury, S.A., and van Leeuwen, S.P. (2011). Perfluoroalkyl and polyfluoroalkyl substances in the environment: terminology, classification, and origins. *Integr. Environ. Assess. Manage.* 7(4): 513–541. doi: 10.1002/ieam.258.

2 Organisation for Economic Co-operation and Development, (2018). Toward a new comprehensive global database of per- and polyfluoroalkyl substances, 2018. http://www.oecd.org/officialdocuments/publicdisplaydocumentpdf/?cote=ENV-JM-MONO(2018)7&doclanguage=en. Accessed 23 February 2021.

3 Company [3M], (2000). Phase-out plan for POSF-based products. USEPA Administrative Record AR226-0600. https://www.regulations.gov/document/EPA-HQ-OPPT-2002-0051-0006. Accessed 14 December 2021.

4 Simon, J.A., Abrams, S., Bradburne, T., Bryant, D., Burns, M., Cassidy, D., Cherry, J., Chiang, S.-Y., Cox, D., Crimi, M., Denly, E., DiGuiseppi, B., Fenstermacher, J., Fiorenza, S., Guarnaccia, J., Hagelin, N., Hall, L., Hesemann, J., Houtz, E., Koenigsberg, S.S., Lauzon, F., Longsworth, J., Maher, T., McGrath, A., Naidu, R., Newell, C.J., Parker, B.L., Singh, T., Tomiczek, P., and Wice, R. (2019). PFAS experts symposium: statements on regulatory policy, chemistry and analytics, toxicology, transport/ fate, and remediation for per- and polyfluoroalkyl substances (PFAS) contamination issues. *Remediation* 29(4): 31–48. doi: 10.1002/rem.21624.

5 Xiao, F. (2017). Emerging poly- and perfluoroalkyl substances in the aquatic environment: a review of current literature. *Water Res.* 124: 482–495. doi: 10.1016/j.watres.2017.07.024.

6 Cui, D., Li, X., and Quinete, N. (2020). Occurrence, fate, sources and toxicity of PFAS: what we know so far in Florida and major gaps. *TrAC, Trends Anal. Chem.* 130: 115976. doi: 10.1016/j.trac.2020.115976.

7 Al Amin, M., Sobhani, Z., Liu, Y., Dharmaraja, R., Chadalavada, S., Naidu, R., Chalker, J.M., and Fang, C. (2020). Recent advances in the analysis of per- and polyfluoroalkyl substances (PFAS) – a review. *Environ. Technol. Innov.* 19: 100879. doi: 10.1016/j.eti.2020.100879.

8 Giesy, J.P. and Kannan, K. (2001). Global distribution of perfluorooctane sulfonate in wildlife. *Environ. Sci. Technol.* 35(7): 1339–1342. doi: 10.1021/es001834k.

9 Kim, M., Li, L.Y., Grace, J.R., and Yue, C. (2015). Selecting reliable physicochemical properties of perfluoroalkyl and polyfluoroalkyl substances (PFASs) based on molecular descriptors. *Environ. Pollut.* 196: 462–472. doi: 10.1016/j.envpol.2014.11.008.

10 Lindstrom, A.B., Strynar, M.J., and Libelo, E.L. (2011). Polyfluorinated compounds: past, present, and future. *Environ. Sci. Technol.* 45(19): 7954–7961. doi: 10.1021/es2011622.

11 Naidu, R., Nadebaum, P., Fang, C., Cousins, I., Pennell, K., Conder, J., Newell, C.J., Longpré, D., Warner, S., Crosbie, N.D., Surapaneni, A., Bekele, D., Spiese, R., Bradshaw, T., Slee, D., Liu, Y., Qi, F., Mallavarapu, M., Duan, L., McLeod, L., Bowman, M., Richmond, B., Srivastava, P., Chadalavada, S., Umeh, A., Biswas, B., Barclay, A., Simon, J., and

Nathanail, P. (2020). Per- and poly-fluoroalkyl substances (PFAS): current status and research needs. *Environ. Technol. Innov.* 19: 100915. doi: 10.1016/j.eti.2020.100915.

12 Prevedouros, K., Cousins, I.T., Buck, R.C., and Korzeniowski, S.H. (2006). Sources, fate and transport of perfluorocarboxylates. *Environ. Sci. Technol.* 40(1): 32–44. doi: 10.1021/es0512475.

13 Glüge, J., Scheringer, M., Cousins, I.T., DeWitt, J.C., Goldenman, G., Herzke, D., Lohmann, R., Ng, C.A., Trier, X., and Wang, Z. (2020). An overview of the uses of per- and polyfluoroalkyl substances (PFAS). *Environ. Sci.: Process. Impacts* 22(12): 2345–2373. doi: 10.1039/D0EM00291G.

14 Domingo, J.L. (2012). Health risks of dietary exposure to perfluorinated compounds. *Environ. Int.* 40: 187–195. doi: 10.1016/j.envint.2011.08.001.

15 Sunderland, E.M., Hu, X.C., Dassuncao, C., Tokranov, A.K., Wagner, C.C., and Allen, J.G. (2019). A review of the pathways of human exposure to poly- and perfluoroalkyl substances (PFASs) and present understanding of health effects. *J. Expo. Sci. Environ. Epidemiol.* 29(2): 131–147. doi: 10.1038/s41370-018-0094-1.

16 Steenland, K. and Winquist, A. (2021). PFAS and cancer, a scoping review of the epidemiologic evidence. *Environ. Res.* 194: 110690. doi: 10.1016/j.envres.2020.110690.

17 U.N.E. Programme, (2010). *New POPs SC-4/17: Listing of Perfluorooctane Sulfonic Acid, Its Salts and Perfluorooctane Sulfonyl Fluoride.* Geneva: United Nations Environment Programme: Stockholm Convention on Persistent Organic Pollutants (POPs).

18 European Commission (2006). European Union Directive 2006/122/ECOF the European Parliament of the Council of 12 December 2006 amending for the 30th time Council Directive 76/769/EEC on the approximation of the laws, regulations and administrative provisions of the member states relating to restrictions on the marketing and use of certain dangerous substances and preparations (perfluorooctane sulfonates). *Off. J.Eur. Commun.* L372: 32–34.

19 European Commission (2012). Water Framework Directive: Proposal for a Directive of the European Parliament and of the Council amending Directives 200/60/EC and 2008/150/EC as regards priority substances in the field of water policy.

20 European Commission (2002). Commission decision of 12 August 2002 implementing Council Directive 96/23/EC concerning the performance of analytical methods and the interpretation of results. *Off. J. Eur. Union* L221/8.

21 European Commission (2004). Regulation (EC) No 850/2004.

22 European Food Safety Authority (2020). Risk to human health related to the presence of perfluoroalkyl substances in food. *EFSA J.* 18(18).

23 European Commission (2013). Directive 2013/39/EU of the European Parliament and of the Council of 12 August 2013 amending Directives 2000/60/EC and 2008/105/EC as regards priority substances in the field of water policy. Text with EEA relevance. *Off. J. Eur. Union.* L 226: 1–17.

24 Environmental Health Canadian, (2006). Perfluorinated carboxylic acids (PFCAs) and precursors: a proposed action plan for assessment and management. http://www.environmentaldefence.ca.Accessed 14 December 2021.

25 United States Environmental Protection Agency (2016). *Drinking Water Health Advisory for Perfluorooctane Sulfonate (PFOS).* 1–103. https://www.epa.gov/sites/production/

files/2016-05/documents/pfos_health_advisory_final-plain.pdf. Accessed 14 December 2021.

26 Benskin, J.P., De Silva, A.O., and Martin, J.W. (2010). Isomer profiling of perfluorinated substances as a tool for source tracking: A review of early findings and future applications. *Rev. Environ. Contam. Toxicol.* 208: 111–160. doi: 10.1007/978-1-4419-6880-7_2.

27 Peng, H., Zhang, S., Sun, J., Zhang, Z., Giesy, J.P., and Hu, J. (2014). Isomer-specific accumulation of Perfluorooctanesulfonate from (N-Ethyl perfluorooctanesulfonamido) ethanol-based Phosphate Diester in Japanese Medaka (Oryzias latipes). *Environ. Sci. Technol.* 48(2): 1058–1066. doi: 10.1021/es404867w.

28 Joudan, S., Liu, R., D'Eon, J.C., and Mabury, S.A. (2020). Unique analytical considerations for laboratory studies identifying metabolic products of per- and polyfluoroalkyl substances (PFASs). *TrAC, Trends Anal. Chem.* 124: 115431. doi: 10.1016/j.trac.2019.02.032.

29 Rand, A.A., Rooney, J.P., Butt, C.M., Meyer, J.N., and Mabury, S.A. (2014). Cellular toxicity associated with exposure to perfluorinated carboxylates (PFCAs) and their metabolic precursors. *Chem. Res. Toxicol.* 27(1): 42–50. doi: 10.1021/tx400317p.

30 Lee, H., De Silva, A.O., and Mabury, S.A. (2012). Dietary bioaccumulation of Perfluorophosphonates and Perfluorophosphinates in Juvenile Rainbow Trout: evidence of metabolism of Perfluorophosphinates. *Environ. Sci. Technol.* 46(6): 3489–3497. doi: 10.1021/es204533m.

31 Joudan, S., Yeung, L.W.Y., and Mabury, S.A. (2017). Biological cleavage of the C-P bond in perfluoroalkyl phosphinic acids in male Sprague-Dawley rats and the formation of persistent and reactive metabolites. *Environ. Health Perspect.* 125(11): 117001. doi: 10.1289/EHP1841.

32 Jackson, D.A. and Mabury, S.A. (2013). Polyfluorinated amides as a historical PFCA source by electrochemical fluorination of alkyl sulfonyl fluorides. *Environ. Sci. Technol.* 47(1): 382–389. doi: 10.1021/es303152m.

33 Lange, C.C. (2000). *The Aerobic Biodegradation of N-EtFOSE Alcohol by the Microbial Activity Present in Municipal Wastewater Treatment Sludge.* U.S. Environmental Protection Agency Docket AR-226.

34 Wong, F., Shoeib, M., Katsoyiannis, A., Eckhardt, S., Stohl, A., Bohlin-Nizzetto, P., Li, H., Fellin, P., Su, Y., and Hung, H. (2018). Assessing temporal trends and source regions of per- and polyfluoroalkyl substances (PFASs) in air under the Arctic Monitoring and Assessment Programme (AMAP). *Atmos. Environ.* 172: 65–73. doi: 10.1016/j.atmosenv.2017.10.028.

35 Young, C.J. and Mabury, S.A. (2010). Atmospheric perfluorinated acid precursors: chemistry, occurrence, and impacts. In: *Reviews of Environmental Contamination and Toxicology*, Volume 208: perfluorinated alkylated substances, De Voogt (ed.), 1–109. New York: Springer. doi: 10.1007/978-1-4419-6880-7_1.

36 Armitage, J.M., MacLeod, M., and Cousins, I.T. (2009). Comparative assessment of the global fate and transport pathways of long-chain Perfluorocarboxylic Acids (PFCAs) and Perfluorocarboxylates (PFCs) emitted from direct sources. *Environ. Sci. Technol.* 43(15): 5830–5836. doi: 10.1021/es900753y.

37 Janousek, R.M., Mayer, J., and Knepper, T.P. (2019). Is the phase-out of long-chain PFASs measurable as fingerprint in a defined area? Comparison of global PFAS concentrations and a monitoring study performed in Hesse, Germany from 2014 to 2018. *TrAC, Trends Anal. Chem.* 120: 115393. doi: 10.1016/j.trac.2019.01.017.

38 Boiteux, V., Dauchy, X., Bach, C., Colin, A., Hemard, J., Sagres, V., Rosin, C., and Munoz, J.-F. (2017). Concentrations and patterns of perfluoroalkyl and polyfluoroalkyl substances in a river and three drinking water treatment plants near and far from a major production source. *Sci. Total Environ.* 583: 393–400. doi: 10.1016/j.scitotenv.2017.01.079.

39 Wang, T., Vestergren, R., Herzke, D., Yu, J., and Cousins, I.T. (2016). Levels, isomer profiles, and estimated riverine mass discharges of perfluoroalkyl acids and fluorinated alternatives at the mouths of Chinese rivers. *Environ. Sci. Technol.* 50(21): 11584–11592. doi: 10.1021/acs.est.6b03752.

40 Munoz, G., Ray, P., Mejia-Avendaño, S., Vo Duy, S., Tien Do, D., Liu, J., and Sauvé, S. (2018). Optimization of extraction methods for comprehensive profiling of perfluoroalkyl and polyfluoroalkyl substances in firefighting foam impacted soils. *Anal. Chim. Acta* 1034: 74–84. doi: 10.1016/j.aca.2018.06.046.

41 Hamid, H., Li, L.Y., and Grace, J.R. (2018). Review of the fate and transformation of per- and polyfluoroalkyl substances (PFASs) in landfills. *Environ. Pollut.* 235: 74–84. doi: 10.1016/j.envpol.2017.12.030.

42 Campo, J., Masia, A., Pico, Y., Farre, M., and Barcelo, D. (2014). Distribution and fate of perfluoroalkyl substances in Mediterranean Spanish sewage treatment plants. *Sci. Total Environ.* 472: 912–922. doi: 10.1016/j.scitotenv.2013.11.056.

43 Ebrahimi, F., Lewis, A.J., Sales, C.M., Suri, R., and McKenzie, E.R. (2021). Linking PFAS partitioning behavior in sewage solids to the solid characteristics, solution chemistry, and treatment processes. *Chemosphere* 271: 129530. doi: 10.1016/j.chemosphere.2020.129530.

44 Ahrens, L. (2011). Polyfluoroalkyl compounds in the aquatic environment: A review of their occurrence and fate. *J. Environ. Monit.* 13(1): 20–31. doi: 10.1039/C0EM00373E.

45 Campo, J., Lorenzo, M., Perez, F., Pico, Y., Farre, M., and Barcelo, D. (2016). Analysis of the presence of perfluoroalkyl substances in water, sediment and biota of the Jucar River (E Spain). Sources, partitioning and relationships with water physical characteristics. *Environ. Res.* 147: 503–512. doi: 10.1016/j.envres.2016.03.010.

46 Lorenzo, M., Campo, J., Suarez-Varela, M.M., and Pico, Y. (2019). Occurrence, distribution and behavior of emerging persistent organic pollutants (POPs) in a Mediterranean wetland protected area. *Sci. Total Environ.* 646: 1009–1020. doi: 10.1016/j.scitotenv.2018.07.304.

47 Chen, H., Han, J., Zhang, C., Cheng, J., Sun, R., Wang, X., Han, G., Yang, W., and He, X. (2017). Occurrence and seasonal variations of per- and polyfluoroalkyl substances (PFASs) including fluorinated alternatives in rivers, drain outlets and the receiving Bohai Sea of China. *Environ. Pollut.* 231: 1223–1231. doi: 10.1016/j.envpol.2017.08.068.

48 Feng, H., Ruan, Y., Zhang, K., and Lam, P.K.S. (2019). Current analytical methodologies and gaps for per- and polyfluoroalkyl substances determination in the marine environment. *TrAC, Trends Anal. Chem.* 121: 115372. doi: 10.1016/j.trac.2018.12.026.

49 Hu, X.C., Andrews, D.Q., Lindstrom, A.B., Bruton, T.A., Schaider, L.A., Grandjean, P., Lohmann, R., Carignan, C.C., Blum, A., Balan, S.A., Higgins, C.P., and Sunderland, E.M. (2016). Detection of Poly- and Perfluoroalkyl Substances (PFASs) in U.S. drinking water linked to industrial sites, military fire training areas, and wastewater treatment plants. *Environ. Sci. Technol. Lett.* 3(10): 344–350. doi: 10.1021/acs.estlett.6b00260.

50 Kaboré, H.A., Vo Duy, S., Munoz, G., Méité, L., Desrosiers, M., Liu, J., Sory, T.K., and Sauvé, S. (2018). Worldwide drinking water occurrence and levels of newly-identified perfluoroalkyl and polyfluoroalkyl substances. *Sci. Total Environ.* 616-617: 1089–1100. doi: 10.1016/j.scitotenv.2017.10.210.

51 Semerád, J., Hatasová, N., Grasserová, A., Černá, T., Filipová, A., Hanč, A., Innemanová, P., Pivokonský, M., and Cajthaml, T. (2020). Screening for 32 per- and polyfluoroalkyl substances (PFAS) including GenX in sludges from 43 WWTPs located in the Czech Republic – evaluation of potential accumulation in vegetables after application of biosolids. *Chemosphere* 261: 128018. doi: 10.1016/j.chemosphere.2020.128018.

52 Wei, C., Wang, Q., Song, X., Chen, X., Fan, R., Ding, D., and Liu, Y. (2018). Distribution, source identification and health risk assessment of PFASs and two PFOS alternatives in groundwater from non-industrial areas. *Ecotoxicol. Environ. Saf.* 152: 141–150. doi: 10.1016/j.ecoenv.2018.01.039.

53 Xu, B., Liu, S., Zhou, J.L., Zheng, C., Weifeng, J., Chen, B., Zhang, T., and Qiu, W. (2021). PFAS and their substitutes in groundwater: occurrence, transformation and remediation. *J. Hazard. Mater.* 125159. doi: 10.1016/j.jhazmat.2021.125159.

54 Kwok, K.Y., Taniyasu, S., Yeung, L.W.Y., Murphy, M.B., Lam, P.K.S., Horii, Y., Kannan, K., Petrick, G., Sinha, R.K., and Yamashita, N. (2010). Flux of perfluorinated chemicals through wet deposition in Japan, the United States, and several other countries. *Environ. Sci. Technol.* 44(18): 7043–7049. doi: 10.1021/es101170c.

55 Nakayama, S.F., Yoshikane, M., Onoda, Y., Nishihama, Y., Iwai-Shimada, M., Takagi, M., Kobayashi, Y., and Isobe, T. (2019). Worldwide trends in tracing poly- and perfluoroalkyl substances (PFAS) in the environment. *TrAC, Trends Anal. Chem.* 121: 115410. doi: 10.1016/j.trac.2019.02.011.

56 Zheng, P., Liu, M., Yin, H., Shen, L., Wang, Y., and Wu, Q. (2020). Analysis of 58 poly-/perfluoroalkyl substances and their occurrence in surface water in a high-technology industrial park. *Environ. Pollut.* 267: 115381. doi: 10.1016/j.envpol.2020.115381.

57 Heydebreck, F., Tang, J., Xie, Z., and Ebinghaus, R. (2015). Alternative and legacy Perfluoroalkyl substances: differences between European and Chinese River/Estuary systems. *Environ. Sci. Technol.* 49(14): 8386–8395. doi: 10.1021/acs.est.5b01648.

58 Kwok, K.Y., Wang, X.-H., Ya, M., Li, Y., Zhang, X.-H., Yamashita, N., Lam, J.C.W., and Lam, P.K.S. (2015). Occurrence and distribution of conventional and new classes of per- and polyfluoroalkyl substances (PFASs) in the South China Sea. *J. Hazard. Mater.* 285: 389–397. doi: 10.1016/j.jhazmat.2014.10.065.

59 Concha-Graña, E., Fernández-Martínez, G., López-Mahía, P., Prada-Rodríguez, D., and Muniategui-Lorenzo, S. (2018). Fast and sensitive determination of per- and polyfluoroalkyl substances in seawater. *J. Chromatogr. A* 1555: 62–73. doi: 10.1016/j.chroma.2018.04.049.

60 Park, H., Choo, G., Kim, H., and Oh, J.E. (2018). Evaluation of the current contamination status of PFASs and OPFRs in South Korean tap water associated with its origin. *Sci. Total Environ.* 634: 1505–1512. doi: 10.1016/j.scitotenv.2018.04.068.

61 Pan, Y., Zhang, H., Cui, Q., Sheng, N., Yeung, L.W.Y., Sun, Y., Guo, Y., and Dai, J. (2018). Worldwide distribution of novel perfluoroether carboxylic and sulfonic acids in surface water. *Environ. Sci. Technol.* 52(14): 7621–7629. doi: 10.1021/acs.est.8b00829.

62 Pozo, K., Harner, T., Wania, F., Muir, D.C.G., Jones, K.C., and Barrie, L.A. (2006). Toward a global network for persistent organic pollutants in air: results from the GAPS study. *Environ. Sci. Technol.* 40(16): 4867–4873. doi: 10.1021/es060447t.

63 Rauert, C., Shoieb, M., Schuster, J.K., Eng, A., and Harner, T. (2018). Atmospheric concentrations and trends of poly- and perfluoroalkyl substances (PFAS) and volatile methyl siloxanes (VMS) over 7 years of sampling in the Global Atmospheric Passive Sampling (GAPS) network. *Environ. Pollut.* 238: 94–102. doi: 10.1016/j.envpol.2018.03.017.

64 Yao, Y., Zhao, Y., Sun, H., Chang, S., Zhu, L., Alder, A.C., and Kannan, K. (2018). Per- and Polyfluoroalkyl Substances (PFASs) in indoor air and dust from homes and various microenvironments in China: implications for human exposure. *Environ. Sci. Technol.* 52(5): 3156–3166. doi: 10.1021/acs.est.7b04971.

65 Ahrens, L., Yeung, L.W.Y., Taniyasu, S., Lam, P.K.S., and Yamashita, N. (2011). Partitioning of perfluorooctanoate (PFOA), perfluorooctane sulfonate (PFOS) and perfluorooctane sulfonamide (PFOSA) between water and sediment. *Chemosphere* 85(5): 731–737. doi: 10.1016/j.chemosphere.2011.06.046.

66 Chen, X., Zhu, L., Pan, X., Fang, S., Zhang, Y., and Yang, L. (2015). Isomeric specific partitioning behaviors of perfluoroalkyl substances in water dissolved phase, suspended particulate matters and sediments in Liao River Basin and Taihu Lake, China. *Water Res.* 80: 235–244. doi: 10.1016/j.watres.2015.04.032.

67 Mejia-Avendaño, S., Munoz, G., Vo Duy, S., Desrosiers, M., Benoît, P., Sauvé, S., and Liu, J. (2017). Novel fluoroalkylated surfactants in soils following firefighting foam deployment during the Lac-Mégantic Railway accident. *Environ. Sci. Technol.* 51(15): 8313–8323. doi: 10.1021/acs.est.7b02028.

68 Lorenzo, M., Campo, J., Farre, M., Perez, F., Pico, Y., and Barcelo, D. (2016). Perfluoroalkyl substances in the Ebro and Guadalquivir river basins (Spain). *Sci. Total Environ.* 540: 191–199. doi: 10.1016/j.scitotenv.2015.07.045.

69 Wang, Q.W., Yang, G.P., Zhang, Z.M., and Zhang, J. (2018). Optimization of sample preparation and chromatography for the determination of perfluoroalkyl acids in sediments from the Yangtze Estuary and East China Sea. *Chemosphere* 205: 524–530. doi: 10.1016/j.chemosphere.2018.04.143.

70 Eriksson, U. and Kärrman, A. (2015). World-wide indoor exposure to Polyfluoroalkyl Phosphate Esters (PAPs) and other PFASs in household dust. *Environ. Sci. Technol.* 49(24): 14503–14511. doi: 10.1021/acs.est.5b00679.

71 Shen, A., Lee, S., Ra, K., Suk, D., and Moon, H.-B. (2018). Historical trends of perfluoroalkyl substances (PFASs) in dated sediments from semi-enclosed bays of Korea. *Mar. Pollut. Bull.* 128: 287–294. doi: 10.1016/j.marpolbul.2018.01.039.

72 Zhang, S., Peng, H., Mu, D., Zhao, H., and Hu, J. (2018). Simultaneous determination of (N-ethyl perfluorooctanesulfonamido ethanol)-based phosphate diester and triester and their biotransformation to perfluorooctanesulfonate in freshwater sediments. *Environ. Pollut.* 234: 821–829. doi: 10.1016/j.envpol.2017.12.021.

73 Codling, G., Hosseini, S., Corcoran, M.B., Bonina, S., Lin, T., Li, A., Sturchio, N.C., Rockne, K.J., Ji, K., Peng, H., and Giesy, J.P. (2018). Current and historical concentrations of poly and perfluorinated compounds in sediments of the northern Great Lakes – Superior, Huron, and Michigan. *Environ. Pollut.* 236: 373–381. doi: 10.1016/j.envpol.2018.01.065.

74 Guo, R., Megson, D., Myers, A.L., Helm, P.A., Marvin, C., Crozier, P., Mabury, S., Bhavsar, S.P., Tomy, G., Simcik, M., McCarry, B., and Reiner, E.J. (2016). Application of a comprehensive extraction technique for the determination of poly- and perfluoroalkyl substances (PFASs) in Great Lakes Region sediments. *Chemosphere* 164: 535–546. doi: 10.1016/j.chemosphere.2016.08.126.

75 Land, M., de Wit, C.A., Bignert, A., Cousins, I.T., Herzke, D., Johansson, J.H., and Martin, J.W. (2018). What is the effect of phasing out long-chain per- and polyfluoroalkyl substances on the concentrations of perfluoroalkyl acids and their precursors in the environment? A systematic review. *Environ. Evid.* 7: 4. doi: 10.1186/s13750-017-0114-y.

76 Chen, H., Han, J., Cheng, J., Sun, R., Wang, X., Han, G., Yang, W., and He, X. (2018). Distribution, bioaccumulation and trophic transfer of chlorinated polyfluoroalkyl ether sulfonic acids in the marine food web of Bohai, China. *Environ. Pollut.* 241: 504–510. doi: 10.1016/j.envpol.2018.05.087.

77 Casal, P., González-Gaya, B., Zhang, Y., Reardon, A.J.F., Martin, J.W., Jiménez, B., and Dachs, J. (2017). Accumulation of perfluoroalkylated substances in oceanic plankton. *Environ. Sci. Technol.* 51(5): 2766–2775. doi: 10.1021/acs.est.6b05821.

78 Munoz, G., Budzinski, H., Babut, M., Drouineau, H., Lauzent, M., Menach, K.L., Lobry, J., Selleslagh, J., Simonnet-Laprade, C., and Labadie, P. (2017). Evidence for the trophic transfer of perfluoroalkylated substances in a temperate macrotidal estuary. *Environ. Sci. Technol.* 51(15): 8450–8459. doi: 10.1021/acs.est.7b02399.

79 Campo, J., Perez, F., Masia, A., Pico, Y., Farre, M., and Barcelo, D. (2015). Perfluoroalkyl substance contamination of the Llobregat River ecosystem (Mediterranean area, NE Spain). *Sci. Total Environ.* 503: 48–57. doi: 10.1016/j.scitotenv.2014.05.094.

80 De Silva, A.O., Spencer, C., Scott, B.F., Backus, S., and Muir, D.C.G. (2011). Detection of a cyclic perfluorinated acid, perfluoroethylcyclohexane sulfonate, in the Great Lakes of North America. *Environ. Sci. Technol.* 45(19): 8060–8066. doi: 10.1021/es200135c.

81 So, M.K., Taniyasu, S., Lam, P.K.S., Zheng, G.J., Giesy, J.P., and Yamashita, N. (2006). Alkaline digestion and solid phase extraction method for perfluorinated compounds in mussels and oysters from South China and Japan. *Arch. Environ. Contam. Toxicol.* 50(2): 240–248. doi: 10.1007/s00244-005-7058-x.

82 Zabaleta, I., Bizkarguenaga, E., Prieto, A., Ortiz-Zarragoitia, M., Fernández, L.A., and Zuloaga, O. (2015). Simultaneous determination of perfluorinated compounds and their potential precursors in mussel tissue and fish muscle tissue and liver samples by liquid chromatography–electrospray-tandem mass spectrometry. *J. Chromatogr. A* 1387: 13–23. doi: 10.1016/j.chroma.2015.01.089.

83 Cui, Q., Pan, Y., Zhang, H., Sheng, N., Wang, J., Guo, Y., and Dai, J. (2018). Occurrence and tissue distribution of novel perfluoroether carboxylic and sulfonic acids and legacy per/polyfluoroalkyl substances in black-spotted frog (Pelophylax nigromaculatus). *Environ. Sci. Technol.* 52(3): 982–990. doi: 10.1021/acs.est.7b03662.

84 So, M.K., Taniyasu, S., Yamashita, N., Giesy, J.P., Zheng, J., Fang, Z., Im, S.H., and Lam, P.K.S. (2004). Perfluorinated compounds in coastal waters of Hong Kong, South China, and Korea. *Environ. Sci. Technol.* 38(15): 4056–4063. doi: 10.1021/es049441z.

85 Ahrens, L., Taniyasu, S., Yeung, L.W.Y., Yamashita, N., Lam, P.K.S., and Ebinghaus, R. (2010). Distribution of polyfluoroalkyl compounds in water, suspended particulate matter and sediment from Tokyo Bay, Japan. *Chemosphere* 79(3): 266–272. doi: 10.1016/j.chemosphere.2010.01.045.

86 Habibullah-Al-Mamun, M., Ahmed, M.K., Raknuzzaman, M., Islam, M.S., Negishi, J., Nakamichi, S., Sekine, M., Tokumura, M., and Masunaga, S. (2016). Occurrence and distribution of perfluoroalkyl acids (PFAAs) in surface water and sediment of a tropical coastal area (Bay of Bengal coast, Bangladesh). *Sci. Total Environ.* 571: 1089–1104. doi: 10.1016/j.scitotenv.2016.07.104.

87 Wang, S., Wang, H., Zhao, W., Cao, Y., and Wan, Y. (2015). Investigation on the distribution and fate of perfluorooctane sulfonate (PFOS) and perfluorooctanoate (PFOA) in a sewage-impacted bay. *Environ. Pollut.* 205: 186–198. doi: 10.1016/j.envpol.2015.05.042.

88 Lai, F.Y., Rauert, C., Gobelius, L., and Ahrens, L. (2019). A critical review on passive sampling in air and water for per- and polyfluoroalkyl substances (PFASs). *TrAC, Trends Anal. Chem.* 121: 115311. doi: 10.1016/j.trac.2018.11.009.

89 Wang, L., Gong, X., Wang, R., Gan, Z., Lu, Y., and Sun, H. (2017). Application of an immobilized ionic liquid for the passive sampling of perfluorinated substances in water. *J. Chromatogr. A* 1515: 45–53. doi: 10.1016/j.chroma.2017.08.001.

90 Gobelius, L., Persson, C., Wiberg, K., and Ahrens, L. (2019). Calibration and application of passive sampling for per- and polyfluoroalkyl substances in a drinking water treatment plant. *J. Hazard. Mater.* 362: 230–237. doi: 10.1016/j.jhazmat.2018.09.005.

91 Yao, Y., Chang, S., Zhao, Y., Tang, J., Sun, H., and Xie, Z. (2017). Per- and poly-fluoroalkyl substances (PFASs) in the urban, industrial, and background atmosphere of Northeastern China coast around the Bohai Sea: occurrence, partitioning, and seasonal variation. *Atmos. Environ.* 167: 150–158. doi: 10.1016/j.atmosenv.2017.08.023.

92 Jahnke, A. and Berger, U. (2009). Trace analysis of per- and polyfluorinated alkyl substances in various matrices – how do current methods perform? *J. Chromatogr. A* 1216(3): 410–421. doi: 10.1016/j.chroma.2008.08.098.

93 Shoeib, M., Harner, T., Lee, S.C., Lane, D., and Zhu, J. (2008). Sorbent-impregnated polyurethane foam disk for passive air sampling of volatile fluorinated chemicals. *Anal. Chem.* 80(3): 675–682. doi: 10.1021/ac701830s.

94 Martin, J.W., Muir, D.C.G., Moody, C.A., Ellis, D.A., Kwan, W.C., Solomon, K.R., and Mabury, S.A. (2002). Collection of airborne fluorinated organics and analysis by gas chromatography/chemical ionization mass spectrometry. *Anal. Chem.* 74(3): 584–590. doi: 10.1021/ac015630d.

95 Karásková, P., Venier, M., Melymuk, L., Bečanová, J., Vojta, Š., Prokeš, R., Diamond, M.L., and Klánová, J. (2016). Perfluorinated alkyl substances (PFASs) in household dust in Central Europe and North America. *Environ. Int.* 94: 315–324. doi: 10.1016/j.envint.2016.05.031.

96 Taniyasu, S., Kannan, K., Horii, Y., Hanari, N., and Yamashita, N. (2003). A survey of perfluorooctane sulfonate and related perfluorinated organic compounds in water, fish, birds, and humans from Japan. *Environ. Sci. Technol.* 37(12): 2634–2639. doi: 10.1021/es0303440.

97 Casal, P., Zhang, Y., Martin, J.W., Pizarro, M., Jiménez, B., and Dachs, J. (2017). Role of snow deposition of Perfluoroalkylated substances at Coastal Livingston Island (Maritime Antarctica). *Environ. Sci. Technol.* 51(15): 8460–8470. doi: 10.1021/acs.est.7b02521.

98 Monteleone, M., Naccarato, A., Sindona, G., and Tagarelli, A. (2012). A rapid and sensitive assay of perfluorocarboxylic acids in aqueous matrices by headspace solid phase microextraction–gas chromatography–triple quadrupole mass spectrometry. *J. Chromatogr. A* 1251: 160–168. doi: 10.1016/j.chroma.2012.06.033.

99 Wang, J., Shi, Y., and Cai, Y. (2018). A highly selective dispersive liquid–liquid microextraction approach based on the unique fluorous affinity for the extraction and detection of per- and polyfluoroalkyl substances coupled with high performance liquid chromatography tandem–mass spectrometry. *J. Chromatogr. A* 1544: 1–7. doi: 10.1016/j.chroma.2018.02.047.

100 Yeung, L.W.Y., Stadey, C., and Mabury, S.A. (2017). Simultaneous analysis of perfluoroalkyl and polyfluoroalkyl substances including ultrashort-chain C2 and C3 compounds in rain and river water samples by ultra performance convergence chromatography. *J. Chromatogr. A* 1522: 78–85. doi: 10.1016/j.chroma.2017.09.049.

101 Yao, X., Zhou, Z., He, M., Chen, B., Liang, Y., and Hu, B. (2018). One-pot polymerization of monolith coated stir bar for high efficient sorptive extraction of perfluoroalkyl acids from environmental water samples followed by high performance liquid chromatography-electrospray tandem mass spectrometry detection. *J. Chromatogr. A* 1553: 7–15. doi: 10.1016/j.chroma.2018.04.014.

102 Tian, Y., Yao, Y., Chang, S., Zhao, Z., Zhao, Y., Yuan, X., Wu, F., and Sun, H. (2018). Occurrence and phase distribution of neutral and ionizable per- and polyfluoroalkyl substances (PFASs) in the atmosphere and plant leaves around landfills: A case study in Tianjin, China. *Environ. Sci. Technol.* 52(3): 1301–1310. doi: 10.1021/acs.est.7b05385.

103 Lu, Z., Lu, R., Zheng, H., Yan, J., Song, L., Wang, J., Yang, H., and Cai, M. (2018). Risk exposure assessment of per- and polyfluoroalkyl substances (PFASs) in drinking water and atmosphere in central eastern China. *Environ. Sci. Pollut. R* 25(10): 9311–9320. doi: 10.1007/s11356-017-0950-x.

104 White, N.D., Balthis, L., Kannan, K., De Silva, A.O., Wu, Q., French, K.M., Daugomah, J., Spencer, C., and Fair, P.A. (2015). Elevated levels of perfluoroalkyl substances in estuarine sediments of Charleston, SC. *Sci. Total Environ.* 521-522: 79–89. doi: 10.1016/j.scitotenv.2015.03.078.

105 Martín, J., Zafra-Gómez, A., Hidalgo, F., Ibáñez-Yuste, A.J., Alonso, E., and Vilchez, J.L. (2017). Multi-residue analysis of 36 priority and emerging pollutants in marine echinoderms (Holothuria tubulosa) and marine sediments by solid-liquid extraction followed by dispersive solid phase extraction and liquid chromatography–tandem mass spectrometry analysis. *Talanta* 166: 336–348. doi: 10.1016/j.talanta.2017.01.062.

106 Lorenzo, M., Campo, J., and Pico, Y. (2018). Analytical challenges to determine emerging persistent organic pollutants in aquatic ecosystems. *TrAC, Trends Anal. Chem.* 103: 137–155. doi: 10.1016/j.trac.2018.04.003.

107 Wang, L., Sun, H., Yang, L., He, C., Wu, W., and Sun, S. (2010). Liquid chromatography/mass spectrometry analysis of perfluoroalkyl carboxylic acids and perfluorooctanesulfonate in bivalve shells: extraction method optimization. *J. Chromatogr. A* 1217(4): 436–442. doi: 10.1016/j.chroma.2009.11.074.

108 Llorca, M., Farré, M., Picó, Y., and Barceló, D. (2009). Development and validation of a pressurized liquid extraction liquid chromatography–tandem mass spectrometry method for perfluorinated compounds determination in fish. *J. Chromatogr. A* 1216(43): 7195–7204. doi: 10.1016/j.chroma.2009.06.062.

109 Gao, Y., Li, X., Li, X., Zhang, Q., and Li, H. (2018). Simultaneous determination of 21 trace perfluoroalkyl substances in fish by isotope dilution ultrahigh performance liquid chromatography tandem mass spectrometry. *J. Chromatogr. B* 1084: 45–52. doi: 10.1016/j.jchromb.2018.03.008.

110 Lacina, O., Hradkova, P., Pulkrabova, J., and Hajslova, J. (2011). Simple, high throughput ultra-high performance liquid chromatography/tandem mass spectrometry trace analysis of perfluorinated alkylated substances in food of animal origin: milk and fish. *J. Chromatogr. A* 1218(28): 4312–4321. doi: 10.1016/j.chroma.2011.04.061.

111 Álvarez-Ruiz, R., Picó, Y., Sadutto, D., and Campo, J. (2021). Development of multi-residue extraction procedures using QuEChERS and liquid chromatography tandem mass spectrometry for the determination of different types of organic pollutants in mussel. *Anal. Bioanal. Chem.* doi: 10.1007/s00216-021-03363-y.

112 Portolés, T., Rosales, L.E., Sancho, J.V., Santos, F.J., and Moyano, E. (2015). Gas chromatography–tandem mass spectrometry with atmospheric pressure chemical ionization for fluorotelomer alcohols and perfluorinated sulfonamides determination. *J. Chromatogr. A* 1413: 107–116. doi: 10.1016/j.chroma.2015.08.016.

113 Ayala-Cabrera, J.F., Javier Santos, F., and Moyano, E. (2018). Negative-ion atmospheric pressure ionisation of semi-volatile fluorinated compounds for ultra-high-performance liquid chromatography tandem mass spectrometry analysis. *Anal. Bioanal. Chem.* 410(20): 4913–4924. doi: 10.1007/s00216-018-1138-z.

114 Gremmel, C., Frömel, T., and Knepper, T.P. (2017). HPLC–MS/MS methods for the determination of 52 perfluoroalkyl and polyfluoroalkyl substances in aqueous samples. *Anal. Bioanal. Chem.* 409(6): 1643–1655. doi: 10.1007/s00216-016-0110-z.

115 Barzen-Hanson, K.A., Roberts, S.C., Choyke, S., Oetjen, K., McAlees, A., Riddell, N., McCrindle, R., Ferguson, P.L., Higgins, C.P., and Field, J.A. (2017). Discovery of 40 classes of per- and polyfluoroalkyl substances in historical aqueous film-forming foams (AFFFs) and AFFF-impacted groundwater. *Environ. Sci. Technol.* 51(4): 2047–2057. doi: 10.1021/acs.est.6b05843.

116 Barzen-Hanson, K.A. and Field, J.A. (2015). Discovery and Implications of C2 and C3 perfluoroalkyl sulfonates in aqueous film-forming foams and groundwater. *Environ. Sci. Technol. Lett.* 2(4): 95–99. doi: 10.1021/acs.estlett.5b00049.

117 Cai, M., Yang, H., Xie, Z., Zhao, Z., Wang, F., Lu, Z., Sturm, R., and Ebinghaus, R. (2012). Per- and polyfluoroalkyl substances in snow, lake, surface runoff water and coastal seawater in Fildes Peninsula, King George Island, Antarctica. *J. Hazard. Mater.* 209-210: 335–342. doi: 10.1016/j.jhazmat.2012.01.030.

118 Liu, Y., Pereira, A.D.S., and Martin, J.W. (2015). Discovery of C5–C17 poly- and perfluoroalkyl substances in water by in-line SPE-HPLC-orbitrap with in-source fragmentation flagging. *Anal. Chem.* 87(8): 4260–4268. doi: 10.1021/acs.analchem.5b00039.

119 Benskin, J.P., Bataineh, M., and Martin, J.W. (2007). Simultaneous characterization of perfluoroalkyl carboxylate, sulfonate, and sulfonamide isomers by liquid chromatography—tandem mass spectrometry. *Anal. Chem.* 79(17): 6455–6464. doi: 10.1021/ac070802d.

120 Ahmed, E., Mohibul Kabir, K.M., Wang, H., Xiao, D., Fletcher, J., and Donald, W.A. (2019). Rapid separation of isomeric perfluoroalkyl substances by high-resolution differential ion mobility mass spectrometry. *Anal. Chim. Acta* 1058: 127–135. doi: 10.1016/j.aca.2019.01.038.

121 Ayala-Cabrera, J.F., Contreras-Llin, A., Moyano, E., and Santos, F.J. (2020). A novel methodology for the determination of neutral perfluoroalkyl and polyfluoroalkyl substances in water by gas chromatography-atmospheric pressure photoionisation-high resolution mass spectrometry. *Anal. Chim. Acta* 1100: 97–106. doi: 10.1016/j.aca.2019.12.004.

122 Zhang, X., Niu, H., Pan, Y., Shi, Y., and Cai, Y. (2010). Chitosan-coated octadecyl-functionalized magnetite nanoparticles: preparation and application in extraction of trace pollutants from environmental water samples. *Anal. Chem.* 82(6): 2363–2371. doi: 10.1021/ac902589t.

123 Liu, Q., Huang, A., Wang, N., Zheng, G., and Zhu, L. (2015). Rapid fluorometric determination of perfluorooctanoic acid by its quenching effect on the fluorescence of quantum dots. *J. Lumin.* 161: 374–381. doi: 10.1016/j.jlumin.2015.01.045.

124 Xia, W., Wan, Y.-J., Wang, X., Li, Y.-Y., Yang, W.-J., Wang, C.-X., and Xu, S.-Q. (2011). Sensitive bioassay for detection of PPARα potentially hazardous ligands with gold nanoparticle probe. *J. Hazard. Mater.* 192(3): 1148–1154. doi: 10.1016/j.jhazmat.2011.06.023.

125 Wang, Z., Cousins, I.T., Scheringer, M., Buck, R.C., and Hungerbühler, K. (2014). Global emission inventories for C4–C14 perfluoroalkyl carboxylic acid (PFCA) homologues from 1951 to 2030, Part I: Production and emissions from quantifiable sources. *Environ. Int.* 70: 62–75. doi: 10.1016/j.envint.2014.04.013.

126 Tran, T., Li, J., Feng, H., Cai, J., Yuan, L., Wang, N., and Cai, Q. (2014). Molecularly imprinted polymer modified TiO_2 nanotube arrays for photoelectrochemical determination of perfluorooctane sulfonate (PFOS). *Sensors Actuators B Chem.* 190: 745–751. doi: 10.1016/j.snb.2013.09.048.

127 Cennamo, N., D'Agostino, G., Porto, G., Biasiolo, A., Perri, C., Arcadio, F., and Zeni, L. (2018). A molecularly imprinted polymer on a plasmonic plastic optical fiber to detect perfluorinated compounds in water. *Sensors* 18: 6. doi: 10.3390/s18061836.

128 Fang, C., Wu, J., Sobhani, Z., Amin, M.A., and Tang, Y. (2019). Aggregated-fluorescent detection of PFAS with a simple chip. *Anal. Methods* 11(2): 163–170. doi: 10.1039/C8AY02382D.

129 Fang, C., Zhang, X., Dong, Z., Wang, L., Megharaj, M., and Naidu, R. (2018). Smartphone app-based/portable sensor for the detection of fluoro-surfactant PFOA. *Chemosphere* 191: 381–388. doi: 10.1016/j.chemosphere.2017.10.057.

130 Yamashita, N., Kannan, K., Taniyasu, S., Horii, Y., Okazawa, T., Petrick, G., and Gamo, T. (2004). Analysis of perfluorinated acids at parts-per-quadrillion levels in seawater using liquid chromatography-tandem mass spectrometry. *Environ. Sci. Technol.* 38(21): 5522–5528. doi: 10.1021/es0492541.

131 Chu, S. and Letcher, R.J. (2017). Side-chain fluorinated polymer surfactants in aquatic sediment and biosolid-augmented agricultural soil from the Great Lakes basin of North America. *Sci. Total Environ.* 607-608: 262–270. doi: 10.1016/j.scitotenv.2017.06.252.

7

High Production Volume Chemicals

Óscar Castro, Eva Pocurull, Francesc Borrull, Rosa Maria Marcé, and Núria Fontanals

Department of Analytical Chemistry and Organic Chemistry, Faculty of Chemistry, Universitat Rovira i Virgili, Tarragona, Spain

7.1 Overview of High Production Volume Chemicals

Constant development has led society to produce tonnes of chemical compounds for many different commercial and industrial products. In the Organisation for Economic Cooperation and Development (OECD) countries, those chemical compounds with levels of production over 1000 tonnes per year in at least one member country are considered High Production Volume Chemicals (HPVCs). Compounds such as phthalate esters (PAEs), benzothiazoles (BTHs), benzotriazoles (BTRs), benzenesulfonamides (BSAs), synthetic phenolic antioxidants (SPAs), benzotriazole ultraviolet stabilizers (BUVs), and organophosphate esters (OPEs) are some of the most common and it is these on which this chapter will mainly focus. Their basic chemical structures are described in Figure 7.1 and the most representative compounds of some groups as well as the abbreviations used in this chapter are included in Table 7.1. Because they are widely and heavily used, and because most of them are not chemically bonded to their substrates, they can easily be released into the environment and are found in all environmental compartments [1–4].

7.1.1 Properties

Phthalate esters (PAEs) are alkyl-substituted derivatives of phthalic acid with disparate log K_{ow} values and volatility. Depending on the chain length, phthalate log K_{ow} values range between 1.65 for dimethyl phthalate (DMP) and 12.06 for ditridecyl phthalate (DTDP). Their main purpose is to act as non-chemically bonded plasticizers in many consumer products and industrial applications [5]. Benzenesulfonamides (BSAs), which are alkyl-substituted benzene rings with an attached N-substituted sulfonamide group, can also be used as plasticizers. Moreover, these compounds also act as intermediate products for producing pesticides, drugs, and sweeteners [6]. Benzotriazoles (BTRs) and benzothiazoles (BTHs) are heterocyclic compounds that have a five-membered ring with

Analytical Methods for Environmental Contaminants of Emerging Concern, First Edition. Edited by Núria Fontanals and Rosa Maria Marcé.

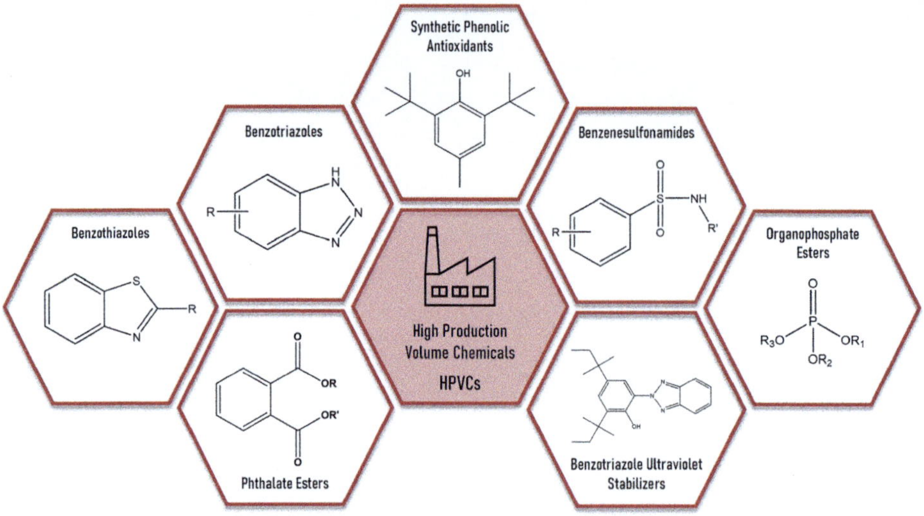

Figure 7.1 General chemical structure of some HPVCs.

Table 7.1 Examples of HPVCs and their abbreviation used in this chapter.

PAEs	**Phthalate and adipate esters**
BBzP	butyl benzyl phthalate
DEHA	di(2-ethylhexyl) adipate
DEHP	di(2-ethylhexyl) phthalate
DEHT	bis(2-ethylhexyl)terephthalate
DEP	diethyl phthalate
DiBP	di-iso-butyl phthalate
DiNP	di-iso-nonyl phthalate
DiDP	di-iso-decyl phthalate
DMP	dimethyl phthalate
DnBP	di-n-butyl phthalate
DnOP	di-n-octyl phthalate
DTDP	ditridecyl phthalate
OPEs	**Organophosphate esters**
DCP	diphenylcresyl phosphate
TBEP	tributoxy ethyl phosphate
TBP	tributyl phosphate
TClEP	tris(2-chloroethyl) phosphate
TClPP	tris(chloroisopropyl) phosphate
TDClPP	tris(1,3-dichloro-2-propyl) phosphate
TPhP	triphenyl phosphate
BTHs	**Benzothiazoles**
BTH	benzothiazole

(Continued)

Table 7.1 (Continued)

ClBTH	2-chlorobenzothiazole
MeSBTH	2-methylthiobenzothiazole
NH_2BTH	2-aminobenzothiazole
OHBTH	2-hydroxybenzothiazole
TCMTBH	2-(thiocyanomethylthio) benzothiazole
SO_3BTH	2-benzothiazole-sulfonic acid
BTRs	**Benzotriazoles**
BTR	1-H-benzotriazole
ClBTR	5-chloro-1H-benzotriazole
4TTR	4-methyl-1H-benzotriazole (4-tolyltriazole)
5TTR	5-methyl-1H-benzotriazole
XTR	5,6-dimethyl-1H-benzotriazole
BSAs	**Benzenesulfonamides**
BSA	benzenosulfonamide
o-TSA	o-toluenesulfonamide
p-TSA	p-toluenesulfonamide
SPAs	**Synthetic phenolic antioxidants and their TP**
BHA	2-tert-butyl-4-methoxyphenol
BHT	2,6-di-tert-butyl-4-methylphenol
BHT-CHO	3,5-di-tert-butyl-4-hydroxybenzaldehyde
BHT-COOH	3,5-di-tert-butyl-4-hydroxybenzoic acid
BHT-OH	2-6-di-tertbutyl-4-(hydroxymethyl)phenol
BHT-Q	2,6-di-*tert*-butyl-1,4-benzoquinone
BHT-quinol	2,6-di-tert-butyl-4-hydroxy-4-methyl-2,5-cyclohexadienone
2,4-DTBP	2,4 di-tert-butylphenol

three nitrogen atoms or a 1,3-thiazole ring attached to benzene, respectively. BTH derivatives include the 2-substituted congeners such as 2-hydroxybenzothiazole (OHBTH), 2-aminobenzothiazole (NH_2BTH), 2-benzothiazole-sulfonic acid (SO_3BTH), and 2-methylthiobenzothiazole (MeSBTH). The log K_{ow} range of these compounds is between −0.99 (SO_3BTH) and 3.22 (MeSBTH). BTH derivatives are widely used as vulcanization accelerators, fungicides, and corrosion inhibitors [2]. As for BTRs, they are highly polar compounds with low K_{ow} and they can function as anti-corrosion additives in antifreeze fluids, washing powders, and many other products which need to inhibit corrosion. They can also be used as antifogging agents and antifungals [7]. Derivatives include the mono- or poly-methylated compounds, 4/5-methyl-1H-benzotriazole (4TTR and 5TTR), and 5,6-dimethyl-1H-benzotriazole (XTR) as well as halogenated congeners such as 5-chlorobenzotriazole (ClBTR). Benzotriazole can be attached to a phenolic group, thus making room for the benzotriazole UV stabilizers (BUVs), also known as tinuvins, which are mainly used as UV light stabilizers in dairy commodities as well as other applications described for BTRs [8].

Synthetic antioxidants play an important role in preventing oxidation. For this reason they are often found in such daily life products as cosmetics, personal care products, food, etc. Depending on their structure, compounds are classified as synthetic phenolic antioxidants (SPAs) or synthetic aromatic antioxidants (SArAs). SArAs comprise compounds such as Irganox 1073 and Irgafos 168 [9]. On the other hand, SPAs are phenolic rings substituted in ortho-position by hindered alkyl moieties. 2,6-di-tert-butyl-4-methylphenol (BHT) is the most used compound. Like PAEs, they are not chemically bonded to their substrates. Other compounds include 2,4-di-tert-butylphenol (2,4-DTBP) and higher molecular weight congeners such as pentaerythitol tetrakis and their transformation products, all of which have K_{ow} values ranging from 1.79 to 19.6 and low volatility [1].

Organophosphate esters (OPEs) took over brominated compounds as novel flame retardants and some of them are also used as plasticizers. As they are used to prevent fire ignition or to modify the properties of plastics, they are present in many areas, from industrial products to electronics, textiles, and even furniture. As for their structure, these compounds consist of a phosphate core which may incorporate alkyl or halogenated moieties. Despite chlorinated organophosphates such as tris(2-chloroethyl) phosphate (TClEP) or tris(chloroisopropyl) phosphate (TClPP) being in the spotlight due to their neurotoxic, carcinogenic, and endocrine disruptor behavior, alkyl compounds such as tributyl phosphate (TBP) are also attracting attention because of their potentially negative health effects [8].

7.1.2 Legislation

Despite the widespread use of HPVCs and the adverse effects that some of them have on human health, only a few are regulated in environmental samples. Of these, some have even been banned or restricted in some products. For instance, in the European Union (EU) six phthalates, di(2-ethylhexyl) phthalate (DEHP), di-n-butyl phthalate (DnBP), butyl benzyl phthalate (BBzP), di-iso-nonyl phthalate (DiNP), di-iso-decyl phthalate (DiDP), di-n-octyl phthalate (DnOP), are restricted in plasticized material in toys and childcare articles and DEHP, BBzP, DnBP, and di-iso-buthyl phthalate (DiBP) in all electrical and electronic equipment, since some PAEs are endocrine-disrupting compounds. In this respect, it should be pointed out that the most regulated HPVCs are PAEs and some of them (e.g. BBzP, DnBP, and DEHP) have been added to the Registration, Evaluation, Authorization and Restriction of Chemicals (REACH Regulation) list as substances of very high concern (SVHC). Six PAEs (DMP, diethyl phthalate (DEP), DnBP, BBzP, DEHP, and DnOP) have been included as priority pollutants on the water lists in the EU, USA, and China. The World Health Organization (WHO) has recommended values of only 8 μg/L for DEHP in drinking water, while USA, Australia, Japan, and New Zealand have recommended maximum values of 6, 9, 100, and 10 μg/L, respectively. The EU also proposed a guideline of 1.3 μg/L for DEHP in inland surface waters while guidelines in Australia and New Zealand proposed trigger values of 5100, 1300, and 64.6 μg/L for DMP, DEP, and BBzP, respectively [4].

In sludge for disposal onto farmland, the EU proposed a limit value of 100 mg/kg for DEPH while the Danish Ministry fixed the value at 50 mg/kg. For sediments, some environmental risk limits have been set for DnBP and DEHP, at 1.82 and 1 mg/kg (d.w.) (dry weight), respectively [4].

Although there has been an increase in the use of OPEs as flame retardants, there is still relatively limited information on their potential health effects, and regulations are still limited. Only Cl-containing OPEs have been proven to be carcinogenic and some of them, such as TClEP, have also been included on the REACH list as SVHC. Triphenyl phosphate (TPhP), diphenylcresyl phosphate (DCP), and TClEP are also considered to be toxic to aquatic organisms and/or (potentially) carcinogenic. TClEP, TClPP, and tris(1,3-dichloro-2-propyl) phosphate (TDClPP) present an important risk to health and the environment due to possible toxicity and persistence [10].

As regards the other HPVCs, some BTHs and BTRs are also included on the REACH list as toxic compounds since BTRs have been associated with estrogenicity and BTHs with mutagenicity and carcinogenicity in humans. Only Australia and Denmark have drinking water standards for some BTRs [7] (the allowable concentration of 5TTR in drinking water is limited to 2400 ng/L in Australia, and 20 μg/L for BTR and 4/5TTR in Denmark).

Some tinuvins are also included on the REACH list as SVHCs. For instance, UV-328 is considered persistent, bioaccumulative, and toxic (PBT) and some others, such as UV-326, UV-327, and UV-234, are under assessment and PBT. But at present they are not regulated in environmental samples.

7.1.3 Reported or Potential Metabolites and/or Transformation Products

Each family of HPVCs has some differences in terms of metabolites and transformation products (TPs). PAEs are rapidly metabolized after they have entered the body. Thus, they are first hydrolyzed, and then monoesters produce the glucuronidated form. Figure 7.2a shows a scheme of the metabolite formation. In fact, primary and secondary metabolites are fully considered biomarkers of phthalate exposure [11]. The scenario is similar for OPEs, since their metabolites are used as biomarkers of exposure. Various studies have shown that they are quickly metabolized to dialkyl, diaryl, chlorinated and hydroxylated metabolites [12]. In addition, for the TPs of BHT, the dominant SPA congener revealed that oxidation occurred. 2,6-di-*tert*-butyl-1,4-benzoquinone (BHT-Q) and 2,6-di-tert-butyl-4-hydroxy-4-methyl-2,5-cyclohexadienone (BHT-quinol) are the result of the oxidation of the phenolic ring, while 2-6-di-tertbutyl-4-(hydroxymethyl)phenol (BHT-OH), 3,5-di-tert-butyl-4-hydroxybenzaldehyde (BHT-CHO), and 3,5-di-tert-butyl-4-hydroxybenzoic acid (BHT-COOH) are the result of the oxidation of the methyl group [1]. Figure 7.2b shows their structure. BUVs metabolites and/or TPs, however, are still largely unknown [13] although BTR and 5TTR have been identified as the most common TPs or the breakdown products of other conjugated derivatives from other compounds in the BTR family [14]. Various studies of the BTR and BTH families have also reported the transformation and chemical, biological, and photolytic degradation products after water treatment processes. In summary, compounds such as phenol, aniline, phenazine, phthalic acid, and carbazole emerge after these treatments [2, 15].

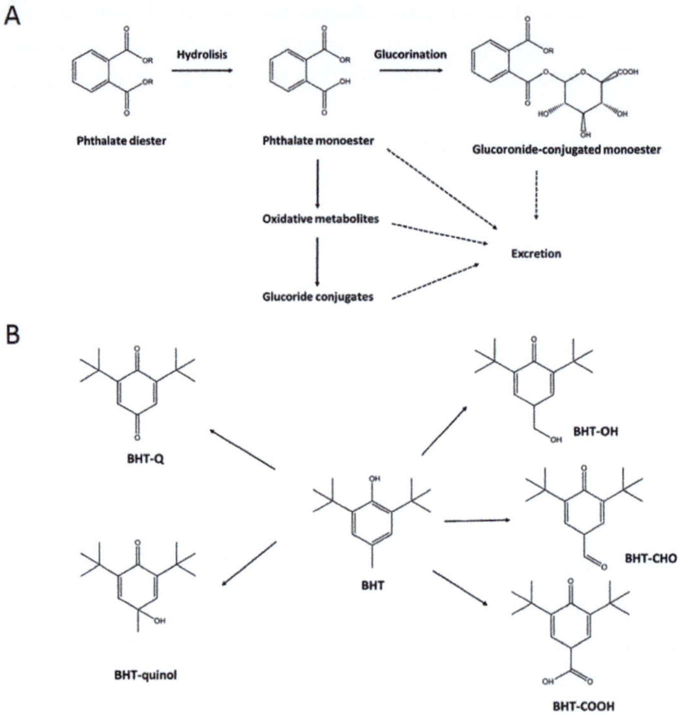

Figure 7.2 Phthalate metabolites pathway (A); structure of BHT and its main transformation products (B).

7.1.4 Occurrence

As discussed above, HPVCs have many different environmental fates. Therefore, it is not unusual for them to be found in many environmental compartments.

Their presence in water depends on the type of water and the family to which they belong. Thus, in influent sewage samples PAEs are present in concentrations ranging from 10 to 200 μg/L, with DEHP followed by DnBP being the most abundant congeners [16, 17]. BTR, 4/5TTR isomers, and p-toluenesulfonamide (p-TSA) have also been reported at concentrations of 10–100 μg/L; and in particular BTR up to 700 μg/L; although other compounds from their families have been reported at lower concentrations (ranging from 100 to 5000 ng/L) [2, 18, 19]. Other UV filters and SPAs occurred at concentrations of 100–1000 ng/L [2, 20]. Among SPAs, it should be pointed out that BHT is usually found in concentrations 10–1000 times higher than the rest of SPAs [20]. In the case of OPEs, total concentrations between 1 and 10 μg/L are usually found, with tributoxy ethyl phosphate (TBEP) and TClPP being the dominant compounds, probably because they are more soluble than other OPEs [21]. The concentration of HPVCs in effluent sewage is lower (10–1000 ng/L) than in influent sewage, which indicates efficient removal (50%–99%) in sewage treatment plants [2, 16, 22]. On the other hand, the levels of o-TSA and BSA increased after treatment, for which no explanation is currently available. Moreover, the occurrence of high concentrations of BTR in effluent sewage samples has been attributed to the lower removal rate in sewage treatment plants (STPs), but also to the fact that BTR is the transformation product of other BTRs [19]. In the case of PAEs, removal involves the formation of monoester metabolites, which are also compounds of concern [17, 23]. OPEs are eliminated at a different rate. In general, alkyl OPEs are easily eliminated while Cl-OPEs are more persistent, with TClPP being the dominant compound in effluent waters [21]. In surface water, HPVCs have been reported at concentrations ranging from not detected to 500 ng/L [1, 13, 17, 19, 21]. PAEs (mainly DEHP and DnBP) have also been reported at concentrations up to 500 ng/L in drinking, mineral, and bottled water because they are released from the containers, and also because of widespread environmental contamination [24]. OPEs have been found in drinking water at one order of magnitude lower than in surface water [21].

The presence of HPVCs has also been studied in a wide variety of air samples: for example, indoor air from laboratories [25], houses, cars, offices [26], and so on, and outdoor air from urban, suburban, and harbor environments. Phenolic antioxidants, BHT and 2,4-DTBP, and two transformation products, BHT-Q and BHT-CHO, aromatic antioxidants, such as Irgafos 168 and Irganox 176, and BUVs have been detected in all PM_{10} samples from a village and Tarragona harbor at concentrations up to 1126 pg/m^3 (BHT-Q), 2860 pg/m^3 (Irgafos 168), and 1854 pg/m^3 (Irganox 1076) [9]. These highest concentrations were from the harbor samples. And the major PAEs found in the samples are DEHP, DnBP, and DiBP [3, 27, 28]. For instance, DEHP was the major PAE in $PM_{2.5}$ and PM_{10} while DnBP and DiBP were the major PAEs in $PM_{2.5}$ in suburban areas of Shanghai, and concentrations were higher in warm periods than in cold ones [27]. DnBP and DEHP were also the most abundant PAEs in another study [28] that found levels 3.5 times higher in urban areas than suburban areas, and higher concentrations in the gas phase than in total suspended particles (TSP).

Of benzothiazoles, BTH is the most abundant in indoor air [26]. Its highest geometric mean concentration is found in bulk air (gas and TSP) from automobiles (148 ng/m^3), followed by homes (49.5 ng/m^3) and offices (18.8 ng/m^3). In PM_{10} from outdoor air in the Tarragona region [3], BTH has also been found in all samples at concentrations up to 492 pg/m^3. OHBTH was found in the highest concentrations (up to 1266 pg/m^3) but not in all samples analyzed. BTR and 4TTR were the benzotriazoles found in highest concentrations. OPEs have been determined at concentrations between 1 and 3 orders of magnitude higher in indoor air than in outdoor air. Their total concentration in outdoor air is of just a few ng/m^3 and there is a consensus that they are mostly distributed in the particle phase rather than the gas phase [21].

Indoor dust is a useful matrix for evaluating the occurrence of low volatility chemical contaminants in environments. As in air samples, BHT,BHT-OH, BHT-CHO, and BHT-Q have been found in almost all samples with the highest concentration being 15178 ng/g [29] and 47.37 μg/g [30] for BHT. As far as UV stabilizers (tinuvins) are concerned, UV-326 has been found at maximum concentrations of 4883 ng/g [31] and 275 ng/g [32] but levels of UV-328 are similar (149 ng/g [31] and 304 ng/g) [32]. Values up to 817 ng/g and 657 ng/g have been found for UV-234 [32] and tinuvin P [31], respectively. Concentrations for other UV stabilizers are lower. TTR and BTR, and BTH and MeSBTH, are the BTRs and BTHs found in the highest concentrations [33]. DEHP is also the predominant phthalate in indoor dust from various European countries [34], with concentrations between 9 and 1957 μg/g, and followed by the plasticizer di(2-ethylhexyl) adipate (DEHA), DiNP, and bis(2-ethylhexyl)terephthalate (DEHT) with variable concentrations among the samples. OPEs are present at low μg/g levels [21].

Solid samples such as soil, sediments, and sludge have been analyzed to determine the occurrence of HPVCs although fewer studies have been made than on water and air samples. Soil studies tend to focus on suburban or agricultural soil. PAEs are usually present in a wide range of concentrations (total concentration between 5 and 1600 ng/g dry weight (d.w.)), with DEHP being found at highest concentrations followed by DnBP [35]. In contrast, BTRs such as BTR and TTR have only been found in industrialized zones up to hundreds of ng/g, but are hardly detected in agricultural soils [7, 36]. The same pattern has been observed for OPEs, with their total concentration usually no higher than 100 ng/g [21].

Samples of sediments from rivers, lakes, and the sea have also been analyzed. Unlike soil samples, total concentrations of PAEs up to thousands of ng/g have been found although the most dominant compounds are the same as for soil [37]. BTRs also tend to be absorbed onto sediments despite their low K_{ow}, although at concentrations up to hundreds of ng/g [38]. SPAs were detected in surface and marine sediments at very low μg/g levels, with the dominant compounds being BHT and BHT-Q. In the case of marine sediments, spatial studies have demonstrated that the concentration of SPAs decreases with the distance from the coast, which suggests that human activities are one of the main source of SPAs [1]. In addition, the concentration of OPEs is similar to that of soil samples [21].

The presence of PAEs has been studied less in sludge samples than in the other solid samples mentioned [4]. However, the presence of BTHs such as OHBTH in sludge has usually been reported at concentrations of up to 200 ng/g, although BTRs have been found more frequently, probably due to their resistance to biodegradation [39]. SPA concentrations of up to 5 μg/g were found in a large-scale study in China, which showed no spatial differences [1]. Information about OPEs in sludge is also scarce but they can be found in total concentrations of up to thousands of ng/g, with TBEP being the most predominant [21].

PAEs are the HPVCs present in fish and seafood at the highest concentrations. Recent literature describes the presence of PAEs such as DEHP and DEP along with some of their metabolites in concentrations up to 987 ng/g d.w. in fish and seafood species like sole, salmon, shrimp, or squid collected from local stores at Tarragona, Spain [40]. Higher concentrations (up to 52.5 μg/g) have also been found in fish species from Hong Kong, with DEHP being the main contributor [41]. Besides DEHP, DEP and DnBP have also been detected in most of the samples analyzed. The use of plastic containers has led to increasing interest in studying the migration of these compounds to the fish themselves. Some studies [42, 43] related the use of these containers to the appearance of PAEs in, among other food, fish samples in markets from Belgium and Norway.

BTHs such as BTH, ClBTH, NH_2BTH, and MeSBTH have been determined in samples of fish and seafood species from local fish stores in Tarragona, Spain, with concentrations ranging between 13 and 49 ng/g d.w. [44]. Higher concentrations of the aforementioned congeners were found in mollusks from the Bohai Sea in China, especially for OHBTH and 2-/thiocyano methylthio)benzothiazole (TCMTB), whose concentrations reached 512 and 829 ng/g d.w., respectively [45]. The occurrence of BTRs in fish and seafood mainly involves the presence of BTR and the 4/5TTR isomers. Low concentrations (0.40–67.9 ng/g d.w.) have been reported in several fish and seafood species from around the world [45–51]. Compounds such as ClBTR and XTR have also been detected in fish species from the Dongjing River [48] and mollusks from the Bohai Sea [45] both in China. Very few studies have been published proving the ubiquity of BUVs in fish or seafood samples. UV-326, UV-327, and UV-328 are the compounds most commonly found, with concentrations ranging from 0.1 to 55 ng/g wet weight (w.w.) in fish species from the Ariake Sea at Japan [52]. On the other hand, UV-326, UV-328, and UV-360 have been found in higher concentrations (up to 34.9 ng/g d.w.) in fish samples from Gran Canaria, Spain [53]. SPAs such as BHT and 2-tert-butyl-4-methoxyphenol (BHA) have been detected in farmed fish species such as salmon, cod, and rainbow trout as a result of these compounds being used in animal feed [54]. The presence of BHT has been reported at a detection frequency of 100% when analyzing 274 mollusk samples [54]. Moreover, BHT-Q was the most common transformation product, with a median concentration of 0.97 μg/g.

7.2 Sample Preparation and Collection

7.2.1 Protocols for Collecting and Preparing Samples

7.2.1.1 Water

The sampling procedure used depends on the type of water that comprehend more complex (for example, influent and effluent sewage) or less complex (surface water such as river, lake, or sea water, groundwater, tap water, and drinking water). In the specific case of PAEs, mineral and bottled water is also analyzed since they might be released by the containers. Water sampling usually consists of active sampling such as grab sampling (samples taken at a particular location at a particular point in time) [22, 55–57] or integrated sampling for measurements over a longer period of time (i.e. 24 hours) [58, 59]. Water is collected in samples of up to 10 L in amber glass containers [20, 22, 56, 60] which, like the rest of the sampling equipment, has been prerinsed with such solvents as ultrapure water or methanol (MeOH) [22, 59–61]. Then, the water sampled should be refrigerated in darkness until analysis [22, 56, 60] and in some instances acidified to prevent biotic degradation [59]. Moreover, before analysis, these aqueous samples are usually filtered through glass fiber or nylon filters to remove suspended particle matter [22, 58, 61]. It should be pointed out that in the case of the determination of PAEs, due to its ubiquity presence, it is recommended to avoid plastic material (except PTFE).

In addition, passive sampling approaches using, for example POCIS (polar organic chemical integrative samplers) or SPMD (semipermeable membrane devices) provide time-weighted average concentrations of the compounds over periods ranging from weeks to months. Accumulated contaminants are desorbed from these samples in the laboratory and analyzed by suitable techniques [62].

7.2.1.2 Air and Dust

To determine HPVCs in outdoor air samples, high-volume air samplers are mainly used [4, 6]. They enable both gas phase and particulate matter to be sampled by retaining compounds in polyurethane foam (PUF) [26] and glass fiber filters (GFF) [26] or quartz fiber filters (QFF) [6, 63], respectively. As regards particulate matter, total suspended particles [28], PM_{10} or PM_5 are analyzed [9, 27, 63]. Sampling is commonly performed for 12–24 hours at flow-rates of 30–60 m^3/h [3, 27, 28], which involves high air volumes. For indoor air, PUF and GFF and QFF are also used, although in low-volume air samplers [26] sampling takes 3–24 hours at about 5 l/min.

When these materials are used, care should be taken that HPVCs do not interfere so, depending on the compounds to be determined, the PUFs should be cleaned [25, 26], and filters baked [27, 28] or cleaned with acetone [63] before use.

Although active sampling is the most commonly used procedure, passive sampling is also used with similar sampling materials to determine the average concentration over a particular period of time [25].

Apart from sampling mentioned, when gas phase is analyzed, a Tenax tube is also used to sample a low volume of air and further thermally desorbed. Other sorbents are less suitable due to the higher retention of the semivolatile organic compounds (SVOCs) and therefore not complete desorption [64]. In this case, care should also be taken with the presence of some compounds in the blanks of the tubes.

Dust is typically sampled with a vacuum cleaner equipped with a nylon sock (25 μm pore size) [34] or paper filter bag [31]. Once the dust samples have been taken, they should be homogenized by being passed through a stainless steel sieve to collect particles <125 μm [30, 65], 500 μm [32, 34, 66], or 2 mm [33] although some authors just include a cotton pad attached on the tube extender and do not sieve the sample [29].

7.2.1.3 Soil, Sediments, and Sludge

The protocols for collecting and preparing solid samples such as soil, sediments, and sludge are quite similar [36, 38, 39, 67]. Composite samples representative of the area under study are usually taken. In the particular case of soil samples, most studies focus on the surface area, which is 0–20 cm deep [36]. Once in the laboratory, they are usually frozen and lyophilized to remove water. In cases of low water content, soil samples are dried at room temperature [36]. In the last step the samples are sieved so that particles are of homogeneous sizes. A wide variety of sieves are used, although a 100–125 μm mesh is the most common. It is important to note that when PAEs and SPAs are determined, it is highly recommended to cover samples with aluminum foil to prevent compounds which may be present in many plastic materials from migrating, and interferences that may hinder the analysis from appearing.

7.2.1.4 Biota

The samples of biota collected usually depend on the purpose of the investigation. Studies involving the occurrence of HPVCs in fish samples require fish specimens to be collected from the studied area when an environmental approach is performed [68], whereas samples for food safety related studies are obtained directly from shops and markets [44]. Samples of biota such as fish and seafood are usually kept at 4°C after collection and during transport. As with other solid samples, aluminum foil is recommended to cover specimens when PAEs are determined [69, 70]. Once the samples reach the laboratory, they are frozen and stored at −20°C until analysis. Depending on the purpose of the study, fish and seafood samples can be analyzed as a whole organism [46] or as constituent parts: i.e., liver [48], muscle [68], bile [71], etc. It is therefore important to dissect the relevant part from the organism. For those species containing shell, it is removed. After the selected part has been obtained, the samples – fresh [72] or lyophilized (using freeze-drying or a dehydrating agent) [44, 68] – are usually homogenized and analyzed. The moisture content of the samples needs to be controlled in order to correctly express the analyte concentrations.

7.2.2 Sample Extraction and Clean-Up

7.2.2.1 Water

Different extraction techniques have been used to extract HPVCs from aqueous samples and Table 7.2 shows selected studies on the extraction techniques. However, as the concentration of some of the HPVCs (i.e. BTR, 4/5TTR, etc.) is in the order of low μg/L in influent sewage, they can also be determined by direct injection. In this respect, in one study [73] of a group of BTRs from environmental waters, BTR and TTR isomers were determined by direct injection of 60 μL of sample, whereas the rest of the BTRs studied were enriched by means of an SPE cartridge.

Of all the extraction techniques (see Table 7.2), solid-phase extraction (SPE) with different sorbents is the most common strategy to concentrate these compounds and simplify the aqueous matrices. SPE is usually performed in the off-line mode using cartridges from 60 mg to 500 mg. Most SPE cartridges are made of polyethylene and polypropylene that can release certain PAEs (DiBP, DnBP, and DEHP). Thus, glass cartridges are recommended for determining PAEs [4]. The main sorbent chosen is a hydrophilic-lipophilic balanced polymeric sorbent such as Oasis HLB [18, 20, 22, 55, 58–60, 73, 74]. For instance, Oasis HLB (500 mg) was used in the extraction of a group of BUVs from 200 mL of effluent and influent sewage samples [58]. Some studies selected the sorbent after making a comparison between different sorbents. Wick et al. [55] compared silica-based sorbents functionalized with C_{18} chains, hydrophilic-lipophilic balanced polymeric sorbents (Oasis HLB, Strata-X, Isolute ENV+), and mixed-mode ion-exchange sorbents (Oasis MCX and Strata-X-C). Oasis HLB was selected since it provided the best recoveries in the extraction of a group of BUVs and BHTs.

The volume of sample percolated is subject to the capacity of the sorbent and the complexity of the aqueous samples. In this respect, up to 1000 mL of less complex aqueous samples (surface, tap, and drinking water) is usually percolated, and reduced volumes of 100–200 mL are percolated for more complex samples (effluent and influent sewage samples). The sample is usually adjusted to an acidic pH although in some studies it is maintained at natural pH [18, 23, 24, 56, 58, 61]. Some studies also included a clean-up step in the SPE protocol, which usually consists of passing ~5 mL of water [22, 58–60] or water with 5–10% MeOH [74]. Depending on the nature of the SPE sorbent, a greater percentage of organic solvent during the washing might elute the compounds. The most commonly used elution solvents are MeOH [18], ethyl acetate (EtAc) [20, 22], dichloromethane (DCM), and their mixtures [23, 55, 58–60, 73, 74]. In the extraction of a group of SPAs, MeOH, ACN, and EtAc were compared as elution solvents, and finally the mixture ACN/EtAc (50/50, v/v) was selected since it provided the largest recoveries, although they only ranged from 59% to 62% for BHT and BHA [60].

Liquid-liquid extraction (LLE) has also been effectively used for the extraction of HPVCs, although to a lesser extent, especially in recent years. For instance, it has been used to extract PAEs from 500 mL of influent and effluent water samples with 3 × 100 mL of DCM [56]. Some protocols require an extra clean-up stage in which the extract is passed through silica or alumina columns [52, 56]. LLE extraction involves a large volume of organic solvent and a time-consuming evaporation step.

Table 7.2 Overview of analytical methods for the determination of HPVCs in environmental waters.

Matrix	Compounds	Extraction	Instrumental analysis	Recovery (%)	MDLs (ng/L)	Ref.
Surface, Influent WW, effluent WW	3 SPAs, 4 SPA metabolites	SPE: Oasis HLB (30 mg) L: 200 mL pH 2.5 W: 3 mL H_2O E: 1 mL EtAc 50 μL MTBSTFA + 50 μL extract	GC-MS (Q) HP-5 MS (30 m × 250 μm, 0.25 μm)	70–115 (except BHT and BHT-Q, 35–61)	2–44	[22]
River, influent WW, effluent WW	4 SPAs	SPE: SupelSectect HLB L: 100 mL pH 2.5 W: 10 mL H_2O E: 12 mL ACN/ EtAc (50/50, v/v) Evap. dryness; 100 μL H_2O/ MeOH (50/50, v/v)	LC-(ESI)MS/MS (QQQ) & LC-(HESI) MS/MS (Q-Orbitrap); Xbridge BEH C_{18} (100 mm × 2.1 mm, 2.5 μm)	60–80	0.14–0.91 (river); 0.11–1 (effluent) 0.15–1.5 (influent)	[60]
Influent WW, effluent WW, sludge	2 SPAs, 5 SPA metabolites	SPE: Oasis HLB (60 mg) L: 100 mL pH 2.5 W: – E: 9 mL EtAc Evap. 1 mL	GC-MS (Q) (only BHT) DB5(15 m × 250 μm, 0.25 μm) LC-MS/MS (QTRAP) (rest comp.) Besatil, C_{18} (100 mm × 2.1 mm, 5 μm)	96–107	0.5–5	[20]
Influent WW, effluent WW, biosolids	5 BUVs	SPE: Oasis HLB (200 mg) L: 100 mL pH 2.5 W: 5 mL 5% MeOH aq + 30 min dry E: 10 mL MTBE/ MeOH (9/1, v/v) Evap. dryness, rediss. in 1 mL ACN	UHPLC-(ESI)MS/ MS (QTRAP) Gemini-NX C_{18} (50 mm × 3 mm, 3 μm)	56–91 (influent); 68–93 (effluent)	0.3–13 (influent); 0.1–5.2 (effluent)	[74]
Tap, river, influent WW, effluent WW	6 BTHs, 5 BTRs, 3 BSAs	SPME: PA fiber 85 μm L: 8 mL (1.6 mL influent diluted to 8 mL) pH 2.5 + 6% NaCl, DI, 40 min W: – TD: 290°C, 10 min	PTV-GC-MS/MS (QqQ) TR-5 M (30 m × 250 μm, 0.25 μm)	60–100	0.0001–1	[75]

(*Continued*)

Table 7.2 (Continued)

Matrix	Compounds	Extraction	Instrumental analysis	Recovery (%)	MDLs (ng/L)	Ref.
Influent WW, effluent WW, sludge	6 BUVs	SPE: Oasis HLB (500 mg) L: 200 mL W: 6 mL H_2O E: 6 mL DCM/MeOH (7/3, v/v) Evap. dryness, rediss. in 1 mL MeOH	LC-(APCI)MS/MS (QqQ) Symmetry Shield C_{18} (150 mm × 4.6 mm, 5 μm)	53–98	1.5–8.5	[58]
Ground water, effluent WW, biosolid	4 BTRs, 3 BUVs	SPE: Oasis HLB (500 mg) L: 1000 mL pH 2 + 2x50 mL 5% MeOH in H_2O W: 5 mL H_2O E: 3 × 2 mL DCM/MeOH (1/1, v/v) Evap. dryness, rediss. 1 mL acetone	GC-MS/MS (QqQ) HP-5 MS (30 m × 220 μm, 0.25 μm)	79–130	0.3–14	[59]
Surface, influent WW, effluent WW, sludge	4 BTHs, 1 BTH metabolite (within 36 EOCs)	SPE: Oasis HLB (200 mg) L: 1000 mL surface, 200 mL effluent WW, 100 mL influent WW pH 6 W: - dried 1 hour E: 4 × 2 mL McOII/Acet (60/40, v/v) Evap. dryness, rediss. 1 mL MeOH/H_2O (50/50, v/v)	LC-(APCI/ESI)MS/MS (QqQ) Synergi fusion C_{18} (150 mm × 3 mm, 4 μm)	18–80 (ESI) 60–113 (APCI)	10–200 (LOQs)	[55]
River, effluent WW, influent WW	4 BTHs, 5 BTRs, 5 BSAs	SPE: Oasis HLB (150 mg) L: 500 mL surface, 250 mL effluent WW, 100 mL influent WW pH 7 W: – E: + florisil (500 mg) 2 × 3 mL MeOH Evap. to 0.1 mL and filled to 1 mL H_2O	LC-(ESI)MS/MS (QqQ) Ascentis Express C_{18} (100 mm × 2.1 mm, 2.7 μm)	75–104	500–10000	[18]

(*Continued*)

Table 7.2 (Continued)

Matrix	Compounds	Extraction	Instrumental analysis	Recovery (%)	MDLs (ng/L)	Ref.
Bottled water	9 PAEs	SPME: PDMS/DVB fiber (65 μm) L: 10 mL,10 or 30% NaCl, HS, 60 min, 90°C W: – TD: 280°C, 5 min	GC-MS (Q) DB-5 MS (30 m × 250 μm, 0.25 μm)	n.d.	3–64	[24]
River, harbor, seawater	7 PAEs	SPME: PDMS/DVB fiber (65 μm) L: 3.5 mL pH not adjusted + 30% NaCl, DI, 30 min, 80°C W: – TD: 250°C, 4.5 min	GC-MS (Q) HP-5 MS (30 m × 250 μm, 0.25 μm)	n.d.	2–27	[61]
Lake,	15 PAEs	LLE: 2 l water and 3 × 100 mL DCM. Evap. dryness, rediss. to 1 mL Hex	GC-MS (Q) DB-5 MS (30 m × 250 μm, 0.25 μm)	78–112	1–9 pg (LODs)	[56]
Lake, influent WW, effluent WW	6 PAEs, 9 PAEMs	*PAEs*: SPE: C_{18} (500 mg) L: 1000 mL lake W: drying E: 5 mL DCM + 5 mL Hex Evap. dryness, recons. 1 mL Hex *PAEMs*: SPE: MAX (500 mg) L: 500 mL W: 5 mL water + 5 mL 5% NH_4OH + 5 mL MeOH E: 6 mL 5% HCOOH in MeOH Evap. dryness, recons. 0.5 mL MeOH	*PAEs*: GC-MS (Q) HP-5 MS (30 m × 250 μm, 0.25 μm) *PAEMs*: UHPLC-(ESI)MS/MS (QqQ) BEH C_{18} (100 mm × 2.1 mm, 1.7 μm)	77–86	0.01–6	[23]
River, influent WW, effluent WW	9 OPEs	SPE: Strata-X (200 mg) L: 500 mL W: – drying 15 min E: 10 mL MeOH/DCM (1/1, v/v) Evap. dryness, rediss. 1 mL MeOH	LC-(ESI)MS/MS (QqQ) Kinetex C_{18} (50 mm × 2.1 mm, 2.7 μm)	94–106	30–250	[99]

(*Continued*)

Table 7.2 (Continued)

Matrix	Compounds	Extraction	Instrumental analysis	Recovery (%)	MDLs (ng/L)	Ref.
River	10 OPEs	SPE: Oasis HLB (150 mg) L: 500 mL W: – E: 15 mL DCM/hex (1/1, v/v) + 15 mL DCM/Acet (1/1, v/v) Evap. dryness, rediss. 250 μL toluene	GC-MS/MS (QqQ) DB-5 MS (15 m × 250 μm, 0.10 μm)	n.d.	0.3–60	[100]
River, influent WW, effluent WW	9 OPEs	SPME: PDMS/DVB fiber (65 μm) L: 22 mL pH neutral + 30% NaCl, DI, 40 min, W: – TD: 270°C, 5 min	GC-NPD DB-5 (30 m × 250 μm, 0.25 μm)	83–110	5–10 (MQLs)	[57]

Acet: acetone; ACN: acetonitrile; APCI: atmospheric pressure chemical ionizations; DCM: dichloromethane; DI: direct immersion; E: elution; ESI: electrospray ionization; EtAc: ethyl acetate; GC: gas chromatography; Hex: hexane; L: loading; LC: liquid chromatography; LLE: liquid-liquid extraction; MeOH: methanol; MS: mass spectrometry; MS/MS: tandem mass spectrometry; MTBE: methyl-tert-butyl ether; MTBSTFA: N-methyl-N-(tertbutyldimethylsilyl)trifluoroacetamide; n.d.: no data; PDMS/DVB: polydimethylsiloxane/divinylbenzene; Q: quadrupole; QqQ: triple quadrupole; SPE: solid-phase extraction; SPME: solid-phase microextraction; PTV: program temperature vaporization; UHPLC: ultra-high-performance liquid chromatography; W: washing; WW: waste water. For abbreviations of compounds, see Table 7.1.

Solid-phase microextraction (SPME) has been also used to extract HPVCs. It is particularly suitable for those chemicals determined by gas chromatography (GC), when the extract fraction can be introduced directly into the GC system and quantitatively desorbed, so making the method more sensitive. Most studies have been performed in the direct immersion mode [61, 75] for less complex samples, although for most volatile compounds such as PAEs, the headspace mode was also feasible [24]. A variety of fiber coatings are available covering a wide range of polarity: PA fibers (polar) have been successfully used to extract a group of BTHs, BTRs, and BSAs [75], whereas apolar fibers (PDMS and PDMS/DVB) are usually selected to extract OPEs or PAEs [24, 61]. Other parameters that have an influence on extraction and desorption should be carefully optimized to enhance extraction. For instance, in SPME by direct immersion of a group of OPEs from river and wastewater samples, parameters such as type of fiber (PDMS, PDMS/DVB, PDMS/CAR/DVB, CW-DVB, PA, CAR-DVB), salt addition (0–30% NaCl), sampling mode (headspace or direct immersion), stirring rate (0–1000 rpm), or extraction time (10–180 min) were optimized [57]. Table 7.2 details the optimum conditions.

Other microextraction techniques such as dispersive liquid-liquid microextraction (DLLME) or liquid-phase microextraction (LPME) have been also assayed in some studies on determining HPVCs [4, 19].

7.2.2.2 Air and Dust

For the extraction of air samples, contaminants from both the gas phase and the particulate matter are retained in a solid material (PUF, filter, etc.), so extraction techniques suitable for solid samples can be applied. Compounds from the gas phase retained in PUF or XAD are extracted using the same techniques as when the particulate matter from air is retained on filters. In these cases, the Soxhlet extraction technique [21, 25, 28] is still used but is gradually being replaced by shaking in an orbital shaker [26], ultrasonic-assisted extraction (USE) [27], pressurized liquid extraction (PLE) [3, 9], and, to a lesser extent, microwave-assisted extraction (MAE) [21, 63], as can be seen in Table 7.3.

As far as solvents are concerned, EtAc is the most used for BTH, BTR, and BSA derivates, plastic additives and OPEs, in both PLE and shaking extraction [3, 9, 26]. To extract PAEs, in some cases DCM or a mixture of hexane/DCM and MeOH are used in Soxhlet extraction [6, 25, 28], while hexane/acetone (4/1, v/v) is used in USE [27]. Good recoveries are reported for the various extraction techniques (see Table 7.3) although the clear advantage of USE, PLE, or MAE is the shorter extraction time and lower volume of solvent. For instance, the extraction of 15 PAEs by Soxhlet takes 24 [25] or 40 hours [28] but when USE is followed by shaking and centrifugation, they can be extracted in 55 minutes [27]. Another example is that PAEs, BTRs, BTHs, BSAs, and OPEs were determined from PM_{10} [3] by PLE, using only one static, 5-minutes cycle using EtAc as solvent and the extract did not need further centrifugation.

In most cases, no clean-up is applied and the extract, after evaporation, is directly injected into a chromatographic instrument. Sometimes, however, a clean-up is used; for instance, when a group of PAEs is extracted from PUF and GFF by Soxhlet, and the 70 mL of cyclohexane is passed through an activated silica gel and further evaporated [28]. In another example, a green approach for determining OPEs from PM_{10} is applied by MAE using a hydroalcoholic mixture followed by SPME [63].

As well as being extracted with solvents, analytes from the sampling tube can also be subject to thermal desorption (TD), when analyzing not only the gaseous phase [64] but also particulate matter [76]. This extraction technique has the advantage that most of the analytes retained on the tube or on the filter in the tube are desorbed and they all enter the GC column. In the case of particulate matter when TD was applied to determine a group of SVOCs including PAEs and OPEs [76], the sampling filter was cut into three parts (one half filter and two quarters), and one quarter was used to determine the most concentrated compounds, such as PAEs. The filter was rolled up and inserted into a glass desorption tube fitted with a glass wool plug and immediately transferred to the thermal desorption (TD) device coupled by a transfer line to the injection system. However, due to the semivolatility of these compounds, TD is not widely used due to the possible carry over effect since this technique is more suitable for volatile compounds.

HPVCs are mainly extracted from dust by shaking a mixture of solvent and dust and them applying centrifugation [29, 33, 65]. USE is also used to decrease the extraction time, but it also requires centrifugation [34, 66]. PLE and MAE are also recommended to improve the extraction efficiency of, for instance, BUVs [31], and OPEs [77], respectively. The amount of dust typically analyzed is between 20 mg [34, 66] and 0.5 g [31, 77], with 100 mg being the preferred amount.

Table 7.4 summarizes the characteristics of these methods, including examples of each family of HPVCs. As mentioned above, the most critical parameter in any extraction is the solvent. For instance, when PLE is used for the extraction of BUVs, the higher the polarity of the solvent is, the

Table 7.3 Analytical methods to determine HPVCs in air samples.

Matrix	Compounds	Extraction	Instrumental analysis	Recovery (%)	MDLs	Ref.
Outdoor gas phase air	7 PAEs, DEHA, 4 OPEs	Thermal desorption of the sampling tube (TENAX) using a cold trap.	GC-MS (Q) ZB-5 (30 m × 0.25 mm, 0.25 μm)	n.d.	0.003–3.3 $\mu g/m^3$	[64]
Indoor gas, particulate matter	4 BTHs	Shaking in an orbital shaker. PUFs: 2 × 100 mL EtAc, 30 min. Quartz filters: EtAc 3 × 5 min. Extracts evap. to 1 mL.	LC-(ESI)MS/MS (QqQ) Betasil C18 column (100 × 2.1 mm, 5 μm) connected to guard column Betasil C_{18} (20 × 2.1 mm)	63.4–122 (air) 70.2–128 (particulate matter)	0.2 ng/mL–2 ng/mL (analytical method LOQs)	[26]
Indoor gas, particulate phase	15 PAEs	PUF disk passive air samples and GFF. Soxhlet: 100 mL n-Hex/DCM (1:1,v/v), 6 cycles/hour, 24 hours. Extract evap. to 2 mL	GC-MS (Q) HP-5 MS (30 m × 0.25 mm, 0.25 μm)	67.9–98.5 (PUF) 76.6–104.3 (GFF)	0.2–5.76 ng/m^3	[25]
Outdoor air particulate matter	7 UV stabilizers (Tinuvins), 2 SArAs, 3 SPAs, 5 degradation products of SPAs	PLE: QFF cut + 1 g DE, EtAc, 80°C, 1500 psi, 10 min, 1 cycle. Evap.	GC-MS (Q) Zebron ZB-50 (30 m × 0.25 mm, 0.25 μm)	68–119	1–32 pg/m^3	[9]
Outdoor air particulate matter	6 PAEs, 4 BTRs, 5 BTHs, 4 BSAs, 7 OPEs	PLE: QFF cut + 1 g DE, EtAc, 70°C, 1500 psi, 5 min, 1 cycle. Evap.	GC-MS (Q) Zebron ZB-50 (30 m × 0.25 mm, 0.25 μm)	45–124	0.6–65 pg/m^3	[3]
Outdoor air gas and particulate matter	14 PAEs	Soxhlet: PUF and GFF. N-Hex/DCM/MeOH (1/1/1) for 40 hours. Activated silica gel column, 70 ml cyclohexane. Evap.	GC-MS (Q) HP-5 MS (30 m × 0.25 mm, 0.25 μm)	Gas phase: 71–105 Particle phase: 82–103	0.4–1.3 pg/m^3	[28]

(*Continued*)

Table 7.3 (Continued)

Matrix	Compounds	Extraction	Instrumental analysis	Recovery (%)	MDLs	Ref.
Indoor air suspended particulate matter	8 PAEs, tributylphosphate, other SVOCs	Thermal desorption Gerstel	GC-MS (Q) Rtx-PCB (60 m × 0.25 mm, 0.25 μm)	n.d.	0.1–1 ng/m^3	[76]
$PM_{2.5}$, PM_{10} in suburban area	9 PAEs	USE: Portion of GFF extract 3 × 4 mL of Hex/Acet(4:1), 30 min, shaking in orbital shaker 20 min each, centrif. 4000 rpm 5 min, evap. of combined extracts under N_2	GC-MS (Q) DB-5 (30 m × 0.25 mm, 0.25 μm)	78–94	0.05 ng/m^3	[27]
PM_{10}	10 OPEs	MAE: filter extracted with 10 mL water-EtOH (50/50, v/v) at 180°C for 1.5 min. 800 μL extract diluted Into 8 mL of water. SPME immersion: DVB/Car/PDMS fiber. 45 min, 80°C, 10% NaCl	GC-MS/MS (QqQ) HP-5 (30 m × 0.25 m, 0.25 μm)	n.d.	0.1 ng/l–0.5 ng/L (analytical method LODs)	[63]

Acet: acetone; DCM: dichloromethane; DE: diatomaceous earth; DVB/Car/PDMS: divinylbenzene/carboxen/polydimethylsiloxane; ESI: electrospray ionization; EtAc: ethyl acetate; EtOH: ethanol; GC: gas chromatography; Hex: N-hexane; LC: liquid chromatography; MAE: microwave-assisted extraction; MeOH: methanol; MS: mass spectrometry; MS/MS: tandem mass spectrometry; n.d.: no data; PLE: pressurized liquid extraction; PUF: polyurethane foam; Q: quadrupole; QFF: quartz fiber filter; QqQ: triple quadrupole; USE: ultrasonic-assisted extraction. For abbreviations of compounds, see Table 7.1.

Table 7.4 Analytical methods to determine HPVCs in dust samples.

Matrix	Compounds	Extraction	Clean-up	Instrumental analysis	Recovery (%)	MDLs	Ref.
Indoor dust	6 PAEs, 1 OPEs, others SVOCs	Thermal desorption Gerstel	-	GC-MS/MS (QqQ) Rtx-PCB (30 m × 0.25 mm, 0.25 μm)	78–151	50–1000 ng/g	[101]
Indoor dust	8 SPAs, 4 degradation products	Shaking: 0.1 g dust sample, 3 extractions ACN, at 4000 rpm 30 min., centrif. Evap. and rediss. in 1 mL MTBE. Centrif.	-	GC-MS (Q) DB-1701 (30 m × 0.25 mm, 0.25 μm)	85–118	0.3–20 ng/g (MQLs)	[29]
Indoor dust	5 BTRs, 5 BTHs, 5 BPs (benzophenone derivates)	Shaking: two extraction with $MeOH/H_2O$ (5/3, v/v), 60 min., centrif. 4500 g 5 min. Extracts combined, evap. to ~ 4 mL and diluted to 10 mL of 0.2% HCOOH in H_2O.	SPE Oasis MCX (60 mg): L: 10 mL sample extract W: 15 mL MeOH/ H_2O (20/80) and 5 mL water. Drying cartridge. E: 5 mL MeOH. Evap. to 1 mL.	LC-(ESI)MS/MS (QqQ) Betasil C_{18} (20 × 2.1 mm) and guard column Betasil (C_{18} 20 × 2.1 mm)	44–113	0.3 ng/g–1 ng/g (MQLs)	[33]
Indoor dust	5 PAEs, 50 other contaminants	USE: 100 mg dust. Vortex 1 min. and USAE 10 min. Hex/Acet (1/1)x 3. Centrif. 3000 rpm 10 min., evap. N_2	Florisil 10 g. Elution 30 mL Hex/Acet (1/1) Evap. to dryness, recons. 1.5 mL	GC-MS/MS (QqQ) ZB-5 MS (30 m × 0.25 mm, 0.25 μm)	n.d.	n.d.	[65]
Indoor dust	5 BUVs	PLE: 0.5 g, 3 static cycles of 10 min, Hex/DCM (8/2), 90°C, 1500 psi	In-cell clean-up with silica (1 g).	GC-MS/MS (IT) BP-5 (30 m × 0.25 mm, 0.25 μm)	82–122	4–10 ng/g (MQLs)	[31]

(Continued)

Table 7.4 (Continued)

Matrix	Compounds	Extraction	Clean-up	Instrumental analysis	Recovery (%)	MDLs	Ref.
Indoor dust	BHT, 4 TPs	USE: ? g of dust, Hex/DCM (1/3) × 3, 15 min. Centrif. at 3000 rpm 5 min. 3 extracts combined and dehydrated with anhydrous Na_2SO_4. Evap. and rediss.	-	GC-MS (Q). HP-5 MS, (30 m × 0.25 mm, 0.25 μm). Derivatization of BHT-COOH and BHT-OH.	71.1–118	0.02–0.34 ng/g	[30]
Indoor dust	5 SPAs, 4 TPs	Shaking: 0.25 g dust, 5 mL DCM/Hex (3/1), shaking 60 min. centrif. at 2000 g 10 min. Extraction twice and extracts combined, concentr. and solvent change to EtAc.	-	GC-MS (Q) for BHT: DB-5 (30 m × 0.25 mm, 0.25 μm). LC-MS/MS (QTRAP) for other SPAs: Betasyl C_{18} (100 × 2.1 mm)	69–106	0.12–2.4 ng/g (MLQs)	[78]
Indoor dust	7 BUVs	0.1 g dust homogenized with anhydrous Na_2SO_4 and extracted using high speed solvent extractor with Acet/Hex (1/1) at 35°C, at 6 ml/min 1 hour. Extract evap. to 10 mL.	Glass column with 4 g of 6% deactivated alumina. Eluted with 50 mL DCM. Evap. and rediss. in 1 mL MeOH	UHPLC-(ESI)MS/MS (QTRAP) Ascentis Express C_{18} (100 × 2.1, 2.7 μm)	67–107	0.6–3.8 ng/g	[32]

Indoor dust	19 OPEs, 20 plasticizers	USE: 20 mg with 2.5 mL of Hex/Acet (3:1, v/v) and 0.5 mL of toluene by a combination of vortexed 1 min and USE 5 min, repeated twice. Centrif. at 300 rpm 3 min. Evap. and rediss. in 1 ml of Hex/toluene (1/1) and vortexed 1 min.	Extracted passed through florisil ENVI (500 mg), washed with 4 mL Acet, 6 mL EtAC and 6 mL Hex, and eluted with 12 mL Hex/DCM (4/1) (F1) and 10 mL EtAc (F2). F2 evap. to dryness and dissol. in 50 μL mL MeOH and divided by LC and GC analysis	LC-(ESI)MS/MS (QqQ) Kinetex biphenyl (100 × 2.1 mm, 2.6 μm) GC-MS (Q) for DEHP and DEHT. HT-8 (25 m × 0.25 mm, 0.25 μm)	n.d.	1–265 ng/g, except 665 (DEHP) and 1100 ng/g (DEHT)	[66]
Indoor dust	9 OPEs	MAE: 0.5 g with 10 mL Acet at 130°C, 30 min. Centrif. 5 min. 300 rpm and supernatant added to 500 mL of water.	SPE: water solution passed Oasis HLB and eluted with 2 mL EtAc Solution passed through 500 mg silica cartridge. Elution with 5 mL EtAc. Evap. to 1 mL	GC-NPD DB-5 (30 m × 0.32 mm, 0.25 μm)	85–104	00.4 μg/g (TiBP, TBP, TCEP, TDPP) and 0.05 μg/g the rest (MQLs)	[77]

Acet: acetone; ACN: acetonitrile; DCM: dichloromethane; ESI: electrospray ionization; EtAC: ethylacetate; GC: gas chromatography; Hex: n-hexane; IT: ion trap; LC: liquid chromatography; MAE: microwave-assisted extraction; MS: mass spectrometry; MS/MS: tandem mass spectrometry; MTBE: methyl-tert-butyl ether; MeOH: methanol; n.d.: no data; NPD: nitrogen-phosphorous detector; PLE: pressurized liquid extraction; Q: quadrupole; QqQ: triple quadrupole; SPE: solid-phase extraction; UHPLC: ultra-high-performance liquid chromatography; USE: ultrasonic-assisted extraction. For abbreviations of compounds, see Table 7.1.

more complex the visual appearance of the extract [31]. However, when the non-polar solvent hexane is used, the most polar analytes are not recovered. Therefore, in most cases a mixture of solvents is preferred. Examples of the solvents used are: for SPAs, ACN [29] and hexane/DCM (1/3, v/v) [30, 78]; for BTRs and BTHs, MeOH/water (5/3,v/v) [33]; for BUVs, hexane/DCM (8/2, v/v) [31] and acetone/hexane (1/1, v/v) [32]; for plasticizers, including PAEs and/or OPEs, hexane/acetone (1/1, v/v) [34, 65], hexane/acetone (3/1) [66] or acetone [77].

Due to the complexity of the sample, a clean-up step using SPE is included in most methods, in which the extract is passed through florisil [34, 65, 66], or alumina cartridges [32], although Oasis MCX [33] and Oasis HLB [77] have also been used for BTRs and BTHs, and OPEs, respectively. In some cases, florisil provided better recoveries than alumina for a wide group of compounds [65]. When PLE was used, an in-cell clean-up with silica was as efficient as florisil, while alumina appeared to be more retentive, although no differences were observed in terms of selectivity [31].

Like particulate matter analysis, another less used option is TD coupled to GC of compounds from dust introduced in a tube. This option has been used to determine a group of SVOCs including PAEs and OPEs [77]. In detail, 2 mg of dust was placed on a quarter of a QFF that was folded, rolled, and inserted into a glass tube and TD was applied in conditions similar to those reported in a previous study that analyzed particulate matter [27], but in this case coupled to GC coupled to tandem mass spectrometry (MS/MS). When the results were compared with those of PLE, limits of quantification (LOQs) were similar although the amount of sample was 100 times lower, and the sample preparation procedure was more environmentally friendly.

7.2.2.3 Soil, Sediments, and Sludge

As stated above for solid samples, HPVCs are usually extracted from soil, sediments, and sludge with USE, MAE, or PLE. Table 7.5 shows information from a variety of studies on this type of sample.

The amount of sample needed varies between 0.5 g and 5 g, with PLE [21, 38, 79] being the technique which requires the least (usually between 0.5 g and 1 g) and USE [35, 67] the one that requires the most (up to 5 g). The selection of the extraction solvent mainly depends on the nature of the contaminants. Polar solvents such as water or MeOH are widely used to extract BTRs, BTHs, and BSAs from any of the solid samples mentioned by MAE [36] or PLE [38, 39, 80]. However, the extraction of PAEs, OPEs, and SPAs requires mixtures of less polar solvents such as hexane, DCM, or EtAc. For instance, hexane:DCM (1/1, v/v) and (1/3, v/v) were used in USE to extract a group of PAEs from an agricultural soil [35] and a group of SPAs from marine sediments [81], respectively. And EtAc/cyclohexane (5/2, v/v) was used to extract a group of OPEs from sediment, sludge, and dust [82]. The volume of extraction solvent needed, independently of the extraction techniques used, varies from 5 mL to 20 mL and a single extraction is usually enough to quantitatively extract the contaminants.

MAE and PLE require a temperature of about 90–120°C to enhance the extraction of the contaminants. As mentioned above, the extraction takes less time in MAE and PLE (about 5–10 min) than in USE, which in most cases takes 30–60 min [21, 35].

Although all the techniques mentioned can satisfactorily extract contaminants from solid samples, co-extraction of interferences is one of their main drawbacks. Thus, a clean-up step is mandatory. For this purpose, SPE is widely used in most of the methods described (Table 7.5). Oasis HLB and silica are the two most commonly used sorbents. Oasis HLB has

Table 7.5 Overview of analytical methods for the determination of HPVCs in soil, sediments, and sludge.

Matrix	Compounds	Extraction	Instrumental analysis	Recovery (%)	MDLs	Ref.
Industrialized area soil	2 BTRs, 3 BTHs, 1 BSAs	Dried at room temperature and sieved at 0.2 mm MAE: 3 g soil + 6 mL MeOH, 200 W, 120°C, 10 min Centrif. at 4500 rpm, 10 min Evap. and rediss. with 0.5 mL MeOH:H_20 (1:1)	LC-UV; MS for confirmation Accucore XL C_{18} (100 × 4.6 mm, 4 μm)	72–119	7–25 μg/kg	[36]
River sediments	3 BTRs	Lyophilized and sieved 120 μm PLE: 1 g sample + 1 g activated aluminum oxide, MeOH or MeOH:H_20 Extract diluted with MeOH to 25 mL Evap. to dryness and rediss. with 500 μL in ACN	LC-(ESI)MS/MS (QTRAP) LiChroCART Purospher STAR RP-18 ec (125 × 2.0 mm, 5 μm)	58–125	0.5–15.5 ng/g (d.w.)	[38]
River sediments	8 BTRs	Lyophilized PLE: 1 g sample + DE; MeOH, 100°C, 1500 psi. Evap. to 0.5 mL and diluted to 10 mL with H_20 Clean-up: SPE Oasis HLB, 6 mL MeOH, evap. to dryness, rediss. 1 mL MeOH:H_20 (20:80)	LC-(ESI)MS/MS (QqQ) CSH C_{18} (100 × 2.1 mm, 1.7 μm)	n.d.	n.d.	[80]
Sewage sludge	5 BTRs, 4 BTHs, 5 BSAs	Lyophilized PLE: 1 g sample + DE, H_2O, 80°C, 1500 psi, 5 min Clean-up: SPE Oasis HLB + fluorisil, 2 × 3 mL MeOH, evap. and rediss. up to 1 mL H_2O	LC-(HESI)HRMS(Orbitrap) Ascentis Express C_{18} fused-core column (100 × 2.1 mm, 2.7 μm)	25–107	0.25–25 ng/g (d.w.)	[39]
Agricultural soil	6 PAEs	Lyophilized and sieved 100 μm USE: 5 g soil + 20 mL Hex:DCM (1:1), 60 min Centrif. 5 min, 2500 rpm Clean-up: SPE silica + Al_2O_3 + Na_2SO_4, 40 mL Acet:Hex (1:4) Evap. dryness, rediss. with 1 mL Hex	GC-MS(Q) DB5-MS (30 m × 0.25 mm, 0.25 μm)	n.d.	n.d.	[35]

(Continued)

Table 7.5 (Continued)

Matrix	Compounds	Extraction	Instrumental analysis	Recovery (%)	MDLs	Ref.
Marine sediments	7 PAEs, 8 OPEs	Freeze-dried, sieved 500 μm USE: 5 mL DCM Centrif. 4000 rpm, 10 min Clean-up: SPE Oasis MAX, 5 mL DCM/EtAc (1:1)	GC-MS(Q) HP-5 MS (30 m × 0.25 mm, 0.25 μm)	66–109 (PAEs); 52–131 (OPEs)	0.0001–0.01 ng	[67]
Marine sediments	14 PAEs	Freeze-dried, sieved 50 μm MAE: 3 g sample, 15 mL MeOH:EtAc (1:1), filtered 0.45 μm. Evap. near dryness, rediss. in 500 μL extraction solvent	GC-MS(Q)	n.d.	n.d.	[37]
Marine sediment	9 SPAs	Freeze-dried, sieved 80 meshes USE: 0.5 g sample + 1 g copper power, 5 mL DCM:Hex (3:1), 20 min Centrif. 3300 rpm. Extract divided in 2 parts. First part, evap. and rediss. with 0.5 mL Hex to be analyzed by GC-MS/MS. Second part, evap. and rediss. with 0.5 mL MeOH to be analyzed by UHPLC-MS/MS.	GC-MS/MS (QqQ) DB-5 capillary column (30 m × 0.25 mm, 0.25 μm) UHPLC-MS/MS (QqQ) CSH Phenyl-hexyl UPLC analytical column (100 × 2.1 mm, 1.7 μm)	n.d.	n.d.	[81]
Sludge	14 SPAs	Freeze-dried, sieved 100 meshes PLE: 0.5 g sample + 15 g Na_2SO_4, DCM:Hex (3:1), 90°C, 1500 psi, 3 × 10 min Clean-up: 8 g silica, elution 120 mL DCM:Hex (1:1), concent. and rediss. with 5 mL MeOH	LC-(ESI)MS/MS (QqQ) Symmetry Shield C_{18} (150 × 2.1 mm, 5 μm)	65–103	n.d.	[79]

Acet: acetone; DCM: dichloromethane; DE: diatomaceous earth; ESI: electrospray ionization; EtAC: ethylacetate; GC: gas chromatography; Hex: n-hexane; HRMS: high-resolution mass spectrometry; LC: liquid chromatography; MAE: microwave-assisted extraction; MeOH: methanol; MS: mass spectrometry; MS/MS: tandem mass spectrometry; n.d.: no data; PLE: pressurized liquid extraction; Q: quadrupole; QqQ: triple quadrupole; SPE: solid-phase extraction; UHPLC: ultra-high-performance liquid chromatography; USE: ultrasonic-assisted extraction; UV: ultraviolet-visible absorption. For abbreviations of compounds, see Table 7.1.

been successfully used to clean-up extracts from PLE [39, 80] and USE [67] when BTRs, BTHs, and BSAs and PAEs are determined in sediments or sludge. In addition, silica has been used as the clean-up sorbent for USE [35] extracts when PAEs are determined in soil.

7.2.2.4 Biota

HPVCs in fish and seafood samples have been determined in the literature using many different extraction approaches. Table 7.6 overviews the analytical methods used to determine HPVCs from biota. PAEs are compounds that have most been studied and the preferred extraction techniques are solid-liquid extraction (SLE) and USE. Sample weights of fish and seafood between 3 and 20 g can be extracted using a mixture of hexane/acetone (1/1, v/v) [83–85] while smaller amounts of sample (from 3–20 g to 2 g) are analyzed with USE [71, 86]. Even though SLE and USE using hexane/acetone mixture as solvent are the most widely used extraction techniques, other studies [72, 87] have shown that matrix solid-phase dispersion (MSPD) and QuEChERS (Quick, Easy, Cheap, Effective, Rugged and Safe) are also efficient extraction techniques, providing comparable or higher recovery rates and replacing hexane with greener solvent options such as EtAc or ACN.

Few methods for determining BTHs and BTRs in fish and seafood have been reported in the literature. Nevertheless, BTHs have been extracted from biota by Trabalón et al. [44], who combined PLE using subcritical water extraction followed by SPME. Recoveries were between 86 and 135% with 1 g of sample and a single cycle with water at 80°C and 1500 psi, followed by the immersion of the aqueous extract in PDMS/DVB fiber. In fact, BTRs are currently best extracted from fish samples by QuEChERS [46–48]. Moreover, the literature also reports methods using PLE with DCM/EtAc (1/1, v/v) and USE with DCM/hexane (1/1, v/v). Wang et al. [88] described determining SPAs such as BHT and their metabolites in mollusks using USE with a DCM/hexane (3:1, v/v) mixture. The presence of BUVs in fish samples can also be determined using USE with an hexane/acetone (1:1, v/v) mixture [89] or even using MAE with ACN [68] as extraction techniques.

The hexane/acetone solvent has also been widely described for extracting OPEs from fish and seafood samples with USE [90–92]. However, PLE or QuEChERS appear to be good alternatives for extracting OPEs from seafood and fish and recoveries are higher. For instance, the determination of 17 OPEs in fish and seafood species using PLE with a single cycle of EtAc as extraction solvent at 100°C yielded recoveries between 77 and 97% [93]. On the other hand, the method developed using QuEChERS for determining OPEs in foodstuff, including fish, provided recoveries of 73–106% [94].

For fish samples, lipids are usually a critical factor when trying to achieve an effective determination. Thus, the use of techniques such as SPE and dispersive SPE (d-SPE) are suitable for this purpose (Table 7.6). As regards the sorbents used for clean-up, florisil [71, 86] and primary secondary amine (PSA) [87] have been proven to be effective at determining PAEs from fish and seafood using SPE and d-SPE, respectively. Oasis HLB cartridges have also been used to purify the extracts obtained from USE for determining BTRs, BTHs [45], and BUVs [88]. BTRs can be determined with a mixture of PSA and C_{18} using d-SPE after a QuEChERS extraction process [46–48]. Approaches such as gel permeation chromatography (GPC), hexane on-cell PLE clean-up, and MSPD using glass carbon black have also been described as good alternatives [44, 84, 85, 89, 95].

Table 7.6 Overview of analytical methods for the determination of HPVCs in biota.

Matrix	Compounds	Extraction	Clean-up	Instrumental analysis	Recovery (%)	MDLs (ng/L)	Ref.
Shrimp, Fish	9 PAEs	SLE: 3–20 g d.w. sample 3x Hex/Acet (1:1) + shaken 30 min + centrif. 10 min + evapor. to 3 mL	LLE Hex/ACN 40 mL Hex/ACN + shake (15 min) + 3 × 2 mL Hex + Hex discarded	GC-MS (Q) DB-5 MS (30 m × 0.25 mm, 0.25 μm)	56–101	2–10 ng g^{-1} w.w.	[83]
Fish	6 PAEs, 9 PAEMs	USE: 2 g d.w. sample 2 × 20 mL Hex/Acetone (8:2) + sonicated 20 min + centrif. 5 min at 970 g + extracts combined	SPE Florisil E: 10 mL Hex/Diethyl Ether (8:2)	GC-MS (Q) ZB-5 (30 m × 0.25 mm, 0.25 μm)	57–87	0.4–10 ng g^{-1} w.w.	[71]
Fish	19 PAEs	QuEChERS: 5 g w.w. sample + 5 mL ACN (1% HCOOH) + vortex 1 min + 2 g NaCl + 1 min shake + sonicated 30 min	d-SPE PSA 3 mL supernatant + 150 mg PSA + 450 mg $MgSO_4$ + vortex 1 min + centrif. 5 min at 3000 rpm	GC- MS/MS (QqQ) HP-5 MS (30 m × 0.25 mm, 0.25 μm)	71–116	0.01–10 ng g^{-1} w.w.	[87]
Fish	5 BTHs	PLE: 1 g d.w. sample + 2 g DE in-cell clean-up 1 cycle H_2O (80°C, 5 min, 1500 psi) Purge cycle (solvent flush 60%) SPME 10 mL extract PDMS/DVB, immersion 40 min, 80°C, 750 rpm	In-cell Hex 2 cycles Hex (60°C, 5 min, 1500 psi, 80% solvent flush)	GC-MS/MS (IT) ZB-5 (30 m × 0.25 mm, 0.25 μm)	86- 135	0.5–10 ng g^{-1} d.w	[44]
Fish	4 BTRs	QuEChERS: 2 g w.w. sample equilibrated at 4°C for 30 min + 5 mL H_2O + vortex 1 min + 10 mL ACN (1% acetic acid) + 6 g $MgSO_4$ + 1.5 g sodium acetate + shaken 1 min + centrif. 10 min	d-SPE PSA + C_{18} extract + 900 mg $MgSO_4$ + 150 mg PSA + 150 mg C_{18} vortex 1 min + centrif. 10 min	LC-ESI(+)-MS/MS (QqQ) Zorbax SB-C18 (100 × 3 mm, 1.8 μm)	51–120	0.02–0.11 ng g^{-1} w.w.	[48]

Mollusk	1 SPA, 4 SPA metabolites	USE: 0.1 g d.w. sample 3 × 10 mL DCM/Hex (3:1) + sonicated 20 min (each step) + shaken 20 min + centrif. 10 min at 3300 rpm + extracts combined + evapor. to dryness + 0.5 mL MeOH + 0.5 mL H_20 + filtered 0.22 um + 9.5 mL H_2O	SPE Oasis HLB L: 7 mL MeOH + 7 mL H_2O W: 10 mL H_2O with 5% MeOH E: 10 mL MeOH P1: evapor. to 0.5 mL P2: solvent exchange to 0.5 mL Hex	GC-MS/MS(QqQ) (SPA) DB-5 (30 m × 0.25 mm, 0.25 μm) LC-ESI (-)-MS/MS (QqQ) (SPA metabolites) Luna C_{18} (150 × 3 mm, 3 μm)	61–92	0.4–1.9 ng g^{-1} d.w.	[88]
Fish	8 BUVs	SLE: 5 g w.w. sample + Hex/Acet (1:1) at 30°C for 30 min + evapor. to < 10 mL and rediss. with Hex	GPC silica 1 mL extract E: 100 mL DCM + evapor. + 1 mL MeOH	LC-(ESI)MS/MS (QqQ) C_{18} (100 × 2, 1 mm, 2.7 μm)	70–112	0.0002–0.009 ng g^{-1} w.	[89]
Fish	6 BUVs	MAE: 0.1 g d.w. sample 2.5 mL ACN + 10 min MAE 200 W	SPE Phree Phospholipid Removal + evapor. + 1 mL ACN	LC-(ESI)MS/MS (QqQ) C_{18} (50 × 2, 1 mm, 1.7 μm)	13–75	0.34–2.87 ng g^{-1} d.w.	[68]
Mollusk	4 BTRs, 4 BTHs	USE: 0.1 g d.w. sample 2 × 5 mL Hex/DCM (1:1) + sonicated 20 min + centrif. 12 min at 3500 rpm + extracts combined + evapor. to dryness + 0.5 mL MeOH + frozen extract at −20°C for 2 hours + centrif. at 12000 rpm for 12 min + rediss. with 10 mL with H_2O 0.03% HCOOH	SPE Oasis HLB W: 6 mL H_2O 10% MeOH E: 6 mL MeOH	GC-MS/MS (QqQ) TG-5 MS (30 m × 0.25 mm, 0.25 μm) + LC-(ESI)MS/MS (QqQ) CSH Phenyl-hexyl (100 × 2.1 mm, 1.7 μm)	51–120	0.607–5.12 ng g^{-1} d.w.	[45]

(*Continued*)

Table 7.6 (Continued)

Matrix	Compounds	Extraction	Clean-up	Instrumental analysis	Recovery (%)	MDLs (ng/L)	Ref.
Fish	16 OPEs	USE: 0.25 g d.w. sample 15 mL Acet/Hex (1:1) + sonicated 15 min x2 + extracts combined + 30 mL of extract evapor. to dryness with N_2 and rediss. with 60 mL of ACN	SPE C_{18} + Alumina E: 60 mL ACN	LC-(ESI)MS/MS (QqQ) Purosphere RP-18 (125 mm × 2 mm, 5 μm)	45–115	1.12–11.6 ng g^{-1}w.	[90]
Fish	20 OPEs	PLE: 1 g d.w. sample + DE, 1 cycle EtAc (100°C, 5 min, 1500 psi) Purge cycle (solvent flush 60%) Extract + evap. to 1 mL	In-cell Hex/DCM 2 cycles Hex/DCM (1:1) (100°C, 5 min, 1500 psi) GPC Alumina + Silica + florisil + PSA 1 g neutral alumina + 2 g silica gel + 2 g florisil + 0.3 g PSA + 2 g Na_2SO_4 E: 20 mL EtAc	GC-MS (Q) DB-5 MS (30 m × 0.25 mm, 0.25 μm)	77–97	0.06–1.29 ng g^{-1} w.w.	[93]
Fish	10 OPEs	QuEChERS: 1 g f.w. sample + 4 mL ACN (5% HCOOH) + vortex + sonicated 10 min + 400 mg NaCl + 400 mg $MgSO_4$ + vortex 1 min + centrif. 5 min at 3000 rpm	d-SPE PSA + C_{18} + Alumina supernatant + 50 mg PSA + 30 mg C_{18} + 30 mg neutral alumina + vortex 1 min + centrif. 5 min at 3000 rpm	LC-(ESI)MS/MS (QqQ) Waters BEH C_{18} (100 mm × 2.1 mm, 1.7 μm)	73–106	0.05–0.42 ng g^{-1} d.w.	[94]

Acet: acetone; ACN: acetonitrile; DCM: dichloromethane; DE: diatomaceous earth; d-SPE: dispersive solid-phase extraction; d.w.: dry weight; E: elution; ESI: electrospray ionization; EtAC: ethylacetate; GC: gas chromatography; GPC: gel permeation chromatography; Hex: hexane; L: loading; LC: liquid chromatography; LLE: liquid-liquid extraction; l.w.: lipid weight; MAE: microwave-assisted extraction; MS: mass spectrometry; MS/MS: tandem mass spectrometry; MeOH: methanol; PDMS/DVB: polydimethylsiloxane/divinylbenzene; PLE: pressurized liquid extraction; PSA: primary secondary amine; Q: quadrupole; QqQ: triple quadrupole; QuEChERS: Quick, Easy, Cheap, Effective, Rugged & Safe extraction; QqQ: triple quadrupole; SLE: solid-liquid extraction; SPE: solid-phase extraction; SPME: solid-phase microextraction; UHPLC: ultra-high-performance liquid chromatography; USE: ultrasonic-assisted extraction; W: washing; w.w.; wet weight. For abbreviations of compounds, see Table 7.1.

7.3 Determination of High Production Volume Chemicals

7.3.1 Development of the Instrumental Method

As HPVCs have a broad range of physical and chemical properties (from very polar to non-polar) different instrumental techniques can be used for their determination. These compounds are usually determined using GC or liquid chromatography (LC) in combination with mass spectrometry (MS)-based detectors.

GC has been used to determine the most apolar compounds (for example, PAEs, SPAs, or OPEs) using mainly non-polar stationary phases such as 5% phenyl dimethyl-polysiloxane (including trademarks such as DB-5 MS, HP-5 MS, or BP-5) [4, 20, 22–25, 27, 28, 31, 35, 56, 59, 61, 67, 75, 81], but also some others, such as the 8% phenyl polycarborane-siloxane (HT-8) [34] or 50% phenyl/50% dimethylsiloxane (Zebron-50) [3, 9], have also been used in a lower extent. The most common column dimensions are 30 m × 0.25 mm (internal diameter) and 0.25 μm (film thickness). For instance, Maceira et al. [3] used a Zebron-50 with these dimensions to simultaneously determine PAEs, OPEs, BTRs, BTHs, and BSAs from the particulate matter in outdoor air. Other columns and dimensions have been used although resolution has been lower and bleeding higher than for the non-polar option [4]. In spite of this, two PAE isomers (i.e. di-n-nonyl phthalate and di-n-decyl phthalate) are more difficult to separate with a capillary column. They can only be separated by LC. For heavier OPEs, and organophosphate monoesters and diesters, LC is more suitable and avoids tailing peaks [21, 95].

In the case of BUVs, BTRs, BTHs, and BSAs, due to their low volatility of these compounds, only few examples determine them by GC. Nevertheless, some authors have selected this technique since it overcomes some of the matrix effect issues encountered in determination by LC-electrospray (ESI)MS/MS [3, 9, 26, 59, 75].

Although PAEs and SPAs are volatile enough to be directly determined by GC, their derivatization might enhance their volatility and so its GC performance. Derivatization is usually required to determine PAE metabolites (PAEMs). Derivatization procedures such as silylation have been conducted with reagents like N-methyl-N-(tertbutyldimethylsilyl) trifluoroacetamide (MTBSTFA), bis(trimethylsilyl)trifluoroacetamide (BSTFA) and N-methyl-N-trimethylsilyl-trifluoroacetamide (MSTFA). For instance, in the determination of BHT, BHA, TBHQ and their main metabolites [22], these three reagents as well as other derivatization conditions were assayed, and results were best with MTBSTFA. However, in another study that evaluated the three derivatization agents for BHT-OH and BHT-COOH, results were best with BSTFA [30].

LC has been widely used in both high-performance and ultra-high-performance modes for determining polar HPVCs such as BTRs, BTHs, BSA, or BUVs, non-volatile OPEs, and metabolites of both OPEs and PAEs. LC separation is commonly achieved with C_{18} apolar reversed-phase columns [18, 20, 23, 26, 29, 32, 33, 39, 55, 58, 60, 74, 80, 95] with different particle technologies – 5 μm, fused-core (2.7 μm) or sub-2 μm – and column dimensions, and additional features such as extra-endcapping to benefit from the separation performance features. For instance, different columns (C_{18} with different particle sizes and column dimensions and a polar embedded reversed-phase column) and LC conditions were evaluated for separating a group of BTRs, BTHs, and BSAs [18], and in particular the isomeric species (4TTR and 5TTR, o-TSA, and p-TSA), which have been separated in very few studies. The optimum conditions for separating all the compounds (including isomeric species) are detailed in Table 7.2. Regarding to mobile phase, binary mobile phases of water with acidic additive (0.1% HCOOH or 0.1% CH_3COOH) or salt (1–10 mm $HCOONH_4$ or CH_3COONH_4) together with polar solvents such as MeOH and ACN (sometimes also with the acidic additive) are usually combined in gradient profile. In the particular case of the separation of isomeric species of BTR and BSA families, 2% ACN had to be added to the aqueous phase [18].

PAEs can also be determined by LC, although sensitivity is lower than with GC-based methods. In fact, LC is more appropriate for determining metabolites such as PAEMs. Thus, when the same study focuses on both PAEs and PAEMs, two chromatographic methods are often used. For instance, Jiang et al. [23] determined a group of PAEs using GC-MS, and the primary and secondary PAEMs using UHPLC-MS/MS. Likewise, in the determination of OPEs and SPAs, some authors determine the parent compounds with GC but the metabolites with LC [20, 66]. LC-MS/MS has also been used to determine a group of 19 OPEs and 20 plasticizers but the isomers DEHP and bis(2-ethylhexyl)terephthalate (DEHT) are better separated by GC-MS [66].

As for detection, MS-based detectors are the usual choice, although nitrogen-phosphorous detector and flame photometric detector are still used for OPEs with limited identification capabilities [21, 77, 95]. The most common instrumental techniques are GC coupled to MS using electron impact (EI) ionization with simple quadrupole (Q) [4, 20, 22–25, 28, 34, 35, 56, 61, 67], ion trap [31], or triple quadrupole (QqQ) [59, 65, 75, 81] analyzers. As expected, when MS/MS using QqQ analyzer was used, the selectivity and sensitivity improve. This was demonstrated in a study in which UV filters and BTRs were determined by GC-(EI)QqQ [59], and the method quantification limits (MQLs) attained were one order of magnitude lower than when GC-(EI)Q was used. In the case of OPEs, MS/MS is strongly recommended since under EI most of the fragments have low masses and the intensity of molecular ions is low in the spectrum and low m/z ions may be interfered by ions from matrix components [21]. As regards ionization, chemical ionization was less effective for most OPEs, except for TClPP, which shows higher sensitivity [96]. Tables 7.2–7.6 show examples of the use of these techniques.

In LC, electrospray ionization (ESI) with MS/MS using QqQ analyzer [18, 21, 32, 33, 55, 60, 66, 79, 80] is the detection of choice. Nevertheless, HPVCs have also been determined with other ionization sources such as atmospheric chemical ionization (APCI) [55, 58] or analyzers such as QTRAP [38, 74], Orbitrap [97], Q-Orbitrap [39, 60], or QTOF [98]. For instance, LC-(ESI)QqQ and LC-(HESI)Q-Orbitrap were compared for the determination of a group of compounds that included some SPAs [60]. As expected, LC-(HESI) Q-Orbitrap provided better limits than LC-(ESI)QqQ. In addition, the matrix effects were lower when the ionization source was HESI, which was attributed to manufacturers' different design of the ESI sources. However, BHA (one of the SPAs studied) was the compound most affected by the matrix effect in both sources. APCI and HESI ionization sources were compared for determining PAEs and PAEMs in biota with PLE followed by LC-Orbitrap [97]. Instrumentally, APCI provided better limits of detection for the target compounds than HESI. Nevertheless, APCI was discarded since it showed considerable ion enhancement for some compounds.

In summary, a variety of instrumental techniques can be used to determine HPVC compounds. Nevertheless, depending on the compounds and their properties, the instrumental technique selected can have a considerable effect on performance.

7.3.2 Figures of Merit

The application of the above-described analytical methods provides sensitive and reliable results in the determination of the different families of HPVCs from environmental samples. Method validation for quantitative analysis is usually performed in terms of linearity, limits of detection and quantification, accuracy (expressed as recoveries), matrix effect, and precision (expressed as repeatability and reproducibility). Tables 7.2–7.6 detail the recovery values and method detection limits in an effort to provide an overview of the figures of merit of these different compounds in such different matrices using the variety of analytical methods described. It should be mentioned that whenever described, the recovery values given comprise the whole method (i.e., apparent recovery or whole recovery) in an attempt to include the efficiency of the extraction (plus the clean-up step) procedures and the signal variation due to the matrix effect when MS is the detection technique. In any case, reading each specific study is highly recommended to learn more about particular data.

7.3.3 Hints and Tips

Owing to the high number of compounds belonging to the set of HPVCs and the wide range of physical and chemical properties to be extracted from samples of different types, several useful hints and tips should be taken into account. They are described below.

- In the case of PAEs and SPAs, due to their ubiquitous presence, care should be taken during the steps of the analytical method to prevent contamination. To this end, plastic should be avoided, especially during sampling and sample preparation, to prevent the migration of these compounds from plastic. Glass materials should be used (including bottles during sampling or cartridges for SPE) and solid samples should be covered with aluminum foil.
- The rapid metabolization of some families of compounds, such as PAEs and OPEs, among others, makes their metabolites also of concern. Therefore, analytical methods for the determination of metabolites and transformation products are also required. However, due to the different chemical and physical properties of the parent compounds compared to their metabolites, it is usually necessary to determine them by means of two separate chromatographic methods. One of them involving GC (for the parent compounds) and another using LC (for metabolites).
- During sampling, the moisture levels in solid samples should be recorded in order to correctly express compound concentrations.
- In biota samples, lipids and fat should be removed as much as possible using clean-up strategies (see Table 7.6 for details) in order to reduce interferences and other coeluted compounds that may interfere in the analysis and hamper correct determination.

7.4 Future Directions and Challenges

Although significant progress has been made in recent years in the development of methods for the determination of HPVCs in general, additional research is needed to improve these analytical methods. Sampling protocols tend to be moving towards passive sampling in order to reduce sample handling and its associated errors, but this should be carefully followed up in the laboratory with adequate equipment and data interpretation. Analytical methods for multi-residue analyses are starting to be used in many studies. However, when dealing with compounds with substantially different chemical (polarity) and physical (volatility) properties, more research in method development is still needed. In addition, it would be interesting to learn more about the metabolization and transformation of HPVCs through suspect and unknown analyses in order to include them in monitoring programs, and eventually in legislation. In fact, although the use of some HPVCs, like some PAEs and OPEs, has already been banned, there is increasing concern about the toxicological effects of their substitutes, which so far are only classified as (emerging) pollutants. Therefore, it is necessary to develop analytical methods to identify, but also quantify, them in environmental samples in order to understand the real scenario of the presence of HPVCs in the environment.

Acknowledgments

The authors would like to thank the *Ministerio de Economía, Industria y Competitividad, the Agencia Estatal de Investigación* and the European Regional Development Fund (ERDF) for the financial support received (Project: PID2020-114587GB-I00).

References

1 Liu, R. and Mabury, S.A. (2020). Synthetic phenolic antioxidants: a review of environmental occurrence, fate, human exposure, and toxicity. *Environ. Sci. Technol.* 54: 11706–11719. doi: 10.1021/acs.est.0c05077.

2 Liao, C., Kim, U.-J., and Kannan, K. (2018). A review of environmental occurrence, fate, exposure, and toxicity of benzothiazoles. *Environ. Sci. Technol.* 52: 5007–5026. doi: 10.1021/acs.est.7b05493.

3 Maceira, A., Pecikoza, I., Marcé, R.M., and Borrull, F. (2020). Multi-residue analysis of several high-production-volume chemicals present in the particulate matter from outdoor air. A preliminary human exposure estimation. *Chemosphere* 252: 126514. doi: 10.1016/j.chemosphere.2020.126514.

4 Net, S., Delmont, A., Sempéré, R., Paluselli, A., and Ouddane, B. (2015). Reliable quantification of phthalates in environmental matrices (air, water, sludge, sediment and soil): a review. *Sci. Total Environ.* 515–516: 162–180. doi: 10.1016/j.scitotenv.2015.02.013.

5 Net, S., Sempéré, R., Delmont, A., Paluselli, A., and Ouddane, B. (2015). Occurrence, fate, behavior and ecotoxicological state of phthalates in different environmental matrices. *Environ. Sci. Technol.* 49: 4019–4035. doi: 10.1021/es505233b.

6 Maceira, A., Marcé, R.M., and Borrull, F. (2020). Analytical methods for determining organic compounds present in the particulate matter from outdoor air. *TrAC -Trends Anal. Chem.*122: 115707. doi: 10.1016/j.trac.2019.115707.

7 Shi, Z.-Q., Liu, Y.-S., Xiong, Q., Cai, W.-W., and Ying, G.-G. (2019). Occurrence, toxicity and transformation of six typical benzotriazoles in the environment: a review. *Sci. Total Environ.* 661: 407–421. doi: 10.1016/j.scitotenv.2019.01.138.

8 Montesdeoca-Esponda, S., Vega-Morales, T., Sosa-Ferrera, Z., and Santana-Rodríguez, J.J. (2013). Extraction and determination methodologies for benzotriazole UV stabilizers in personal-care products in environmental and biological samples. *TrAC – Trends Anal. Chem.* 51: 23–32. doi: 10.1016/j.trac.2013.05.012.

9 Maceira, A., Borrull, F., and Marcé, R.M. (2019). Occurrence of plastic additives in outdoor air particulate matters from two industrial parks of Tarragona, Spain: human inhalation intake risk assessment. *J. Hazard. Mater.* 373: 649–659. doi: 10.1016/j.jhazmat.2019.04.014.

10 van der Veen, I. and de Boer, J. (2012). Phosphorus flame retardants: properties, production, environmental occurrence, toxicity and analysis. *Chemosphere* 88: 1119–1153. doi: 10.1016/j.chemosphere.2012.03.067.

11 Tsochatzis, E.D., Tzimou-Tsitouridou, R., and Gika, H.G. (2017). Analytical methodologies for the assessment of phthalate exposure in humans. *Crit. Rev. Anal. Chem.* 47: 279–297. doi: 10.1080/10408347.2016.1273754.

12 Den Eede, N., de Meester, I., Maho, W., Neels, H., and Covaci, A. (2016). Biotransformation of three phosphate flame retardants and plasticizers in primary human hepatocytes: untargeted metabolite screening and quantitative assessment. *J. Appl. Toxicol.* 36: 1401–1408. doi: 10.1002/jat.3293.

13 Carve, M., Nugegoda, D., Allinson, G., and Shimeta, J. (2021). A systematic review and ecological risk assessment for organic ultraviolet filters in aquatic environments. *Environ. Pollut.* 268: 115894. doi: 10.1016/j.envpol.2020.115894.

14 Molins-Delgado, D., Távora, J., Díaz-Cruz, M.S., and Barceló, D. (2017). UV filters and benzotriazoles in urban aquatic ecosystems: the footprint of daily use products. *Sci. Total Environ.*601–602: 975–986. doi: 10.1016/j.scitotenv.2017.05.176.

15 Careghini, A., Mastorgio, A.F., Saponaro, S., and Sezenna, E. (2015). Bisphenol A, nonylphenols, benzophenones, and benzotriazoles in soils, groundwater, surface water, sediments, and food: a review. *Environ. Sci. Pollut. Res.* 22: 5711–5741. doi: 10.1007/s11356-014-3974-5.

16 Gao, D.-W. and Wen, Z.-D. (2016). Phthalate esters in the environment: a critical review of their occurrence, biodegradation, and removal during wastewater treatment processes. *Sci. Total Environ.* 541: 986–1001. doi: 10.1016/j.scitotenv.2015.09.148.

17 Wu, Q., Lam, J.C.W., Kwok, K.Y., Tsui, M.M.P., and Lam, P.K.S. (2017). Occurrence and fate of endogenous steroid hormones, alkylphenol ethoxylates, bisphenol A and phthalates in municipal sewage treatment systems. *J. Environ. Sci.* 61: 49–58. doi: 10.1016/j.jes.2017.02.021.

18 Herrero, P., Borrull, F., Pocurull, E., and Marcé, R.M. (2013). Efficient tandem solid-phase extraction and liquid chromatography-triple quadrupole mass spectrometry method to determine polar benzotriazole, benzothiazole and benzenesulfonamide contaminants in environmental water samples. *J. Chromatogr. A* 1309: 22–32. doi: 10.1016/j.chroma.2013.08.018.

19 Herrero, P., Borrull, F., Pocurull, E., and Marcé, R.M. (2014). An overview of analytical methods and occurrence of benzotriazoles, benzothiazoles and benzenesulfonamides in the environment. *TrAC Trends Anal. Chem.* 62: 46–55. doi: 10.1016/j.trac.2014.06.017.

20 Wang, W. and Kannan, K. (2018). Inventory, loading and discharge of synthetic phenolic antioxidants and their metabolites in wastewater treatment plants. *Water Res* 129: 413–418. doi: 10.1016/j.watres.2017.11.028.

21 Wang, X., Zhu, Q., Yan, X., Wang, Y., Liao, C., and Jiang, G. (2020). A review of organophosphate flame retardants and plasticizers in the environment: analysis, occurrence and risk assessment. *Sci. Total Environ.* 731: 139071. doi: 10.1016/j.scitotenv.2020.139071.

22 Rodil, R., Quintana, J.B., Basaglia, G., Pietrogrande, M.C., and Cela, R. (2010). Determination of synthetic phenolic antioxidants and their metabolites in water samples by downscaled solid-phase extraction, silylation and gas chromatography-mass spectrometry. *J. Chromatogr. A* 1217: 6428–6435. doi: 10.1016/j.chroma.2010.08.020.

23 Jiang, J., Mu, D., Ding, M., Zhang, S., Zhang, H., and Hu, J. (2018). Simultaneous determination of primary and secondary phthalate monoesters in the Taihu Lake: exploration of sources. *Chemosphere* 202: 17–24. doi: 10.1016/j.chemosphere.2018.03.070.

24 Cao, X.-L. (2008). Determination of phthalates and adipate in bottled water by headspace solid-phase microextraction and gas chromatography/mass spectrometry. *J. Chromatogr. A* 1178: 231–238. doi: 10.1016/j.chroma.2007.11.095.

25 Feng, Y.-X., Feng, N.-X., Zeng, L.-J., Chen, X., Xiang, L., Li, Y.-W., Cai, Q.-Y., and Mo, C.-H. (2020). Occurrence and human health risks of phthalates in indoor air of laboratories. *Sci. Total Environ.* 707: 135609. doi: 10.1016/j.scitotenv.2019.135609.

26 Wan, Y., Xue, J., and Kannan, K. (2016). Benzothiazoles in indoor air from Albany, New York, USA, and its implications for inhalation exposure. *J. Hazard. Mater.* 311: 37–42. doi: 10.1016/j.jhazmat.2016.02.057.

27 Ma, J., Chen, L., Guo, Y., Wu, Q., Yang, M., Wu, M., and Kannan, K. (2014). Phthalate diesters in Airborne PM2.5 and PM10 in a suburban area of Shanghai: seasonal distribution and risk assessment. *Sci. Total Environ.* 497–498: 467–474. doi: 10.1016/j.scitotenv.2014.08.012.

28 Wang, P., Wang, S.L., and Fan, C.Q. (2008). Atmospheric distribution of particulate- and gas-phase phthalic esters (PAEs) in a Metropolitan City, Nanjing, East China. *Chemosphere* 72: 1567–1572. doi: 10.1016/j.chemosphere.2008.04.032.

29 Liu, R. and Mabury, S.A. (2019). Synthetic phenolic antioxidants and transformation products in dust from different indoor environments in Toronto, Canada. *Sci. Total Environ.* 672: 23–29. doi: 10.1016/j.scitotenv.2019.03.495.

30 Zhang, R., Li, C., Li, Y., Cui, X., and Ma, L.Q. (2018). Determination of 2,6-di-tert-butyl-hydroxytoluene and its transformation products in indoor dust and sediment by gas chromatography–mass spectrometry coupled with precolumn derivatization. *Sci. Total Environ.* 619–620: 552–558. doi: 10.1016/j.scitotenv.2017.11.115.

31 Carpinteiro, I., Abuín, B., Rodríguez, I., Ramil, M., and Cela, R. (2010). Pressurized solvent extraction followed by gas chromatography tandem mass spectrometry for the determination of benzotriazole light stabilizers in indoor dust. *J. Chromatogr. A* 1217: 3729–3735. doi: 10.1016/j.chroma.2010.04.022.

32 Kim, J.-W., Isobe, T., Malarvannan, G., Sudaryanto, A., Chang, K.-H., Prudente, M., and Tanabe, S. (2012). Contamination of benzotriazole ultraviolet stabilizers in house dust from the Philippines: implications on human exposure. *Sci. Total Environ.* 424: 174–181. doi: 10.1016/j.scitotenv.2012.02.040.

33 Wang, L., Asimakopoulos, A.G., Moon, H.-B., Nakata, H., and Kannan, K. (2013). Benzotriazole, benzothiazole, and benzophenone compounds in indoor dust from the

United States and East Asian countries. *Environ. Sci. Technol.* 47: 4752–4759. doi: 10.1021/es305000d.

34 Christia, C., Poma, G., Harrad, S., de Wit, C.A., Sjostrom, Y., Leonards, P., Lamoree, M., and Covaci, A. (2019). Occurrence of legacy and alternative plasticizers in indoor dust from various EU countries and implications for human exposure via dust ingestion and dermal absorption. *Environ. Res.* 171: 204–212. doi: 10.1016/j.envres.2018.11.034.

35 Wei, L., Li, Z., Sun, J., and Zhu, L. (2020). Pollution characteristics and health risk assessment of phthalate esters in agricultural soil and vegetables in the Yangtze River Delta of China. *Sci. Total Environ.* 726: 137978. doi: 10.1016/j.scitotenv.2020.137978.

36 Speltini, A., Sturini, M., Maraschi, F., Porta, A., and Profumo, A. (2016). Fast low-pressurized microwave-assisted extraction of benzotriazole, benzothiazole and benezenesulfonamide compounds from soil samples. *Talanta* 147: 322–327. doi: 10.1016/j.talanta.2015.09.074.

37 Zhang, B.-T., Gao, Y., Lin, C., Yang, W., Liu, T., Liu, X., and Wang, Y. (2020). Spatial distribution of phthalate acid esters in sediments of the Laizhou Bay and its relationship with anthropogenic activities and geochemical variables. *Sci. Total Environ.* 722: 137912. doi: 10.1016/j.scitotenv.2020.137912.

38 Mizukawa, A., Molins-Delgado, D., de Azevedo, J.C.R., Fernandes, C.V.S., Díaz-Cruz, S., and Barceló, D. (2017). Sediments as a sink for UV filters and benzotriazoles: the case study of Upper Iguaçu watershed, Curitiba (Brazil). *Environ. Sci. Pollut. Res.* 24: 18284–18294. doi: 10.1007/s11356-017-9472-9.

39 Herrero, P., Borrull, F., Marcé, R.M., and Pocurull, E. (2014). A pressurised hot water extraction and liquid chromatography–high resolution mass spectrometry method to determine polar benzotriazole, benzothiazole and benzenesulfonamide derivates in sewage sludge. *J. Chromatogr. A* 1355: 53–60. doi: 10.1016/j.chroma.2014.05.086.

40 Hidalgo-Serrano, M., Borrull, F., Marcé, R.M., and Pocurull, E. (2021). Simple method for determining phthalate diesters and their metabolites in seafood species using QuEChERS extraction and liquid chromatography-high resolution mass spectrometry. *Food Chem* 336: 127722. doi: 10.1016/j.foodchem.2020.127722.

41 Wang, W., Leung, A.O.W., Chu, L.H., and Wong, M.H. (2018). Phthalates contamination in China: status, trends and human exposure-with an emphasis on oral intake. *Environ. Pollut.* 238: 771–782. doi: 10.1016/j.envpol.2018.02.088.

42 Van Holderbeke, M., Geerts, L., Vanermen, G., Servaes, K., Sioen, I., De Henauw, S., and Fierens, T. (2014). Determination of contamination pathways of phthalates in food products sold on the Belgian market. *Environ. Res.* 134: 345–352. doi: 10.1016/j.envres.2014.08.012.

43 Sakhi, A.K., Lillegaard, I.T.L., Voorspoels, S., Carlsen, M.H., Løken, E.B., Brantsæter, A.L., Haugen, M., Meltzer, H.M., and Thomsen, C. (2014). Concentrations of phthalates and bisphenol A in Norwegian foods and beverages and estimated dietary exposure in adults. *Environ. Int.* 73: 259–269. doi: 10.1016/j.envint.2014.08.005.

44 Trabalón, L., Nadal, M., Borrull, F., and Pocurull, E. (2017). Determination of benzothiazoles in seafood species by subcritical water extraction followed by solid-phase microextraction-gas chromatography-tandem mass spectrometry: estimating the dietary intake. *Anal. Bioanal. Chem.* 409: 5513–5522. doi: 10.1007/s00216-017-0487-3.

45 Jia, J., Zhu, Q., Liu, N., Liao, C., and Jiang, G. (2019). Occurrence of and human exposure to benzothiazoles and benzotriazoles in mollusks in the Bohai Sea, China. *Environ. Int.* 130: 104925. doi: 10.1016/j.envint.2019.104925.

46 Jakimska, A., Huerta, B., Bargańska, Z., Kot-Wasik, A., Rodríguez-Mozaz, S., and Barceló, D. (2013). Development of a liquid chromatography-tandem mass spectrometry procedure for determination of endocrine disrupting compounds in fish from Mediterranean rivers. *J. Chromatogr. A* 1306: 44–58. doi: 10.1016/j.chroma.2013.07.050.

47 Álvarez-Muñoz, D., Rodríguez-Mozaz, S., Maulvault, A.L., Tediosi, A., Fernández-Tejedor, M., Van Den Heuvel, F., Kotterman, M., Marques, A., and Barceló, D. (2015). Occurrence of pharmaceuticals and endocrine disrupting compounds in macroalgaes, bivalves, and fish from coastal areas in Europe. *Environ. Res.* 143: 56–64. doi: 10.1016/j.envres.2015.09.018.

48 Yao, L., Zhao, J.L., Liu, Y.S., Yang, Y.Y., Liu, W.R., and Ying, G.G. (2016). Simultaneous determination of 24 personal care products in fish muscle and liver tissues using QuEChERS extraction coupled with ultra pressure liquid chromatography-tandem mass spectrometry and gas chromatography-mass spectrometer analyses. *Anal. Bioanal. Chem.* 408: 8177–8193. doi: 10.1007/s00216-016-9924-y.

49 Yao, L., Zhao, J.L., Liu, Y.S., Zhang, Q.Q., Jiang, Y.X., Liu, S., Liu, W.R., Yang, Y.Y., and Ying, G.G. (2018). Personal care products in wild fish in two main Chinese rivers: bioaccumulation potential and human health risks. *Sci. Total Environ.* 621: 1093–1102. doi: 10.1016/j.scitotenv.2017.10.117.

50 Álvarez-Muñoz, D., Rambla-Alegre, M., Carrasco, N., López de Alda, M., and Barceló, D. (2019). Fast analysis of relevant contaminants mixture in commercial shellfish. *Talanta*205: 119884. doi: 10.1016/j.talanta.2019.04.085.

51 Terechovs, A.K.E., Ansari, A.J., McDonald, J.A., Khan, S.J., Hai, F.I., Knott, N.A., Zhou, J., and Nghiem, L.D. (2019). Occurrence and bioconcentration of micropollutants in Silver Perch (*Bidyanus bidyanus*) in a reclaimed water reservoir. *Sci. Total Environ.* 650: 585–593. doi: 10.1016/j.scitotenv.2018.08.431.

52 Nakata, H., Murata, S., and Filatreau, J. (2009). Occurrence and concentrations of benzotriazole UV stabilizers in marine organisms and sediments from the Ariake Sea, Japan. *Environ. Sci. Technol.* 43: 6920–6926. doi: 10.1021/es900939j.

53 Montesdeoca-Esponda, S., Torres-Padrón, M.E., Sosa-Ferrera, Z., and Santana-Rodríguez, J.J. (2021). Fate and distribution of benzotriazole UV filters and stabilizers in environmental compartments from Gran Canaria Island (Spain): a comparison study. *Sci. Total Environ.* 756: 144086. doi: 10.1016/j.scitotenv.2020.144086.

54 Lundebyea, A.K., Hovea, H., Mågea, A., Bohneb, V.J.B., and Hamrea, K. (2010). Levels of synthetic antioxidants (ethoxyquin, butylated hydroxytoluene and butylated hydroxyanisole) in fish feed and commercially farmed fish. *Food Addit. Contam. – Part A Chem. Anal. Control. Expo. Risk Assess.* 27: 1652–1657. doi: 10.1080/19440049.2010.508195.

55 Wick, A., Fink, G., and Ternes, T.A. (2010). Comparison of electrospray ionization and atmospheric pressure chemical ionization for multi-residue analysis of biocides, UV-filters and benzothiazoles in aqueous matrices and activated sludge by liquid chromatography–tandem mass spectrometry. *J. Chromatogr. A* 1217: 2088–2103. doi: 10.1016/j.chroma.2010.01.079.

56 Zeng, F., Cui, K., Xie, Z., Liu, M., Li, Y., Lin, Y., Zeng, Z., and Li, F. (2008). Occurrence of phthalate esters in water and sediment of urban lakes in a subtropical city, Guangzhou, South China. *Environ. Int.* 34: 372–380. doi: 10.1016/j.envint.2007.09.002.

57 Rodríguez, I., Calvo, F., Quintana, J.B., Rubí, E., Rodil, R., and Cela, R. (2006). Suitability of solid-phase microextraction for the determination of organophosphate flame retardants and plasticizers in water samples. *J. Chromatogr. A* 1108: 158–165. doi: 10.1016/j.chroma.2006.01.008.

58 Song, S., Ruan, T., Wang, T., Liu, R., and Jiang, G. (2014). Occurrence and removal of benzotriazole ultraviolet stabilizers in a wastewater treatment plant in China. *Environ. Sci. Process. Impacts.* 16: 1076–1082. doi: 10.1039/C3EM00483J.

59 Liu, Y.-S., Ying, -G.-G., Shareef, A., and Kookana, R.S. (2011). Simultaneous determination of benzotriazoles and ultraviolet filters in ground water, effluent and biosolid samples using gas chromatography–tandem mass spectrometry. *J. Chromatogr. A* 1218: 5328–5335. doi: 10.1016/j.chroma.2011.05.100.

60 Chen, W., Huang, H., Chen, C.-E., Qi, S., Price, O.R., Zhang, H., Jones, K.C., and Sweetman, A.J. (2016). Simultaneous determination of 20 trace organic chemicals in waters by solid-phase extraction (SPE) with triple-quadrupole mass spectrometer (QqQ-MS) and hybrid quadrupole Orbitrap high resolution MS (Q-Orbitrap-HRMS). *Chemosphere* 163: 99–107. doi: 10.1016/j.chemosphere.2016.07.080.

61 Peñalver, A., Pocurull, E., Borrull, F., and Marcé, R.M. (2001). Comparison of different fibers for the solid-phase microextraction of phthalate esters from water. *J. Chromatogr. A* 922: 377–384. doi: 10.1016/S0021-9673(01)00920-7.

62 Fontanals, N., Pocurull, E., Marcé, R.M., and Borrull, F. (2019). Water analysis. Organic compounds. In: *Encyclopedia of Analytical Science* (3rd edn.), P. Worsfold, C. Poole, A. Townshend, and M. Miró (ed.), 286–298. Amsterdam: Elsevier. doi: 10.1016/B978-0-12-409547-2.14497-X.

63 Naccarato, A., Tassone, A., Moretti, S., Elliani, R., Sprovieri, F., Pirrone, N., and Tagarelli, A. (2018). A green approach for organophosphate ester determination in airborne particulate matter: microwave-assisted extraction using hydroalcoholic mixture coupled with solid-phase microextraction gas chromatography-tandem mass spectrometry. *Talanta* 189: 657–665. doi: 10.1016/j.talanta.2018.07.077.

64 Aragón, M., Borrull, F., and Marcé, R.M. (2013). Thermal desorption-gas chromatography–mass spectrometry method to determine phthalate and organophosphate esters from air samples. *J. Chromatogr. A* 1303: 76–82. doi: 10.1016/j.chroma.2013.06.025.

65 Velázquez-Gómez, M., Hurtado-Fernández, E., and Lacorte, S. (2018). Comprehensive method for the analysis of multi-class organic micropollutants in indoor dust. *Sci. Total Environ.* 635: 1484–1494. doi: 10.1016/j.scitotenv.2018.04.186.

66 Christia, C., Tang, B., Yin, -S.-S., Luo, X.-J., Mai, B.-X., Poma, G., and Covaci, A. (2019). Simultaneous determination of legacy and emerging organophosphorus flame retardants and plasticizers in indoor dust using liquid and gas chromatography–tandem mass spectrometry: method development, validation, and application. *Anal. Bioanal. Chem.* 411: 7015–7025. doi: 10.1007/s00216-019-02078-5.

67 Alkan, N., Alkan, A., Castro-Jiménez, J., Royer, F., Papillon, L., Ourgaud, M., and Sempéré, R. (2021). Environmental occurrence of phthalate and organophosphate esters in sediments across the Gulf of Lion (NW Mediterranean Sea). *Sci. Total Environ.* 760: 143412. doi: 10.1016/j.scitotenv.2020.143412.

68 Montesdeoca-Esponda, S., Torres-Padrón, M.E., Novák, M., Krchová, L., Sosa-Ferrera, Z., and Santana-Rodríguez, J.J. (2020). Occurrence of benzotriazole UV stabilizers in coastal fishes. *J. Environ. Manage.* 269. doi: 10.1016/j.jenvman.2020.110805.

69 Guerranti, C., Cau, A., Renzi, M., Badini, S., Grazioli, E., Perra, G., and Focardi, S.E. (2016). Phthalates and perfluorinated alkylated substances in Atlantic bluefin tuna (Thunnus thynnus) specimens from Mediterranean Sea (Sardinia, Italy): levels and risks for human consumption. *J. Environ. Sci. Heal. – Part B Pestic. Food Contam. Agric. Wastes.* 51: 661–667. doi: 10.1080/03601234.2016.1191886.

70 Hu, X., Gu, Y., Huang, W., and Yin, D. (2016). Phthalate monoesters as markers of phthalate contamination in wild marine organisms. *Environ. Pollut.* 218: 410–418. doi: 10.1016/j.envpol.2016.07.020.

71 Valton, A.S., Serre-Dargnat, C., Blanchard, M., Alliot, F., Chevreuil, M., and Teil, M.J. (2014). Determination of phthalates and their by-products in tissues of roach (Rutilus rutilus) from the Orge river (France). *Environ. Sci. Pollut. Res.* 21: 12723–12730. doi: 10.1007/s11356-014-3213-0.

72 Wang, Z.S.L., Zhao, H., Yu, G., and Shao, L., (2011) Study on contents of phthalates in seafood by liquid chromatography spectrometry with matrix solid- phase dispersion, in: 5th Int. Conf. Bioinforma. Biomed., 1–4.

73 Weiss, S. and Reemtsma, T. (2005). Determination of benzotriazole corrosion inhibitors from aqueous environmental samples by liquid chromatography-electrospray ionization-tandem mass spectrometry. *Anal. Chem.* 77: 7415–7420. doi: 10.1021/ac051203e.

74 Lu, Z., Smyth, S.A., Peart, T.E., and De Silva, A.O. (2017). Occurrence and fate of substituted diphenylamine antioxidants and benzotriazole UV stabilizers in various Canadian wastewater treatment processes. *Water Res* 124: 158–166. doi: 10.1016/j.watres.2017.07.055.

75 Naccarato, A., Gionfriddo, E., Sindona, G., and Tagarelli, A. (2014). Simultaneous determination of benzothiazoles, benzotriazoles and benzosulfonamides by solid phase microextraction-gas chromatography-triple quadrupole mass spectrometry in environmental aqueous matrices and human urine. *J. Chromatogr. A* 1338: 164–173. doi: 10.1016/j.chroma.2014.02.089.

76 Mercier, F., Glorennec, P., Blanchard, O., and Le Bot, B. (2012). Analysis of semi-volatile organic compounds in indoor suspended particulate matter by thermal desorption coupled with gas chromatography/mass spectrometry. *J. Chromatogr. A* 1254: 107–114. doi: 10.1016/j.chroma.2012.07.025.

77 García, M., Rodríguez, I., and Cela, R. (2007). Microwave-assisted extraction of organophosphate flame retardants and plasticizers from indoor dust samples. *J. Chromatogr. A* 1152: 280–286. doi: 10.1016/j.chroma.2006.11.046.

78 Wang, W., Asimakopoulos, A.G., Abualnaja, K.O., Covaci, A., Gevao, B., Johnson-Restrepo, B., Kumosani, T.A., Malarvannan, G., Minh, T.B., Moon, H.-B., Nakata, H., Sinha, R.K., and Kannan, K. (2016). Synthetic phenolic antioxidants and their metabolites in indoor dust from homes and microenvironments. *Environ. Sci. Technol.* 50: 428–434. doi: 10.1021/acs.est.5b04826.

79 Liu, R., Song, S., Lin, Y., Ruan, T., and Jiang, G. (2015). Occurrence of synthetic phenolic antioxidants and major metabolites in municipal sewage sludge in China. *Environ. Sci. Technol.* 49: 2073–2080. doi: 10.1021/es505136k.

80 Cantwell, M.G., Sullivan, J.C., Katz, D.R., Burgess, R.M., Bradford Hubeny, J., and King, J. (2015). Source determination of benzotriazoles in sediment cores from two urban estuaries on the Atlantic Coast of the United States. *Mar. Pollut. Bull.* 101: 208–218. doi: 10.1016/j.marpolbul.2015.10.075.

81 Wang, X., Hou, X., Zhou, Q., Liao, C., and Jiang, G. (2018). Synthetic phenolic antioxidants and their metabolites in sediments from the coastal area of Northern China: spatial and vertical distributions. *Environ. Sci. Technol.* 52: 13690–13697. doi: 10.1021/acs.est.8b04448.

82 Cristale, J. and Lacorte, S. (2013). Development and validation of a multiresidue method for the analysis of polybrominated diphenyl ethers, new brominated and organophosphorus flame retardants in sediment, sludge and dust. *J. Chromatogr. A* 1305: 267–275. doi: 10.1016/j.chroma.2013.07.028.

83 Guo, Y., Zhang, Z., Liu, L., Li, Y., Ren, N., and Kannan, K. (2012). Occurrence and profiles of phthalates in foodstuffs from China and their implications for human exposure. *J. Agric. Food Chem.* 60: 6913–6919. doi: 10.1021/jf3021128.

84 Fierens, T., Vanermen, G., Van Holderbeke, M., De Henauw, S., and Sioen, I. (2012). Effect of cooking at home on the levels of eight phthalates in foods. *Food Chem. Toxicol.* 50: 4428–4435. doi: 10.1016/j.fct.2012.09.004.

85 Fierens, T., Servaes, K., Van Holderbeke, M., Geerts, L., De Henauw, S., Sioen, I., and Vanermen, G. (2012). Analysis of phthalates in food products and packaging materials sold on the Belgian market. *Food Chem. Toxicol.* 50: 2575–2583. doi: 10.1016/j.fct.2012.04.029.

86 Teil, M.J., Tlili, K., Blanchard, M., Chevreuil, M., Alliot, F., and Labadie, P. (2012). Occurrence of polybrominated diphenyl ethers, polychlorinated biphenyls, and phthalates in freshwater fish from the orge river (Ile-de France). *Arch. Environ. Contam. Toxicol.* 63: 101–113. doi: 10.1007/s00244-011-9746-z.

87 Xu, Y., Weng, R., Lu, Y., Wang, X., Zhang, D., Li, Y., Qiu, J., and Qian, Y. (2018). Evaluation of phthalic acid esters in fish samples using gas chromatography tandem mass spectrometry with simplified QuEChERS technique. *Food Anal. Methods.* 11: 3293–3303. doi: 10.1007/s12161-018-1313-z.

88 Wang, X., Hou, X., Hu, Y., Zhou, Q., Liao, C., and Jiang, G. (2018). Synthetic phenolic antioxidants and their metabolites in Mollusks from the Chinese Bohai Sea: occurrence, temporal trend, and human exposure. *Environ. Sci. Technol.* 52: 10124–10133. doi: 10.1021/acs.est.8b03322.

89 Kim, J.W., Ramaswamy, B.R., Chang, K.H., Isobe, T., and Tanabe, S. (2011). Multiresidue analytical method for the determination of antimicrobials, preservatives, benzotriazole UV stabilizers, flame retardants and plasticizers in fish using ultra high performance liquid chromatography coupled with tandem mass spectrometry. *J. Chromatogr. A* 1218: 3511–3520. doi: 10.1016/j.chroma.2011.04.006.

90 Santín, G., Eljarrat, E., and Barceló, D. (2016). Simultaneous determination of 16 organophosphorus flame retardants and plasticizers in fish by liquid chromatography-tandem mass spectrometry. *J. Chromatogr. A* 1441: 34–43. doi: 10.1016/j.chroma.2016.02.058.

91 Giulivo, M., Capri, E., Eljarrat, E., and Barceló, D. (2016). Analysis of organophosphorus flame retardants in environmental and biotic matrices using on-line turbulent flow chromatography-liquid chromatography-tandem mass spectrometry. *J. Chromatogr. A* 1474: 71–78. doi: 10.1016/j.chroma.2016.10.042.

92 Garcia-Garin, O., Vighi, M., Sala, B., Aguilar, A., Tsangaris, C., Digka, N., Kaberi, H., Eljarrat, E., and Borrell, A. (2020). Assessment of organophosphate flame retardants in Mediterranean Boops boops and their relationship to anthropization levels and microplastic ingestion. *Chemosphere* 252: 1–8. doi: 10.1016/j.chemosphere.2020.126569.

93 Bekele, T.G., Zhao, H., Wang, Q., and Chen, J. (2019). Bioaccumulation and trophic transfer of emerging organophosphate flame retardants in the marine food webs of Laizhou Bay, North China. *Environ. Sci. Technol.* 53: 13417–13426. doi: 10.1021/acs.est.9b03687.

94 Ding, J., Deng, T., Xu, M., Wang, S., and Yang, F. (2018). Residuals of organophosphate esters in foodstuffs and implication for human exposure. *Environ. Pollut.* 233: 986–991. doi: 10.1016/j.envpol.2017.09.092.

95 Pantelaki, I. and Voutsa, D. (2019). Organophosphate flame retardants (OPFRs): a review on analytical methods and occurrence in wastewater and aquatic environment. *Sci. Total Environ.* 649: 247–263. doi: 10.1016/j.scitotenv.2018.08.286.

96 Gustavsson, J., Ahrens, L., Nguyen, M.A., Josefsson, S., and Wiberg, K. (2017). Development and comparison of gas chromatography–mass spectrometry techniques for analysis of flame retardants. *J. Chromatogr. A* 1481: 116–126. doi: 10.1016/j.chroma.2016.12.028.

97 Hidalgo-Serrano, M., Borrull, F., Pocurull, E., and Marcé, R.M. (2020). Pressurised liquid extraction and liquid chromatography–high resolution mass spectrometry for the simultaneous determination of phthalate diesters and their metabolites in seafood species. *Food Anal. Methods* 13: 1442–1453. doi: 10.1007/s12161-020-01759-7.

98 Langford, K.H., Reid, M.J., Fjeld, E., Øxnevad, S., and Thomas, K.V. (2015). Environmental occurrence and risk of organic UV filters and stabilizers in multiple matrices in Norway. *Environ. Int.* 80: 1–7. doi: 10.1016/j.envint.2015.03.012.

99 Lorenzo, M., Campo, J., and Picó, Y. (2016). Ultra-high-pressure liquid chromatography tandem mass spectrometry method for the determination of 9 organophosphate flame retardants in water samples. *MethodsX* 3: 343–349. doi: 10.1016/j.mex.2016.04.006.

100 Cristale, J., García Vázquez, A., Barata, C., and Lacorte, S. (2013). Priority and emerging flame retardants in rivers: occurrence in water and sediment, Daphnia magna toxicity and risk assessment. *Environ. Int.* 59: 232–243. doi: 10.1016/j.envint.2013.06.011.

101 Mercier, F., Gilles, E., Soulard, P., Mandin, C., Dassonville, C., and Le Bot, B. (2020). On-line coupling of thermal extraction with gas chromatography/tandem mass spectrometry for the analysis of semivolatile organic compounds in a few milligrams of indoor dust. *J. Chromatogr. A* 1615: 460768. doi: 10.1016/j.chroma.2019.460768.

8

Musk Fragrances

Irene Aparicio, Julia Martín, Juan Luis Santos, and Esteban Alonso

Departamento de Química Analítica, Escuela Politécnica Superior, Universidad de Sevilla, Sevilla, Spain

8.1 Overview of Musk Fragrances

Synthetic musk fragrances appeared in the twentieth century as a low-cost alternative to natural perfumes to accommodate increasing demand [1]. They are extensively used in cosmetics, perfumes, toiletries, air fresheners, detergents and other household products. Synthetic musks are classified according to their chemical structure in polycyclic musks (PCMs), nitro musks (NMs), macrocylic musks (MCMs) and alicyclic musks (ACMs). Their chemical structures and abbreviations can be seen in Table 8.1. NMs were the first synthetised musks. They were widely used as musk fragrances but they began to be removed from the market in the 1990s when the toxicological and environmental effects associated with the presence of a nitroaromatic group in their structure began to be known [2, 3]. Their use in cosmetics has been banned [2] as they have suspected or confirmed carcinogenicity [4]. PCMs began to be synthetised in 1951 [5]. Their production is continuously increasing, constituting the predominant group of synthetic musks [2]. PCMs constitute more than 90% of all synthetic musks [6] as they are relatively inexpensive to synthetise and safer that NMs [2, 5]. Among them, HHCB and AHTN are the most-used ones [7]. MCMs are expected to replace PCMs because of their intense smell and because they are more degradable that other musk fragrances [3, 8]. Nevertheless, they are not so widely used as PCMs because of their high production cost [2, 9]. ACMs began to be used in the 1990s but their use in personal care products is still very scarce [3]. Nevertheless, they are considered the future of synthetic musk fragrances because of their low cost and biodegradability [8].

8.1.1 Properties

Table 8.1 shows the boiling points and water–octanol partition coefficients (Log K_{ow} values) of the most-used synthetic musk fragrances. Most of these compounds have Log K_{ow} values higher than 4 and boiling points in the range from 300 to 400°C. Solubilities of

Analytical Methods for Environmental Contaminants of Emerging Concern, First Edition. Edited by Núria Fontanals and Rosa Maria Marcé.

Table 8.1 Classification and physicochemical properties of synthetic musk fragrances.*

Musk fragrance	Chemical structure	BP (°C, 760 mm Hg)	Log K_{ow}
Polycyclic musks (PCMs)			
Cashmeran (DPMI)	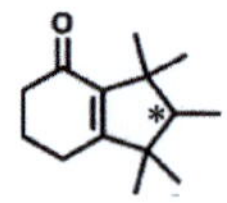	286.1	4.9[a]
Celestolide (ADBI)	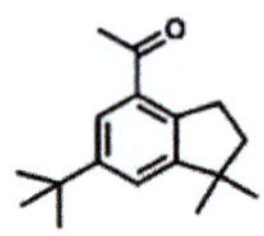	309	6.6[a]
Galaxolide (HHCB)	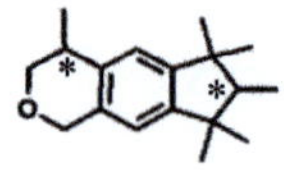	326	5.9[a]
Phantolide (AHMI)	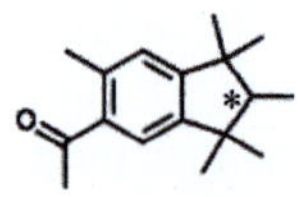	336.6	6.7[a]
Tonalide (AHTN)	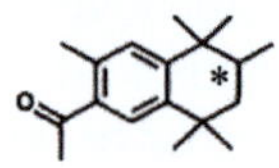	356.8	5.7[a]
Traseolide (ATTI)	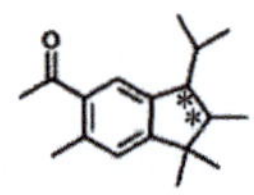	350	6.3[a]
Galaxolidone (HHCB-lactone) (degradation product)	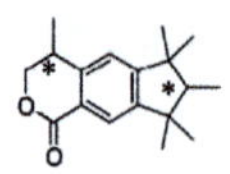	–	–
Nitro musks (NMs)			
Musk ambrette (MA)	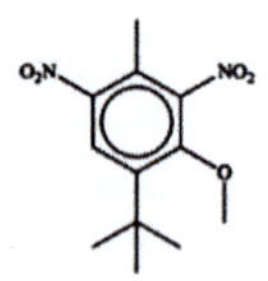	–	3.7[a]
Musk ketone (MK)	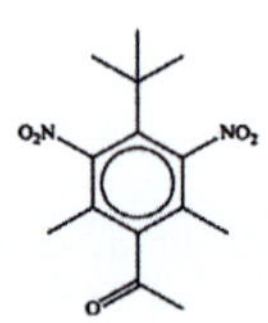	369	4.3[a]

(Continued)

Table 8.1 (Continued)

Musk fragrance	Chemical structure	BP (°C, 760 mm Hg)	Log K_{ow}
Musk moskene (MM)	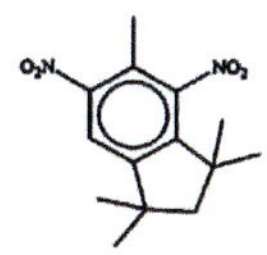	351.1	5.8[a]
Musk tibetene (MT)	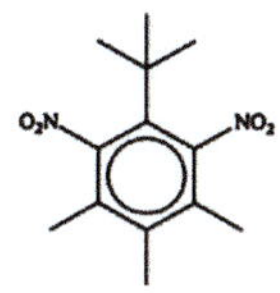	–	5.9[b]
Musk xylene (MX)	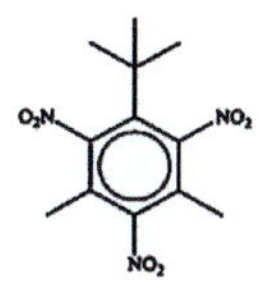	392.3	4.8[a]
Macrocyclic musk fragrances (MCMs)			
Ambrettolide	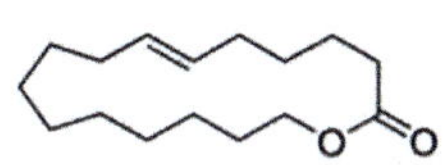	378.7[c]	5.52[c]
Civetone	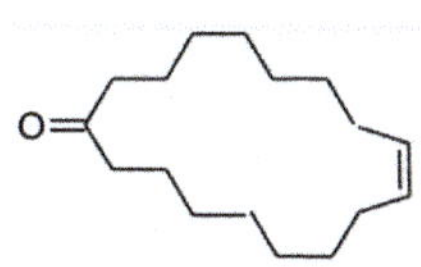	371.4[c]	6.31[c]
Ethylene brassylate	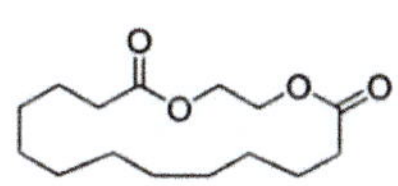	330.0	4.7[b]
Exaltolide (Thibetolide)	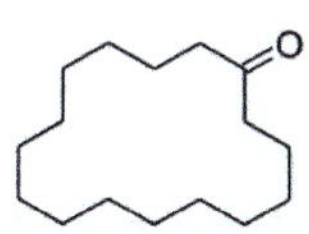	344.8[c]	6.10[c]
Exaltone	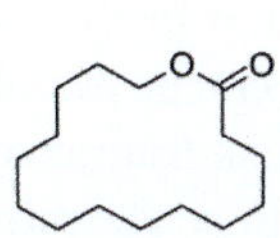	338.3[c]	5.84[c]

(*Continued*)

Table 8.1 (Continued)

Musk fragrance	Chemical structure	BP (°C, 760 mm Hg)	Log K_{ow}
Habanolide (Globalide)	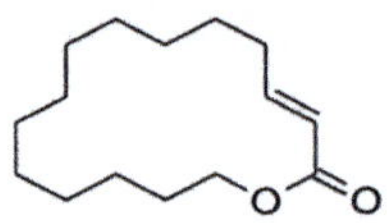	388.4[c]	5.53[c]
Muscone	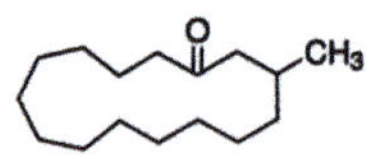	329.5[c]	6.33[c]
Musk MC4	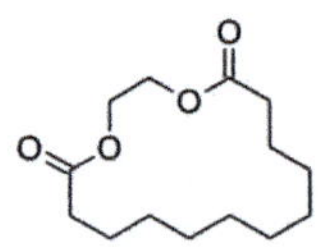	464.5[c]	2.33[c]
Musk-NN	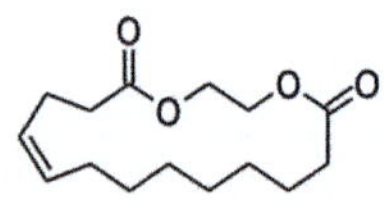	330.5[c]	2.90[c]
Alicyclic musks (ACMs)			
Romandolide	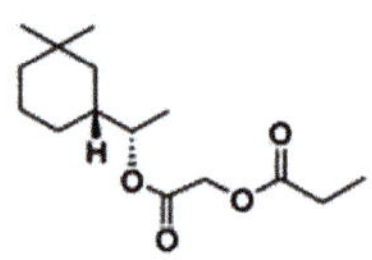	334.9	–
Heveltolide	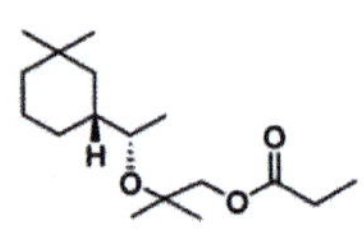	346.3	–

Data not found; *Chiral centre; BP: boiling point;[a][47];[b][51];[c][31].

PCMs are in the range from 0.015 mg L^{-1} (ADBI) to 1.75 mg L^{-1} (HHCB) whereas solubilities of MA, MK and MX are 2.11, 1.9 and 0.40 mg L^{-1} [10]. More physical-chemical properties of PCMs and NMs can be found in the work from McDonough et al. [11]. Detailed information about their toxicity can be found in the review from Rainieri et al. [2]. The high Log K_{ow} values of most musk fragrances reveal not only a high sorption potential onto sewage sludge generated in wastewater treatment plants (WWTPs) [12] but also a high bioaccumulation tendency in living organisms. Synthetic musks are considered relatively volatile compounds. Therefore, they can evaporate from aqueous or solid environmental matrices to the air [11, 13]. Volatilisation fluxes of HHCB and AHTN from surface waters to air in the range from 11 to 341 ng m^{-2} day^{-1} and −3 to 28 ng m^{-2} day^{-1}, respectively, have been reported by McDonough et al. [11].

8.1.2 Legislation

The production or use of some musk fragrances is restricted, or even banned, because of their toxicity or environmental persistence. Regarding NMs, in Europe, the use of MA, MM and MT is forbidden is cosmetic products by the Regulation (EC) No 1223/2009 whereas the use of MK is forbidden in oral products and limited to up to 1.4% in other cosmetics [14]. MX is banned under REACH (Registration, Evaluation, Authorization and Restriction of Chemicals), after being classified as a very persistent and very bioaccumulative (vPvB) substance.

In China, the use of MA, MA and MT in cosmetics products is also banned whereas the use of MK and MX is restricted [13]. In addition, the European Chemicals Agency (ECHA) has limited to the production/importation of DPMI, ADBI, AHMI and ATTI (1–10 tonnes per year), MK (100–1000 tonnes per year), and HHCB and AHTN (1000–10,000 tonnes per year) in Europe.

8.1.3 Reported or Potential Metabolites and/or Transformation Products

Musk fragrances are introduced into the aquatic environment mainly through urban wastewater discharges from WWTPs. Metabolites and transformation products (TPs) can be generated by biotic and abiotic processes, enzymatic reactions and photodegradation in WWTPs, in the environment and/or in organisms [3]. HHCB-lactone is the predominant TP and the one most widely evaluated in the environment because of its concentration and stability. Other TPs are 2-amino MX, 4-amino MX and 2-amino MK [2, 3]. They have been proven to be more toxic than their parent compounds [8].

HHCB-lactone has been reported to be generated by biological degradation in WWPTs operating with activated sludge [15]. For instance, Tasselli and Guzzella [16] reported an increase of the concentration of HHCB-lactone from biological tank inlet (1447–1883 ng L^{-1}) to biological tank outlet (4397–5695 ng L^{-1}). HHCB-lactone can also be generated in freshwater environments [15, 17]. Martin et al [17], reported the formation of 6 TPs of HHCB, including HHCB-lactone, in freshwater caused by two mitosporic fungi. They also identified the formation of 10 TPs of AHTN. A study carried out by Lange et al. [15] revealed that the transformation of HHCB into HHCB-lactone in surface water was temperature conditioned. They obtained the result that the ratio HHCB-lactone/HHCB in surface water increased from 1 : 1 to 5 : 1 when surface water temperature increased from 8°C to 15°C [15]. Water treatments with ozone and UV irradiation can also generate TPs of PCMs. Herrera-Lõpez et al. [18] identified 18 TPs of HHCB, generated during water treatments with ozone, UV and Xe, with HHCB-lactone being the most stable one. NMs can suffer photochemical transformation in the environment. Gao et al. [4] evaluated the mechanisms, eco-toxicity and health effects of the indirect photochemical transformation of NMs using MX as a representative NM. They concluded that TPs of MX consisted of cyclic, aldehyde and demethylation products with carcinogenic activity although it was lower than that of NX. It has been reported that MX can be also metabolised in humans to 2-amino MX and 4-amino MX whereas MK can be metabolised to 2-amino MK [2].

8.1.4 Occurrence in the Environment

Musk fragrances most frequently monitored in the environment are PCMs, mainly HHCB and AHTN but also DPMI, ADBI, AHMI and ATTI and NMs, mainly MK and MX [10, 11, 19]. In addition, HHCB and AHNT are the PCMs most frequently detected and the ones at the highest concentration levels in wastewater [20–22], sediments [23, 24], air [11, 25], seafood [26], freshwater species [27, 28], vegetables [29] and leaves [30]. This fact can be explained because they are the ones most frequently used [7] and the ones at the highest concentrations in personal care products [6]. There are only a few papers reporting the presence of MCMs. Vallecillos et al. [31] determined 8 MCMs in sewage sludge from 3 WWTPs. The highest concentrations corresponded to ambrettolide (up to 0.85 ng g^{-1} dry weight (dw), n = 24) and musk-NN (up to 0.89 ng g^{-1} dw, n = 24). There is also scarce information about ACMs in the environment, probably due to the fact that their use is still limited and they are more biodegradable than the other groups of synthetic polycyclic musks [2].

8.1.4.1 Occurrence in Wastewater and Sewage Sludge

Some musk fragrances are mainly used in detergents and in cleaning products so, after their use, they are released to the sewer system through domestic wastewaters. In WWTPs, due to their high Log K_{OW} values, these pollutants can be sorbed onto sewage sludge generated during wastewater treatment, can be converted into TPs or can be released to the receiving streams through wastewater effluents [32]. As the conventional processes applied in WWTPs are not capable of removing synthetic musks, if removal from wastewater is observed, it is mainly due to sorption onto sewage sludge [33]. Several studies have evaluated the occurrence of musk fragrances in wastewater [16, 20, 21, 33–35] and in sewage sludge [16, 21, 33, 34, 36]. The occurrence of PCMs in wastewater and sewage sludge has been recently reviewed by Liu et al. [13].

PCMs are the synthetic musk fragrances most frequently detected in WWTPs and the ones at the highest concentrations [33]. Tasselli and Guzzella [16] reported concentrations of PCMs in effluent wastewater (ADBI: 29–37 ng L^{-1}; HHCB: 6698–7324 ng L^{-1}; AHTN: 285–328 ng L^{-1}) similar to those in influent wastewater (ADBI: 25–39 ng L^{-1}, HHCB: 5255–6404 ng L^{-1}; AHTN: 339–518 ng L^{-1}) whereas HHCB-lactone concentration increased before wastewater treatment (from 616–1129 ng L^{-1} to 4786–5386 ng L^{-1}). Wastewater treatments do not produce significant changes in enantiomeric fractions of chiral musk fragrances [32].

Once sorbed into sludge, their removal by anaerobic digestion is poor for HHCB, AHTN, ATTI and ADBI) (removals in the range from 0 to 10%) [34, 36] whereas MK is efficiently removed (94%) [34]. Concentrations of AHTN and HHCB up to 11.5 and 1.6 μg g^{-1}, respectively, have been reported in sewage sludge from Spain [37]. More information about the presence and fate of PCMs in sewage sludge can be found in the review from Venegas et al. [12]. Their review reported concentrations of AHTN, HHCB, AHMI and ATTI in different types of sewage sludge in the ranges 0.4–11 ng g^{-1}, 3.8–36 ng g^{-1}, 0.2–1.8 ng g^{-1} and 0.2–1.0 ng g^{-1}, respectively. Detailed information about the concentration levels, phase distributions and efficiency of removal of synthetic musks in WWTPs can be seen in the review from Homem et al. [33].

8.1.4.2 Occurrence in Surface Water, Soils, Sediments and Air

Musk fragrances have been reported in surface water [6, 15, 20, 38–40], seawater [41], soils [20, 34], sediments [20, 23, 24] and air [11, 25]. The occurrence, distribution, ecological effects and behaviour of PCMs in the environment has been recently reviewed by Liu et al. [13]. Although most of the studies have been focused in the determination of PCMs in water samples, in recent years there has been an increasing concern about their occurrence in sediments and organisms whereas the interest in their occurrence in sludge has decreased [13]. Their occurrence in the air has been scarcely evaluated in spite of the fact that they are semi-volatile organic contaminants which can evaporate from aqueous or solid environmental matrices to the air. McDonough et al. [11] reported mean concentrations in air from four sampling points of the Lower Great Lakes in Canada in the ranges 0.6–2.8 pg m^{-3}, 2.2–24 pg m^{-3}, 47–493 pg m^{-3}, $<$ method detection limit (MDL)-152 pg m^{-3} and 5.4–302 pg m^{-3} for ADBI, AHMI, ATII, HHCB and AHNT, respectively. Weinberg et al. [25] detected synthetic musk fragrances in all air samples collected from two landfields with a total concentration in the range from 146 to 990 pg m^{-3} whereas particle-bound musk fragrances were only detected in two of the sampling sites and at a lower concentration level (6–103 pg m^{-3}).

8.1.4.3 Occurrence in Biota

Most of the studies have been carried out in aquatic organisms and have been focused on PCMs and NMs. Their occurrence in seafood and microalgae [42–44], marine bivalves [45, 46] and freshwater aquatic species [15, 27] has been reported. Some of the studies have also included the monitoring of HHCB-lactone [42–44]. Detailed information about the occurrence of musk fragrances in different marine species and aquatic organisms can be found in the reviews from Rainieri et al. [2] and Liu et al. [13], respectively. Nevertheless, only a few papers have reported their occurrence in leaves [30] and vegetables [29].

A study carried out by Cunha et al. [42] in five European hotspot revealed that HHCB and AHTN were the most frequent musk fragrances and the ones at the highest concentration levels (up to 34.52 ng g^{-1} dw and 12.99 ng g^{-1} dw, respectively) in different seafood (mussels, flounder, mullet and clams) and macroalgae (*L. digitata* and *S. Latissima*). The highest concentration levels for HHBC, AHTN and HHCB-lactone were found in mussels (up to 34.52, 12.99 and 63.51 ng g^{-1} dw, respectively) whereas the highest concentrations of DPMI were found in flounder (9.11 ng g^{-1} dw). The monitoring study carried out by Aminot et al. [45] revealed a widespread occurrence of HHCB (mean: 2.27 ng g^{-1} dw), AHTN (mean: 0.724 ng g^{-1} dw), MX (mean: 0.032 ng g^{-1} dw) and MK (mean: 0.066 ng g^{-1} dw) and a decrease trend just for MX in marine bivalves from French coastal areas from 2010 to 2019. A decrease temporal trend was observed just for MX. The monitoring study of 6 PCMs and MX carried out by Tavera Busso et al. [30] in leaves from 28 sampling points located in urban, industrial and peri-urban areas revealed concentrations in industrial areas four times higher than in urban areas and eight times higher than in peri-urban sites. HHCB, AHMI and DPMI were the musk fragrances most frequently detected (61%, 50% and 43%, respectively) [30].

The bioaccumulation of HHCB in aquatic organisms has been evidenced to be not only species-specific [45] but also season- [28] and tissue-dependent [26]. A different metabolisation pattern could explain the significantly lower levels of HHCB in oysters than in mussels [45]. Vimalkumar et al. [28] reported that bioconcentration factors (BCFs) for HHCB and AHTN in fish were season-dependent. BCFs of HHCB were higher during wet season (November to February, 25–35°C) whereas BCFs of AHTN were higher during dry season (March to July, 35–42°C). Saraiva et al. [26] reported higher concentrations of PCMs and NMs in liver and gonads than in muscle tissues. The highest concentrations corresponded to HHCB (mean values: 96.4 ng g^{-1} in gonads and 6.4 ng g^{-1} in liver) and AHTN (mean values: 11.7 ng g^{-1} in gonads and 3.95 ng g^{-1} in liver).

8.2 Sample Preparation and Collection

Analytical methods reported for the determination of synthetic musk fragrances in environmental samples and biota have been reviewed by Vallecillos et al. [8] and Marchal and Beltran [3]. In sections below, an overview of such methods, and methods reported after that, are included.

8.2.1 Protocols for Collecting and Preparing Samples

As synthetic musks are extensively used in everyday products, it is mandatory to be extremely careful to prevent sample contamination during laboratory process. Detergents should be replaced by alcohol-based detergents whereas acetone can be used for washing glassware [27]. The use of personal care products containing musk fragrances such as perfumes, soaps, hand cream, detergents, etc. should be limited. Other precautions, such as using musk-free gloves and preparing samples in a fume cupboard [9], have been also proposed. In addition, blank samples should be frequently analysed in each batch of samples as part of the routine quality control [16, 27, 30].

8.2.1.1 Air Samples

Polyethylene passive samplers [11] and high-volume active samplers [25] have been reported for further determination of PCMs and NMs in air. Polyethylene passive samplers, located inside a protective metal bowl, have been used for monitoring emissions of PCMs from a lake to the air. Passive samplers were secured in a protective chamber 2 m above the water surface on a buoy [11]. Weinberg et al. [25] monitored 6 PCMs and 2 NMs in air from landfills using high-volume samplers operating at a sampling rate of 350 m^3 d^{-1}. They obtained 1-day and 3-days composite samples.

8.2.1.2 Water Samples

Water samples are commonly collected in pre-cleaned amber glass bottles and stored in the dark at 4°C for short-term conservation [9, 16, 47–50] or at −20°C [22]. Automatic samplers allowing flow proportional composite samples are commonly used for collecting wastewater samples [16, 47]. In a lower extent, passive samplers have been also reported for sample collection [10, 11]. Polyethylene passive samplers, enclosed within a perforated metal cage and secured to a buoy about 4 m below water surface, have been used for sampling PCMs from lake water [11] whereas passive samplers based on diffusive gradients in thin films (DGT) have been reported for the determination of 6 PCMs and 3 NMs in wastewater [10]. The DGT device consisted of a 0.5 mm binding gel (XAD-2 resin gel), a 0.8 mm diffusion gel, and a 0.14 mm filter membrane fixed with an ABS-based plastic DGT moulding (DGT Research Ltd., UK). Musk fragrances were linearly accumulated for at least 5 days, except HHCB and DPMI which were linearly accumulated within 7 and 10 days, respectively.

Once in the laboratory, it has been recommended to analyse samples without filtration to avoid underestimation of the synthetic musks which tend to be sorbed onto particles [15]. This practice has been reported by several authors [28, 49]. Nevertheless, most of them filter aqueous samples prior to treatment through a 0.45 μm cellulose or nylon filter [9, 47, 50], or through two consecutive filters (1 μm and 0.7 μm glass microfiber filters) [16], or remove suspended particles by centrifugation [51].

8.2.1.3 Sludge, Soil and Sediment Samples

After sampling, solid samples are stored frozen at −20°C prior pretreatment [19]. Pretreatment of solid samples commonly consists on frozen, freeze-drying in a lyophiliser, crushing in a mortar and sieving [16, 19, 29, 52].

8.2.1.4 Biota

Pretreatment operations commonly applied to seafood involve rinsing with fresh water, homogenisation, frozen and lyophilisation [26, 42, 45]. Some authors homogenise samples again after lyophilisation [44, 46]. Fishes are commonly dissected to separately analyse liver, muscle and gonads to evaluate the tissue-dependent accumulation of musk fragrances [26, 53]. Pretreatment of vegetables commonly consists on cutting and freeze-drying in a lyophiliser [29]. In addition, some authors calculate moisture content of leaves and vegetables [54] or lipid content of seafood and fishes [26].

8.2.2 Sample Extraction and Clean-up

8.2.2.1 Air Samples

Extraction of PCMs and NMs from PE passive samplers require long-time solid-liquid extractions. For instance, their extraction from polyethylene required extraction with pentane for 18–24 hours [11] whereas their extraction from PUF/XAD-2/PUF cartridges required 3 extractions with hexane:acetone (1 : 1, v/v) (approximately 450 mL) for 1 hour, 1 hour and 3 hours, respectively [25].

8.2.2.2 Water Samples

A wide variety of extraction methods have been proposed for the extraction of musk fragrances from water samples [3, 8, 13] such as liquid-liquid extraction (LLE) [49], solid-phase extraction (SPE) [16, 55] and microextraction methods such as dispersive liquid-liquid microextraction (DLLME) [50], ultrasound-assisted dispersive liquid-liquid microextraction (UA-DLLME) [22, 52], dispersive solid-phase extraction (d-SPE) [56], membrane-assisted solvent extraction (MASE) [47], needle trap (NT) microextraction [48], ionic liquid-based headspace single drop microextraction (IL-HS-SDME) [9] and stir-bar sorptive extraction (SBSE) [57]. Some of them, such as NT microextraction, IL-HS-SDME and SBSE, allow direct injection of extracted compounds by thermal desorption in the GC system. An overview of such methods can be seen in Table 8.2.

Table 8.2 Overview of analytical methods for the determination of musk fragrances in environmental samples.

Matrix	Analytes	Extraction + clean-up	Chromatographic determination	R (%)	MDLs	Reference
Air and water	5 PCMs, 2 NMs	Polyethylene passive sampler-SLE (18–24 hour, pentane) Water: SLE + SPE	GC-MS (EI, Q) DB-5 MS column (30 m × 0.25 mm i.d.) Flow-rate: 1.9 mL min^{-1}	79–145	Air: 0.9–0.1 ng m^{-3} Water: 0.6–0.04 ng L^{-1}	[11]
Water samples	6 PCMs, 2 NM, *HHCB-lactone*	IL-HS-SDME	GC-MS/MS (TD-splitless, EI, ion trap) Fused silica capillary column (3 m × 0.25 mm) (guard column) connected to a ZB-50 (30 m × 0.25 mm, 0.25 μm). Run-time: 19 minutes; Flow-rate: 1 mL min^{-1}	n.d.	10–30 ng L^{-1}	[9]
Water samples, wastewater	6 PCMs, 5 NMs, 5 MCMs, 2 ACMs	SBSE	GC-MS (TD-splitless, EI, Q) HP- 5 ms column (30 m × 0.25 mm, 0.2 μm) Run-time: 32 minutes	n.d.	5–80 ng L^{-1} (MQL)	[57]
Water samples	5 PCMs, 5 NMs, 2 MCMs (+ 6 UV filters)	UAE + DLLME	GC-MS/MS (Splitless, EI, QqQ) CP-Sil 8 CB column (50 m × 0.25 mm, 0.12 μm) Flow-rate: 1.0 mL min^{-1}	83–122	0.1–10.9 ng L^{-1}	[52]
Water samples, wastewater	Enantiomers of 5 PCMs	SPE (Oasis HLB)	GC-MS/MS (Splitless, EI, QqQ) Achiral HP-5 MS (30 m × 0.25 mm, 0.25 μm) connected to a Cyclosil-B column (30 m × 0.25 mm, 0.25 μm) Run-time: 240 minutes; Flow-rate: 1.2 mL min^{-1}	84–116	1.0–2.0 ng L^{-1}	[55]

River water, wastewater	Enantiomers of 5 PCMs	LLE (CH_2Cl_2-hexane)	GC-MS/MS (Splitless, EI, QqQ) Cyclosil-B column (25 m × 0.25 mm, 0.25 μm) Run-time: 49 minutes; Flow-rate: 1.0 mL min^{-1}	n.d.	18–34 ng L^{-1}	[49]
Surface water, wastewater	6 PCMs, 1 NM	MASE	GC-MS (PTV-LVI, EI, Q) HP-5 MS column (30 m × 0.25 mm, 0.25 μm) Flow-rate: 1.3 mL min^{-1}	47–126	4–25 ng L^{-1}	[47]
Wastewater	6 PCMs, 3 NMs	DGT device (in situ)	GC-MS (Splitless, EI, Q) SH-Rxi-5Sil MS column (30 m × 0.25 mm, 0.25 μm) Run-time: 14 minutes; Flow-rate: 1.1 mL min^{-1}	n.d.	0.02–0.09 ng L^{-1} (for 7 day sampling period)	[10]
Wastewater	2 PCMs (+16 fragrance allergens)	HS-SPME	GC-MS (Splitless, EI, ion trap) TG-5SIL MS column (30 m × 0.25 mm i.d.; 0.25 μm) Run-time: 26 minutes; Flow-rate: 1 mL min^{-1}	30–71	40 ng L^{-1}	[35]
Wastewater	6 PCMs, 3 NMs	HS-SPME	GC-MS (TD-split 1 : 5, EI) DB-1 MS column (30 m × 0.25 mm, 0.25 μm) Run-time: 24.2 minutes; Flow-rate: 1 mL min^{-1}	79–107	1–2 ng L^{-1}	[6]

(*Continued*)

Table 8.2 (Continued)

Matrix	Analytes	Extraction + clean-up	Chromatographic determination	R (%)	MDLs	Reference
Wastewater	6 PCMs, 2 NMs, *HHCB-lactone*	NT microextraction	GC-MS/MS (TD-PTV-splitless, EI, ion trap) Fused silica capillary column (3 m × 0.25 mm) (guard column) connected to a 50% diphenyl/50% dimethyl-polysiloxane analytical column (30 m × 0.25 mm, 0.25 μm). Run-time: 17.5 minutes; Flow-rate: 1 mL min^{-1}		2.5–10 ng L^{-1}	[48]
Wastewater	4 PCMs, *HHCB-lactone*	SPE (C18 cartridges)	GC-MS/MS (Splitless, EI, QqQ) SLB-5 ms column (60 m × 0.25 mm, 0.25 μm) Run-time: 30 minutes; Flow-rate: 1.3 mL min^{-1}	71–95	1.7–80 ng L^{-1}	[16]
Soil	6 PCMs, 2 NMs	SBSE	GC-MS (TD-PTV-split, EI, Q) HP5-MS column (30 m × 0.25 mm, 0.25 μm) Flow-rate: 1.3 mL min^{-1}	74–126	0.01–1.1 ng g^{-1} dw	[29]
Sewage sludge	8 MCMs	HS-SPME	GC-MS (PTV-splitless, EI, ion trap) ZB-50 column (30 m × 0.25 mm, 0.25 μm) Run-time: 10 minutes; Flow-rate: 1 mL min^{-1}	n.d.	10–25 pg g^{-1} dw	[31]
Sewage sludge	Enantiomers of HHCB, *HHCB-lactone*, AHTN	Soxhlet (72 hour)	GC-MS/MS (Splitless, EI, QqQ) Hydrodex-β-6TBDM column (25 m × 0.25 mm, 0.25 μm) Run-time: 270 minutes; Flow-rate: 1 mL min^{-1}	n.d.	10–45 ng L^{-1} (IDL)	[60]
Sewage sludge and sediments	6 PCMs	MA-HS-SPME	GC-MS (Splitless, EI, Q) DB-5 MS column (30 m × 0.25, 0.25 μm) Run-time: 21 minutes	67–89	0.04–0.1 ng g^{-1} fresh weight	[59]

Sewage sludge	4 PCMs, *HHCB-lactone*	UAE + d-SPE	GC-MS (PTV-split 1 : 8, EI, ion trap) SLB- 5 ms column (30 m × 0.25 mm, 0.25 μm) Run-time: 22 minutes; Flow-rate: 1.3 mL min^{-1}	>77%	0.1–210 ng g^{-1}	[16]
Sediments	6 PCMs, 2 NMs	In-cell clean-up PLE	GC-MS (Splitless, EI, Q) DB-WAX column (30 m × 0.25 mm, 0.25 μm) Run-time: 19 minutes; Flow-rate: 1 mL min^{-1}	86.0–104	0.03–0.05 ng g^{-1}	[19]

ACM: alicyclic musks; DGT: diffusive gradients in thin films; DLLME: dispersive liquid-liquid microextraction; dw: dry weight; EI: electron impact ionisation source; GC-MS: gas chromatography-mass spectrometry; GC-MS/MS: gas chromatography tandem mass spectrometry; HS-SPME: headspace solid-phase microextraction; IL-HS-SDME: ionic liquid-based headspace single drop microextraction; LLE: liquid-liquid extraction; LVI: large volume injection mode; MA-HS-SPME: microwave-assisted headspace solid-phase microextraction; MASE: membrane-assisted solvent extraction; MCMs: macrocylic musks; MDL: method detection limit; MQL: method quantitation limit; n.d.: no data; NMs: nitro musks; PCMs: polycyclic musks; PLE: pressurised liquid extraction; PTV: programmable temperature vaporisation injector; Q: quadrupole; QqQ: triple quadrupole; R: recovery; SBSE: stir-bar sorptive extraction; SLE: solid-liquid extraction; SPE: solid-phase extraction; SPME: solid-phase microextraction; TD: thermal desorption unit; UAE: ultrasound-assisted extraction.

8.2.2.3 Sludge, Soil and Sediment Samples

Extraction of solid samples has been carried out by pressurised liquid extraction (PLE) [58], PLE using in-cell clean-up [19], stir-bar sorptive extraction (SBSE) [29], headspace solid-phase microextraction (HS-SPME) [31], microwave-assisted headspace solid-phase microextraction (MA-HS-SPME) [59], ultrasound-assisted extraction (UAE) [16], QuEChERS method [58] and Soxhlet extraction [60]. In Table 8.2, it can be seen an overview of extraction methods applied to the determination of synthetic musk fragrances in solid environmental samples.

Che et al. [19] compared the extraction of 6 PCMs and 2 NMs from lake sediments by means of in-cell clean-up PLE, Soxhlet extraction and ultrasound-assisted extraction (UAE). They reported that in-cell clean-up PLE provided the best extraction efficiency (from 86 to 104%), the lowest solvent consumption and extraction time and it was the extraction method easiest to perform. Some of them, such as SBSE and HS-SPME, allow direct injection of extracted compounds by thermal desorption in the GC system. HS-SPME is a solventless, easy-to-perform and fast extraction technique, proposed in 2013 for the first-time determination of 8 MCMs in sewage sludge [31]. Extraction was carried out exposing a polydimethylsiloxane-divinylbenzene PDMS/DVB 65 μm fibre for 45 minutes in the headspace of a 20 mL vial containing sewage sludge (0.25 g dry weight (dw)) mixed with 0.5 mL of water at 80°C and stirred at 750 rpm. They achieved method detection limits (MDLs) in the range from 10 to 25 pg g^{-1} dw. In the one-step in situ MA-HS-SPME method proposed for the determination of 6 PCMs in sewage sludge and sediments [60], the solid sample (5 g dw) was mixed with 20-mL deionised water (containing 3 g of NaCl in a 40-mL sample-vial) and extracted by a PDMS-DVB fibre placed in the headspace while the slurry was microwave irradiated at 80 W for 5 minutes. Recoveries and MDLs were in the range 85–96% and 0.1–0.3 ng g^{-1} fresh weight, respectively. SBSE was applied for the first time by Aguirre et al. [29] to the extraction of musks from soils and vegetables. Samples (0.5 g) were mixed with water:MeOH (80 : 20, v/v) and a PDMS stir-bar was introduced. The mixture was stirred for 3 hours at 40°C and then the stir bars were thermally desorbed into a thermal desorption unit connected to the GC injector. This method requires low solvent consumption and it is fast and easy-to-perform. Nevertheless, some recent reported methods are still based on traditional methods, such as Soxhlet, requiring long extraction times and high solvent volumes. For instance, the method reported by Gao et al. [60] required 72 hours to extract samples whereas the PLE method reported by Che et al. [19], and the UAE method reported by Tasselli and Guzzella [16], required extraction times of just 10 minutes and 30 minutes, respectively.

8.2.2.4 Biota

Several extraction methods have been reported for the determination of musk fragrances in biota: QuEChERS method has been reported for the determination of PCMs and NMs in seafood [26, 43, 44, 46]; solid-phase microextraction Arrow (SPME Arrow) has been reported for the determination of 6 PCMs and 2 NMs in fish; SBSE has been used for the determination of 6 PCMs and 2 NMs in vegetables [29]; UAE has been applied for the extraction of AHTN, HHCB and MK from fish muscle tissue [28] and 6 PCMs and 5 NMs from pine needles [54]; PLE [45] and in-cell clean-up PLE [42] have been reported for the determination of PCMs and NMs in seafood and macroalgae. Table 8.3 shows an overview of such extraction methods.

Table 8.3 Overview of analytical methods for the determination of musk fragrances in biota.

Matrix	Analytes	Extraction + clean-up	Chromatographic determination	R (%)	MDLs	Reference
Mussels	3 PCMs, 4 NMs	QuEChERS + DLLME	GC-MS (Splitless, EI, Q) DB-5 MS column (30 m × 0.25 mm, 0.25 μm) Run-time: 28 minutes; Flow-rate: 1 mL min^{-1}	59–73	0.5–15.0 ng g^{-1}	[46]
Seafood	5 PCMs, 4 NMs	QuEChERS	GC-MS (PTV-LVI, EI, Q) HP-5 MS column (30 m × 0.25 mm, 0.25 μm) Run-time: 28 minutes; Flow-rate: 1.3 mL min^{-1}	46–120	0.001–1.91 ng g^{-1}	[26]
Seafood	6 PCMs, 3 NMs, *HHCB-lactone*	QuEChERS	GC-MS/MS (PTV-LVI, EI, ion trap) ZB-50 column (30 m × 0.25 mm, 0.25 μm) Run-time: 15 minutes; Flow-rate: 1 mL min^{-1}	47–117	1–10 ng g^{-1} dw	[44]
Seafood and macroalgae	6 PCMs, 3 NM, *HHCB-lactone*	In-cell clean-up PLE	GC-MS/MS (PTV-LVI, EI, ion trap) Fused silica guard column (3 m × 0.25 mm) + ZB-50 column (30 m × 0.25 mm, 0.25 μm) Run-time: 16 minutes; Flow-rate: 1 mL min^{-1}	n.d.	0.25–10 ng g^{-1} dw	[42]
Marine bivalves	2 PCMs, 2 NMs	PLE + GPC	GC-MS (EI: HHCB and AHTN; NCI: MX and MK) GC-MS/MS (APCI, QqQ): HHCB, AHTN and MK DB-5 MS column (40 m × 0.18 mm, 0.18 μm)	95 ± 4	0.006–0.54 ng g^{-1} dw	[45]

(*Continued*)

Table 8.3 (Continued)

Matrix	Analytes	Extraction + clean-up	Chromatographic determination	R (%)	MDLs	Reference
Fish	6 PCMs, 2 NMs	SPME Arrow	GC-MS/MS (PTV, EI, ion trap) ZB-50 column (30 m × 0.25 mm, 0.25 μm) Run-time: 16 minutes; Flow-rate: 1 mL min^{-1}	n.d.	0.5–2.5 ng g^{-1} dw	[53]
Pine needles	6 PCMs, 5 NMs (BFRs, OCPs, PAHs and PCBs)	UAE + SPE + GPC	GC-MS (Splitless, EI, ion trap) Fused silica deactivated retention gap (5 m × 0.25 mm) + CP-Sil 8CB column (50 m × 0.25 mm, 0.2 μm)	48–107	0.0004–0.129 ng g^{-1} fresh weight	[54]
Leaves	6 PCMs, 5 NMs	UAE + SPE + GPC	GC-MS (Splitless, EI, Q) DB-5 MS column (30 m × 0.25 mm, 0.25 μm)	91 ± 14	0.003–0.114 ng g^{-1} dw	[30]

BFRs: brominated flame retardants; DLLME: dispersive liquid-liquid microextraction; dw: dry weight; EI: electron impact ionisation source; GC-MS: gas chromatography-mass spectrometry; GC-MS/MS: gas chromatography tandem mass spectrometry; GPC: gel permeation chromatography; HS-SPME: headspace solid-phase microextraction; LVI: large volume injection mode; MDL: method detection limit; n.d.: no data; NMs: nitro musks; OCPs: organochlorine pesticides; PAHs: polycyclic aromatic hydrocarbons; PCBs: polychlorinated biphenyls; PCMs: polycyclic musks; PLE: pressurised liquid extraction; PTV: programmable temperature vaporisation injector; Q: quadrupole; QqQ: triple quadrupole; R: recovery; SPE: solid-phase extraction; SPME: solid-phase microextraction; UAE: ultrasound-assisted extraction.

After extraction, extract clean-up is commonly required. After PLE, clean-up first by gel permeation chromatography (GPC) and then with Florisil and silica/alumina columns was applied by Aminot et al. [45] to extracts from mussels and oysters. This clean-up step was avoided by Cunha [42] in the extraction of 6 PCMs, HHCB-lactone and 3 NMs from seafood and microalgae by adding a clean-up sorbent (Florisil) to PLE extraction cells (in-cell clean-up PLE). Castro et al. [46] applied DLLME after QuEChERS method to achieve higher enrichment factors and, therefore good MDLs (from 0.5 to 15 ng g^{-1} dw), in the determination of 3 PCMs and 4 NMs in wild mussels.

SPME Arrow was applied in 2019 for the first-time determination of PCMs and NMs in fish samples [53]. The method combined the advantages of SBSE, with its large sorption phases, with the advantages of conventional SPME. In the SPME Arrow method proposed by Castro et al. [53], the SPME Arrow was exposed for 45 minutes to the vial headspace containing fish sample heated at 100°C and stirred at 750 rpm. After that, arrow desorption was carried out in the GC injection port heated at 250°C. Aguirre et al. [29] tested three different SBSE configurations for the extraction of 6 PCMs and 2 NMs from vegetables and soils. The configurations were: (a) headspace mode by placing the stir bar in the headspace of a vial containing the solid sample (0.5 g) and heated at 80°C; (b) headspace mode by placing the stir bar in the headspace of a vial containing the solid sample (0.5 g) and water (6 mL) at 80°C; and (c) immersion mode by immersing the stir bar inside a mixture a vial containing the solid sample (0.5 g) and water (6 mL) at room temperature. They reported that immersion mode provided the best extraction recoveries for all the compounds.

8.3 Determination of Musk Fragrances

8.3.1 Chromatography

GC is the analytical technique most commonly applied for the determination of synthetic musk fragrances [3, 8] as they are semi-volatile and thermally-stable compounds. Synthetic musk fragrances are separated mainly in low polar 5% phenyl-95% methylpolysiloxane (DB-5 MS, HP-5 MS, CP-Sil 8 CB, SLB®-5 ms) [11, 16, 22, 29, 61] but also in mid-polarity 50% phenyl-50% dimethylpolysiloxane GC columns (ZB-50) [44, 53]. Column dimensions are usually 30 m × 0.25 mm internal diameter, 0.25 μm film thickness [6, 10, 11, 19, 29, 35, 52, 53, 61]. Run times for the separation of PCMs and NMs are commonly lower than 20 min [9, 10, 44, 53]. Methods including the simultaneous determination of PCMs, NMs and MCMs are scarce. The method reported by Homem et al. [22] allowed the separation of 5 PCMs, 5 NMs and 2 MCMs in 40 min using a 50 m column. The longest run times (up to 270 min for the separation of enantiomers of HHCB, HHCB-lactone and AHTN [60]) correspond to methods developed for the enantioseparation of chiral musk fragrances. Their separation requires chiral columns, commonly cyclodextrin-based columns, and slow temperature gradients, which results in long run times [49, 55, 60]. Wang et al. [55] tested four chiral GC columns for the enantioseparation of 5 chiral

PCMs but finally they had to apply two-dimensional gas chromatography (GC × GC) to achieve a proper separation of enantiomers. Split/splitless injector, in splitless mode, is the one most commonly used [3, 8, 11, 22]. The use of splitless mode is required to allow the determination of musks fragrances at low concentration levels. The use of programmable temperature vaporisation injectors (PTV), in splitless mode or in large volume injection mode (LVI), has been also reported. LVI mode allows sensitivity to be improved [16, 44, 53] by venting the solvent and, therefore, concentrating the analytes. Thermal desorption units (TD) have been reported when SBSE [3, 29, 57] or HS-SPME [6] have been applied. LC has been scarcely used for the determination of synthetic musk fragrances. It has been used for the identification of transformations products of HHCB generated from ozone and UV treatments [18], for the simultaneous determination of musk fragrances and other eight groups of pollutants in sediments [58] and PCMs in wastewater effluents and sludge [62]. An overview of chromatographic conditions applied to the determination of synthetic musk fragrances in environmental samples and biota can be seen in Tables 8.2 and 8.3.

8.3.2 Detection

Detection is commonly carried out by mass spectrometry [11, 22, 28, 29] in electron impact ionisation (EI) mode (70 ev) using quadrupole (Q) [10, 11, 29, 46, 57] and ion trap [16, 22, 44, 53] analysers. Recently, the use of triple quadrupole (QqQ) analysers has been also reported [16, 42, 53, 60]. In the determination of musk fragrances by LC-MS/MS, Chiaia-Hernandez et al. [58] compared the use of electrospray ionisation (ESI) and atmospheric pressure photoionisation (APPI) sources. APPI is an ionisation technique suitable for the analysis of poorly ionisable and less polar compounds that allowed sensitivity and sensibility 2 orders of magnitude higher than ESI and the differentiation between HHCB and AHTN, which could not be chromatographically separated, by means of their different ionisation behaviour [58]. APPI was also applied by Herrera-Lõpez et al. [18] for the LC/ESI-QTOF-MS identification of 18 TPs of HHCB generated during water treatments.

8.4 Future Directions and Challenges

To the date, few methods have been reported for the determination of HHCB-lactone, MCMs, ACMs and TPs of musk fragrances in environmental samples; for the determination of NMs and PCMs in air, soils and vegetables; and for the enantiodetermination of chiral PCMs. Future research is expected to be directed to those topics. In addition, it is of special relevance the development of analytical methods that allow their determination in agricultural soils and in edible vegetables to prevent them to enter into the food chain by vegetable consumption. In addition, reduction of sample treatment or chromatographic run times constitutes a significant analytical challenge in the case of the determination of musk fragrances in air samples and analytical enantiodetermination of chiral PCMs, respectively.

References

1 Chisvert, A., López-Nogueroles, M., Miralles, P., and Salvador, A. (2018). *Perfumes in Cosmetics: Regulatory Aspects and Analytical Methods*, 2nd edn. Elsevier B.V. doi: 10.1016/B978-0-444-63508-2.00010-2.

2 Rainieri, S., Barranco, A., Primec, M., and Langerholc, T. (2017). Occurrence and toxicity of musks and UV filters in the marine environment. *Food Chem. Toxicol.* 104: 57–68. doi: 10.1016/j.fct.2016.11.012.

3 Marchal, M. and Beltran, J. (2016). Determination of synthetic musk fragrances. *Int. J. Environ. Anal. Chem.* 96: 1213–1246. doi: 10.1080/03067319.2016.1249479.

4 Gao, Y., Li, G., Qin, Y., Ji, Y., Mai, B., and An, T. (2019). New theoretical insight into indirect photochemical transformation of fragrance nitro-musks: mechanisms, eco-toxicity and health effects. *Environ. Int.* 129: 68–75. doi: 10.1016/j.envint.2019.05.020.

5 David, O.R.P. (2020). A chemical history of polycyclic musks. *Chem. Eur. J.* 26: 7537–7555. doi: 10.1002/chem.202000577.

6 Hong, J.-H., Lee, J.-Y., Ha, H.-J., Lee, J.-H., Oh, S.-R., Lee, Y.-M., Lee, M.-Y., and Zoh, K.-D. (2021). Occurrence and sources of synthetic musk fragrances in the sewage treatment plants and the Han River, Korea. *Water* 13: 392. doi: 10.3390/w13040392.

7 Ehiguese, F.O., Alam, M.R., Pintado-Herrera, M.G., Araújo, C.V.M., and Martin-Diaz, M.L. (2020). Potential of environmental concentrations of the musks galaxolide and tonalide to induce oxidative stress and genotoxicity in the marine environment. *Mar. Environ. Res.* 160: 105019. doi: 10.1016/j.marenvres.2020.105019.

8 Vallecillos, L., Borrull, F., and Pocurull, E. (2015). Recent approaches for the determination of synthetic musk fragrances in environmental samples. *TrAC – Trends Anal. Chem.* 72: 80–92. doi: 10.1016/j.trac.2015.03.022.

9 Vallecillos, L., Pocurull, E., and Borrull, F. (2012). Fully automated ionic liquid-based headspace single drop microextraction coupled to GC-MS/MS to determine musk fragrances in environmental water samples. *Talanta* 99: 824–832. doi: 10.1016/j.talanta.2012.07.036.

10 Ren, S., Tan, F., Wang, Y., Zhao, H., Zhang, Y., Zhai, M., Chen, J., and Wang, X. (2020). In situ measurement of synthetic musks in wastewaters using diffusive gradients in thin film technique. *Water Res.* 185: 116239. doi: 10.1016/j.watres.2020.116239.

11 McDonough, C.A., Helm, P.A., Muir, D., Puggioni, G., and Lohmann, R. (2016). Polycyclic musks in the air and water of the lower great lakes: spatial distribution and volatilization from surface waters. *Environ. Sci. Technol.* 50: 11575–11583. doi: 10.1021/acs.est.6b03657.

12 Venegas, M., Leiva, A.M., Reyes-Contreras, C., Neumann, P., Piña, B., and Vidal, G. (2021). Presence and fate of micropollutants during anaerobic digestion of sewage and their implications for the circular economy: a short review. *J. Environ. Chem. Eng.* 9. doi: 10.1016/j.jece.2020.104931.

13 Liu, J., Zhang, W., Zhou, Q., Zhou, Q., Zhang, Y., and Zhu, L. (2020). Polycyclic musks in the environment: a review of their concentrations and distribution, ecological effects and behavior, current concerns and future prospects. *Crit. Rev. Environ. Sci. Technol.* 51: 323–377. doi: 10.1080/10643389.2020.1724748.

14 EC, (2009). Regulation EC No 1223/2009 of the European Parliament and of the Council of 30 November 2009 on cosmetic products.

15 Lange, C., Kuch, B., and Metzger, J.W. (2015). Occurrence and fate of synthetic musk fragrances in a small German river. *J. Hazard. Mater.* 282: 34–40. doi: 10.1016/j.jhazmat.2014.06.027.

16 Tasselli, S. and Guzzella, L. (2020). Polycyclic musk fragrances (PMFs) in wastewater and activated sludge: analytical protocol and application to a real case study. *Environ. Sci. Pollut. Res.* 27: 30977–30986. doi: 10.1007/s11356-019-06767-7.

17 Martin, C., Moeder, M., Daniel, X., Krauss, G., and Schlosser, D. (2007). Biotransformation of the polycyclic musks HHCB and AHTN and metabolite formation by fungi occurring in freshwater environments. *Environ. Sci. Technol.* 41: 5395–5402. doi: 10.1021/es0711462.

18 Herrera-Lõpez, S., Hernando, M.D., Gõmez, M.J., Santiago-Morales, J., Rosal, R., and Fernández-Alba, A.R. (2013). Investigation of Galaxolide degradation products generated under oxidative and irradiation processes by liquid chromatography/hybrid quadrupole time-of-flight mass spectrometry and comprehensive two-dimensional gas chromatography/time-of-flight mass spect. *Rapid Commun. Mass Spectrom.* 27: 1237–1250. doi: 10.1002/rcm.6575.

19 Che, J.S., Yu, R.P., Song, Q.J., Wang, L.P., and Wu, S.F. (2011). Determination of synthetic musks in the sediment of the Taihu lake by using accelerated solvent extraction (ASE) and GC/MS. *Int. J. Environ. Anal. Chem.* 91: 387–399. doi: 10.1080/03067311003782633.

20 Chase, D.A., Karnjanapiboonwong, A., Fang, Y., Cobb, G.P., Morse, A.N., and Anderson, T.A. (2012). Occurrence of synthetic musk fragrances in effluent and non-effluent impacted environments. *Sci. Total Environ.* 416: 253–260. doi: 10.1016/j.scitotenv.2011.11.067.

21 Hu, Z., Shi, Y., Zhang, S., Niu, H., and Cai, Y. (2011). Assessment of synthetic musk fragrances in seven wastewater treatment plants of Beijing, China. *Bull. Environ. Contam. Toxicol.* 86: 302–306. doi: 10.1007/s00128-011-0215-1.

22 Homem, V., Alves, A., Alves, A., and Santos, L. (2016). Ultrasound-assisted dispersive liquid-liquid microextraction for the determination of synthetic musk fragrances in aqueous matrices by gas chromatography-mass spectrometry. *Talanta* 148: 84–93. doi: 10.1016/j.talanta.2015.10.049.

23 Boehler, S., Strecker, R., Heinrich, P., Prochazka, E., Northcott, G.L., Ataria, J.M., Leusch, F.D.L., Braunbeck, T., and Tremblay, L.A. (2017). Assessment of urban stream sediment pollutants entering estuaries using chemical analysis and multiple bioassays to characterise biological activities. *Sci. Total Environ.* 593–594: 498–507. doi: 10.1016/j.scitotenv.2017.03.209.

24 Huang, W., Xie, Z., Yan, W., Mi, W., and Xu, W. (2016). Occurrence and distribution of synthetic musks and organic UV filters from riverine and coastal sediments in the Pearl River estuary of China. *Mar. Pollut. Bull.* 111: 153–159. doi: 10.1016/j.marpolbul.2016.07.018.

25 Weinberg, I., Dreyer, A., and Ebinghaus, R. (2011). Landfills as sources of polyfluorinated compounds, polybrominated diphenyl ethers and musk fragrances to ambient air. *Atmos. Environ.* 45: 935–941. doi: 10.1016/j.atmosenv.2010.11.011.

26 Saraiva, M., Cavalheiro, J., Lanceleur, L., and Monperrus, M. (2016). Synthetic musk in seafood products from south Europe using a quick, easy, cheap, effective, rugged and safe extraction method. *Food Chem* 200: 330–335. doi: 10.1016/j.foodchem.2016.01.017.

27 Lyu, Y., Ren, S., Zhong, F., Han, X., He, Y., and Tang, Z. (2021). Occurrence and trophic transfer of synthetic musks in the freshwater food web of a large subtropical lake. *Ecotoxicol. Environ. Saf.* 213: 112074. doi: 10.1016/j.ecoenv.2021.112074.

28 Vimalkumar, K., Nikhil, N.P., Arun, E., Mayilsamy, M., and Babu-Rajendran, R. (2021). Synthetic musks in surface water and fish from the rivers in India: seasonal distribution and toxicological risk assessment. *J. Hazard. Mater.* 414: 125558. doi: 10.1016/j.jhazmat.2021.125558.

29 Aguirre, J., Bizkarguenaga, E., Iparraguirre, A., Fernández, L.Á., Zuloaga, O., and Prieto, A. (2014). Development of stir-bar sorptive extraction-thermal desorption-gas chromatography-mass spectrometry for the analysis of musks in vegetables and amended soils. *Anal. Chim. Acta.* 812: 74–82. doi: 10.1016/j.aca.2013.12.036.

30 Tavera Busso, I., Tames, F., Silva, J.A., Ramos, S., Homem, V., Ratola, N., and Carreras, H. (2018). Biomonitoring levels and trends of PAHs and synthetic musks associated with land use in urban environments. *Sci. Total Environ.* 618: 93–100. doi: 10.1016/j.scitotenv.2017.10.295.

31 Vallecillos, L., Borrull, F., and Pocurull, E. (2013). An automated headspace solid-phase microextraction followed by gas chromatography-mass spectrometry method to determine macrocyclic musk fragrances in wastewater samples. *Anal. Bioanal. Chem.* 405: 9547–9554. doi: 10.1007/s00216-013-7375-2.

32 Wang, L., Wijekoon, K.C., Nghiem, L.D., and Khan, S.J. (2014). Removal of polycyclic musks by anaerobic membrane bioreactor: biodegradation, biosorption, and enantioselectivity. *Chemosphere* 117: 722–729. doi: 10.1016/j.chemosphere.2014.10.004.

33 Homem, V., Silva, J.A., Ratola, N., Santos, L., and Alves, A. (2015). Long lasting perfume. A review of synthetic musks in WWTPs. *J. Environ. Manage.* 149: 168–192. doi: 10.1016/j.jenvman.2014.10.008.

34 Biel-Maeso, M., Corada-Fernández, C., and Lara-Martín, P.A. (2019). Removal of personal care products (PCPs) in wastewater and sludge treatment and their occurrence in receiving soils. *Water Res* 150: 129–139. doi: 10.1016/j.watres.2018.11.045.

35 Godayol, A., Besalú, E., Anticó, E., and Sanchez, J.M. (2015). Monitoring of sixteen fragrance allergens and two polycyclic musks in wastewater treatment plants by solid phase microextraction coupled to gas chromatography. *Chemosphere* 119: 363–370. doi: 10.1016/j.chemosphere.2014.06.072.

36 Gonzalez-Gil, L., Papa, M., Feretti, D., Ceretti, E., Mazzoleni, G., Steimberg, N., Pedrazzani, R., Bertanza, G., Lema, J.M., and Carballa, M. (2016). Is anaerobic digestion effective for the removal of organic micropollutants and biological activities from sewage sludge? *Water Res.* 102: 211–220. doi: 10.1016/j.watres.2016.06.025.

37 Taboada-Santos, A., Braz, G.H.R., Fernandez-Gonzalez, N., Carballa, M., and Lema, J.M. (2019). Thermal hydrolysis of sewage sludge partially removes organic micropollutants but does not enhance their anaerobic biotransformation. *Sci. Total Environ.* 690: 534–542. doi: 10.1016/j.scitotenv.2019.06.492.

38 Bai, X., Lutz, A., Carroll, R., Keteles, K., Dahlin, K., Murphy, M., and Nguyen, D. (2018). Occurrence, distribution, and seasonality of emerging contaminants in urban watersheds. *Chemosphere* 200: 133–142. doi: 10.1016/j.chemosphere.2018.02.106.

39 Vecchiato, M., Cremonese, S., Gregoris, E., Barbaro, E., Gambaro, A., and Barbante, C. (2016). Fragrances as new contaminants in the Venice lagoon. *Sci. Total Environ.* 566–567: 1362–1367. doi: 10.1016/j.scitotenv.2016.05.198.

40 Villa, S., Assi, L., Ippolito, A., Bonfanti, P., and Finizio, A. (2012). First evidences of the occurrence of polycyclic synthetic musk fragrances in surface water systems in Italy: spatial and temporal trends in the Molgora River (Lombardia Region, Northern Italy). *Sci. Total Environ.* 416: 137–141. doi: 10.1016/j.scitotenv.2011.11.027.

41 Vecchiato, M., Gregoris, E., Barbaro, E., Barbante, C., Piazza, R., and Gambaro, A. (2017). Fragrances in the seawater of Terra Nova Bay, Antarctica. *Sci. Total Environ.* 593–594: 375–379. doi: 10.1016/j.scitotenv.2017.03.197.

42 Cunha, S.C., Fernandes, J.O., Vallecillos, L., Cano-Sancho, G., Domingo, J.L., Pocurull, E., Borrull, F., Maulvault, A.L., Ferrari, F., Fernandez-Tejedor, M., Van Den Heuvel, F., and Kotterman, M. (2015). Co-occurrence of musk fragrances and UV filters in seafood and macroalgae collected in European hotspots. *Environ. Res.* 143: 65–71. doi: 10.1016/j.envres.2015.05.003.

43 Cunha, S.C., Trabalón, L., Jacobs, S., Castro, M., Fernandez-Tejedor, M., Granby, K., Verbeke, W., Kwadijk, C., Ferrari, F., Robbens, J., Sioen, I., Pocurull, E., Marques, A., Fernandes, J.O., and Domingo, J.L. (2018). UV-filters and musk fragrances in seafood commercialized in Europe Union: occurrence, risk and exposure assessment. *Environ. Res.* 161: 399–408. doi: 10.1016/j.envres.2017.11.015.

44 Trabalón, L., Cano-Sancho, G., Pocurull, E., Nadal, M., Domingo, J.L., and Borrull, F. (2015). Exposure of the population of Catalonia (Spain) to musk fragrances through seafood consumption: risk assessment. *Environ. Res.* 143: 116–122. doi: 10.1016/j.envres.2015.04.007.

45 Aminot, Y., Munschy, C., Héas-Moisan, K., Pollono, C., and Tixier, C. (2021). Levels and trends of synthetic musks in marine bivalves from French coastal areas. *Chemosphere* 268. doi: 10.1016/j.chemosphere.2020.129312.

46 Castro, M., Fernandes, J.O., Pena, A., and Cunha, S.C. (2018). Occurrence, profile and spatial distribution of UV-filters and musk fragrances in mussels from Portuguese coastline. *Mar. Environ. Res.* 138: 110–118. doi: 10.1016/j.marenvres.2018.04.005.

47 Posada-Ureta, O., Olivares, M., Navarro, P., Vallejo, A., Zuloaga, O., and Etxebarria, N. (2012). Membrane assisted solvent extraction coupled to large volume injection-gas chromatography-mass spectrometry for trace analysis of synthetic musks in environmental water samples. *J. Chromatogr. A* 1227: 38–47. doi: 10.1016/j.chroma.2011.12.104.

48 Vallecillos, L., Borrull, F., Sanchez, J.M., and Pocurull, E. (2015). Sorbent-packed needle microextraction trap for synthetic musks determination in wastewater samples. *Talanta* 132: 548–556. doi: 10.1016/j.talanta.2014.08.016.

49 Lee, I., Gopalan, A.I., and Lee, K.P. (2016). Enantioselective determination of polycyclic musks in river and wastewater by GC/MS/MS. *Int. J. Environ. Res. Public Health.* 13: 1–9. doi: 10.3390/ijerph13030349.

50 López-Nogueroles, M., Chisvert, A., Salvador, A., and Carretero, A. (2011). Dispersive liquid-liquid microextraction followed by gas chromatography-mass spectrometry for the determination of nitro musks in surface water and wastewater samples. *Talanta* 85: 1990–1995. doi: 10.1016/j.talanta.2011.07.048.

51 Ramos, S., Homem, V., and Santos, L. (2019). Simultaneous determination of synthetic musks and UV-filters in water matrices by dispersive liquid-liquid microextraction followed by gas chromatography tandem mass-spectrometry. *J. Chromatogr. A* 1590: 47–57. doi: 10.1016/j.chroma.2019.01.013.

52 Vallecillos, L., Pocurull, E., and Borrull, F. (2013). A simple and automated method to determine macrocyclic musk fragrances in sewage sludge samples by headspace solid-phase microextraction and gas chromatography-mass spectrometry. *J. Chromatogr. A* 1314: 38–43. doi: 10.1016/j.chroma.2013.09.033.

53 Castro, Ó., Trabalón, L., Schilling, B., Borrull, F., and Pocurull, E. (2019). Solid phase microextraction arrow for the determination of synthetic musk fragrances in fish samples. *J. Chromatogr. A* 1591: 55–61. doi: 10.1016/j.chroma.2019.01.032.

54 Silva, J.A., Ratola, N., Ramos, S., Homem, V., Santos, L., and Alves, A. (2015). An analytical multi-residue approach for the determination of semi-volatile organic pollutants in pine needles. *Anal. Chim. Acta* 858: 24–31. doi: 10.1016/j.aca.2014.12.042.

55 Wang, L., McDonald, J.A., and Khan, S.J. (2013). Enantiomeric analysis of polycyclic musks in water by chiral gas chromatography-tandem mass spectrometry. *J. Chromatogr. A* 1303: 66–75. doi: 10.1016/j.chroma.2013.06.006.

56 Maidatsi, K.V., Chatzimitakos, T.G., Sakkas, V.A., and Stalikas, C.D. (2015). Octyl-modified magnetic graphene as a sorbent for the extraction and simultaneous determination of fragrance allergens, musks, and phthalates in aqueous samples by gas chromatography with mass spectrometry. *J. Sep. Sci.* 38: 3758–3765. doi: 10.1002/jssc.201500578.

57 Arbulu, M., Sampedro, M.C., Unceta, N., Gómez-Caballero, A., Goicolea, M.A., and Barrio, R.J. (2011). A retention time locked gas chromatography-mass spectrometry method based on stir-bar sorptive extraction and thermal desorption for automated determination of synthetic musk fragrances in natural and wastewaters. *J. Chromatogr. A* 1218: 3048–3055. doi: 10.1016/j.chroma.2011.03.012.

58 Chiaia-Hernandez, A.C., Krauss, M., and Hollender, J. (2013). Screening of lake sediments for emerging contaminants by liquid chromatography atmospheric pressure photoionization and electrospray ionization coupled to high resolution mass spectrometry. *Environ. Sci. Technol.* 47: 976–986. doi: 10.1021/es303888v.

59 Wu, S.F. and Ding, W.H. (2010). Fast determination of synthetic polycyclic musks in sewage sludge and sediments by microwave-assisted headspace solid-phase microextraction and gas chromatography-mass spectrometry. *J. Chromatogr. A* 1217: 2776–2781. doi: 10.1016/j.chroma.2010.02.067.

60 Gao, S., Tian, B., Zeng, X., and Yu, Z. (2019). Enantiomeric analysis of polycyclic musks AHTN and HHCB and HHCB-lactone in sewage sludge by gas chromatography/tandem mass spectrometry. *Rapid Commun. Mass Spectrom.* 33: 607–612. doi: 10.1002/rcm.8390.

61 Iparraguirre, A., Prieto, A., Vallejo, A., Moeder, M., Zuloaga, O., Etxebarria, N., and Paschke, A. (2017). Tetraphasic polar organic chemical integrative sampler for the determination of a wide polarity range organic pollutants in water. The use of performance reference compounds and in-situ calibration. *Talanta* 164: 314–322. doi: 10.1016/j.talanta.2016.11.034.

62 Sun, P., Casteel, K., Dai, H., Wehmeyer, K.R., Kiel, B., and Federle, T. (2014). Distributions of polycyclic musk fragrance in wastewater treatment plant (WWTP) effluents and sludges in the United States. *Sci. Total Environ.* 493: 1073–1078. doi: 10.1016/j.scitotenv.2014.04.038.

9

Disinfection Byproducts in Water

Cristina Postigo[1], Joshua M. Allen[2], Amy A. Cuthbertson[3], María José Farré[4,5], and Susana Y. Kimura[6]

[1] *Water, Environmental and Food Chemistry Unit (ENFOCHEM), Department of Environmental Chemistry, Institute of Environmental Assessment and Water Research, (IDAEA-CSIC), Barcelona, Spain*
[2] *Department of Chemistry, High Point University, North Carolina, U.S.A.*
[3] *Department of Civil and Environmental Engineering, University of California, Berkeley, California, U.S.A.*
[4] *Catalan Institute for Water Research (ICRA), Girona, Spain*
[5] *University of Girona, 17071 Girona, Spain*
[6] *Department of Chemistry, University of Calgary, Calgary, Canada*

9.1 Overview of Main DBP Classes

Disinfection byproducts (DBPs) form after the reaction of natural and anthropogenic organic matter and other inorganic substances present in water with the disinfectants used to inactivate pathogens [1, 2]. To date, more than 700 DBPs have been identified in disinfected water. The use of more advanced analytical technologies (based on high-resolution mass spectrometry (HRMS)) has boosted DBP discovery in recent years [3]. Less than two-tenths of DBPs are currently regulated or included in drinking water guidelines, even though many of the newly discovered DBPs have been proven to be more toxic in *in vitro* studies than the regulated ones [4, 5]. This may be attributed to the lack of knowledge on their formation and occurrence in disinfected waters in most cases. To advance in this field and properly inform water management authorities and policy-makers, analytical methods that allow a reliable determination of DBPs in water are needed. One big limitation for DBP analysis is the commercial unavailability of analytical standards for newly discovered DBPs.

The chemical nature of DBPs is very wide. They can be classified according to the presence of halogen atoms (halogenated *vs.* non-halogenated DBPs) and/or nitrogen (N-DBPs *vs.* C-DBPs) in their structure, their volatility (volatile *vs.* non-volatile), or their specific chemical class (e.g., halomethanes, haloacids, haloamides, halonitriles, nitrosamines, etc.). Some of these DBPs are usually present at the μg/L level in disinfected waters, but others are only formed at the sub-ppb level. Although multi-residue methods for the simultaneous determination of several DBP classes in water exist, the wide variability in nature and levels of DBPs in water requires the design and optimization of specific

Analytical Methods for Environmental Contaminants of Emerging Concern, First Edition. Edited by Núria Fontanals and Rosa Maria Marcé.

methods for their reliable quantification. In this context, this chapter aims to provide insights into the most common DBP classes regarding their chemical properties, environmental occurrence, and the most suitable methodologies for their reliable determination in water. The chemical classes covered in this chapter include the halogenated DBP classes trihalomethanes (THMs), haloacetic acids (HAAs), haloacetaldehydes (HALs), halobenzoquinones (HBQs), haloacetonitriles (HANs), halonitromethanes (HNMs), haloacetamides (HAMs), and the non-halogenated DBP class nitrosamines.

9.1.1 Properties

For halogenated DBPs, the type and number of halogen substitutions in the molecule determine its chemical properties, which are used to separate and extract them from water. A summary of the main physical-chemical properties of the halogenated DBP classes covered are provided in Tables 9.1–9.7. Overall, hydrophobicity increases and solubility decreases with increasing number of halogens, while volatility increases with increasing number of chlorine atoms. Furthermore, those containing solely chlorine (Cl) are more polar, and soluble in water, than those containing bromine (Br) or iodine (I).

Chlorine and bromine-containing regulated trihalomethanes (THM4) were the first DBPs identified in chlorinated water [6, 7]. THMs are chemical compounds in which three of the four hydrogen atoms of the methane molecule (CH_4) are substituted by halogen atoms (Cl, Br, or I). They are low water-soluble substances characterized by not having ionizable atoms (Table 9.1).

Together with trihalomethanes, HAAs are the most abundant class of DBPs in chlorinated water. These substances are extremely polar and soluble in water (Table 9.2). Furthermore, their acidic properties (pK_a values between 0.7 and 3.1) hinder their extraction from water. Some of them, viz., dichloroacetic acid (DCAA), trichloroacetic acid (TCAA), dibromoacetic acid (DBAA), and bromochloroacetic acid (BCAA), are classified as possibly carcinogenic to humans (group 2B) [8].

The HAL class includes all halogenated derivatives of ethanal or acetaldehyde. Acetaldehyde is one of the most important naturally occurring aldehydes and is also produced on a large scale in the industry. Its halogenated derivatives are less polar and soluble in water than the original compound (log P of −0.38 and solubility of 303.73 mg/mL at pH of 7.4, according to ChemAxon Chemicalize). HALs react with water to form a hydrate known as a geminal-diol. HALs and their corresponding hydrate are in equilibrium, where the hydrated species is favored with increasing polarity (Cl > Br > I) and the number of halogens substituted at the alfa-carbon. For example, hydration constants for mono-substituted HALs are 2–4 orders magnitude lower than tri-substituted HALs (Table 9.3). Tri-substituted HALs will then exist in water primarily as their hydrated species [9].

The HBQ class includes all compounds consisting of a benzoquinone ring (a quinone with a single benzene ring) with its hydrogen atoms substituted by one or several halogens, and in some forms, by alkyl and/or hydroxyl groups (Table 9.4). Because of their benzoquinone ring, HBQs have unique redox properties that need to be taken into consideration for their analysis. HBQs may get sequentially reduced via one electron and two electrons to a semiquinone radical and hydroquinone, respectively. This redox property is reversible, and hydroquinone can get oxidized back to benzoquinone. In this

Table 9.1 Main physical-chemical properties of THMs.

Analyte CAS NR SMILEs	Abbrev.	Molecular structure	Molecular formula	Molar mass (g/mol)	Log P*	Water solubility* (mg/mL)	Vapor pressure (mmHg 25°C)°
Trichloromethane (chloroform) 67-66-3 ClC(Cl)Cl	TCM	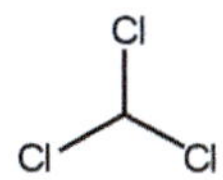	$CHCl_3$	119.37	1.83	2.57	200
Bromodichloromethane 75-27-4 ClC(Cl)Br	BDCM	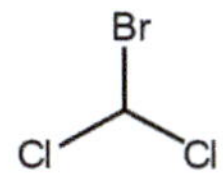	$CHBrCl_2$	163.82	1.981	2.23	65.3
Dibromochloromethane 124-48-1 ClC(Br)Br	DBCM	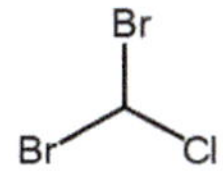	$CHBr_2Cl$	208.28	2.13	1.93	21.0
Tribromomethane (bromoform) 2909-52-6 BrC(Br)Br	TBM	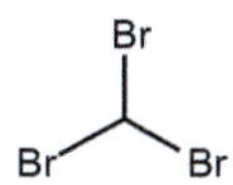	$CHBr_3$	252.73	2.28	1.70	5.2
Dichloroiodomethane 594-04-7 ClC(Cl)I	DCIM	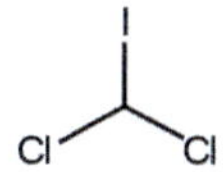	$CHCl_2I$	210.82	2.31	0.67	13.2
Dibromoiodomethane 593-94-2 BrC(Br)I	DBIM	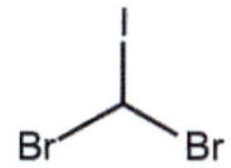	$CHBr_2I$	299.73	2.07	0.54	0.9
Bromochloroiodomethane 34970-00-8 ClC(Br)I	BCIM	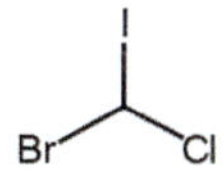	CHBrClI	255.28	2.19	0.59	3.6
Chlorodiiodomethane 638-73-3 ClC(I)I	CDIM	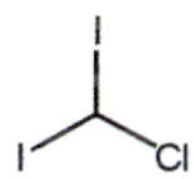	$CHClI_2$	302.28	4.14	0.19	0.7
Bromodiiodomethane 557-95-9 BrC(I)I	BDIM	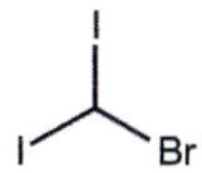	$CHBrI_2$	346.73	4.32	0.18	0.2
Triiodomethane (iodoform) 75-47-8 IC(I)I	TIM	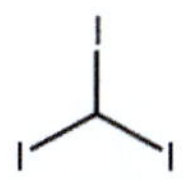	CHI_3	393.73	4.65	0.07	0.0

*Properties estimated with ChemAxon https://chemicalize.com/app. Octanol-water partition coefficient (log P) and water solubility values are provided at pH = 7.4.
°Predicted using ACD/Labs Percepta Platform – Physchem Module (source: ChemSpider).

Table 9.2 Main physical-chemical properties of the HAAs investigated to date in water.

Analyte CAS NR SMILEs	Abbrev.	Molecular structure	Molecular formula	Molar mass (g/mol)	Log P*	Water solubility* (mg/mL)	pK_a*	Vapor pressure (mmHg 25°C)°
2-chloroacetic acid 79-11-8 OC(=O)CCl	CAA	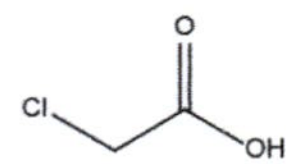	$C_2H_3ClO_2$	94.49	−3.15	8615.57	3.06	0.3
2-bromoacetic acid 79-08-3 OC(=O)CB	BAA	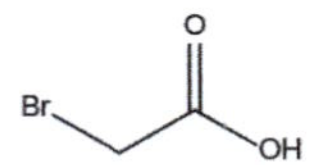	$C_2H_3BrO_2$	138.95	−3.01	7691.58	2.64	0.1
2-iodoacetic acid 64-69-7 OC(=O)CI	IAA	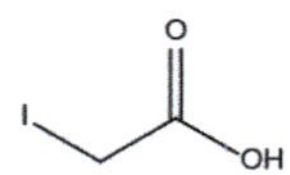	$C_2H_3IO_2$	185.95	−2.75	2285.45	3.05	0.0
2,2-dichloroacetic acid 79-43-6 OC(=O)C(Cl)Cl	DCAA	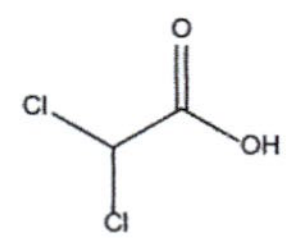	$C_2H_2Cl_2O_2$	128.94	−2.46	2521.34	2.30	0.2
2,2-dibromoacetic acid 631-64-1 OC(=O)C(Br)Br	DBAA	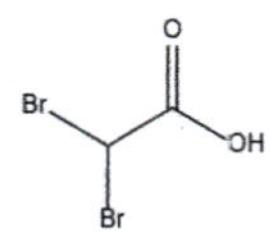	$C_2H_2Br_2O_2$	217.84	−2.89	1883.22	1.60	0.0
2-bromo,2-chloroacetic acid 5589-96-8 OC(=O)C(Cl)Br	BCAA	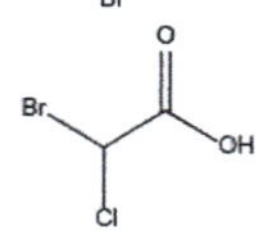	$C_2H_2BrClO_2$	157.39	−2.80	2175.34	1.97	0.1
2-chloro,2-iodoacetic acid 53715-09-6 OC(=O)C(Cl)I	CIAA	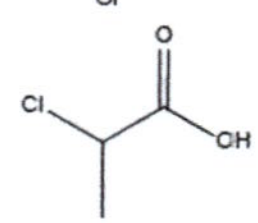	$C_2H_2ClIO_2$	220.39	−1.73	650.87	2.28	0.0

2-bromo,2-iodoacetic acid 71815-43-5 OC(=O)C(Br)I	BIAA	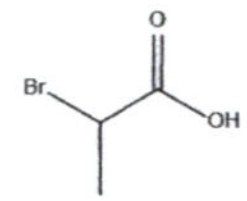	$C_2H_2BrIO_2$	264.84	−1.55	581.17	1.94	0.0
2,2-diiodoacetic acid 598-89-00 OC(=O)C(I)I	DIAA	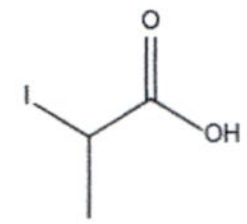	$C_2H_2I_2O_2$	311.85	−1.32	311.85	2.29	0.0
Trichloroacetic acid 76-03-9 OC(=O)C(Cl)(Cl)Cl	TCAA	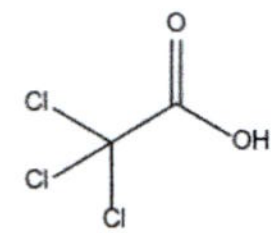	$C_2HCl_3O_2$	163.38	−2.00	190.97	1.72	0.2
Tribromooacetic acid 75-96-7 OC(=O)C(Br)(Br)Br	TBAA	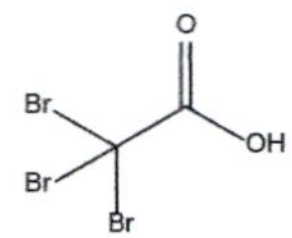	$C_2HBr_3O_2$	296.74	−1.55	296.74	0.71	0.0
2-bromo-2,2-dichloroacetic acid 71133-14-7 OC(=O)C(Cl)(Cl)Br	BDCAA	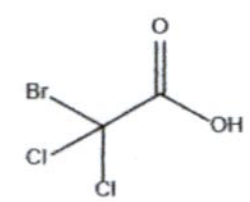	$C_2HBrCl_2O_2$	207.83	−1.85	207.83	1.45	0.1
2,2-dibromo-2-chloroacetic acid 5278-95-5 OC(=O)C(Cl)(Br)Br	DBCAA	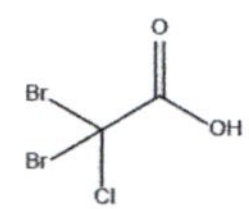	$C_2HBr_2ClO_2$	252.29	−1.70	252.29	1.12	0.1

*Properties estimated with ChemAxon https://chemicalize.com/app. Octanol-water partition coefficient (log P) and water solubility values are provided at pH = 7.4.
°Predicted using ACD/Labs Percepta Platform –Physchem Module (source: ChemSpider).

Table 9.3 Main physical-chemical properties of the HALs investigated to date in water.

Analyte CAS NR SMILEs	Abbrev.	Molecular structure	Molecular formula	Molar mass (g/mol)	Log P*	Water solubility* (mg/mL)	pK_a*	Vapor pressure (mmHg 25°C)°	$K_{Hydration}$ Δ
2-chloro-acetaldehyde 107-20-0 [H]C(=O)CCl	CAL	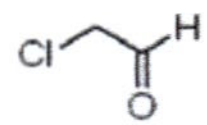	C_2H_3ClO	78.50	0.16	112.66	14.07	70.6	17.8
2-bromo-acetaldehyde 17157-48-1 [H]C(=O)CBr	BAL	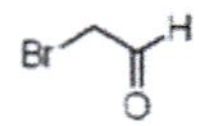	C_2H_3BrO	122.95	0.34	104.41	14.76	48.8	11.0
2-iodo-acetaldehyde 57782-51-62 [H]C(=O)CI	IAL	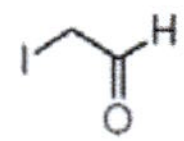	C_2H_3IO	169.95	0.56	31.19	14.89	5.2	4.37
2,2-dichloro-acetaldehyde 79-02-7 [H]C(=O)C(Cl)Cl	DCAL	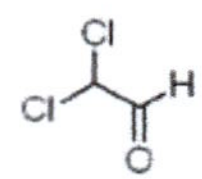	$C_2H_2Cl_2O$	112.94	0.90	34.08	11.95	56.0	1950
2,2-dibromo-acetaldehyde 3039-13-2 [H]C(=O)C(Br)Br	DBAL	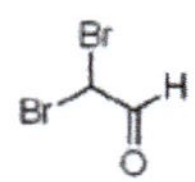	$C_2H_2Br_2O$	201.85	0.43	25.58	13.19	7.4	1580
2-bromo,2-chloro-acetaldehyde 98136-99-3 [H]C(=O)C(Cl)Br	BCAL	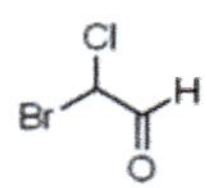	C_2H_2BrClO	157.39	0.52	29.70	12.57	25.1	1700

2,2,2-trichloro-acetaldehyde 75-87-6 [H]C(=O)C(Cl)(Cl)Cl	TCAL	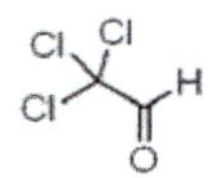	C_2HCl_3O	147.38	1.37	2.61	n.i.	41.1	32400
2,2,2-tribromo-acetaldehyde 115-17-3 [H]C(=O)C(Br)(Br)Br	TBAL	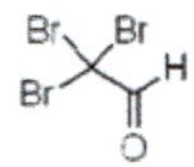	C_2HBr_3O	280.74	1.82	1.77	n.i.	1.2	11500
2-bromo-2,2-dichloro-acetaldehyde 34619-29-9 [H]C(=O)C(Cl)(Cl)Br	BDCAL	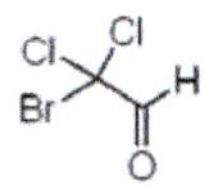	C_2HBrCl_2O	191.83	1.52	2.24	n.i.	30.4	43700
2,2-dibromo-2-chloro-acetaldehyde 64316-11-6 [H]C(=O)C(Cl)(Br)Br	DBCAL	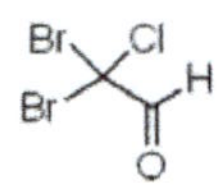	C_2HBr_2ClO	236.29	1.67	1.96	n.i.	10.0	20000

*Properties estimated with ChemAxon https://chemicalize.com/app. Octanol-water partition coefficient (log P) and water solubility values are provided at pH = 7.4.
°Predicted using ACD/Labs Percepta Platform –Physchem Module (source: ChemSpider).
$^{\Delta}$Hydration constant extracted from [9].n.i. not ionizable atoms found

Table 9.4 Main physical-chemical properties of the HBQs investigated to date in water.

Analyte CAS NR SMILEs	Abbrev.	Molecular structure	Molecular formula	Molar mass (g/mol)	Log P*	Water solubility* (mg/mL)	pK_a*	Vapor pressure (mmHg 25°C)°
2,5-dichloro-1,4-benzoquinone 615-93-0 ClC1 = CC(=O)C(Cl) = CC1 = O	25DCBQ	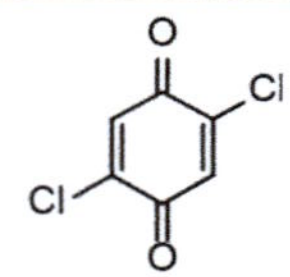	$C_6H_2Cl_2O_2$	176.98	1.80	0.245	6.3**	0.0
2,6-dichloro-1,4-benzoquinone 697-91-6 ClC1 = CC(=O)C = C(Cl)C1 = O	26DCBQ	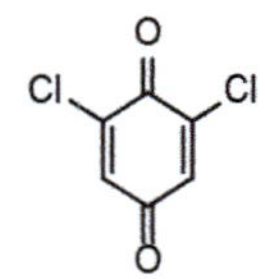	$C_6H_2Cl_2O_2$	176.98	1.80	0.245	6.3**	0.0
2,5-dibromo-1,4-benzoquinone 1633-14-3 BrC1 = CC(=O) C(Br) = CC1 = O	25DBBQ	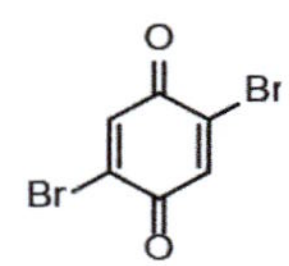	$C_6H_2Br_2O_2$	265.89	2.13	0.19	6.0**	0.0
2,6-dibromo-1,4-benzoquinone 19643-45-9 BrC1 = CC(=O)C = C(Br) C1 = O	26DBBQ	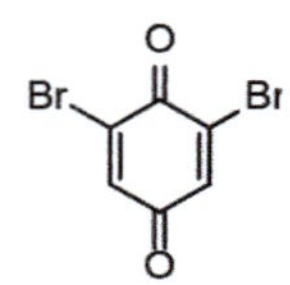	$C_6H_2Br_2O_2$	265.89	2.13	0.19	6.0**	0.0

2,3-diiodo-1,4-benzoquinone 222417-90-5 IC1 = C(I)C(=O)C = CC1 = O	23DIBQ	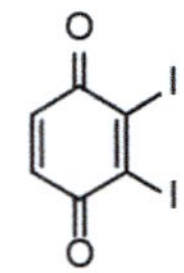	$C_6H_2I_2O_2$	359.89	2.45	0.021	na	
2,3,5-trichloro-1,4-benzoquinone 634-85-5 ClC1 = CC(=O)C(Cl) = C(Cl)C1 = O	235TCBQ	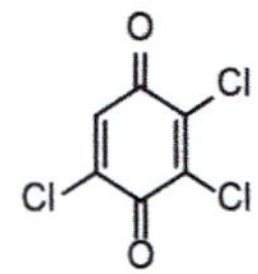	$C_6HCl_3O_2$	211.42	2.19	0.083	5.7**	0.0
3,4,5,6-tetrachloro-1,2-benzoquinone 2435-53-2 ClC1 = C(Cl)C(=O)C(=O) C(Cl) = C1Cl	TetraC-12 BQ	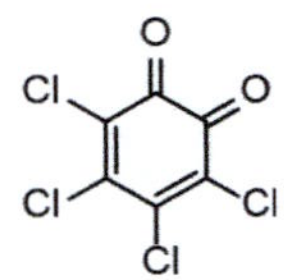	$C_6Cl_4O_2$	245.86	2.58	0.029	na	1.9
2,3,5,6-tetrachloro-1,4-benzoquinone 142655-99-0 ClC1 = C(Cl)C(=O)C(Cl) = C(Cl) C1 = O	TetraC-BQ	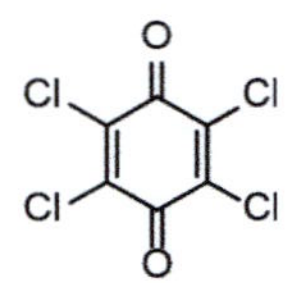	$C_6Cl_4O_2$	245.86	2.58	0.029	na	0.0
3,4,5,6-tetrabromo-1,2-benzoquinone 2435-54-3 BrC1 = C(Br)C(=O)C(=O) C(Br) = C1Br	TetraB-12 BQ		$C_6Br_4O_2$	423.68	3.24	0.027	na	0.4

(*Continued*)

Table 9.4 (Continued)

Analyte CAS NR SMILEs	Abbrev.	Molecular structure	Molecular formula	Molar mass (g/mol)	Log P*	Water solubility* (mg/mL)	pK$_a$*	Vapor pressure (mmHg 25°C)°
2,3,5,6-tetrachloro-1,4-benzoquinone 488-48-2 BrC1 = C(Br)C(=O)C(Br) = C(Br) C1 = O	TetraB-BQ		$C_6Br_4O_2$	423.68	3.24	0.027	na	0.0
2,6-dichloro-3-methyl-1,4-benzoquinone 40100-98-9 CC1 = C(Cl)C(=O)C(Cl) = CC1 = O	26DC3MBQ	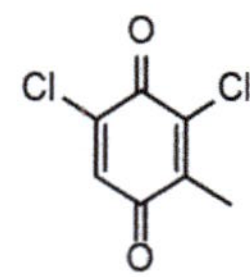	$C_7H_4Cl_2O_2$	191.01	2.19	0.275	5.9**	n.a.
2,6-dicbromo-3-chloro-5-methyl-1,4-benzoquinone 351005-10-2 CC1 = C(Br)C(=O)C(Br) = C(Cl) C1 = O	26DB3C5MBQ	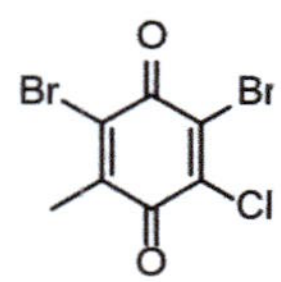	$C_7H_3Br_2ClO_2$	314.36	2.91	0.079	na	0.0
2,3-dicbromo-5,6-dimethyl-methyl-1,4-benzoquinone 38969-08-3 CC1 = C(C)C(=O)C(Br) = C(Br) C1 = O	23DB56DMBQ	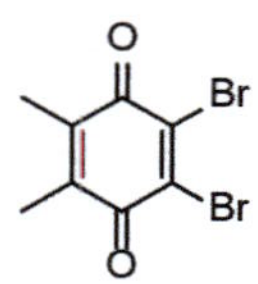	$C_8H_6Br_2O_2$	293.94	2.92	0.25	na	0.0

2,6-dichloro-3-hydroxy-1,4-benzoquinone CAS NR n.a OC1 = C(Cl)C(=O)C(Cl) = CC1 = O	OH-DCBQ	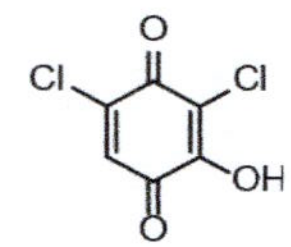	$C_6H_2Cl_2O_3$	192.98	−1.92	192.9	3.52	n.a.
2,6-dibromo-3-hydroxy-1,4-benzoquinone CAS NR n.a. OC1 = C(Br)C(=O)C(Br) = CC1 = O	OH-DBBQ	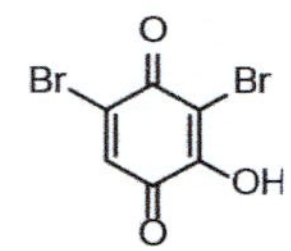	$C_6H_2Br_2O_3$	281.89	−1.60	281.9	3.49	n.a.
2,3,5-trichloro-6-hydroxy-1,4-benzoquinone 877-13-4 OC1 = C(Cl)C(=O)C(Cl) = C(Cl)C1 = O	OH-TCBQ	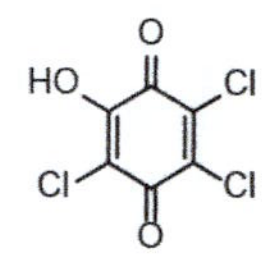	$C_6HCl_3O_3$	227.42	−1.79	227.4	2.65	0.0
2,6-dichloro-3-methyl-5-hydroxy-1,4-benzoquinone CAS NR n.a CC1 = C(Cl)C(=O)C(Cl) = C(O)C1 = O	OH-DCMBQ	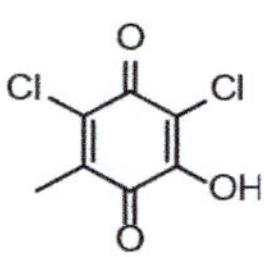	$C_7H_4Cl_2O_3$	207.01	−1.47	207.0	3.62	n.a.

*Properties estimated with ChemAxon https://chemicalize.com/app. Octanol-water partition coefficient (log P) and water solubility values are provided at pH = 7.°Predicted using ACD/Labs Percepta Platform –Physchem Module (source: ChemSpider).**Extracted from [145].na. Not available redox process, HBQs may acquire up to six redox states (Figure 9.1) [10]. Furthermore, HBQs may also undergo the Michael addition, this is the 1,4-addition of a doubly stabilized carbon nucleophile to an α,β-unsaturated carbonyl compound. For instance, the addition of oxygen nucleophiles such as oxygen (O_2) or hydrogen peroxide (H_2O_2) to the –C = C– bond would result in the formation of HBQ epoxides. On the contrary, the reductive addition of sulfur nucleophile would result in a thioether derivative.

The N-containing halogenated DBP classes covered include HANs, HAMs, and HNMs. The HANs is a DBP chemical class that contains a nitrile group (CN) and up to three substituted halogen atoms (Cl, Br, and/or I) attached to the α-carbon. HANs are semi-volatile compounds with limited water solubility (Table 9.5). The HAMs were first identified as DBPs in drinking water in the United States [1]. These compounds are highly hydrophilic, and water-soluble with low volatility (Table 9.6). The HNMs are polar halogenated derivatives of nitromethane and are a highly toxic emerging class of nitrogen-containing DBPs (Table 9.7).

N-nitrosamines are carcinogenic compounds that can be formed during disinfection processes as byproducts in drinking and recycled water systems. Among them, *N*-nitrosodimethylamine (NDMA) is of particular interest, especially in systems employing chloramines (or chlorine in the presence of ammonia), because it is among the most widely detected nitrosamine [11]. Apart from NDMA other relevant *N*-nitrosamines can be present in drinking water (Table 9.8). In particular, *N*-nitrosodiethylamine (NDEA) specifically is around 10 times more genotoxic than NDMA [12].

9.1.2 Legislation

The sum of chloroform (TCM), bromodichloromethane (BDCM), dibromochloromethane (DBCM), and bromoform (TBM), known as THM4, has a parametric value in drinking water of 80 μg/L in the US [13] and 100 μg/L in European countries [14]. This parameter is often used as a DBP exposure indicator and therefore, the DBP regulatory limits adopted in most countries focus on THM4. Although iodine-containing trihalomethanes (I-THMs) have been proved to be more toxic than Cl and Br-containing THMs, they have not been regulated yet.

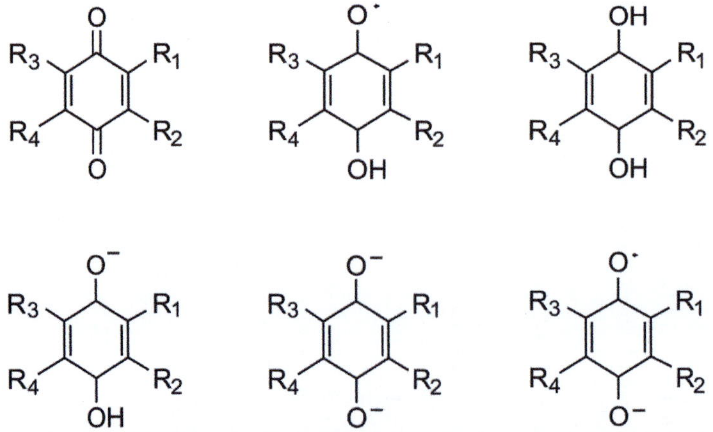

Figure 9.1 The redox states of halobenzoquinones.

Table 9.5 Main physical-chemical properties of HANs investigated to date in water

Analyte CAS NR SMILEs	Abbrev.	Molecular structure	Molecular formula	Molar mass (g/mol)	Log P*	Water solubility* (mg/mL)	Vapor pressure (mmHg 25°C)*
Chloroacetonitrile 107-14-2 N#CCCl	CAN	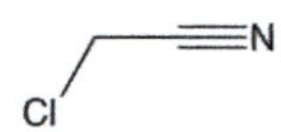	$ClCH_2CN$	75.50	0.18	35	11.6
Bromoacetonitrile 590-17-0 N#CCBr	BAN	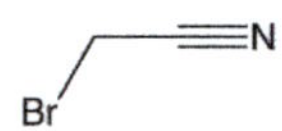	$BrCH_2CN$	119.95	0.51	29	3.18
Iodoacetonitrile 624-75-9 N#CCI	IAN	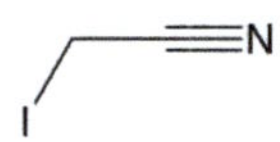	ICH_2CN	166.95	0.79	32	0.71
Dichloroacetonitrile 3018-12-0 N#CC(Cl)Cl	DCAN	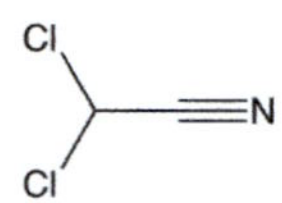	Cl_2CHCN	109.94	0.94	13	21.7
Bromochloroacetonitrile 83463-62-1 N#CC(Br)Cl	BCAN	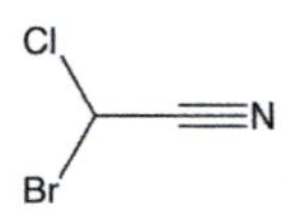	ClBrCHCN	154.39	1.22	9.7	14.8
Dibromoacetonitrile 3252-43-5 N#CC(Br)Br	DBAN	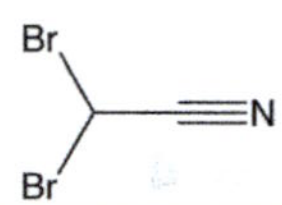	Br_2CHCN	198.84	1.57	6.2	2.10

(Continued)

Table 9.5 (Continued)

Analyte CAS NR SMILEs	Abbrev.	Molecular structure	Molecular formula	Molar mass (g/mol)	Log P*	Water solubility* (mg/mL)	Vapor pressure (mmHg 25°C)*
Trichloroacetonitrile 545-06-2 N#CC(Cl)(Cl)Cl	TCAN	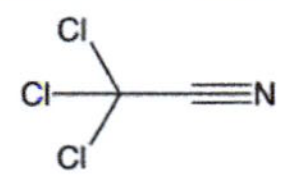	Cl_3CHCN	144.39	2.15	3.3	68.6
Bromodichloroacetonitrile 60523-73-1 N#CC(Br)(Cl)Cl	BDCAN	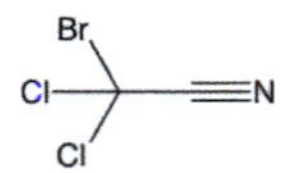	$BrCl_2CHCN$	188.84	2.39	2.3	25.6
Dibromochloroacetonitrile 144772-39-4 N#CC(Br)(Br)Cl	DBCAN	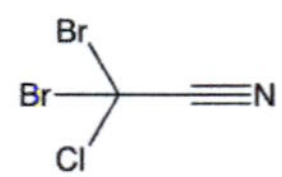	$Br_2ClCHCN$	233.29	2.69	1.5	16.8
Tribromoacetonitrile 75519-19-6 N#CC(Br)(Br)Br	TBAN	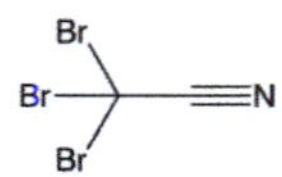	Br_3CHCN	277.74	3.06	0.86	10.0

°Predicted using ACD/Labs Percepta Platform – Physchem Module (source: SciFinder) at pH of 7 at 25°C.

Table 9.6 Main physical-chemical properties of HAMs investigated to date in water.

Analyte CAS NR SMILEs	Abbrev.	Molecular structure	Molecular formula	Molar mass (g/mol)	Log P*	Water solubility* (mg/mL)	Vapor pressure (mmHg 25°C)°
2-Chloroacetamide 79-07-2 ClCC(=O)N	CAM	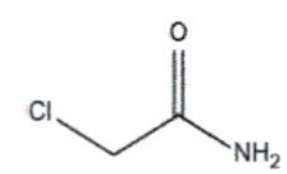	C_2H_4ClNO	93.51	−0.51	71	0.0158
2-Bromoacetamide 683-57-8 BrCC(=O)N	BAM	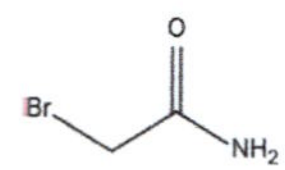	C_2H_4BrNO	137.96	−0.52	70	0.00638
2-Iodoacetamide 144-48-9 ICC(=O)N	IAM	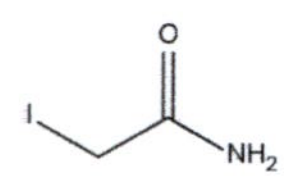	C_2H_4INO	184.96	−0.19	70	0.00138
2,2-Dichloroacetamide 2909-52-6 ClC(Cl)C(=O)N	DCAM	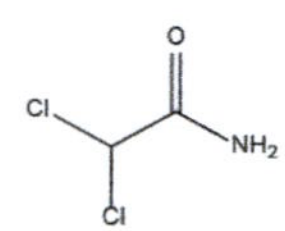	$C_2H_3Cl_2NO$	127.96	0.07	29	0.0532
2-Bromo-2-chloroacetamide 62872-34-8 ClC(Br)C(=O)N	BCAM	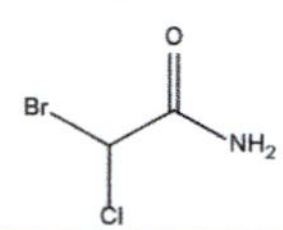	$C_2H_3BrClNO$	172.41	0.03	28	0.0116

(Continued)

Table 9.6 (Continued)

Analyte CAS NR SMILEs	Abbrev.	Molecular structure	Molecular formula	Molar mass (g/mol)	Log P*	Water solubility* (mg/mL)	Vapor pressure (mmHg 25°C)°
2,2-Dibromoacetamide 598-70-9 BrC(Br)C(=O)N	DBAM	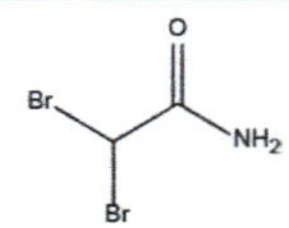	$C_2H_3Br_2NO$	216.86	0.04	22	0.00457
2-Chloro-2-iodoacetamide 62872-35-9 ClC(I)C(=O)N	CIAM	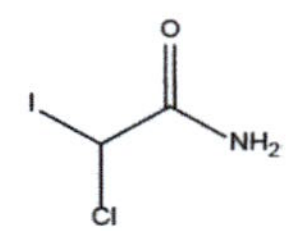	C_2H_3ClINO	219.409	0.41	15	0.00328
2-Bromo-2-iodoacetamide 62872-36-0 BrC(I)C(=O)N	BIAM	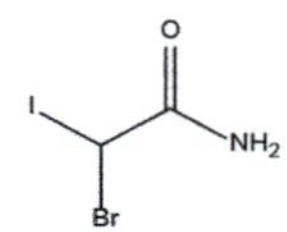	C_2H_3BrINO	263.86	0.5	7.4	0.00137
2,2,2-Diiodoacetamide 5875-23-0 IC(I)C(=O)N	DIAM	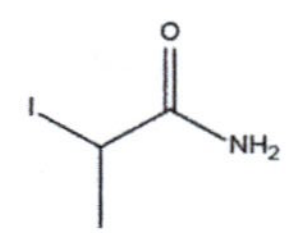	$C_2H_3I_2NO$	310.86	0.62	19	0.000187

2,2,2-Trichloroacetamide 594-65-0 ClC(Cl)(Cl)C(=O)N	TCAM	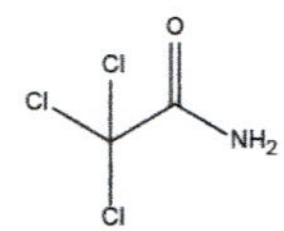	$C_2H_2Cl_3NO$	162.4	1.11	8.1	0.0411
2-Bromo-2,2-dichloroacetamide 98137-00-9 BrC(Cl)(Cl)C(=O)N	BDCAM	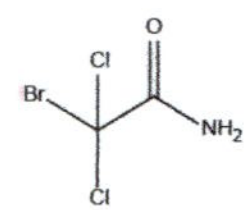	$C_2H_2BrCl_2NO$	206.85	1.02	7.2	0.0555
2,2-Dibromo-2-chloroacetamide 855878-13-6 BrC(Br)(Cl)C(=O)N	DBCAM	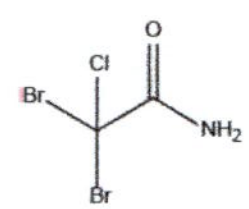	$C_2H_2Br_2ClNO$	251.30	1.00	6.0	0.0262
2,2,2-Tribromoacetamide 594-47-8 BrC(Br)(Br)C(=O)N	TBAM	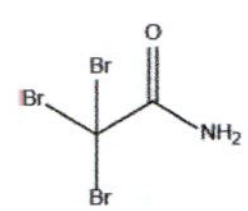	$C_2H_2Br_3NO$	295.76	1.03	4.4	0.0105

*Properties estimated with ChemAxon https://chemicalize.com/app. Octanol-water partition coefficient (log P) and water solubility values are provided at pH = 7.4.
°Predicted using ACD/LAbs Percepta Platform – Physchem Module (source: ChemSpider).

Table 9.7 Main physical-chemical properties of the HNMs investigated to date in water.

Analyte CAS NR SMILEs	Abbrev.	Molecular structure	Molecular formula	Molar mass (g/mol)	Log P*	Water solubility* (mg/mL)	pK_a*	Vapor pressure (mmHg 25°C)°
Chloronitromethane 1794-84-9 [O-][N +](=O)CCl	CNM	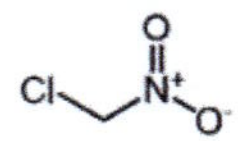	CH_2ClNO_2	95.48	0.56	22.43	8.09	13.9
Bromonitromethane 563-70-2 [O-][N +](=O)CBr	BNM	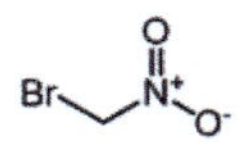	CH_2BrNO_2	139.94	0.63	19.18	8.20	4.4
Dichloronitromethane 7119-89-3 [O-][N +](=O)C(Cl)Cl	DCNM	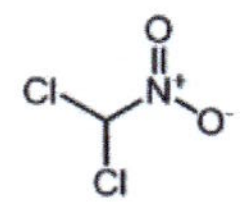	$CHCl_2NO_2$	129.92	0.92	14.40	7.13	27.6
Dibromonitromethane 598-91-4 [O-][N +](=O)C(Br)Br	DBNM	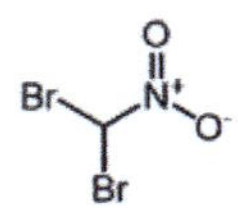	$CHBr_2NO_2$	218.83	1.38	7.44	7.41	3.5
Bromochloronitromethane 135531-25-8 [O-][N +](=O)C(Cl)Br	BCNM	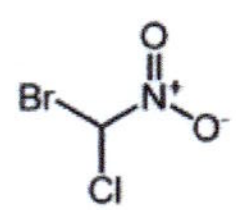	$CHBrClNO_2$	174.38	1.15	10.19	7.28	8.8

Trichloronitromethane (Chloropicrin) 76-06-2 [O-][N +](=O)C(Cl)(Cl)Cl	TCNM	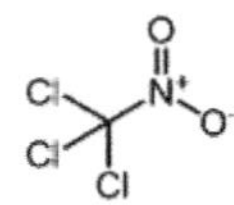	CCl_3NO_2	164.37	2.48	0.38	n.i.	22.2
Bromodichloronitro-methane 918-01-4 [O-][N +](=O)C(Cl)(Cl)Br	BDCNM	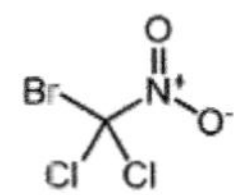	$CBrCl_2NO_2$	208.82	2.63	0.33	n.i.	19.0
Dibromochloronitro-methane 1184-89-0 [O-][N +](=O)C(Cl)(Br)Br	DBCNM	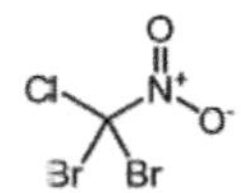	CBr_2ClNO_2	253.27	2.78	0.29	n.i.	7.9
Tribromonitromethane (Bromopicrin) 464-10-8 [O-][N +](=O)C(Br)(Br)Br	TBNM	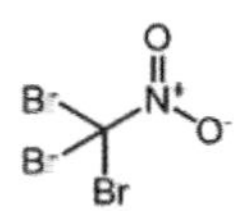	CBr_3NO_2	297.73	2.93	0.27	n.i.	3.0

*Properties estimated with ChemAxon https://chemicalize.com/app. Octanol-water partition coefficient (log P) and water solubility values are provided at pH = 7.4.
°Predicted using ACD/Labs Percepta Platform –Physchem Module (source: ChemSpider).n.i. not ionizable atoms found

Table 9.8 Main physical-chemical properties of nitrosamines detected in water.

Analyte CAS NR SMILEs	Abbrev.	Molecular structure	Molecular formula	Molar mass (g/mol)	Log P*	Water solubility (mg/mL)*	Vapor pressure (mmHg 25°C)
N-nitrosodimethylamine 62-75-9 CN(C)N = O	NDMA	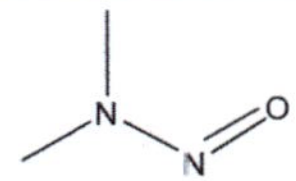	$C_2H_6N_2O$	74.08	0.039	83.60	4.6 ± 0.2
N-Nitrosodiethylamine 55-18-5 CCN(CC)N = O	NDEA	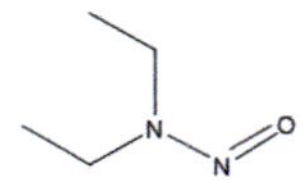	$C_4H_{10}N_2O$	102.14	0.752	20.92	1.7 ± 0.3
N-Nitrosodi-n-propylamine 621-64-7 CCCN(CCC)N = O	NDPA	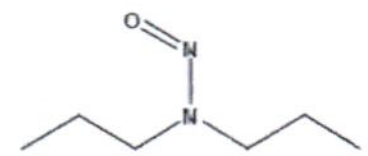	$C_6H_{14}N_2O$	130.19	1.797	3.93	68.3 ± 0.2
N-Nitrosodi-n-butylamine 924-16-3 CCCCN(CCCC)N = O	NDBA	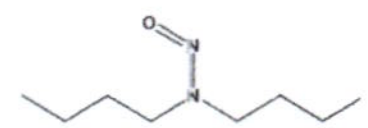	$C_8H_{18}N_2O$	158.25	2.686	0.41	0.0 ± 0.5
N-Nitrosomethylethylamine N/A CCNCN = O	NMEA	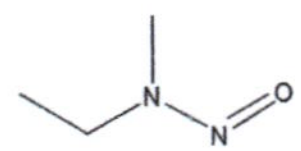	$C_3H_8N_2O$	88.11	0.062	82.40	4.1 ± 0.3
N-Nitrosopiperidine 68374-62-9 O = NN1CCCCC1	NPIP	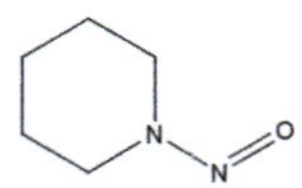	$C_5H_{10}N_2O$	114.15	0.889	9.79	0.1 ± 0.4
N-Nitrosopyrrolidine 930-55-2 O = NN1CCCC1	NPYR	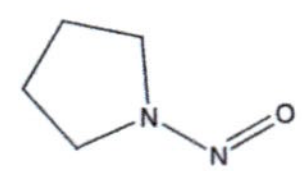	$C_4H_8N_2O$	100.12	0.444	17.67	0.2 ± 0.4
N-Nitrosomorpholine 59-89-2 O = NN1CCOCC1	NMOR	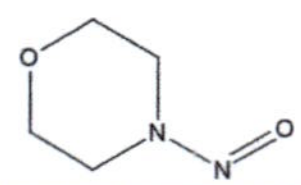	$C_4H_8N_2O_2$	116.12	−0.18	31.60	0.1 ± 0.4

*Properties estimated with ChemAxon https://chemicalize.com/app. Octanol-water partition coefficient (log P) and water solubility values are provided at pH = 7.4.
°Predicted using ACD/Labs Percepta Platform –Physchem Module (source: ChemSpider).

The sum of the concentrations of monochloroacetic acid (CAA), monobromoacetic acid (BAA), DCAA, DBAA, and TCAA, known as HAA5, has a parametric value in drinking water of 60 μg/L in the US [13]. Furthermore, the recent proposal for the revision of the old European drinking water Directive [14] will also enforce this parametric value in the EU countries shortly.

Regulatory and guideline values for THMs and HAAs are summarized in Tables 9.9 and 9.10, respectively. For instance, according to Chinese regulations, a maximum concentration of 100 μg/L of TCM has been established in drinking water in the whole territory of China, and maximum values for THM4 (1 mg/L), BDCM (60 μg/L), DBCM (100 μg/L), and TBM (100 μg/L) should be adopted in specific regions or under certain scenarios [15]. The World Health Organization (WHO) has issued guideline values for individual THM4 (TCM (300 μg/L), BDCM (60 μg/L), DBCM (100 μg/L), and TBM (100 μg/L)) and HAAs (CAA (20 μg/L), DCAA (50 μg/L), TCAA (200 μg/L)) in drinking water [16]. Similar values have been adopted by different countries worldwide.

As for HALs, a guideline value of 10 μg/L of trichloroacetaldehyde (TCAL) in drinking water was set by the WHO in 2004 [17]. However, since it commonly occurs at concentrations well below those of health concern, no guideline value is currently established for this compound [16]. A maximum concentration of 10 μg/L of TCAL has been established in drinking water of specific regions and under specific scenarios in China [15]. A guideline value of 100 μg/L has been set for TCAL in Australia [18]. Apart from these, no other regulations or guideline values exist for the other HALs.

There are currently no drinking water regulatory limits or guideline values for HBQs, HAMs, and HNMs. Although HANs are also not regulated in drinking water, there is some concern about their potential human health risks [19]. For this reason, the WHO has

Table 9.9 Parametric (*) and guideline values established for THMs in drinking water in μg/L.

	THM4	TCM	BDCM	DBCM	TBM	Ref
United States *	80					[13]
Europe *	100					[14]
Canada	100		16			[193]
WHO	1§	300	60	100	100	[16]
China*	1§	100	60	100	100	[15]
India*		200	60	100	100	[194]
Mexico*	200					[195]
Egypt	100	300	100		100	[196]
Japan*	100	60	30	100	90	[197]
Australia	250					[18]
New Zealand*	1§	400	60	150	100	[198]

§Maximum result after the sum of the ratio of the concentration of each THM4 (Ci) to its corresponding guideline value (GVi) (Maximum result after the sum of the ratio of the concentration of each THM4 (Ci) to its corresponding guideline value (GVi) (Ci/GVi, where i = TCM, BDCM, DBCM, TBM).Ci/GVi, where i = TCM, BDCM, DBCM, TBM).

Table 9.10 Parametric (*) and guideline values established for HAAs in drinking water in μg/L.

	HAA5	CAA	DAA	TCAA	Ref
United States *	60				[13]
Europe *	60				[14]
Canada	80				[193]
WHO		20	50	200	[16]
China*			50	100	[15]
Japan*		20	30	30	[197]
Australia		150	100	100	[18]
New Zealand*		20	50	200	[198]

provided guideline values for two HANs: a guideline value of 70 μg/L for dibromoacetonitrile (DBAN) and a provisional guideline value of 20 μg/L for dichloroacetonitrile (DCAN) [16].

As for nitrosamines, NDMA is included in the Fourth Contaminant Candidate List of the United States Environmental Protection Agency (USEPA) [20]. California's Department of Public Health has established 10 ng/L as a notification level for NDMA in drinking water [21], while California's Office of Environmental Health Hazard Assessment [22] has defined 3 ng/L as a public health goal. Massachusetts has a guideline value of 10 ng/L [23], Health Canada [24] has set a 40 ng/L national guideline value, whereas the Ontario Ministry of the Environment [25] has a maximum allowable concentration for NDMA of 9 ng/L in drinking water. The WHO guideline was established at 100 ng/L for NDMA in drinking water [16], which was adopted as a target value by Japan and Australia. Apart from NDMA, the USEPA has included 4 other *N*-nitrosamines, namely *N*-nitrosodiethylamine (NDEA), *N*-Nitrosodiphenylamine (NDPhA), *N*-nitrosodipropylamine (NDPA), and *N*-nitrosopyrrolidine (NPYR) in the Fourth Contaminant Candidate List [20].

THMs have been also regulated in swimming pool water, although in very few countries. Germany has set a maximum value of 20 μg/L for THM4 (calculated as the main THM4 species formed, i.e., usually TCM in chlorine-disinfected swimming pools), while Denmark has established a maximum value for THM4 of 25 μg/L in indoor swimming pools or 50 μg/L in outdoor swimming pools and spas [26]. Other countries in Europe, namely France, United Kingdom, and Finland, have set a maximum concentration of THM4 of 100 μg/L [27].

9.1.3 Potential Metabolites and/or Transformation Products

Hydrolysis and photochemical oxidation have been recognized as the main removal mechanism for halomethanes in natural waters; however, the THM half-lives under these reactions are of the order of years [28]. THM hydrolysis rates are significantly enhanced when the water temperature is raised to its boiling point [29]. THMs also degrade under anaerobic conditions [28], while brominated THMs are effectively degraded by UV irradiation at 254 nm, resulting in the formation of bromide and chloride ions [30].

As for HAAs, bromine and iodine-containing trihaloacetic acids decompose via decarboxylation in water to form the corresponding THMs. The decomposition rates are influenced by the electronegativity of the halide group, increasing with the increasing number of bromine and iodine atoms in the molecule [31]. Dehalogenation of highly substituted HAAs to produce less halogenated HAAs can also occur. The rate of dehalogenation usually increases with increasing pH and with decreasing halogen electronegativity [32]. Dihaloacetic acids are very stable in water [33].

Like HAAs, HALs may undergo decarboxylation to form halomethanes during hydrolysis. The HAL hydrolysis rate increases with increasing water pH values and is higher for bromine- and iodine-containing HALs than for chloro-HALs [34, 35]. The dehalogenation of trihalogenated HALs in the presence of sulfite over a range of pHs has also been reported [35].

HBQs in water may undergo redox, photochemical, and nucleophilic reactions, resulting in the formation of semiquinone radicals, haloquinones (HQs), benzene-1,2,4-triols, and hydroxyl-quinones as final products. HBQ concentrations are observed to decrease as their corresponding hydroxyl analogs (OH-HBQs) increase with the distance to the drinking water treatment plant (DWTP) in the distribution systems [36]. OH-HBQs are more stable at neutral pH than HBQs.

HANs are unstable compounds that can undergo two types of reactions in the environment: a nucleophilic attack (i.e., water) on the nitrile carbon and/or removal of a hydrogen at the α-position by a base to produce a resonance-stabilized species [37]. For the first reaction, water can hydrolyze HANs to form HAAs with HAMs as an intermediate [38]. Hydrolysis of HANs has been observed to be faster under alkaline conditions and with increasing halogen substitution at the α-position. On the contrary, HANs have been shown to be stable under acidic conditions.

During hydrolysis, HAMs mainly decompose to form the corresponding HAAs via a nucleophilic reaction between the carbonyl carbon and hydroxide [39]. Due to the decreasing electron-withdrawing inductive effects from chorine to iodine, chlorinated HAMs are more unstable than their brominated analogs, followed by iodinated HAMs. During chlorination reactions, HAMs react with hypochlorite to form *N*-chloro-HAMs, which further degrade to form HAAs via hypochlorous acid addition [39]. HAM hydrolysis and chlorination rates increase with increasing pH and the number of halogens substituted on the methyl group. HAMs decomposition, when exposed to hypochlorite, is two and nine orders of magnitude faster than HAM basic and neutral hydrolysis reactions, respectively. In contrast, HAMs reactions with chloramines were found to be insignificant [39]. Dehalogenation of highly substituted HAMs to form less halogenated HAMs can also occur in the presence of sulfite and zero-valent iron [40, 41].

One study discovered that trichloronitromethane, also known as chloropicrin, (TCNM) could undergo dehalogenation in the presence of pipe corrosion solids within drinking water distribution systems to form dichloronitromethane (DCNM) and chloronitromethane (CNM) [42]. Fang et al. observed that monohalogenated and dihalogenated HNMs, and TCNM to a lesser extent, undergo UV photolysis at neutral and alkaline pH values to form halides and nitrite [43]. HNMs also degrade through solar photolysis [44] and in the presence of hydroxyl radicals ($HO^{\bullet}$) and hydrated electrons [45].

NDMA may transform under UV photolysis to dimethylamine, methylamine, nitrite, nitrate, and formate. A half-life between 8 and 38 hours was estimated in aquatic systems over 1 m depth and for relatively low levels of solar irradiation. Other nitrosamines (NMEA, NDEA, NDPA; NDBA, NPYR, NPIP) showed photolysis rates in the laboratory similar to NDMA, resulting in half-lives less than 20 minutes for the irradiation conditions tested. Overall, nitrosamine quantum yields are 10–100 times larger than that of some other photosensitive environmental contaminants, like naproxen or nonylphenol [46].

9.1.4 Occurrence in the Environment

THM4 and HAA5 formation is enhanced in the presence of chlorine compared to chloramine. These DBPs are the most abundant DBPs in chlorinated drinking water. Although THM4 prevail, HAAs may form at equal or even greater concentrations than THM4 in specific scenarios, occurring in any case at the μg/L range. On the contrary, I-THMs and iodine-containing haloacetic acids (I-HAAs) are usually formed at sub-ppb levels [47, 48], with dichloroiodomethane and bromochloroiodomethane being the main I-THMs species formed in chloraminated waters, with maximum concentrations of 5.4 and 5.7 μg/L, respectively [48]. While the formation of iodo-DBPs requires the presence of iodide or an iodine source in the source water [49], THM4 and HAA9 speciation is highly dependent on the presence of bromide in source waters, being more abundant bromine-substituted THMs than TCM and bromine-substituted HAAs than DCAA and TCAA in bromide-containing waters [50, 51]. BCAA, DCAA, and TCAA were continuously measured in tap water in the US at maximum concentrations of 11 μg/L, 30 μg/L, and 36 μg/L, respectively [52]. The concentrations of 11 HAAs in the tap water of Barcelona and its distribution system ranged between 0.4 and 45 μg/L [53], while those of THM4 ranged between 36 and 51 μg/L [51]. The lowest values of THMs and HAAs were observed in the water coming from a desalination plant [51, 53]. It is important to highlight that HAAs are not the only haloacids present in drinking water; haloacids with longer carbon chains (e.g., 3,3-dichloropropenoic acid) may also form [54]. Levels of THM4 and HAA9 in finished drinking water can be reduced by applying enhanced coagulation processes, eliminating the pre-chlorination step, and/or optimizing the final chlorine dose in the drinking water treatment trains [55].

Concentrations of THM4 in swimming pools are usually higher than in drinking water, with TCM being the main THM4 species formed, except in those swimming pools disinfected with bromo-chlorodimethylhydantoin and in seawater swimming pools, where TBM is the main contributor to total THM4 levels. THM4 levels reached up to 311 μg/L in chlorinated outdoor swimming pool waters in Canada, 996 μg/L in indoor seawater swimming pools in France, or 223 μg/L in indoor swimming pools in England [56]. In a study recently conducted in Shanghai (China), HAA5 concentrations in swimming pool water were one order of magnitude higher than those measured in drinking water (407 μg/L *vs.* 35 μg/L) [57].

Following THMs and HAAs, HALs was ranked as the third-largest DBP class by weight in a US nationwide DBP occurrence study, with dichloroacetaldehyde (DCAL) presenting the highest concentrations (up to 16 μg/L) [58]. TCAL, present in water in its hydrated form (chloral hydrate), is by far the most investigated HAL in disinfected waters. Its concentration in chlorinated swimming pool waters ranged between 5 and 35 μg/L [59]. TCAL in drinking waters was measured from sub-μg/L levels to maximum concentrations of 263 μg/L, the latter at the outlet of a Canadian DWTP using pre-ozonation and subsequent chlorination [60]. Pre-ozonation may increase the overall formation potential of this class of DBPs of the water because it results in the formation of acetaldehyde that has been pointed out as the main precursor of HALs [61]. IAL (up to 4.5 μg/L) has been only detected in chloraminated drinking waters, regardless of being treated with ozone [9]. The formation of brominated HALs was observed in disinfected bromide-containing waters [9, 51, 60]. Tribromoacetaldehyde (TBAL) was the HAL species that with 12.6 μg/L contributed the most (44%) to total HAL concentrations in chloraminated water with high bromide content (540 μg/L) [9].

The occurrence of HBQs has been investigated in drinking water and swimming pool water. The most abundant and ubiquitous HBQs found are 26DCBQ and 26DBBQ, the latter being preferably formed during the chlorination of high bromide content source water. Overall, chlorination favors the formation of HBQs over other disinfection treatments (chloramination or ozonation). 26DCBQ and 26DBBQ were quantified in chlorinated drinking water at a maximum concentration of 275 ng/L [62] and 254 ng/L [63], respectively. Comparatively lower concentrations of DCMBQ (up to 6.5 ng/L) and TCBQ (up to 9.1 ng/L) were also found in chlorinated waters. The remaining HBQs investigated in water were not detected in disinfected waters. The four transformation products of HBQs investigated to date, viz. OH-DCBQ, OH-DBBQ, OH-TCBQ, and OH-DCMBQ, were measured in drinking water networks at maximum concentrations ranging between 7 ng/L (OH-DCMBQ) and 20 ng/L (OH-DCBQ and OH-DBBQ).

HANs have been detected in drinking waters treated with chlorine, chloramine, ozone, and chlorine dioxide [64]. In the U.S. EPA ICR (Information Collection Rule), four HANs were monitored in 296 public utilities for about 1.5 years and a maximum total HAN concentration of 41 μg/L was reported [65]. Another nationwide occurrence study that monitored a total of nine HANs in the U.S. reported a maximum level of total HANs of 14 μg/L [54, 58]. Dichloroacetonitrile (DCAN) has been reported with the highest concentration and is the most frequently detected of all HAN species, followed by bromochloroacetonitrile (BCAN) and dibromoacetonitrile (DBAN). Trichloroacetonitrile (TCAN), chloroacetonitrile (CAN), and bromoacetonitrile (BAN) are detected with less frequency and lower concentration levels compared to dihalogenated acetonitriles [54, 58, 66]. Like THMs, HAAs, and HALs, the presence of high bromide concentration in source waters enhances the formation of brominated HANs.

HAMs were first detected in drinking water in the United States during a nationwide occurrence study conducted between 2000 and 2002 [1, 54]. Since then, a varying suite of HAMs has been measured in drinking water around the globe, including England [67], Japan [68], Australia [69, 70], and China [71]. The most commonly measured HAMs are the dihalogenated compounds, with dichloroacetamide (DCAM) being the most prevalent measured in drinking waters [41, 58, 67, 68]. For example, in an occurrence study of six HAMs in Japan (CAM, BAM, DCAM, BCAM, DBAM, TCAM), HAMs concentration ranged from 0.3 to 3.8 μg/L, with DCAM and bromochloroacetamide (BCAM) the most frequently detected [68]. Some studies have measured a larger suite including mixed brominated and iodinated HAMs [72–74]. Studies have shown that the formation of HAMs may be higher upon chloramination [1, 54, 58, 75]. Bromine substitution also increases with chloramination [67].

HNM formation has been investigated in a variety of treated waters, including drinking water, wastewater effluent, reclaimed wastewater, and swimming pools [54, 73, 74, 76–83]. TCNM is the most widely measured HNM and was the first of this class to be identified as a DBP [84]. HNMs are typically found in drinking water at concentrations ranging from 0.1 to 3 μg/L, and trihalogenated HNMs are more commonly detected at quantifiable levels (> 0.1 μg/L) than monohalogenated and dihalogenated HNMs [73, 76, 77]. HNM concentrations in swimming pools typically fall within a similar range [78, 85], though a recent study reported tribromonitromethane (TBNM) concentrations in an indoor pool up to 6.1 μg/L [78]. No iodinated HNM DBPs have been reported. HNMs have been reported in source waters impacted by wastewater effluent [82, 83, 86], especially those associated with oil and gas extraction processes [82], with dibromochloronitromethane (DBCNM) particularly found at elevated concentrations up to 9 μg/L. This is most likely due to the presence of additional organic and inorganic precursors, including phenol and bromide, introduced from the extraction procedure. Though many disinfection scenarios lead to HNM formation, it is well-established that ozonation followed by chlorination or chloramination can significantly increase the overall formation of this DBP class [79, 87, 88]. One study demonstrated that ozonation of *N*-methylamine drugs potentially present in wastewater, such as ephedrine and methamphetamine, yielded nitromethane at >50%, which subsequently forms TCNM upon chlorination [83]. Further, the presence of bromide during pre-ozonation greatly enhances brominated HNM formation [58, 87]. Both granular activated carbon and biological activated carbon are effective for removing organic DBP precursors, but not halides, which enhances brominated HNM formation because of the increased bromide-to-organic matter ratio [73, 74, 89]. Other known HNM precursors include nitrite, primary amines, and ion exchange resins [87, 88, 90].

The first report that documents the presence of NDMA in water appeared in the 1970s at a location adjacent to an industrial facility near Baltimore, Maryland USA that produced unsymmetrical dimethylhydrazine (UDMH) [91, 92]. Later, NDMA was identified as a DBP, with reported occurrence in chlorinated drinking water supplies in California and Canada [92]. Since these discoveries, the research on NDMA and other nitrosamines has increased around the world. More recently, researchers have reported that alternative disinfectants, including chlorine dioxide (ClO_2) and ozone (O_3), can also produce NDMA [11, 93]. It is believed that NDMA precursors are mainly of anthropogenic origin, in contrast to other DBPs, such as THMs and HAAs, which are derived from organic matter of different environmental sources [94]. Therefore, NDMA formation is mostly related to wastewater impacted drinking water sources waters as their precursors are high in wastewater effluents. However, drinking water practices such as coagulation with PolyDADMAC have also been related to NDMA formation [95]. Algal blooms, and to a lesser degree agricultural or stormwater runoff, may also be an important source of nitrosamine precursors, but likely only in very adverse conditions when a major fraction of the water supply is affected [93].

In previously published occurrence surveys, NDMA was found at relatively high concentration (>10 ng/L) in DWTPs in the US, Australia, China, and Spain [11, 96–102], whereas in the UK and Japan it has been seldomly detected and at generally lower concentration levels [11, 103, 104]. One study investigated the occurrence of nitrosamines at one specific DWTP in Spain employing peroxidation with potassium permanganate followed by coagulation, sedimentation, and filtration before disinfection with chloramines [102]. The average NDMA concentration in the four samples taken in the distribution system was 7.1 ng/L, with maximum values measured up to 20 ng/L. Planas et al. [96] also measured NDMA in another specific Spanish DWTP using ozonation and found up to 10 ng/L after this treatment. This value was reduced after granular activated carbon and the concentration measured in the final treated waters was not higher than 5 ng/L. More recently, Farré et al. investigated the presence of NDMA and other nitrosamines in drinking waters from central Spain [105]. The actual concentration of NDMA in final treated waters and samples taken from the distribution system was never above 4.2 ± 0.2 ng/L.

When other nitrosamines are reported, NDMA tends to be the highest nitrosamine concentration [11]. However, in the US a concentration of NDEA above 5 ng/L was found in 2% of 1198 public water systems, while in China this proportion was up to 15% [97]. NDPhA was found at concentrations between 0.1 and 0.4 ng/L in distribution systems of Canada and the US [106]. Another study reported concentrations of different *N*-nitrosamines in drinking water that ranged from 4.6 to 20.5 ng/L for NDMA (found in 7/12 samples), 1.9 to 16.3 ng/L for NDEA (9/12), 0.4 to 3.4 ng/L for *N*-Nitrosodi-n-butylamine (NDBA) (6/12), 1.1 ng/L for *N*-Nitrosomethylethylamine (NMEA) (1/12), and 3.3 ng/L for NDPhA (1/12), respectively [107]. In the Spanish study [105], NDEA and NDBA were also detected in almost all samples, however, their concentrations did not exceed 1.5 ng/L in any case. Much higher concentrations have been detected in China's drinking water [108, 109]. Jurado-Sánchez et al. [102] measured 11, 9.4, 1.8 ng/L average concentrations of NDBA, NDPhA, and NDEA respectively, in samples taken from a specific distribution system in Spain. Planas et al. [96] reported values up to 12.9 ng/L of NDEA in treated water from a specific Spanish plant employing ozonation.

Yoon et al. [110] reported NDMA concentrations measured in treated wastewater effluents from less than 10 ng/L to 73 ng/L, depending on the treatment used in Japan. Sanchís et al. reported concentrations ranging from 20 and 40 ng/L in treated secondary wastewater effluent [111]. This agrees with the first data published by Sedlak et al. [112], who reported average concentrations of NDMA in secondary effluent before disinfection of 46 ng/L from different wastewater treatment plants (WWTPs) in the USA. Concentrations measured in another US study found values between < 2–320 ng/L in secondary and tertiary treated effluents [113]. In contrast, the values were higher and up to 1161 ng/L in a secondary effluent from WWTPs of Shanghai [114], which may have been caused by high NDMA concentration at the inlet of the WWTP from industrial wastewater discharge. NMOR [115], *N*-nitrosodiethanolamine [116], and other tobacco-specific nitrosamines [117] are also relevant in wastewater and water reclamation practices as they have been found at relevant concentrations too.

9.2 Sample Preparation and Collection

9.2.1 Protocols for Collecting and Preparing Samples

Water samples are collected in headspace-free amber vials or bottles to avoid loss of the most volatile DBPs such as THMs or TCAN through volatilization and photodegradation of light-sensitive DBPs like HBQs [118] or nitrosamines [119]. Both borosilicate glass and polytetrafluoroethylene (PTFE) bottles have been used to collect and store the water samples.

Grab samples represent the chemical composition of the water at a specific moment in time. For this reason, quenchers are commonly added to grab samples to stop oxidants (i.e., chlorine, chloramine) from generating additional DBPs during storage before analysis. However, quenchers can also degrade DBPs. While sodium sulfite (24 mg/L) has been commonly used to quench residual free chlorine and preserve THM concentrations [120], sulfur-based quenchers (sodium sulfite and sodium thiosulfate) can catalyze the degradation of many DBPs (HANs, HALs, TCNM, and even HAAs at pH 8) [121] and should be avoided. On the contrary, these sulfur-based quenchers are the most common quenchers employed for nitrosamine preservation [92, 122]. Ascorbic acid (36 ppm:1 p.p.m. of free chlorine, 1.6 g/L or 12.5 mg/L) is an appropriate quencher for both chlorine and chloramines to preserve THMs [53, 123–125], HAAs [121, 123], HALs [9, 126], and HAMs [127]; however, it rapidly degrades trihalogenated HANs such as tribromoacetonitrile (TBAN) and TCAN and trihalogenated HNMs like TCNM [76, 121]. The addition of ammonium chloride (100 p.p.m. [128–130] or 1 mol per 1.3 mol of free chlorine [76]) to chlorinated water only interferes with BCAL and DBAL [121]. Ammonium chloride is the most suitable quenching agent for HAMs [39], HANs, and HNMs [76, 80, 131]. Ammonium chloride reacts with chlorine to produce chloramines, a weaker oxidant. However, the addition of ammonium chloride to chloramines does not affect chloramine concentration and therefore no quencher is preferred in chloraminated water. In this case, a pH adjustment has been reported to work well to preserve some DBP classes. For instance, HANs degrade less than 30% after 3 days under cold storage at a water pH of 5 [76, 80, 132], and HAL concentrations were also preserved by lowering the sample pH to 3.5 [126, 133]. Formic acid added to the water sample in a proportion of 0.1% [63] or 0.25% (v/v) provides the best results in terms of HBQ stabilization for up to 5 days at 4°C. Trihalogenated HNMs undergo base-catalyzed hydrolysis in alkaline conditions, so pH should be adjusted between 3 and 7 for adequate preservation [131].

Samples are stored at 4°C until extraction, which is usually conducted within a maximum of 24 hours (for HAAs) [123] or 48 hours of sample collection [76, 134]. In the case of nitrosamine analysis, sample storage is accepted up to 14 days, protected from light, with the addition of sodium thiosulfate [135]. Sample filtration of drinking water samples is usually not needed. For direct injection of drinking water samples into the ion chromatograph (IC) or liquid chromatograph (LC) to determine HAAs, samples are usually filtered (0.22 μm). In the particular case of LC methods, samples need to be acidified to pH 3 (usually with 0.1% formic acid) [53, 123, 136]. Water matrices other than drinking water (e.g., surface water, treated wastewater, etc.) are usually done with 0.45 μm membranes. For the analysis of nitrosamines, 0.45 μm nitrocellulose filters are used.

9.2.2 Sample Extraction and Clean-up

Liquid-liquid extraction (LLE) is the most widely used extraction method due to its simplicity (i.e., no specialized equipment required), low cost, and effectiveness in simultaneously extracting various DBP classes [76, 80, 128, 137, 138]. LLE with methyl-tert-butyl ether (MTBE) is indeed the approach recommended by the US. Environmental Protection

Agency (US EPA) for the multi-residue extraction of DBPs (method 551.1 [128]) and HAAs (method 552.2 [139]). In method 551.1, designed for the extraction of THM4, BCAN, DBAN, DCAN, TCAN, TCAL, TCM, and two haloketones among other organic compounds, 10 mL of the water sample fortified with the surrogate standard (1,2-dibromopropane) solution (50 μL of 10 ng/mL) is mixed sequentially with 3 mL of MTBE and 20 g of anhydrous sodium sulfate (Na_2SO_4). Then the mixture is vigorously shaken by hand for 4 minutes. After shaking, the two phases are allowed to separate for few minutes and the upper MTBE layer is recovered and transferred into a 2-mL GC vial. In method 552.2, designed for the extraction of HAAs, 40 mL of the water sample is acidified with 2 mL concentrated sulfuric acid (H_2SO_4), then 16 g of Na_2SO_4 is added to the solution, and finally 4 mL of MTBE is added as extracting solvent [139]. After vortex or manual shaking for few minutes, the supernatant is recovered and dried using a Na_2SO_4 column. Variations of these methods have been used by different authors to extract THMs, HAAs, HALs, HANs, HNMs [51], and HAMs from water [47, 51, 76].

For instance, for multi-residue extraction 100 mL of sample volume, 1 mL or 5 mL of H_2SO_4 and 30 g of Na_2SO_4 have been used [51, 76]. Lowering the pH below 1 maximizes the co-extraction of additional DBPs that have a very low pK_a (e.g., HAAs). In other LLE approaches, a sample volume of 50 mL is used, and the pH is maintained at 4.5 and instead of Na_2SO_4, NaCl (16 g) is added, for the specific extraction of trihalogenated HALs [126], or the pH is adjusted to 4 and 20 g of Na_2SO_4 is added for the simultaneous extraction of N-containing halogenated DBPs (HANs, HNMs and HAMs) [80] or the specific extraction of HAMs [131]. In preparation for injection-port silylation of HAMs another method used 10 mL sample (pH 7) with 3 g of Na_2SO_4 and 4 mL of ethyl acetate [140].

Although MTBE is the extracting solvent used in most LLE approaches, ethyl acetate has been also used for LLE of trihalogenated HALs [141] and HAM [140]. After adding the extracting solvent, LLE is done using a mechanical wrist action shaker for several minutes (from 3 to 30 minutes), or manually for a couple of minutes. When treating low sample volumes, LLE can be assisted with a vortex. While most methods employ single extraction with MTBE (4 mL and 4 minutes [80], 3 mL and 4 minutes [139], 3 mL and 3 minutes [126], 2.5 mL and 30 minutes [51], 5 mL and 4 minutes [131]), methods that employ multiple extractions (up to three) can recover more analyte from water samples that lead to higher recovery rates and lower detection limits (3 × 5 mL and 15 minutes each time [76, 137], 3 × 3 mL and 10 minutes each time [138]). For complex wastewater samples, a small sample volume (10 mL) is usually extracted to minimize matrix effects [138].

The remaining water molecules in the extract are removed with anhydrous Na_2SO_4 before transfer into the analytical vial, and for methods that obtain a large volume extract (>15 mL), the extract is blown down to 200 μL under N_2 to increase method sensitivity [131, 137]. For injection-port silylation of HAMs, the final extract was evaporated to dryness and re-dissolved in 100 μL. The internal standard (1,2-dibromopropane in many cases) is added either in the MTBE extracting solvent, in the sample right before extraction, or in the final evaporated extract.

Liquid-liquid microextraction (LLME) approaches are similar to LLE with the difference that LLME utilizes a low sample and solvent volume, and thus minimizes amounts of solvents and reagents used, and requires a centrifuge to induce phase separation. Ma et al. developed an LLME method that extracts seven HANs from 8 mL of sample with

40 μL of methylene chloride [142]. LLME approaches were also applied by Serrano et al. [133] and Laflamme et al. [141] to extract HALs. In these methods, 9 mL of sample at pH 3–3.5 [133] or 10 mL of sample containing 75 μL of 0.1 M H_2SO_4 [141] is extracted with 200 μL [133] or 500 μL [141] of ethyl acetate, respectively. Then, 3 g of Na_2SO_4 is added and vortexed for 1 minute [133] or 5 minutes [141]. After phase separation, 60 μL [133] or 300 μL [141] of the supernatant is transferred into a vial containing 10 mg of Na_2SO_4, and vortexed. Finally, 30 μL [133] or 100 μL [141] of the supernatant is transferred into the analytical vial. An LLME approach has also been applied for HNM extraction [143]. In the optimized method, 9 mL of sample at pH 3.4 is spiked with 4 μg/L fluorobenzene, used as internal standard, and extracted with 200 μL of ethyl acetate and 3 g of Na_2SO_4. The sample is vortexed for 1 minute, and after phase separation, ~50 μL of the organic extract is transferred into a vial containing 10 mg of Na_2SO_4 to remove water.

Solid-phase extraction (SPE) has also been employed for the extraction of multiple DBP classes and is the technology preferred for HBQ and nitrosamine extraction. SPE can also achieve high analyte recovery using low quantities of solvent with the option of automation. However, SPE requires one-time use cartridges and specialized equipment that increase the cost of analysis.

Roumiguieres et al. evaluated five commonly used SPE cartridges for environmental contaminants and found that Bakerbond SDB-1 (40–150 μm, 200 mg, *Interchim*) previously conditioned with 5 mL of methanol and 10 mL of ultrapure water had the highest recoveries for the majority of the tested DBPs (THMs, HANs, HNMs, haloketones, and TBAL) [134]. Final extraction conditions included 1 L of sample, pH 8, 5 mL of ultrapure water for cartridge rinse, and 2 × 0.8 mL of MTBE as eluent. Finally, the organic extract is transferred into a 300 μL GC vial. Hladik et al. also developed a method for the analysis of THMs, HALs, HANs, and HNM in river water containing wastewater discharge [82]. This method uses Oasis HLB SPE cartridges (6 mL, 500 mg; Waters Corporation, Milford, MA), which are pre-cleaned with 10 mL of MTBE, 10 mL of methanol, and 10 mL of water. Filtered river water (1 L) is loaded onto the SPE cartridge at 5 mL/min, and the cartridges are subsequently dried under N_2 for ~15 minutes. The samples are eluted with 10 mL of MTBE, which is then concentrated to 400 μL under N_2 followed by the addition of internal standard (1-chlorooctane). The final extract is placed in a vial with no headspace. SPE cartridges containing styrene divinyl benzene polymeric beads (Varian Bond Elut-PPL, 200 mg, 3 mL, Palo Alto, CA) have also been used for the simultaneous extraction of HNMs, THMs, HANs, and HALs from drinking water [144]. The SPE cartridge is first conditioned with 8 mL methanol, and then loaded with two sample aliquots of 18 mL each and eluted with 1.6 mL MTBE.

SPE of HAAs has been successfully achieved with Bakerbond SDB (200 mg) cartridges [124]. In the optimal approach, acidified water sample (1 L, pH < 2, using H_2SO_4) is transferred into the cartridge, previously conditioned with methanol and acidified water (10% H_2SO_4, v/v). After sample loading, the cartridge is washed with acidified water and allowed to dry for 10 minutes. Finally, HAAs are eluted with 5 mL of acidified methanol (10% H_2SO_4, v/v) and 3 mL of MTBE [124].

Trihalogenated HALs are extracted from 100 mL of water using Oasis HLB cartridges previously conditioned with 6 mL of MeOH, 6 mL of MTBE, and 6 mL of Milli-Q water [9]. After sample loading, and without drying the cartridge, analytes are gravity-eluted dropwise with 3 mL of MTBE, allowing the MTBE to soak the cartridge every 10 drops. The eluate is mixed with 300 ng of 1,2-dibromopropane (used as internal standard) and transferred to a 2 mL vial after removing the water molecules that may be present using a Na_2SO_4 column.

HBQ extraction has been accomplished using offline and online SPE. In offline SPE approaches the sample volume (500 mL) is passed at a flow rate of 2–3 mL/min through an Oasis HLB cartridge (6 mL, 200 mg; Waters, Milford, MA) that has been previously conditioned with 6 mL of acidified methanol (0.25% FA) and 6 mL acidified water (0.25% FA). After sample loading, the cartridge is washed with 6 mL of acidified water (0.25% FA), and 6 mL of acidified methanol/water solution (50/50, v/v) (0.25% FA). Then the cartridge is vacuum-dried for 10 minutes. The HBQs retained on the cartridge are eluted with 6 mL of acidified methanol (0.25% FA). The eluent is concentrated to 100 μL under nitrogen and reconstituted to 500 μL with acidified water (0.25% FA) for injection [145, 146]. While the volumes indicated were designed for the extraction of drinking water samples, the optimum extraction of more complex water samples; i.e., swimming pool waters or surface waters require the use of higher wash volumes (18 mL of acidified water + 6 mL of acidified methanol/water solution) [147]. Improved versions have doubled the volume of the conditioning and elution solvents [36].

In the online SPE approach [63], the water sample (5 mL) is carried from the sample loop into the SPE column (Oasis HLB dual online SPE column, 2.1 × 20 mm, 25 μm) with acidified water (0.25% formic acid) at 1 mL/min for 12 minutes (3 times the volume of the sample loop). HBQs are eluted from the SPE column using the LC gradient. After elution, the sample loop and the syringe are flushed twice with 5 mL of a solvent mixture (water 25%, acetonitrile 25%, methanol 25%, and isopropanol 25%) and then twice with 5 mL of LC-grade water to avoid cross-contamination and carryover.

SPE methods have been also specifically designed for the extraction of HAMs. Zhou et al. used a Bond Elut-PPL cartridge (Agilent CA, USA), conditioned twice with 3 mL ethyl acetate and 3 mL water, loaded 10 mL of sample, washed with 2 mL water (5% methanol), and eluted with 3 mL ethyl acetate [148]. Chu et al. used an Oasis HLB cartridge and eluted it with 0.5 mL water and 5 mL methanol, with the final volume concentrated to 0.5 mL [149].

An alternative method used to extract DBPs from water is headspace solid-phase microextraction (HS-SPME). In this approach, extracting solvents are not required because analytes are directly sorbed onto a fiber. HS-SPME can be automated; however, it requires specialized equipment and solid-phase microextraction (SPME) fibers that can be expensive to operate. This technology, limited to compounds with high vapor pressure, has been used for the extraction of THMs, HANs, and HNMs. The fiber coatings that have the highest performance for THM4 are carboxen/polydimethylsiloxane (CAR/PDMS), while for the six I-THMs, the carbowax/divinylbenzene (Carbowax/DVB) coating works best. The whole suite of THMs can be simultaneously extracted using a carboxen/polydimethylsiloxane/divinylbenzene (CAR/PDMS/DVB, 50/30 μm) fiber. For this, 10 mL of sample, spiked with the surrogate standard (1,2-dibromopropane), is mixed with 5.5 g of anhydrous Na_2SO_4, and then heated for 15 minutes at 70°C so that THMs can adsorb onto the fiber in the headspace. Analyte desorption is achieved in 8 minutes using a programmed temperature vaporizer (PTV) inlet (160°C for 3 minutes, and raised at 1°C/s up to 270°C, temperature held for 3 minutes to regenerate the fiber) [120]. A cryogenic trap that allows setting the gas chromatograph oven at −20°C is also needed so that the analytes can be retained on the head of the gas chromatography (GC) column. Kristiana et al. developed an HS-SPME method to extract six HANs and two long chain nitriles [132]. For this, a 50/30 μm DVB/CAR/PDMS fiber is exposed to a sample volume of 12 mL at pH 5, containing 30% w/v of salt, during 15 minutes and at a temperature of 40°C. Then, the HS-SPME fiber is desorbed inside the GC inlet at 200°C. Luo et al. also used the same HS-SPME fiber with slightly different conditions (10 mL sample, 40% w/v of salt, extraction temperature of 40°C for 30 minutes, desorption at 170°C) for the extraction of HANs [150]. The DVB/CAR/PDMS fiber was also found to effectively extract TCNM and haloketones from water [151]. Brominated HANs were found to adsorb well onto a novel polydimethylsiloxane/divinylbenzene-*N*-vinylpyrrolidone fiber [151]. A HS-single drop microextraction (SDME) method was specifically developed for the analysis of the whole suite of HNMs [151–153]. In this method, a 2.5 μL volume of 1-hexanol was exposed to the headspace of the sample vial using a microsyringe. The sample (10 mL containing 3 g of Na_2SO_4 and at a pH of 3.2, adjusted with 0.1 M H_2SO_4) was stirred at 600 rpm for 20 minutes at 30°C [152].

An approach commonly used for the analysis of THM4 in commercial labs is purge and trap, which allows measuring the total amount of the volatile analytes in the sample. The purged THMs are focused on a cold or an adsorbent trap before being introduced into the column. The excessive water purged in the process results in peak distortion in the early part of the chromatogram. To avoid this, devices that include a solvent elimination system or a drying step have been developed [154]. In the EPA 524.2 method [155], a sample volume of 25 mL or 5 mL is purged at room temperature for 11 minutes, and purged volatiles are trapped into a trap that combines 2,6-diphenylene oxide polymer, silica gel, and coconut charcoal as adsorbents. The desorption of the analytes is done at 180°C with the desorption gas at a flow rate suitable for the column being used for about 4 minutes. If a cryogenic interface is available, this should be set at −150°C or lower, and desorption is done by rapidly heating the trap to 180°C while backflushing with the inert gas for about 5 minutes [155]. Additional methods described in the literature to extract THMs from water have been reviewed in detail by Perez-Pavon et al. [154].

The analysis of highly polar DBPs (e.g., HAAs or monohalogenated and dihalogenated HALs) with GC requires previous derivatization. In the case of HAAs, this is done either with acidic methanol (3 mL of the extract is mixed with 1 mL of 10% H_2SO_4 methanolic solution and left to react at 50°C for 2 hours) [139, 156] or diazomethane (0.5 mL of the extract is mixed with 0.25 mL of freshly prepared diazomethane gas and left to react at room temperature for 1 hour) [51]. In the case of monohalogenated and dihalogenated HALs they are derivatized with O-(2,3,4,5,6-pentafluorobenzyl)hydroxylamine (PFBHA) at 35°C. For this, 100 mL of the water sample is mixed with 50 ng of fluorobenzaldehyde (used as surrogate standard), 0.1 mL of the derivatization quenching agent solution ($(NH_4)_2SO_4$, 1.7 M), 5 mL of potassium hydrogen phthalate (KHP)/sodium hydroxide (NaOH) buffer (pH = 6), and 5 mL of the PFBHA solution (15 mg/mL). Upon derivatization (under a continuous gentle stir at 35°C for 2 hours), samples are left to cool down at room temperature, and the derivatization reaction is quenched by adding 0.5 mL of

H_2SO_4. After the addition of 20 ng of 1,2 dibromopropane in hexane (used as internal standard), the oximes formed are extracted in triplicate with hexane (10 mL) for 3 minutes in a mechanical shaker. The recovered organic layers are combined and dried with a Na_2SO_4 column, and the dried extract is evaporated under N_2 to a final volume of 0.3 mL.

To avoid the extraction step and derivatization of the extremely polar and water-soluble HAAs, analytical methods based on the direct injection of water samples onto ion chromatography(IC)-MS and LC-MS systems after uncomplicated pretreatment steps (such as filtration, acidification, and/or internal standard addition) have been recently developed. In these methods, the adjustment of the sample pH to 2 or 3 removes the matrix effects originated by the ions present in the tap water that strongly affect the separation and detection of HAAs of small molecular weight (e.g., CAA, BAA, DCAA or BCAA) [53, 123, 136]. Direct aqueous injection into the GC system has also been used for THM4 determination using electron capture detection (ECD) [157] and mass spectrometry (MS) [158] systems. For this, capillary columns with a thick film of an apolar phase are used, and the sample is injected under cold conditions. However, direct injection is not a recommended approach when using GC systems as it deteriorates the GC column coating. This problem can be reduced by placing deactivated guard columns that protect the GC column

Direct analysis of nitrosamines and especially NDMA has been difficult mainly because low detection limits are required (low ng/L range). Therefore, samples between 500 mL and 1 L are concentrated 1000 fold to reach the required levels of detection.

SPE is the most commonly used method for the determination of nitrosamines in water [104, 106, 115, 135, 159–167]. The USEPA method 521 [135] for the determination of seven nitrosamines in drinking water proposes an SPE using coconut charcoal as the sorbent, dichloromethane as the elution solvent, and concentration of the extract to less than 1 mL. The other nitrosamines analyzed with this method are NMEA, NDEA, NDPA, NDBA, NPYR, and *N*-Nitrosopiperidine (NPIP). Before extraction, deuterated surrogates are added to quantify the losses and an internal standard is also added after concentration to 1 mL. Several different SPE sorbents have been evaluated for nitrosamine extraction and concentration. This was extensively reviewed by Boyd et al. [168] and an excellent update since 2010 has been recently published by Bian et al. [169]. Some of the most recent sorbents used in the analysis of nitrosamines in water are LiChrolut EN or Ambersorb 572 (500 mg) [106], coconut charcoal cartridges [164, 166], Sep-Pak cartridges (NH2 cartridge connected to an AC-2 cartridge) [170], Sep-Pak Plus AC-2 cartridges (400 mg, 85 μm) [171], combined cartridges (200 mg of HLB on the top and 500 mg of UCT at the bottom) [172]. Dichloromethane is the solvent of preference for the elution of nitrosamines. After elution, dichloromethane is dried up with sodium sulfate before GC injection.

LLE has also been employed to analyze nitrosamines although its use is less extensive [173, 174] because the recoveries for nitrosamines were found to be low. Few LLE methods have been published in the past few years, as they have largely been replaced with more efficient SPE methods. SPME [175–177] or dispersive SPE [178] have also been employed to extract nitrosamines. A major issue for SPME analysis of nitrosamines has been the sensitivity required to detect low ng/L concentrations.

9.3 Determination of DBPs

The determination of halogenated DBPs in water has been traditionally done with GC-ECD [126, 128, 139]. However, GC-ECD methods require two different chromatographic columns for analyte confirmation. Because of this, MS is preferred in recent years because it has a higher method specificity.

One of the main limitations of employing GC-MS to determine highly volatile DBPs like TCM and TCAN is their low retention in most GC columns with splitless injection mode. These compounds elute in the peak solvent front (TCM) or very close to it (TCAN), which affects their reliable quantitation. The use of PTV injection methods, or headspace extraction methods can be used to overcome this limitation. Specific GC columns are also capable of resolving TCM from the solvent peak (e.g., ZB1701, 30 m × 0.25 mm × 0. 25 μm, from Phenomenex [125]).

Although GC-negative chemical ionization (NCI)-MS has been employed for the determination of HAAs in water [51], more specific methods based on GC, IC, or LC coupled to tandem mass spectrometry (MS/MS) have been recently developed. In any case, the analysis of these compounds using GC requires their derivatization to form their corresponding methyl esters (haloacetic acid methyl ester, HAAME), which are more volatile than the original HAAs. To avoid this step, the determination of these compounds using IC- or LC-based methodologies has gained popularity in recent years.

The GC-MS determination of HBQs is limited to (semi-)volatile, thermally stable, non-ionic, and non-polar HBQs. To overcome this limitation, HBQs analogs have been successfully derivatized with N-methyl-N(tert-butyldimethylsilyl)-trifluoroacetamide (MTBSTFA), 1,2-phenylenediamine, heptafluorobutyrylimidazole (HFBI), and acetic anhydride with zinc powder. However, to avoid the tedious derivatization process, LC-MS/MS-based methods have been developed.

GC-MS analysis has shown itself to be an excellent method for the detection of NDMA and other semi-volatile and thermally stable nitrosamines, due to the high sensitivity and selectivity offered by the selected ion monitoring (SIM) capability of the MS detector and the ruggedness needed to detect these substances routinely below the guideline concentrations. However, the use of GC precludes the direct analysis of thermally unstable or non-volatile nitrosamines, such as NDPhA. Methods employing LC have also been developed [168].

9.3.1 Development of the Instrumental Method

9.3.1.1 Chromatography

Table 9.11 shows the various GC columns and conditions used for the separation of THMs, HALs, HANs, HNMs, and HAMs in water samples. Helium gas is always used as the mobile phase or carrier gas at a constant flow rate (0.7–1.3 mL/min) or pressure. The injection volume of sample extracts is typically 1 μL, but it can be higher (up to 1 mL) if LLME or SDME is used.

In most approaches, the sample extracts are introduced into the GC in splitless mode with and without a PTV inlet. Recent methods have preferred PTV inlets to enhance DBP separation and retention of the most volatile DBPs, and minimize DBP degradation at high temperatures. For instance, higher response was found for nine HANs with a PTV inlet compared to a normal split/splitless inlet, particularly for CAN, BAN, and TCAN [80]. Trihalogenated HNMs, thermally unstable under commonly used GC injection-port temperatures (200–250°C) [79], are preserved when injection is done at cooler temperatures (33–40°C) and then ramped to higher temperatures. Inlets without ramping capabilities are typically maintained at 170°C or below to avoid thermal degradation of trihalogenated HNMs. Transfer line temperatures also need to be reduced to 200–225°C to prevent HNM decomposition, although 250°C has been also used [82]. Injection-port silylation has been used to reduce HAM polarity and enhance the detection of these DBPs with GC-MS [140].

In methods that use HS-SPME for sample extraction, a constant inlet temperature is used to completely desorb analytes from the SPME fiber [132]. Additionally, the GC initial temperature should be low enough (0°C with cryogenic cooling) to minimize DBP separation before being fully desorbed in the inlet.

The most commonly used GC columns for halogenated DBPs are low polarity phenyl methyl polysiloxane phases of low bleed capability that provide good separation, high signal-to-ratio, and sensitivity (e.g., ZB-5MS/Rtx-5MS/TG-5MS/HP-5MS) [76, 80, 134, 137, 138] (Table 9.11). The Rtx-200MS is a mid-polarity phase column (trifluoropropylmethyl polysiloxane) that also provides good separation and high sensitivity for halogenated DBPs, and better peak shape (no peak tailing) for HNMS and HAMs than the aforementioned phenyl methyl polysiloxane phases [76].

LC has been also the technique of choice to separate polar halogenated DBPs like HAAs, HBQs, and HAMs. The columns and conditions used for the analysis of HAAs with selected GC, IC, and LC methods are summarized in Table 9.12. A C18 column modified to increase its affinity with polar analytes (Luna Omega Polar C18) in LC methods and anion exchange columns in IC approaches have been successfully used to retain and resolve HAAs. LC separation of HBQs is achieved using a stationary reversed-phase and acidified mobile phase. In offline approaches for HBQ analysis, the injection volume is 20 μL while it increases to 5 mL in an online method (Table 9.13). LC separation of HAMs after injection of 10 μL of the extract has been achieved using an Hypersil GOLD C18 column (100 × 2.1 mm, ID 5 μm) at 40°C and a linear organic gradient of a water-methanol mobile phase at a constant flow rate of 0.3 mL/min [149].

The most common methodology to separate nitrosamines before analysis is GC [105, 177, 179–183]. For this, various GC columns (e.g., 5Sil MS (Restek), DB35 MS, DB624 (Agilent), SLB-IL111 (Supelco), ZB-5MS (Zebron), and TG-1701MS (Thermo)) have been used. Table 9.14 summarizes the conditions used for GC separation of nitrosamines. Few methods employ capillary or binary HPLC or ultra-performance LC (UPLC) using reversed-phase (C8 or C18) columns to analyze nitrosamines [104, 161–163, 184–188]. An HPLC-photochemical reaction-chemiluminescence that does not need sample clean-up has been also proposed for very clean samples such as reverse osmosis permeated [180]. With this method, a detection limit from 0.2 to 2 ng/L was obtained.

Table 9.11 Columns and conditions used for GC separation of halogenated DBPs.

		Sample injection		Temperature gradient	Ref
DBP class	Stationary phase	Mode	Injector T	GC oven T	
THMs HANs	DB-1 (30 m × 0.25 mm ID × 1 μm film thickness) *Agilent* *Santa Clara, CA* + Rtx-1301 (30 m × 0.25 mm ID × 1 μm film thickness) *Restek Corporation, Bellefonte, PA*	Splitless mode	200°C	35°C (for 22 min) 10°C/min to 145°C (for 2 min) 20°C to 225°C (for 15 min) 10°C to 260°C (for 30 min)	[128] *
THMs HALs HANs HNMs HAMs	Rtx-200 MS (30 m × 0.25 mm ID × 0.25 μm film thickness) *Restek Corporation*	Pulsed splitless (MMI)	35°C for 0.1 min 360°C/min to 220°C (for 5 min) 720°C/min to 280°C	35°C for 5 min 9°C/min to 200°C (for 5 min) 9°C/min 280°C (for 20 min)	[76]
HALs	Rtx-200 MS (30 m × 0.25 mm ID × 0.25 μm film thickness) *Restek Corporation, Bellefonte, PA*	Splitless (flow = 50 mL/min, time = 1.5 min)	200°C	30°C (for 5 min) 9°C/min to 165°C 20°C/min to 285°C (for 10 min)	[51]
HALs	DB-5 (30 m × 0.32 mm ID × 1 μm film thickness) *Agilent*	PTV	80°C for 0.1 min 140°C/min to 240°C (for 20 min)	35°C (for 5 min) 2°C/min to 55°C (for 1 min) 5°C/min to 100°C (for 1 min) 40°C/min to 200°C (for 5 min)	[126]

HALs	ZB-5 (30 m × 0.25 mm ID × 0.25 μm film thickness) *Phenomenex, Torrance, CA*	Splitless (flow = 10 mL/min, time = 1 min)	250°C	35°C (for 2 min) 9°C/min to 160°C (for 1 min) 5°C/min to 180°C (for 1 min) 22°C/min to 285°C (for 10 min)	[9]
HALs	HP-5 MS (30 m × 0.25 mm ID × 0.25 μm film thickness) *Agilent*	Pulsed splitless (PTV) (flow = 60 mL/min, time = 0.01 min)	45°C for 0.01 min 250°C/min to 200°C (for 20 min)	40°C (for 4 min) 1°C/min to 50°C 10°C/min to 100°C 50°C/min to 200°C (for 2 min)	[133]
HANs	Rtx-200 MS (30 m × 0.25 mm ID × 0.25 μm film thickness) *Restek Corporation*	Pulsed splitless (PTV)	40°C for 0.1 min 360°C/min to 170°C (for 5 min) 720°C/min to 280°C	35°C (for 5 min) 9°C/min to 220°C 20°C/min to 280°C (for 20 min)	[137, 138]
THMs	ZB-5 (30 m × 0.25 mm ID × 1 μm film thickness) *Phenomenex*	Fiber desorption using PTV (Splitless mode)	160°C for 3 min 1°C/min to 270°C (for 3 min)	-20°C (for 7.65 min) 50°C/min to 80°C (for 6 min) 10°C/min to 220°C (for 1 min) 50°C/min to 300°C (for 3 min)	[120]

(*Continued*)

Table 9.11 (Continued)

		Sample injection		Temperature gradient	Ref
DBP class	**Stationary phase**	**Mode**	**Injector T**	**GC oven T**	
THMs	HP 5 MS (30 m × 0.25 mm ID × 0.25 μm film thickness) *Agilent*	Splitless (PTV) THM4: 1 μL I-THM: 50 μL (flow = 50 mL/min, time = 1 min)	THM4: 200°C for 0.1 min 150°C/min to 300°C	30°C (for 5 min) 60°C/min to 150°C (for 2 min) 30°C/min to 180°C (for 0.1 min) 50°C/min to 300°C (for 1 min) 35°C (for 0 min) 15°C/min to 45°C 3°C/min to 75°C 8°C/min to 100°C 100°C/min to 280°C (for 1.5 min)	[47]
THMs HANs	TG 5 MS (30 m × 0.25 mm ID × 0.25 μm film thickness) *Thermo Fisher, Waltham, MA*	Splitless (PTV) (flow = 50 mL/min, time = 2.35 min)	I-THMs 35°C (for 9 min) 200°C/min to 170°C for 1.5 min 200°C/min to 300°C 60°C for 0 min 300°C/min to 220°C (for 0.2 min)	35°C (for 3.6 min) 20°C/min to 210°C 60°C/min to 280°C (for 1.5 min)	[134]
THMs	ZB1701, (30 m × 0.25 mm ID × 0.25 μm film thickness) *Phenomenex*	Splitless	200°C	40°C for 25 min, 5°C/min to 145°C (for 2 min) 20°C/min to 260°C (for 10 min)	[125]

HANs	ZB-5 MS (30 m × 0.25 mm ID × 1.0 μm film thickness) *Phenomenex*	Splitless Mode	200°C	0°C for 1 min (cryogenic) 5 °C/min to 200 °C 15 °C/min to 300 °C 300 °C (for 5 min).	[132]
HANs	Rtx-5 MS (30 m × 0.25 mm ID × 0.25 μm film thickness) *Restek Corporation*	Split/ Splitless (PTV)	30°C for .01 min (Split) 150°C/min to 170°C (Splitless for 0.99 min) 300°C/min to 250°C (purge) (No line spacing, this refers to 250ºC (urge) for 5 min)	35°C (for 4 min) 15 °C/min to 180 °C 180 °C (for 5 min)	[142]
HANs HNMs HAMs	ZB-5 MS (30 m × 0.25 mm ID × 1.0 μm film thickness) *Phenomenex*	Splitless PTV (100 mL/min, 1.5 min)	40°C for 1 min 300°C/min to 160°C (for 5 min) 300°C/min to 270°C (for 5 min)	40°C (for 6 min) 5 °C/min to 100 °C (for 1 min) 15 °C/min to 130 °C (for 1 min) 20 °C/min to 300 °C (for 5 min)	[80]
HNMs	HP-5 MS (30 m × 0.25 mm ID × 0.25 μm film thickness) *Agilent*	Split (split ratio 1 : 20, 1 min)	170°C	40°C (for 3 min) 40°C/min to 140°C (for 2 min) 40°C/min to 180°C (for 3 min)	[153]
HNMs	SLB-5 MS (30 m × 0.25 mm ID × 0.25 μm film thickness) *Supelco*	Solvent vent mode (PTV) (flow = 60 mL/min, 0.01 min, carrier flow = 1 mL/min)	45°C for 0.01 min 200°C/min to 170°C (for 20 min)	40°C (for 4 min) 0.5°C/min to 50°C 50°C/min to 100°C 80°C/min to 180°C (for 2 min)	[143]

(Continued)

Table 9.11 (Continued)

		Sample injection		Temperature gradient	Ref
DBP class	Stationary phase	Mode	Injector T	GC oven T	
HNMs	HP-1 MS (30 m × 0.25 mm ID × 1 μm film thickness) *Agilent*	Splitless	90°C	33°C (for 14 min) 4°C/min to 60°C 10°C/min to 200°C 20°C/min to 300°C	[82]
HAMs	ZB-5 MS (30 m × 0.25 mm ID × 0.25 μm film thickness) *Phenomenex*	Splitless mode	140°C	32°C (for 5 min) 3°C/min to 54°C 5°C/min to 150°C 25°C/min to 285°C for 4 min	[131]
HAMs	DB-5 (30 m × 0.25 mm ID × 1 μm film thickness) *Agilent*	Injection-port silyation		50°C (for 4 min) 20°C/min to 170°C 10°C/min to 234°C 25°C/min 280 (for 2 min)	[140]
HAMs	DB-5 (30 m × 0.25 mm ID × 1 μm film thickness) *Agilent*	Splitless mode	220°C	60°C (for 1 min) 5°C/min to 160°C (for 5 min) 30°C/min to 220°C	[148]

ID: internal diameter, MMI: Multimode inlet, PTV: programmable temperature vaporization.
*For MTBE-based LLE extracts and intended for GC-ECD determination of THMs.

Table 9.12 Columns and conditions used for the chromatographic separation of HAAs in waters

	Stationary phase	Eluent or mobile phase: Type	Flow	T, organic, or ion exchanger gradient	Injection volume	Ref
GC	TraceGold TG-5 MS (30 m × 0.25 mm ID, 0.25 μm) *ThermoFisher*	Helium	1.4 mL/min	35°C (hold 5 min) 10°C/min to 110°C 20°C/min to 200°C (hold 17 min)	1.5 μL (PTV injection splitless mode)	[124]
	Zebron ZB-5MS (30 m × 0.25 mm ID, 0.25 μm) *Phenomenex*	Helium	1 mL/min	35°C (hold 5 min) 1°C/min to 40°C (hold 1 min) 30°C/min to 200°C	3 μL (Splitless mode)	[156]
IC	Ion Pac AS24 (250 mm × 2 mm ID) *Dionex*	KOH	0.3 mL/min	0–15.1 min (7 mm KOH) 15.1–25.8 min (7–15.5 mm KOH) 25.9–46 (60 mm KOH)	120 μL	[130]
	IonPac AS16 (250 × 2 mm) *Dionex*	Aqueous methylamine (1 M): Acetonitrile (30 : 70, v/v)	0.3 mL/min	Isocratic (30 : 70, v/v)	10 μL	[129]
LC	Luna Omega Polar C18 (100 × 4.6 mm, 3 μm) *Phenomenex*	H_2O: Acetonitrile (both with 0.1% FA)	1 mL/min	0 min: 5% (hold 3 min) 5 min: 20% 7 min: 100% (hold 3 min) 14.5 min: 5% (hold 4.5 min)	200 μL	[123]
	Acquity UPLC HSS T3 (2.1 mm × 50 mm, 1.8 μm) *Waters*	H_2O: Methanol (both with 0.125 mm FA)	0.2 mL/min	0 min: 1% (hold 1.5 min) 4 min: 50% (hold 1 min) 5 min: 1 % (hold 3 min)	10 μL	[136]

FA: formic acid, ID: internal diameter, PTV: programmable temperature vaporization.

Table 9.13 Columns and conditions for LC separation of HBQs.

Stationary phase		**Mobile phase**			**Ref**
Column type	*T [°C]*	*Composition*	*Flow [mL/min]*	*Organic gradient*	
Luna C18 (100 × 2 mm, 3 μm) *Phenomenex*	Room T	H_2O: MeOH (both with 0.1% FA)	0.17	0 min: 20% 22 min: 60% 29 min: 90% (hold 7 min) 36.1 min: 20% (hold 15 min) 0 min: 20% 20 min: 90% (hold 5 min) 25.1 min: 20% (hold 5 min)	[153, 43]
Uptisphere® Strategy C18 100 A° (50 × 3 mm, 2.6 μm) *Interchim, Montlucon, France*	30°C	H_2O: AcN (both with 0.1% FA)	0.3	0 min: 20% 40 min: 90% (hold 5 min) 45.1 min: 20% (hold 5 min)	[63]
Acquity UPLC BEH C18 (100 × 2.1 mm, 1.7 μm) *Waters*		H_2O: MeOH (both with 0.1% FA)	0.4	0–1 min: 20% 12 min: 90% (hold 2 min) 14.1 min: 20% (hold 4 min)	[146]

MeOH: methanol, ACN: acetonitrile, FA: formic acid.

9.3.1.2 Detection

While the determination of halogenated DBPs with GC-ECD relies only on the compound's retention time, the use of MS allows registering also characteristic ions. For selective MS determination, single quadrupole or triple quadrupole mass analyzers are typically employed. The later increases method specificity, as they grant the acquisition of selected reaction monitoring (SRM) transitions [130]. HRMS analyzers have increased the selectivity and sensitivity of the MS methods. They have been used for HAN [137] and HAA [123] determination. The reported TOF-MS detector for HAN analysis is a relatively newer technology that contains a short flight path that provides a mass resolution of 600 p.p.m. and high sensitivity [137].

In GC-MS approaches, DBPs are ionized under electron ionization [76, 80, 131, 134] using an energy of 70 eV. Some DBPs, e.g., HNMs, do not exhibit a molecular ion in their mass spectra. Detection is done in either selected ion monitoring (SIM) mode (with single quadrupole or ion traps analyzers) [9, 76, 80, 120, 126, 131, 133, 140, 143] or SRM mode (with triple quadrupole analyzers) [47, 105, 124, 134, 138]. A minimum of two SRM transitions per analyte are registered (one for quantification and at least one as a

Table 9.14 Columns and conditions used for GC separation of nitrosamines.

	Sample injection		Temperature gradient	Ref
Stationary phase	**Mode**	**Injector T**	**GC oven T**	
Rtx 5SiL MS (30 m, 0.25 mm ID, 1.0 μm) *Restek*	Splitless mode	250°C	37°C (0.72 min hold) 100°C/min to 250°C, (40 min hold)	[135]
Rtx 5SiL MS (30 m, 0.25 mm ID, 1.0 μm). *Restek*	Splitless mode	250°C	35°C (6 min hold), 4°C/min to 90°C 2°C/min to 130°C 10°C/min to 260°C	[183]
TG-1701 MS (30 m, 0.25 mm ID, 0.50 μm) *Thermo*	Splitless mode	240°C	40°C (1 min hold), 5°C/min to 65°C, 10°C/min to 110°C (1 min hold), 25°C/min to 240°C (1 min hold)	[105]
DB-624 (30 m, 0.25 mm ID, 1.4 mm) *Agilent*	Splitless mode	250°C	45°C (2 min hold), 10°C /min to 90°C (1 min hold), 15°C /min to 200°C (12 min hold), 20°C /min to 230°C (1 min hold)	[177]
SLB-IL111 (30 m, 0.25 mm ID, 0.25 μm) *Supelco*	Splitless mode	270°C	65°C (2 min hold), 15°C/min to 120°C, 5°C/min to 265°C (20 min hold)	[181]

qualifying ion/transition). Positive identification is done according to the chromatographic retention time of the analyte and the presence of two characteristic SIM ions or SRM transitions, and the peak ratio between them. In a GC-MS method that utilizes injection-port silylation for HAMs determination, the tert-butyl dimethylsilyl (TBDMS) derivatives of HAMs are analyzed using selected-ion-storage (SIS) [140].

Nitrosamines are also commonly analyzed by GC-MS. To this end, both triple quadrupole and Orbitrap-MS analyzers have been utilized for the determination of nitrosamine in water and wastewater [104, 161–163]. A soft ionization source such as ammonia positive chemical ionization (PCI) is recommended in the literature [160]. The USEPA method 521 [135] employs GC coupled to tandem MS detection by chemical ionization source operating in positive (PCI) mode and using an ion trap mass analyzer (GC-PCI-MS/MS). Pozzi et al. [159] compared EI and chemical ionization (CI) with methanol using three ion analysis modes, full scan, SIS, and MS/MS. Results showed that CI and SIS were the best combinations to analyze nitrosamines. Farré et al. used EI and SRM to analyze NDMA and other nitrosamines [105].

For the determination of HAAs with IC-MS and LC-MS approaches, ionization is always achieved by operating the electrospray ionization (ESI) source in the negative mode [123, 129, 130]. In the IC-MS approach, acetonitrile is added to the column eluate before entering the electrospray source to improve gas phase ion generation [130]. The precursor ions of monohalogenated and dihalogenated HAAs result after the deprotonation of the carboxylic group $[M\text{-}H]^-$, while those of trihalogenated HAAs are formed after decarboxylation of the molecule in the ionization source $[M\text{-}COOH]^-$. Formic acid adduct ions $[M\text{-}H + HCOOH]^-$ are occasionally reported as the major precursor ions of monohalogenated and dihalogenated HAAs [53]. The main product ions formed after HAA fragmentation are the least electronegative halogen of the pseudomolecular ion, and the ion that results after decarboxylation.

In LC-MS methods, HAMs are ionized using atmospheric pressure chemical ionization (ESI in positive mode) and analyzed using SRM [149].

Overall, the best LC-MS signal for HBQs is obtained under negative ESI [36, 63, 146, 147]. Under this ionization condition, the HBQs, which are present mainly in their neutral form in acidified solutions, get reduced (C = O converted into C–OH) to dihydrohaloquinones [M + 2H] that are rapidly deprotonated to produce $[M + H]^-$. However, high source temperatures (400°C vs 700°C) favor the formation of the $M^{-\bullet}$ ion. The main product ions are the halogens themselves (Cl or Br) or result from the loss of HX, CO, or both moieties. While the transition $[M + H]^-$ to X is usually the most abundant SRM transition, used for quantification, in the case of 23 DB-56DM-BQ only $M^{-\bullet}$ to Br is formed. Although the most stable and intense signal of HBQs under negative electrospray ionization corresponds with $[M + 2H]^-$, in the case of their analogs OH-HBQs the best precursor ion for their monitoring is compound-dependent [36].

Methods employing LC-MS to analyze nitrosamines use ESI in positive mode [104, 161–163, 184–188]. Zhao et al. [161] tested ESI and atmospheric pressure chemical ionization (APCI) for ionization of the target nitrosamines. The ESI produces characteristic ions of the parent compounds and product ions for all nitrosamines, whereas the APCI cannot generate the parent ion of thermally unstable nitrosamines such as NDPhA. The detection of both the parent and product ions is important for the specific determination of nitrosamines at trace concentrations in water. Therefore, ESI is preferred as an interface of the LC with the tandem MS [161].

9.3.2 Figures of Merit

9.3.2.1 Linearity

Linearity is obtained in most analytical approaches with the internal standard calibration method. The linearity of the different methods developed to analyze DBPs expands between 2 to 3 orders of magnitude, e.g., from the limit of quantification (LOQ) to 100 μg/L for THM4 and HAAs [124, 156], up to 2 or 20 μg/L for I-THMs [47, 120], and 8 or 10 μg/L for monohalogenated and dihalogenated HALs [9]. The linearity range for GC-MS methods are usually narrower than for LC-MS methods, but this aspect is also compound-specific.

9.3.2.2 Precision and Accuracy

The analysis of halogenated DBPs in water samples is performed in many occasions with multi-residue generic methods, and therefore the conditions adopted may compromise the performance of the method for some DBPs, particularly the most volatile ones. TCM and TCAN are two DBPs that are often affected by methods that require an evaporation step. The absolute extraction recoveries (total amount of DBP recovered from the initial spike to the final extract) for TCM and TCAN have been reported below 30% [76]. Despite these low extraction recoveries, the accuracy of these LLE methods is still high (e.g., average trueness of 100% for TCM or 83% for TCAN) because calibration curves are prepared by adding standard spikes to water which are processed with the same extraction method as samples. Furthermore, high precision values (relative standard values (RSD) of 6% for TCM or TCAN) ensures the reliability of the measurements with these multi-residue LLE methods [76, 137]. An artifact also occurs during multi-residue LLE that result in elevated analyte recoveries, such as trihalogenated HALs (164–217%) [76]. However, HALs show an average accuracy between 71 and 124 %, with RSD values below 13% [76]. With these exceptions, the performance of multi-residue methods optimized to accommodate a wide variety of DBPs classes does not differ significantly compared to methods developed exclusively for individual DBP classes.

A good agreement between measured and theoretical spiked concentrations (bias below 12% and precision below 16%) is obtained for THMs in different water matrices when concentrating 1 L of water with SPE [134]. The HS-SPME extraction of all ten THMs provides recoveries between 84 and 103%, with RSD values below 15% for all THMs except for TCM. An, RSD value of 23% was observed for TCM, most likely due to its high volatility [120].

Since analytical standards are not commercially available for the oximes of monohalogenated and dihalogenated HALs, the recovery of the derivatization and the extraction process for these HALs cannot be evaluated [9]. However, HAL conversion during PFBHA derivatization has been consistently observed to be over 75% [54], and quantification by the internal standard method can correct any artifact that may occur during the derivatization step. The extraction recoveries of trihalogenated HALs using SPE decrease in the order TBAL (>82%) > dibromochloroacetaldehyde (58–72%) > bromodichloroacetaldehyde (43–68%) > TCAL (23–30%), in line with the decreasing polarity of these compounds [9]. The RSD values of concentrations measured in n = 6 fortified waters are below 12% [9].

In the case of HANs, absolute extraction recoveries with LLE approaches are between 20–153%, where TCAN has the lowest values (20–34%) [76, 138]. HANs were evaluated in different matrices with overall good accuracy (70–130%) [138]. Using the multiple LLE approach [76], absolute recoveries at 10 μg/L of the seven HNMs studied (which excluded monohalogenated HNMs) ranged from 26 to 51% in ultrapure water, 40–70% in chloraminated tap water, and 50–94% in chlorinated tap water. For a single LLE method [80], all 9 HNMs were evaluated, and spiked recoveries at 5 μg/L ranged from 63–116% in tap water, 58–127% in pool water, and 76–130% in spa water (with RSD values below 20%). Recoveries of HNMs in tap water after LLME at 5 and 20 μg/L ranged between 92 to 99%, with RSD below 9% [143]. Recoveries of 8 HNMs (excluding CNM) using the SPE approach with matrix spikes ranged from 71 to 87% (spiked concentration not reported) [82].

Most methods employed to determine HAMs present good accuracy (trueness) (60–100%) [131, 140, 149], although the absolute extraction recoveries of HAMs with a multi-residue LLE method are in general low [76].

As for the analysis of HAAs, direct injection methods provide figures of merit in terms of trueness and sensitivity comparable to those obtained with SPE and LLE preconcentration methods. The worst performance of direct injection LC-MS methods has been consistently reported for CAA, which shows trueness values around 75% or below and RSD values up to 25% for the lowest concentration [53, 123]. This could be attributed to the presence of matrix effects since CAA is a very small molecule with a very short retention time. For the remaining compounds, trueness values in direct injection LC-MS/MS methods are in the range 91–120% and RSD values are below 17% [123]. Improved RSD values (<12%) are reported for all target HAAs with direct injection IC-MS/MS [129]. LLE and subsequent derivatization of HAAs also provides satisfactory accuracy and precision figures for HAA9 (all Cl and Br-containing HAAs) (trueness values between 81 and 107% and RSD values below 19%) [156]. The performance of SPE-based approaches depends strongly on the sorbent selected for extraction. The Bakerbond Carbon sorbent shows good retention capacity for monohalogenated HAAs (100%) but performs worse in the case of trihalogenated HAAs (47–69%) [124].

The HBQ recoveries in offline SPE approaches at 10 and 20 ng/L are between 46 and 102% [36, 146]. Recoveries below 60% are observed for 25DCBQ and TetraB-BQ isomers. Matrix effects in the form of ion suppression effect are between 7 (TetraCBQ) and 87% (23DB56DMBQ) [146]. In the online SPE approach, the recoveries of HBQs at their LOQ are between 90 and 117% on average in bottled water and between 45 and 133% in tap water, due to matrix effects. RSD values of the results are below 12% in both offline and online approaches.

The most challenging issue for the analysis of this NDMA is the percentage of the recovery that should be between 70 and 130% [135]. To quantify this recovery and avoid matrix effects, an internal standard (NDMA-d_6) is used.

9.3.2.3 Sensitivity

While GC-ECD offers higher sensitivity than GC-MS, it provides lower specificity, and therefore is not ideal for analyzing complex multi-residue mixtures.

The LOQs of THMs in LLE-based methods vary from 30 to 100 ng/L when the final LLE extract is concentrated under nitrogen or between 40 and 390 ng/L when the final extract is not evaporated. The HS-SPME approach can reliably quantify THMs between 4 ng/L (iodoform) and 68 ng/L (TCM). SPE of 1 L of water provided LOQs for THMs between 3 and 500 ng/L.

Direct injection LC-MS/MS approaches allow the quantification of most HAAs between 0.06 and 6 μg/L in water samples [53, 123]. LOQs in the sub-ppb level (<0.41 μg/L) for all targeted HAAs are reported with IC-MS/MS methods [129, 130]. Overall, the worst sensitivity in direct injection methods is reported for CAA and trihalogenated HAAs. LOQs similar to those previously indicated can be also obtained with LLE and GC-MS/MS [156]. The concentration of 1 L of water using SPE and subsequent GC-MS/MS analysis provides LOQs between 0.01 and 0.5 μg/L [124].

Monohalogenated and dihalogenated HALs can be quantified between 0.1 and 0.5 μg/L in water using the PFBHA derivatization method [9]. The sensitivity for trihalogenated HALs achieved with the method that use multiple LLE cycles and an extract concentration step (LOQs = 0.03–0.1 μg/L) [76] is slightly better than those methods using one single LLE cycle (LOQs = 0.25–0.5 μg/L) [51] or SPE extraction (LOQs = 0.25–1 μg/L) [9]. The LLME approach allows the detection of dihalogenated and trihalogenated HALs in water at concentrations between 0.006 and 0.02 μg/L.

LOQs of HBQs are lower in offline approaches than in online approaches, because in the latter a conservative approach to avoid false positives is adopted. Offline methods allow the quantification of the target HBQs at levels below 8.4 ng/L. In online approaches, dihalogenated HBQs can be quantified at similar levels (≤5 ng/L) whereas highly halogenated compounds and methylated species present LOQs between 50 and 1000 ng/L.

Most analytical methods for HAN quantification have low method detection limits (MDLs). MDLs are much lower for methods that utilize HS-SPME (0.9–80 ng/L) [132, 150], triple LLE (3.2–100 ng/L) [76, 137, 138], or single LLME (1.2–44 ng/L) [142] compared to SPE (10–2000 ng/L) [134] and single LLE (80–1700 ng/L) [80].

MDLs with LLE and nitrogen evaporation were 0.008 μg/L for dihalogenated HNMs and TCNM, 0.05 μg/L for BDCNM, and 0.1 μg/L for DBCNM and TBNM. The LOQ was lowest for TCNM (0.015 μg/L) and highest for TBNM (0.5 μg/L) [76]. Montesinos et al. reported detection limits of 0.03–0.6 μg/L for headspace analysis and 0.06–1.2 μg/L using HS-SDME [152, 153]. Though SPE produced comparable MDLs to LLE, the analysis requires a high sample volume (1 L) and can be time-consuming. Additionally, headspace methods may not be suitable for less volatile DBPs and demand specialized instrumentation for analysis.

The sensitivity of HAMs in LLE-based approaches varies from 8 to 2000 ng/L, being slightly higher in those methods using only one extraction as compared to those using multiple extractions. All methods that utilized LLE, concentrated the final extract under a gentle stream of nitrogen. SPE coupled with LC-MS offered better detection limits for these compounds (20–60 ng/L) than GC-MS (usually above 100 ng/L), particularly the less halogenated HAMs (CAM, BAM, iodoacetamide). This is likely because these compounds are highly hydrophilic, which also explains the poor recoveries for some HAMs using LLE.

The NDMA limits of detection of reported methods are at nanogram per liter in drinking water. Table 9.15 shows a summary of method detection limits for different works published in literature.

9.3.3 Hints and Tips

To avoid the loss of the most volatile DBPs, water samples should be collected in headspace-free bottles. To preserve original DBP levels, the residual chlorine should be removed by adding a quenching agent such as ammonium chloride or ascorbic acid and/or adjusting the sample pH. In any case, analysis of water samples within 48 hours of collection is recommended.

Table 9.15 Limit of detection of selected methodologies published for the determination of nitrosamine in water.

Sample prep. & detection technique (reference)	Nitrosamines (ng/L)							
	NDMA	NMEA	NDEA	NDPA	NDBA	NPIP	NPYR	NMOR
SPE-GC-EI-MS/MS (TQ) [166]	1.1	1.8	1.6	2.3	3.1	2.2	1.7	1.9
GC-PCI-MS/MS (IT) [135]	0.3	0.3	0.3	0.3	0.4	0.7	0.4	n/a
SPE-GC-PCI-MS/MS (TQ) [164]	0.5	0.6	0.9	0.8	1.7	0.9	1.2	0.7
SPE LC-ESI-MS/MS (Orbitrap) [163]	0.4	0.9	0.7	0.8	3.3	0.1	1	0.2
SPE LC-ESI-MS/MS (QTRAP) [161]	3.1	2.4	10.6	0.2	3.1	0.9	2.1	0.2
SPE-GC-EI-MS (Q) [102]	10	n/a	3	n/a	13	4	12	7
SPE GC-PCI-MS/MS (IT) [159]	1	1.9	1.4	1.4	2.7	1.1	1.5	0.8
SPME GC-PCI-MS/MS (TQ) [175]	30	90	64	59	79	59	n/a	138
SPE-GC-AEI- MS/MS (TQ) [105]	0.1	0.1	0.1	0.5	0.1	0.1	0.5	n/a

IT: ion trap, TQ: triple quadrupole, Q: quadrupole, QTRP: linear ion trap, n/a: not analyzed.

In LLE approaches, the surrogate standard can be either added to the water sample as a small aliquot of a concentrated MTBE solution (50 μL or 100 μL) or dissolved in the extraction solvent [125]. When several samples are being sequentially prepared for LLE, after the addition of Na_2SO_4 to each vial, the sample vial should be shaken and placed in the horizontal position to avoid solidification of the salt at the bottom of the vial, which will not dissolve afterward. The use of $MgSO_4$ as a salting-out agent results in the degradation of brominated acetaldehydes into their corresponding THMs, due to its exothermic hydration. Because of the exothermic reaction of sulfuric acid with water, in those approaches that require lowering the water pH, adding the sulfuric acid before the salting-out agent favors the dissolution of salts.

As for LLME-based methods, manual and vortex shaking provides better analyte recoveries than ultrasonic or magnetic stirring. In these methods, the salting-out agent needs to be added at the end of the LLE process.

The use of PTV inlets that allow to start the injection at low temperature may help retain TCM and TCAN in the column and avoid their elution in the chromatographic solvent peak or just after it. SPME fibers or specific GC columns may be also used to address this constraint. Low injection temperatures are also required to avoid the thermal degradation of trihalogenated HNMs (injection-port temperature ≤170°C and transfer line temperature ≤225°C).

Due to their polarity, the derivatization of HAAs is required for their GC-MS determination. A common derivatization technique is acidic methanol, which converts HAAs to the corresponding methyl ester (HAAMEs). The use of acidic methanol over diazomethane is recommended since it is a less toxic methylating agent alternative. However, contradictory results are provided regarding the process efficiency using acidic methanol as indicated in the US EPA method 552.2 [139], especially for trihalogenated HAAs [189, 190]. Furthermore, one study indicated that HAMs can interfere with HAAs analysis by generating the same corresponding HAAMEs that are generated when HAAs react with methyl alcohol [191].

The concentration of the LLE and SPE extracts under nitrogen may increase the method sensitivity; however, drying the extract completely during the evaporation step results in the loss of the target DBPs.

Because of general instability of some DBPs, preparing stock standard solutions monthly is advised for correct calibration. The construction of calibration curves in water and processing the calibration solutions as if they were samples may result in more accurate results compared to solvent-based calibration curves. When analyzing a suite of 61 DBPs from various classes, TBNM, BDCNM, and DBCNM must be calibrated separately from other DBPs due to their reactivity [76]. Calibration using the internal standard method or the isotope dilution method is recommended. In the absence of isotopically labelled standards, quantification in LC-MS approaches should be done with the standard addition method to effectively remove matrix effects.

Certain DBPs, particularly *N*-nitrosamines, are potential carcinogens for humans and animals. Safety precautions have to be taken when handling these compounds, and waste disposal must follow proper safety procedures. If nitrite is present in solution, NDMA may form during SPE with activated carbon cartridges [192].

9.4 Future Directions and Challenges

While most research on known DBPs has been studied in drinking water, the formation mechanisms and occurrence of these compounds in reclaimed wastewater are not well characterized. This field is of increasing scientific interest because wastewater reuse is key to achieve a circular water economy. The composition of treated wastewater effluents and source drinking water is expected to be different, and so the DBP mixtures formed after disinfection processes may also be different.

Known DBPs represent only a small fraction of the halogenated material formed during the disinfection process. In this regard, HRMS has been demonstrated to be a valuable analytical tool to fully characterize DBP mixtures [3], although the extraction and detection of the most polar components still represents a challenge. The confirmation of the identity of newly discovered DBPs requires analytical standards. This is sometimes difficult because they may not be commercially available. However, a big effort should be directed towards the production of pure standards so that the occurrence and toxicity of newly discovered compounds can be properly assessed.

The development of target multi-class methods is encouraged to obtain valuable information at once for many DBPs, minimizing economical and lab-effort resources.

Acknowledgements

CP acknowledges the funding from the second call of the ComFuturo Programme by the FGCSIC. MJF acknowledges her Ramón y Cajal fellowship (RyC-2015-17108), from the AEI-MICIU. CP would also like to acknowledge the support of grant CEX2018-000794-S, funded by MCIN/AEI/10.13039/501100011033.

References

1 Richardson, S.D., Plewa, M.J., Wagner, E.D., Schoeny, R., and DeMarini, D.M. (2007). Occurrence, genotoxicity, and carcinogenicity of regulated and emerging disinfection by-products in drinking water: a review and roadmap for research. *Mutat. Res. – Rev. Mutat. Res.* 636: 178–242. doi: 10.1016/j.mrrev.2007.09.001.

2 Doederer, K., Gernjak, W., Weinberg, H.S., and Farré, M.J. (2014). Factors affecting the formation of disinfection by-products during chlorination and chloramination of secondary effluent for the production of high quality recycled water. *Water Res.* 48: 218–228. doi: 10.1016/j.watres.2013.09.034.

3 Postigo, C. and Richardson, S.D. (2021). Non-target screening and novel methods based on mass spectrometry detection for identification of unknown disinfection byproducts. In: *Analysis and Formation of Disinfection Byproducts in Drinking Water*. Compr. Anal. Chem. Ser. (T. Manasfi and J.-L. Boudenne, ed.), 1–29. Elsevier. doi: 10.1016/bs.coac.2021.01.001.

4 Wagner, E.D. and Plewa, M.J. (2017). CHO cell cytotoxicity and genotoxicity analyses of disinfection by-products: an updated review. *J. Environ. Sci. (China)* 58: 64–76. doi: 10.1016/j.jes.2017.04.021.

5 Plewa, M.J., Wagner, E.D., Muellner, M.G., Hsu, K.-M., and Richardson, S.D. (2008). Comparative mammalian cell toxicity of N-DBPs and C-DBPs. *Crop Sci.* doi: 10.1021/bk-2008-0995.ch003.

6 Bellar, T.A., Lichtenberg, J.J., and Kroner, R.C. (1974). Occurrence of organohalides in chlorinated drinking waters. *J. Am. Water Work. Assoc.* 66: 703–706. http://www.jstor.org/stable/41267646. Accessed 14 December 2021.

7 Rook, J.J. (1974). Formation of haloforms during chlorination of natural waters. *J. Water Treat. Exam.* 23: 234–243.

8 International Agency for Research on Cancer – IARC monographs on the identification of carcinogenic hazards to humans – list of classifications. Retrieved from https://bit.ly/3w8a9z7. Accessed in November 2020.

9 Jeong, C.H., Postigo, C., Richardson, S.D., Simmons, J.E., Kimura, S.Y., Mariñas, B.J., Barcelo, D., Liang, P., Wagner, E.D., and Plewa, M.J. (2015). Occurrence and comparative toxicity of haloacetaldehyde disinfection byproducts in drinking water. *Environ. Sci. Technol.* 49: 13749–13759. doi: 10.1021/es506358x.

10 Wang, W., Moe, B., Li, J., Qian, Y., Zheng, Q., and Li, X.-F. (2016). Analytical characterization, occurrence, transformation, and removal of the emerging disinfection byproducts halobenzoquinones in water. *TrAC – Trends Anal. Chem.* 85: 97–110. doi: 10.1016/j.trac.2016.03.004.

11 Krasner, S.W., Mitch, W.A., McCurry, D.L., Hanigan, D., and Westerhoff, P. (2013). Formation, precursors, control, and occurrence of nitrosamines in drinking water: a review. *Water Res.* 47: 4433–4450. doi: 10.1016/j.watres.2013.04.050.

12 Scientific Committee on Consumer Products SCCP. Opinion on the presence and release of nitrosamines and nitrosatable compounds from rubber balloons SCCP/1132/07. Retrieved from https://bit.ly/3u5xxeC. Accessed in November 2020.

13 US EPA, Comprehensive Disinfectants and Disinfection Byproducts Rules (Stage 1 and Stage 2): Quick Reference Guide. EPA816-F-10-080, Office of Water. Retrieved from https://bit.ly/2xXmFHP. Accessed in March 2020.

14 COM/2017/0753 final – 2017/0322 (COD) – Proposal for a Directive of the European Parliament and of the Council on the quality of water intended for human consumption (recast). Retrieved from: https://bit.ly/3fmCHyO. Accessed in November 2020.

15 National Standard of the People's Republic of China, GB 5749-2006 Standards for Drinking Water Quality. Retrieved from https://bit.ly/3aoXlcp. Accessed in March 2020.

16 World Health Organization. (2017). Guidelines for drinking-water quality: fourth edition incorporating the 1st addendum. Geneva. https://www.who.int/publications/i/item/9789241549950. Accessed 14 December 2021.

17 World Health Organization. (2004). Guidelines for Drinking Water Quality, 3rd edition. Vol 1. Recommendations. Geneve. Retrieved from https://bit.ly/3fsAD8k. Accessed in November 2020.

18 National Health and Medical Research Council (NRMMC). (2011). Australian drinking water guidelines 6: National Water Quality Management Strategy. Version 3.5. updated August 2018. Commonwealth of Australia.

19 World Health Organization. (2004). Halogenated acetonitriles in drinking-water: background document for development of WHO guidelines for drinking-water quality.

20 US EPA. (2016). Fourth drinking water contaminant candidate list – CCL4. Retrieved from https://bit.ly/3fwkTBh. Accessed in November 2020.

21 California State Water Resources and Control Board. NDMA and other Nitrosamines 2014. https://www.waterboards.ca.gov/drinking_water/certlic/drinkingwater/NDMA.html. Accessed in November 2020.

22 California Office of Environmental Health Hazard Assessment. (2006). Public health goal for N-nitrosodimethylamine in drinking water. Retrieved from https://bit.ly/39ssOMg. Accessed in November 2020.

23 Massachusetts Department of Environmental Protection. (2014). Standards and guidelines for contaminants in Massachusetts drinking waters. Retrieved from https://bit.ly/3wamDpH. Accessed in November 2020.

24 Health Canada. Guidelines for Canadian drinking water quality – summary table. Retreived from https://bit.ly/2PDd2ae. Accessed in November 2020.

25 Ontario Ministry of the Environment. (2006). Technical support document for Ontario drinking water standards, objectives and guidelines. Retrieved from https://bit.ly/3sDBtTF. Accessed in November 2020.

26 Umwelt Bundesamt. Wasserqualität der Schwimm- und Badebecken. Retrieved from: https://www.umweltbundesamt.de/themen/wasser/schwimmen-baden/schwimm-badebecken/wasserqualitaet-3kZ1GqY. Accessed November 2020.

27 Simard, S., Tardif, R., and Rodriguez, M.J. (2013). Variability of chlorination by-product occurrence in water of indoor and outdoor swimming pools. *Water Res.* 47: 1763–1772. doi: 10.1016/j.watres.2012.12.024.

28 Bouwer, E.J., Rittmann, B.E., and McCarty, P.L. (1981). Anaerobic degradation of halogenated 1- and 2-carbon organic compounds. *Environ. Sci. Technol.* 15: 596–599. doi: 10.1021/es00087a012.

29 Zhang, X., Yang, H., Wang, X., Karanfil, T., and Xie, Y.F. (2015). Trihalomethane hydrolysis in drinking water at elevated temperatures. *Water Res.* 78: 18–27. doi: 10.1016/j.watres.2015.03.027.

30 Nicole, I., De Laat, J., Dore, M., Duguet, J.P., and Suty, H. (1991). Etude de la degradation des trihalo-methanes en milieu aqueux dilue par irradiation UV - determination du rendement quantique de photolyse a 253,7 nm. *Environ. Technol.* 12: 21–31. doi: 10.1080/09593339109384978.

31 Zhang, X. and Minear, R.A. (2002). Decomposition of trihaloacetic acids and formation of the corresponding trihalomethanes in drinking water. *Water Res.* 36: 3665–3673. doi: 10.1016/S0043-1354(02)00072-6.

32 Hua, G. and Reckhow, D.A. (2012). Effect of alkaline pH on the stability of halogenated DBPs. *J. AWWA* 104: E107–E120. doi: 10.5942/jawwa.2012.104.0025.

33 Reckhow, D.A., Singer, P.C., and Malcolm, R.L. (1990). Chlorination of humic materials: byproduct formation and chemical interpretations. *Environ. Sci. Technol.* 24: 1655–1664. doi: 10.1021/es00081a005.

34 Chen, B. (2011). Hydrolytic stabilities of halogenated disinfection byproducts: review and rate constant quantitative structure–property relationship analysis. *Environ. Eng. Sci.* 28: 385–394. doi: 10.1089/ees.2010.0196.

35 Xie, Y. and Reckhow, D.A. (1996). Hydrolysis and dehalogenation of trihaloacetaldehydes. *Disinfect. By-Products Water Treat. Chem. Their Form. Control* 283–291. doi: 10.4324/9781315141350.

36 Wang, W., Qian, Y., Li, J., Moe, B., Huang, R., Zhang, H., Hrudey, S.E., and Li, X.-F.X.-F. (2014). Analytical and toxicity characterization of halo-hydroxyl-benzoquinones as stable halobenzoquinone disinfection byproducts in treated water. *Anal. Chem.* 86: 4982–4988. doi: 10.1021/ac5007238.

37 Jones, M.J. (2005). *Organic Chemistry*, 3rd ed. New York: Norton & Co.

38 Glezer, V., Harris, B., Tal, N., Iosefzon, B., and Lev, O. (1999). Hydrolysis of haloacetonitriles: linear free energy relationship, kinetics and products. *Water Res.* 33: 1938–1948. doi: 10.1016/S0043-1354(98)00361-3.

39 Ding, S., Chu, W., Krasner, S.W., Yu, Y., Fang, C., Xu, B., and Gao, N. (2018). The stability of chlorinated, brominated, and iodinated haloacetamides in drinking water. *Water Res.* 142: 490–500. doi: 10.1016/j.watres.2018.06.024.

40 Ding, S., Wang, F., Chu, W., Cao, Z., Pan, Y., and Gao, N. (2018). Rapid degradation of brominated and iodinated haloacetamides with sulfite in drinking water: degradation kinetics and mechanisms. *Water Res.* 143: 325–333. doi: 10.1016/j.watres.2018.06.049.

41 Chu, W., Li, X., Bond, T., Gao, N., Bin, X., Wang, Q., and Ding, S. (2016). Copper increases reductive dehalogenation of haloacetamides by zero-valent iron in drinking water: reduction efficiency and integrated toxicity risk. *Water Res.* 107: 141–150. doi: 10.1016/j.watres.2016.10.047.

42 Lee, J.-Y., Pearson, C.R., Hozalski, R.M., and Arnold, W.A. (2008). Degradation of trichloronitromethane by iron water main corrosion products. *Water Res.* 42: 2043–2050. doi: 10.1016/j.watres.2007.12.012.

43 Fang, J.-Y., Ling, L., and Shang, C. (2013). Kinetics and mechanisms of pH-dependent degradation of halonitromethanes by UV photolysis. *Water Res.* 47: 1257–1266. doi: 10.1016/j.watres.2012.11.050.

44 Chen, B., Lee, W., Westerhoff, P.K., Krasner, S.W., and Herckes, P. (2010). Solar photolysis kinetics of disinfection byproducts. *Water Res.* 44: 3401–3409. doi: 10.1016/j.watres.2010.03.014.

45 Mezyk, S.P., Helgeson, T., Cole, S.K., Cooper, W.J., Fox, R.V., Gardinali, P.R., and Mincher, B.J. (2006). Free radical chemistry of disinfection-byproducts. 1. Kinetics of hydrated electron and hydroxyl radical reactions with halonitromethanes in water. *J. Phys. Chem. A* 110: 2176–2180. doi: 10.1021/jp054962+.

46 Plumlee, M.H. and Reinhard, M. (2007). Photochemical attenuation of N-nitrosodimethylamine (NDMA) and other nitrosamines in surface water. *Environ. Sci. Technol.* 41: 6170–6176. doi: 10.1021/es070818l.

47 Ioannou, P., Charisiadis, P., Andra, S.S., and Makris, K.C. (2016). Occurrence and variability of iodinated trihalomethanes concentrations within two drinking-water distribution networks. *Sci. Total Environ.* 543: 505–513. doi: 10.1016/j.scitotenv.2015.10.031.

48 Richardson, S.D., Fasano, F., Ellington, J.J., Crumley, F.G., Buettner, K.M., Evans, J.J., Blount, B.C., Silva, L.K., Waite, T.J., Luther, G.W., McKague, B.A., Miltner, R.J., Wagner, E.D., and Plewa, M.J. (2008). Occurrence and mammalian cell toxicity of iodinated disinfection byproducts in drinking water. *Environ. Sci. Technol.* 42: 8330–8338. doi: 10.1021/es801169k.

49 Postigo, C. and Zonja, B. (2019). Iodinated disinfection byproducts: formation and concerns. *Curr. Opin. Environ. Sci. Heal.* 7: 19–25. doi: 10.1016/J.COESH.2018.08.006.

50 Chang, H.H., Tung, H.H., Chao, C.C., and Wang, G.S. (2010). Occurrence of haloacetic acids (HAAs) and trihalomethanes (THMs) in drinking water of Taiwan. *Environ. Monit. Assess.* 162: 237–250. doi: 10.1007/s10661-009-0792-1.

51 Postigo, C., Emiliano, P., Barceló, D., and Valero, F. (2018). Chemical characterization and relative toxicity assessment of disinfection byproduct mixtures in a large drinking water supply network. *J. Hazard. Mater.* 359: 166–173. doi: 10.1016/j.jhazmat.2018.07.022.

52 Wang, L., Chen, Y., Chen, S., Long, L., Bu, Y., Xu, H., Chen, B., and Krasner, S. (2019). A one-year long survey of temporal disinfection byproducts variations in a consumer's tap and their removals by a point-of-use facility. *Water Res.* 159: 203–213. doi: 10.1016/j.watres.2019.04.062.

53 Planas, C., Palacios, Ó., Ventura, F., Boleda, M.A., Martín, J., and Caixach, J. (2019). Simultaneous analysis of 11 haloacetic acids by direct injection-liquid chromatography-electrospray ionization-triple quadrupole tandem mass spectrometry and high resolution mass spectrometry: occurrence and evolution in chlorine-treated water. *Anal. Bioanal. Chem.* 411: 3905–3917. doi: 10.1007/s00216-019-01864-5.

54 Weinberg, H.S., Krasner, S.W., Richardson, S.D., and Thruston, A.D. (2002). The occurrence of disinfection by-products (DBPs) of health concern in drinking water: results of a nationwide DBP occurrence study. EPA/600/R02/068, 1–460.

55 Mazhar, M.A., Khan, N.A., Ahmed, S., Khan, A.H., Hussain, A., Changani, R.F., Yousefi, M., Ahmadi, S., and Vambol, V. (2020). Chlorination disinfection by-products in municipal drinking water – a review. *J. Clean. Prod.* 273: 123159. doi: 10.1016/j.jclepro.2020.123159.

56 Yang, L., Chen, X., She, Q., Cao, G., Liu, Y., Chang, V.W.-C., and Tang, C.Y. (2018). Regulation, formation, exposure, and treatment of disinfection by-products (DBPs) in swimming pool waters: a critical review. *Environ. Int.* 121: 1039–1057. doi: 10.1016/j.envint.2018.10.024.

57 Zhao, H., Yang, L., Li, Y., Xue, W., Li, K., Xie, Y., Meng, S., and Cao, G. (2020). Environmental occurrence and risk assessment of haloacetic acids in swimming pool water and drinking water. *RSC Adv.* 10: 28267–28276. doi: 10.1039/d0ra02389b.

58 Krasner, S.W., Weinberg, H.S., Richardson, S.D., Pastor, S.J., Chinn, R., Sclimenti, M.J., Onstad, G.D., Thruston, A.D. Jr. (2006). Occurrence of a new generation of disinfection byproducts. *Environ. Sci. Technol.* 40: 7175–7185. doi: 10.1021/es060353j.

59 Lee, J., Jun, M.-J., Lee, M.-H., Lee, M.-H., Eom, S.-W., and Zoh, K.-D. (2010). Production of various disinfection byproducts in indoor swimming pool waters treated with different disinfection methods. *Int. J. Hyg. Environ. Health* 213: 465–474. doi: 10.1016/j.ijheh.2010.09.005.

60 Koudjonou, B., LeBel, G.L., and Dabeka, L. (2008). Formation of halogenated acetaldehydes, and occurrence in Canadian drinking water. *Chemosphere* 72: 875–881. doi: 10.1016/j.chemosphere.2008.03.062.

61 Mao, Y., Qi, S., Zhao, E., Yang, H., and Xie, Y.F. (2019). Mechanism of ozonation enhanced formation of haloacetaldehydes during subsequent chlorination. *Chemosphere* 236: 124361. doi: 10.1016/j.chemosphere.2019.124361.

62 Zhao, Y., Anichina, J., Lu, X., Bull, R.J., Krasner, S.W., Hrudey, S.E., and Li, X.-F. (2012). Occurrence and formation of chloro- and bromo-benzoquinones during drinking water disinfection. *Water Res.* 46: 4351–4360. doi: 10.1016/j.watres.2012.05.032.

63 Cuthbertson, A.A., Bach, C., Richardson, S.D., and Dauchy, X. (2020). A novel automated method for the quantification of ten halobenzoquinones in drinking water using online solid-phase extraction coupled with liquid chromatography tandem mass spectrometry. *J. Chromatogr. A* 1612. doi: 10.1016/j.chroma.2019.460642.

64 Richardson, S.D. (2011). Disinfection by-products: formation and occurrence in drinking water. DOI: 10.1016/B978-0-444-52272-6.00276-2.

65 McGuire, M.J., McLain, J.L., and Obolensky, A. (2002). *Information Collection Rule Data Analysis*. Denver, CO: AwwaRF and AWWA.

66 Mian, H.R., Hu, G., Hewage, K., Rodriguez, M.J., and Sadiq, R. (2018). Prioritization of unregulated disinfection by-products in drinking water distribution systems for human health risk mitigation: a critical review. *Water Res.* 147: 112–131. doi: 10.1016/j.watres.2018.09.054.

67 Bond, T., Templeton, M.R., Mokhtar Kamal, N.H., Graham, N., and Kanda, R. (2015). Nitrogenous disinfection byproducts in English drinking water supply systems: occurrence, bromine substitution and correlation analysis. *Water Res.* 85: 85–94. doi: 10.1016/j.watres.2015.08.015.

68 Kosaka, K., Ohkubo, K., and Akiba, M. (2016). Occurrence and formation of haloacetamides from chlorination at water purification plants across Japan. *Water Res.* 106: 470–476. doi: 10.1016/j.watres.2016.10.029.

69 Kristiana, I., Liew, D., Henderson, R.K., Joll, C.A., and Linge, K.L. (2017). Formation and control of nitrogenous DBPs from Western Australian source waters: investigating the impacts of high nitrogen and bromide concentrations. *J. Environ. Sci.* 58: 102–115. doi: 10.1016/j.jes.2017.06.028.

70 Liew, D., Linge, K.L., and Joll, C.A. (2016). Formation of nitrogenous disinfection by-products in 10 chlorinated and chloraminated drinking water supply systems. *Environ. Monit. Assess.* 188: 518. doi: 10.1007/s10661-016-5529-3.

71 Bai, X., Ma, X., Xu, F., Li, J., Zhang, H., and Xiao, X. (2015). The drinking water treatment process as a potential source of affecting the bacterial antibiotic resistance. *Sci. Total Environ.* 533: 24–31. doi: 10.1016/j.scitotenv.2015.06.082.

72 Plewa, M.J., Muellner, M.G., Richardson, S.D., Fasano, F., Buettner, K.M., Woo, Y.-T., Mckague, A.B., and Wagner, E.D. (2008). Occurrence, synthesis, and mammalian cell cytotoxicity and genotoxicity of haloacetamides: an emerging class of nitrogenous drinking water disinfection byproducts. *Environ. Sci. Technol.* 42: 955–961. doi: 10.1021/es071754h.

73 Cuthbertson, A.A., Kimura, S.Y., Liberatore, H.K., Summers, R.S., Knappe, D.R.U., Stanford, B.D., Maness, J.C., Mulhern, R.E., Selbes, M., and Richardson, S.D. (2019). Does granular activated carbon with chlorination produce safer drinking water? From disinfection byproducts and total organic halogen to calculated toxicity. *Environ. Sci. Technol.* 53: 5987–5999. doi: 10.1021/acs.est.9b00023.

74 Cuthbertson, A.A., Kimura, S.Y., Liberatore, H.K., Knappe, D.R.U., Stanford, B., Summers, R.S., Dickenson, E.R., Maness, J.C., Glover, C., Selbes, M., and Richardson, S.D. (2020). GAC to BAC: does it make chloraminated drinking water safer? *Water Res.* 172: 115432. doi: 10.1016/j.watres.2019.115432.

75 Chu, W., Gao, N., Yin, D., and Krasner, S.W. (2013). Formation and speciation of nine haloacetamides, an emerging class of nitrogenous DBPs, during chlorination or chloramination. *J. Hazard. Mater.* 260: 806–812. doi: 10.1016/j.jhazmat.2013.06.044.

76 Cuthbertson, A.A., Liberatore, H.K., Kimura, S.Y., Allen, J.M., Bensussan, A.V., and Richardson, S.D. (2020). Trace analysis of 61 emerging Br-, Cl-, and I-DBPs: new methods to achieve part-per-trillion quantification in drinking water. *Anal. Chem.* 92: 3058–3068. doi: 10.1021/acs.analchem.9b04377.

77 Allen, J.M., Cuthbertson, A.A., Liberatore, H.K., Kimura, S.Y., Mantha, A., Edwards, M.A., and Richardson, S.D. (2017). Showering in Flint, MI: is there a DBP problem? *J. Environ. Sci.* 58: 271–284. doi: 10.1016/j.jes.2017.06.009.

78 Allen, J.M., Plewa, M.J., Wagner, E.D., Wei, X., Bollar, G.E., Quirk, L.E., Liberatore, H.K., and Richardson, S.D. (2021). Making swimming pools safer: does copper–silver ionization with chlorine lower the toxicity and disinfection byproduct formation? *Environ. Sci. Technol.* 55: 2908–2918. doi: 10.1021/acs.est.0c06287.

79 Plewa, M.J., Wagner, E.D., Jazwierska, P., Richardson, S.D., Chen, P.H., and McKague, A.B. (2004). Halonitromethane drinking water disinfection byproducts: chemical characterization and mammalian cell cytotoxicity and genotoxicity. *Environ. Sci. Technol.* 38: 62–68. doi: 10.1021/es030477l.

80 Carter, R.A.A., Liew, D.S., West, N., Heitz, A., and Joll, C.A. (2019). Simultaneous analysis of haloacetonitriles, haloacetamides and halonitromethanes in chlorinated waters by gas chromatography-mass spectrometry. *Chemosphere* 220: 314–323. doi: 10.1016/j.chemosphere.2018.12.069.

81 Carter, R.A.A., Allard, S., Croué, J.-P., and Joll, C.A. (2019). Occurrence of disinfection by-products in swimming pools and the estimated resulting cytotoxicity. *Sci. Total Environ.* 664: 851–864. doi: 10.1016/j.scitotenv.2019.01.428.

82 Hladik, M.L., Focazio, M.J., and Engle, M. (2014). Discharges of produced waters from oil and gas extraction via wastewater treatment plants are sources of disinfection by-products to receiving streams. *Sci. Total Environ.* 466–467: 1085–1093. doi: 10.1016/j.scitotenv.2013.08.008.

83 Shi, J.L. and McCurry, D.L. (2020). Transformation of N-methylamine drugs during wastewater ozonation: formation of nitromethane, an efficient precursor to halonitromethanes. *Environ. Sci. Technol.* 54: 2182–2191. doi: 10.1021/acs.est.9b04742.

84 Hoigné, J. and Bader, H. (1988). The formation of trichloronitromethane (chloropicrin) and chloroform in a combined ozonation/chlorination treatment of drinking water. *Water Res.* 22: 313–319. doi: 10.1016/S0043-1354(88)90120-0.

85 Carter, R.A.A. and Joll, C.A. (2017). Occurrence and formation of disinfection by-products in the swimming pool environment: a critical review. *J. Environ. Sci.* 58: 19–50. doi: 10.1016/j.jes.2017.06.013.

86 Song, H., Addison, J.W., Hu, J., and Karanfil, T. (2010). Halonitromethanes formation in wastewater treatment plant effluents. *Chemosphere* 79: 174–179. doi: 10.1016/j.chemosphere.2010.01.001.

87 Hu, J., Song, H., and Karanfil, T. (2010). Comparative analysis of halonitromethane and trihalomethane formation and speciation in drinking water: the effects of disinfectants, pH, Bromide, and Nitrite. *Environ. Sci. Technol.* 44: 794–799. doi: 10.1021/es902630u.

88 Shah, A.D. and Mitch, W.A. (2012). Halonitroalkanes, halonitriles, haloamides, and N-nitrosamines: a critical review of nitrogenous disinfection byproduct formation pathways. *Environ. Sci. Technol.* 46: 119–131. doi: 10.1021/es203312s.

89 Krasner, S.W., Lee, T.C.F., Westerhoff, P., Fischer, N., Hanigan, D., Karanfil, T., Beita-Sandí, W., Taylor-Edmonds, L., and Andrews, R.C. (2016). Granular activated carbon treatment may result in higher predicted genotoxicity in the presence of bromide. *Environ. Sci. Technol.* 50: 9583–9591. doi: 10.1021/acs.est.6b02508.

90 Kemper, J.M., Westerhoff, P., Dotson, A., and Mitch, W.A. (2009). Nitrosamine, dimethylnitramine, and chloropicrin formation during strong base anion-exchange treatment. *Environ. Sci. Technol.* 43: 466–472. doi: 10.1021/es802460n.

91 Fine, D.H., Rounbehler, D.P., Rounbehler, A., Silvergleid, A., Sawicki, E., Krost, K., and DeMarrais, G.A. (1977). Determination of dimethylnitrosamine in air, water, and soil by thermal energy analysis: measurements in Baltimore, Md. *Environ. Sci. Technol.* 11: 581–584. doi: 10.1021/es60129a003.

92 Mitch, W.A., Sharp, J.O., Trussell, R.R., Valentine, R.L., Alvarez-Cohen, L., and Sedlak, D.L. (2003). N-Nitrosodimethylamine (NDMA) as a drinking water contaminant: a review. *Environ. Eng. Sci.* 20: 389–404. doi: 10.1089/109287503768335896.

93 Sgroi, M., Vagliasindi, F.G.A., Snyder, S.A., and Roccaro, P. (2018). N-Nitrosodimethylamine (NDMA) and its precursors in water and wastewater: a review on formation and removal. *Chemosphere* 191: 685–703. doi: 10.1016/j.chemosphere.2017.10.089.

94 Bond, T., Goslan, E.H., Parsons, S.A., and Jefferson, B. (2012). A critical review of trihalomethane and haloacetic acid formation from natural organic matter surrogates. *Environ. Technol. Rev.* 1: 93–113. doi: 10.1080/09593330.2012.705895.

95 Krasner, S.W., Lee, C.F.T., Mitch, W.A., and Von Gunten, U. (2018). Impact of combined chlorination and chloramination conditions on N-nitrosodimethylamine formation. *J. AWWA* 110: 11–24. doi: 10.1002/awwa.1128.

96 Planas, C., Palacios, Ó., Ventura, F., Rivera, J., and Caixach, J. (2008). Analysis of nitrosamines in water by automated SPE and isotope dilution GC/HRMS: occurrence in the different steps of a drinking water treatment plant, and in chlorinated samples from a reservoir and a sewage treatment plant effluent. *Talanta* 76: 906–913. doi: 10.1016/j.talanta.2008.04.060.

97 Bei, E., Shu, Y., Li, S., Liao, X., Wang, J., Zhang, X., Chen, C., and Krasner, S. (2016). Occurrence of nitrosamines and their precursors in drinking water systems around mainland China. *Water Res.* 98: 168–175. doi: 10.1016/j.watres.2016.04.013.

98 Liew, D., Culbert, J., Linge, K., Farré, M.J., Knight, N., Morran, J., Halliwell, D., Newcombe, G., and Charrois, J.W.A. (January 1, 2015). National occurrence of N-Nitrosodimethylamine (NDMA). In: *Recent Adv. Disinfect. By-Products, 135-149 SE–8*. American Chemical Society. doi: 10.1021/bk-2015-1190.ch008.

99 Wang, W., Yu, J., An, W., and Yang, M. (2016). Occurrence and profiling of multiple nitrosamines in source water and drinking water of China. *Sci. Total Environ.* 551–552: 489–495. doi: 10.1016/j.scitotenv.2016.01.175.

100 US EPA. (2014). Technical Fact Sheet – N-Nitrosodimethylamine (NDMA). Retrieved from https://bit.ly/3lJFiUx. Accessed in November, 2020.

101 Charrois, J.W.A., Boyd, J.M., Froese, K.L., and Hrudey, S.E. (2007). Occurrence of N-nitrosamines in Alberta public drinking-water distribution systems. *J. Environ. Eng. Sci.* 6: 103–114. doi: 10.1139/s06-031.

102 Jurado-Sánchez, B., Ballesteros, E., and Gallego, M. (2012). Occurrence of aromatic amines and N-nitrosamines in the different steps of a drinking water treatment plant. *Water Res.* 46: 4543–4555. doi: 10.1016/j.watres.2012.05.039.

103 Templeton, M.R. and Chen, Z. (2010). NDMA and seven other nitrosamines in selected UK drinking water supply systems. *J. Water Supply Res. Technol.* 59: 277–283. doi: 10.2166/aqua.2010.077.

104 Asami, M., Oya, M., and Kosaka, K. (2009). A nationwide survey of NDMA in raw and drinking water in Japan. *Sci. Total Environ.* 407: 3540–3545. doi: 10.1016/j.scitotenv.2009.02.014.

105 Farre, M.J. (2020). Occurrence of: N -nitrosamines and their precursors in Spanish drinking water treatment plants and distribution systems. *Environ. Sci. Water Res. Technol.* 6: 210–220.

106 Qian, Y., Wu, M., Wang, W., Chen, B., Zheng, H., Krasner, S.W., Hrudey, S.E., and Li, X.-F. (2015). Determination of 14 nitrosamines at nanogram per liter levels in drinking water. *Anal. Chem.* 87: 1330–1336. doi: 10.1021/ac504104k.

107 Wang, W., Ren, S., Zhang, H., Yu, J., An, W., Hu, J., and Yang, M. (2011). Occurrence of nine nitrosamines and secondary amines in source water and drinking water: potential of secondary amines as nitrosamine precursors. *Water Res.* 45: 4930–4938. doi: 10.1016/j.watres.2011.06.041.

108 Zhao, C., Lu, Q., Gu, Y., Pan, E., Sun, Z., Zhang, H., Zhou, J., Du, Y., Zhang, Y., Feng, Y., Liu, R., Pu, Y., and Yin, L. (2019). Distribution of N-nitrosamines in drinking water and human urinary excretions in high incidence area of esophageal cancer in Huai'an, China. *Chemosphere* 235: 288–296. doi: 10.1016/j.chemosphere.2019.06.124.

109 Chen, Z., Yang, L., Huang, Y., Spencer, P., Zheng, W., Zhou, Y., Jiang, S., Ye, W., Zheng, Y., and Qu, W. (2019). Carcinogenic risk of N-nitrosamines in Shanghai drinking water: indications for the use of ozone pretreatment. *Environ. Sci. Technol.* 53: 7007–7018. doi: 10.1021/acs.est.8b07363.

110 Yoon, S., Nakada, N., and Tanaka, H. (2013). Occurrence and fate of N-nitrosamines and their formation potential in three wastewater treatment plants in Japan. *Water Sci. Technol.* 68: 2118–2126. doi: 10.2166/wst.2013.458.

111 Sanchís, J., Gernjak, W., Munné, A., Catalán, N., Petrovic, M., and Farré, M.J. (2020). Fate of N-nitrosodimethylamine and its precursors during a wastewater reuse trial in the Llobregat River (Spain). *J. Hazard. Mater.* 124346. doi: 10.1016/j.jhazmat.2020.124346.

112 Sedlak, D.L., Deeb, R.A., Hawley, E.L., Mitch, W.A., Durbin, T.D., Mowbray, S., and Carr, S. (2005). Sources and fate of nitrosodimethylamine and its precursors in municipal wastewater treatment plants. *Water Environ. Res.* 77: 32–39. doi: 10.2175/106143005X41591.

113 Zeng, T., Plewa, M.J., and Mitch, W.A. (2016). N-Nitrosamines and halogenated disinfection byproducts in U.S. full advanced treatment trains for potable reuse. *Water Res.* 101: 176–186. doi: 10.1016/j.watres.2016.03.062.

114 Wang, L., Li, Y., Shang, X., and Shen, J. (2014). Occurrence and removal of N-nitrosodimethylamine and its precursors in wastewater treatment plants in and around Shanghai. *Front. Environ. Sci. Eng.* 8: 519–530. doi: 10.1007/s11783-013-0610-4.

115 Schreiber, I.M. and Mitch, W.A. (2006). Occurrence and fate of nitrosamines and nitrosamine precursors in wastewater-impacted surface waters using boron as a conservative tracer. *Environ. Sci. Technol.* 40: 3203–3210. doi: 10.1021/es052078r.

116 Dai, N., Zeng, T., and Mitch, W.A. (2015). Predicting N-nitrosamines: N-nitrosodiethanolamine as a significant component of total N-nitrosamines in recycled wastewater. *Environ. Sci. Technol. Lett.* 2: 54–58. doi: 10.1021/acs.estlett.5b00005.

117 Wu, M., Qian, Y., Boyd, J.M., Leavey, S., Hrudey, S.E., Krasner, S.W., and Li, X.-F. (2014). Identification of tobacco-specific nitrosamines as disinfection byproducts in chloraminated water. *Environ. Sci. Technol.* 48: 1828–1834. doi: 10.1021/es405075g.

118 Qian, Y., Wang, W., Boyd, J.M., Wu, M., Hrudey, S.E., and Li, X.-F. (2013). UV-induced transformation of four halobenzoquinones in drinking water. *Environ. Sci. Technol.* 47: 4426–4433. doi: 10.1021/es305044k.

119 Ho, T.-F.L., Bolton, J.R., and Lipczynska-Kochany, E. (1996). Quantum yields for the photodegradation of pollutants in dilute aqueous solution: phenol, 4-chlorophenol and N-nitrosodimethylamine. *J. Adv. Oxid. Technol.* 1: 170–178. doi: 10.1515/jaots-1996-0210.

120 Allard, S., Charrois, J.W.A., Joll, C.A., and Heitz, A. (2012). Simultaneous analysis of 10 trihalomethanes at nanogram per liter levels in water using solid-phase microextraction and gas chromatography mass-spectrometry. *J. Chromatogr. A* 1238: 15–21. doi: 10.1016/j.chroma.2012.03.020.

121 Kristiana, I., Lethorn, A., Joll, C., and Heitz, A. (2014). To add or not to add: the use of quenching agents for the analysis of disinfection by-products in water samples. *Water Res.* 59: 90–98. doi: 10.1016/j.watres.2014.04.006.

122 Sacher, F., Schmidt, C.K., and Lee, C. (2008). *Strategies for Minimizing Nitrosamine Formation during Disinfection*. Denver, CO: American Water Works Association.

123 Postigo, C., Emiliano, P., and Valero, F. (2020). High-throughput and reliable determination of 13 haloacetic acids and dalapon in water and evaluation of control strategies. *Environ. Sci. Water Res. Technol.* 6: 2499–2509. doi: 10.1039/D0EW00296H.

124 Kinani, A., Olivier, J., Roumiguières, A., Bouchonnet, S., and Kinani, S. (2018). A sensitive and specific solid-phase extraction–gas chromatography–tandem mass spectrometry method for the determination of 11 haloacetic acids in aqueous samples. *Eur. J. Mass Spectrom.* 24: 375–383. doi: 10.1177/1469066718781302.

125 Sanchís, J., Jaén-Gil, A., Gago-Ferrero, P., Munthali, E., and Farré, M.J. (2020). Characterization of organic matter by HRMS in surface waters: effects of chlorination on molecular fingerprints and correlation with DBP formation potential. *Water Res.* 176:. doi: 10.1016/j.watres.2020.115743.

126 Koudjonou, B.K. and LeBel, G.L. (2006). Halogenated acetaldehydes: analysis, stability and fate in drinking water. *Chemosphere* 64: 795–802. doi: 10.1016/j.chemosphere.2005.10.063.

127 Huang, H., Zhu, H., Gan, W., Chen, X., and Yang, X. (2017). Occurrence of nitrogenous and carbonaceous disinfection byproducts in drinking water distributed in Shenzhen, China. *Chemosphere* 188: 257–264. doi: 10.1016/j.chemosphere.2017.08.172.

128 US EPA. Method 551.1 determination of chlorination disinfection byproducts, chlorinated solvents, and halogenated pesticides/herbicides in drinking water by liquid-liquid extraction and gas chromatography with electron-capture detection. Cincinnati, OH.

129 Cheng, S., Wu, Y.-P., Young, T.R., Dodd, M.C., Wu, J., Zhang, H., Huo, Z.-L., Qian, Y.-T., Li, Y., Li, W.-T., and Li, A.-M. (2021). Rapid determination of trace haloacetic acids in water and wastewater using non-suppressed ion chromatography with electrospray ionization-tandem mass spectrometry. *Sci. Total Environ.* 754: 142297. doi: 10.1016/j.scitotenv.2020.142297.

130 Bruzzoniti, M.C., Rivoira, L., Meucci, L., Fungi, M., Bocina, M., Binetti, R., and Castiglioni, M. (2019). Towards the revision of the drinking water directive 98/83/EC. Development of a direct injection ion chromatographic-tandem mass spectrometric method for the monitoring of fifteen common and emerging disinfection by-products along the drinking water suppl. *J. Chromatogr. A.* doi: 10.1016/j.chroma.2019.07.004.

131 Liew, D., Linge, K.L., Joll, C.A., Heitz, A., and Charrois, J.W.A. (2012). Determination of halonitromethanes and haloacetamides: an evaluation of sample preservation and analyte stability in drinking water. *J. Chromatogr. A* 1241: 117–122. doi: 10.1016/j.chroma.2012.04.037.

132 Kristiana, I., Joll, C., and Heitz, A. (2012). Analysis of halonitriles in drinking water using solid-phase microextraction and gas chromatography–mass spectrometry. *J. Chromatogr. A* 1225: 45–54. doi: 10.1016/j.chroma.2012.01.005.

133 Serrano, M., Silva, M., and Gallego, M. (2011). Micro liquid–liquid extraction combined with large-volume injection gas chromatography–mass spectrometry for the determination of haloacetaldehydes in treated water. *J. Chromatogr. A* 1218: 8295–8302. doi: 10.1016/j.chroma.2011.09.048.

134 Roumiguières, A., Kinani, A., Bouchonnet, S., and Kinani, S. (2018). Development and validation of a multiclass method for the determination of organohalogen disinfectant by-products in water samples using solid phase extraction and gas chromatography-tandem mass spectrometry. *J. Chromatogr. A* 1579: 89–98. doi: 10.1016/j.chroma.2018.10.016.

135 Munch, J.W. (2005). Method 521: determination of nitrosamines in drinking water by solid phase extraction and capillary column gas chromatography with large volume injection and chemical ionization tandem mass spectrometry (MS/MS). US EPA, Washington DC. Re.

136 Duan, J., Li, W., Si, J., and Mulcahy, D. (2011). Rapid determination of nine haloacetic acids in water using ultra-performance liquid chromatography-tandem mass spectrometry in multiple reactions monitoring mode. *Anal. Methods* 3: 1667–1673. doi: 10.1039/C1AY05122A.

137 Kimura, S.Y., Cuthbertson, A.A., Byer, J.D., and Richardson, S.D. (2019). The DBP exposome: development of a new method to simultaneously quantify priority disinfection by-products and comprehensively identify unknowns. *Water Res.* 148: 324–333. doi: 10.1016/j.watres.2018.10.057.

138 Ortega-Hernandez, A., Acayaba, R., Verwold, C., Montagner, C.C., and Kimura, S.Y. (2021). Emerging investigator series: emerging disinfection by-product quantification method for wastewater reuse: trace level assessment using tandem mass spectrometry. *Environ. Sci. Water Res. Technol.* 7: 285–297. doi: 10.1039/D0EW00947D.

139 US EPA. Method 552.2 Determination of haloacetic acids and dalapon in drinking water by liquid-liquid extraction, derivatization and gas chromatography with electron capture detector. Revision 1.0. Cincinnati, OH. Retrieved from https://www.google.com/url?sa=t&rct=j&q=&esrc=s&source=web&cd=&ved=2ahUKEwj475mPyt_1AhWEy4UKHckxCL8QFnoECAMQAQ&docs/epa-method-552.2.pdf&usg=AOvVaw0YLP5-vKQCzKjLfDwmGZDY. Accessed 14 December 2021.

140 Chen, T.-L., Tzing, S.-H., and Ding, W.-H. (2015). Rapid screening of haloacetamides in water using salt-assisted liquid–liquid extraction coupled injection-port silylation gas chromatography–mass spectrometry. *J. Chromatogr. A* 1422: 340–344. doi: 10.1016/j.chroma.2015.10.047.

141 Laflamme, O., Sérodes, J.-B., Simard, S., Legay, C., Dorea, C., and Rodriguez, M.J. (2020). Occurrence and fate of ozonation disinfection by-products in two Canadian drinking water systems. *Chemosphere* 260: 127660. doi: 10.1016/j.chemosphere.2020.127660.

142 Ma, H., Li, Y., Zhang, H., Shah, S.M., and Chen, J. (2014). Salt-assisted dispersive liquid–liquid microextraction coupled with programmed temperature vaporization gas chromatography–mass spectrometry for the determination of haloacetonitriles in drinking water. *J. Chromatogr. A* 1358: 14–19. doi: 10.1016/j.chroma.2014.06.021.

143 Montesinos, I. and Gallego, M. (2012). Solvent-minimized extraction for determining halonitromethanes and trihalomethanes in water. *J. Chromatogr. A* 1248: 1–8. doi: 10.1016/j.chroma.2012.05.067.

144 Chinn, R., Lee, T., Krasner, S., Dale, M., Richardson, S.D., Pressman, J.G., Speth, T.F., Miltner, R.J., and Simmons, J.E. (2007) Solid-phase extraction of 35 DBPs with analysis by GC/ECD and GC/MS. Proceedings, 2007 Water Qual. Technol. Conf. Charlotte, NC, Novemb. 04-08, 2007, 1–20, American Water Works Association, Denver, CO.

145 Zhao, Y., Qin, F., Boyd, J.M., Anichina, J., and Li, X.-F. (2010). Characterization and determination of chloro- and bromo-benzoquinones as new chlorination disinfection byproducts in drinking water. *Anal. Chem.* 82: 4599–4605. doi: 10.1021/ac100708u.

146 Huang, R., Wang, W., Qian, Y., Boyd, J.M., Zhao, Y., and Li, X.-F. (2013). Ultra pressure liquid chromatography-negative electrospray ionization mass spectrometry determination of twelve halobenzoquinones at ng/L levels in drinking water. *Anal. Chem.* 85: 4520–4529. doi: 10.1021/ac400160r.

147 Wang, W., Qian, Y., Boyd, J.M., Wu, M., Hrudey, S.E., and Li, X.-F. (2013). Halobenzoquinones in swimming pool waters and their formation from personal care products. *Environ. Sci. Technol.* 47: 3275–3282. doi: 10.1021/es304938x.

148 Zhou, R., Xu, Z., Zhu, J., Liu, W., Meng, Y., Zhu, P., Zhou, W., Huang, C., and Ding, X. (2020). Determination of 10 Haloacetamides in drinking water by gas chromatography with automated solid phase extraction. *J. Chromatogr. B* 1150: 122191. doi: 10.1016/j.jchromb.2020.122191.

149 Chu, W., Gao, N., Yin, D., Krasner, S.W., and Templeton, M.R. (2012). Trace determination of 13 haloacetamides in drinking water using liquid chromatography triple quadrupole mass spectrometry with atmospheric pressure chemical ionization. *J. Chromatogr. A* 1235: 178–181. doi: 10.1016/j.chroma.2012.02.074.

150 Luo, Q., Chen, X., Wei, Z., Xu, X., Wang, D., and Wang, Z. (2014). Simultaneous and high-throughput analysis of iodo-trihalomethanes, haloacetonitriles, and halonitromethanes in drinking water using solid-phase microextraction/gas chromatography-mass spectrometry: an optimization of sample preparation. *J. Chromatogr. A* 1365: 45–53. doi: 10.1016/j.chroma.2014.09.003.

151 Kermani, F.R., Tugulea, A.-M., Hnatiw, J., Niri, V.H., and Pawliszyn, J. (2013). Application of automated solid-phase microextraction to determine haloacetonitriles, haloketones, and chloropicrin in Canadian drinking water. *Water Qual. Res. J.* 48: 85–98. doi: 10.2166/wqrjc.2013.012.

152 Montesinos, I. and Gallego, M. (2012). Headspace gas chromatography–mass spectrometry for rapid determination of halonitromethanes in tap and swimming pool water. *Anal. Bioanal. Chem.* 402: 2315–2323. doi: 10.1007/s00216-011-5516-z.

153 Montesinos, I., Cardador, M.J., and Gallego, M. (2011). Determination of halonitromethanes in treated water. *J. Chromatogr. A* 1218: 2497–2504. doi: 10.1016/j.chroma.2011.02.056.

154 Pérez Pavón, J.L., Herrero Martín, S., García Pinto, C., and Moreno Cordero, B. (2008). Determination of trihalomethanes in water samples: a review. *Anal. Chim. Acta* 629: 6–23. doi: 10.1016/j.aca.2008.09.042.

155 US EPA. Method 524.2 measurement of purgeable organic compounds in water by capillary column gas chromatography-mass spectrometry. Revision 4.0. Cincinnati, OH. Retrieved from https://bit.ly/3m9xxHF. Accessed in January 2021.

156 Franco, E.S., Pádua, V.L., Rodriguez, M.D.V.R., Silva, D.F., Libânio, M., Pereira, M.C., Silva, P.H.G., Santanta Júnior, I.C., Rocha, B.A., Camargo, J.A., Mourão, A.O., and Rodrigues, J.L. (2019). A simple liquid-liquid extraction-gas chromatography-mass spectrometry method for the determination of haloacetic acids in environmental samples: application in water with Microcystis aeruginosa cells. *Microchem. J* 150: 104088. doi: 10.1016/j.microc.2019.104088.

157 Polkowska, Z. (2004). Determination of volatile organohalogen compounds in urban precipitation in Tricity area (Gdańsk, Gdynia, Sopot). *Chemosphere* 57: 1265–1274. doi: 10.1016/j.chemosphere.2004.08.044.

158 Aeppli, C., Berg, M., Hofstetter, T.B., Kipfer, R., and Schwarzenbach, R.P. (2008). Simultaneous quantification of polar and non-polar volatile organic compounds in water samples by direct aqueous injection-gas chromatography/mass spectrometry. *J. Chromatogr. A* 1181: 116–124. doi: 10.1016/j.chroma.2007.12.043.

159 Pozzi, R., Bocchini, P., Pinelli, F., and Galletti, G.C. (2011). Determination of nitrosamines in water by gas chromatography/chemical ionization/selective ion trapping mass spectrometry. *J. Chromatogr. A* 1218: 1808–1814. doi: 10.1016/j.chroma.2011.02.009.

160 Charrois, J.W.A., Arend, M.W., Froese, K.L., and Hrudey, S.E. (2004). Detecting N-nitrosamines in drinking water at nanogram per liter levels using ammonia positive chemical ionization. *Environ. Sci. Technol.* 38: 4835–4841. doi: 10.1021/es049846j.

161 Zhao, -Y.-Y., Boyd, J., Hrudey, S.E., and Li, X.-F. (2006). Characterization of new nitrosamines in drinking water using liquid chromatography tandem mass spectrometry. *Environ. Sci. Technol.* 40: 7636–7641. doi: 10.1021/es061332s.

162 Zhou, W.-J., Boyd, J.M., Qin, F., Hrudey, S.E., and Li, X.-F. (2009). Formation of N-nitrosodiphenylamine and two new N-containing disinfection byproducts from chloramination of water containing diphenylamine. *Environ. Sci. Technol.* 43: 8443–8448. doi: 10.1021/es901935v.

163 Krauss, M. and Hollender, J. (2008). Analysis of nitrosamines in wastewater: exploring the trace level quantification capabilities of a hybrid linear ion trap/orbitrap mass spectrometer. *Anal. Chem.* 80: 834–842. doi: 10.1021/ac701804y.

164 McDonald, J.A., Harden, N.B., Nghiem, L.D., and Khan, S.J. (2012). Analysis of N-nitrosamines in water by isotope dilution gas chromatography–electron ionisation tandem mass spectrometry. *Talanta* 99: 146–154. doi: 10.1016/j.talanta.2012.05.032.

165 Lashgari, M., Yamini, Y., Basheer, C., and Lee, H.K. (2017). Ordered mesoporous carbon as sorbent for the extraction of N-nitrosamines in wastewater and swimming pool water. *J. Chromatogr. A* 1513: 35–41. doi: 10.1016/j.chroma.2017.07.046.

166 Chen, W., Li, X., Huang, H., Zhu, X., Jiang, X., Zhang, Y., Cen, K., Zhao, L., Liu, X., and Qi, S. (2017). Comparison of gas chromatography-mass spectrometry and gas chromatography-tandem mass spectrometry with electron ionization for determination of N-nitrosamines in environmental water. *Chemosphere* 168: 1400–1410. doi: 10.1016/j.chemosphere.2016.11.109.

167 Lee, M., Lee, Y., Soltermann, F., and Von Gunten, U. (2013). Analysis of N-nitrosamines and other nitro(so) compounds in water by high-performance liquid chromatography with post-column UV photolysis/Griess reaction. *Water Res.* 47: 4893–4903. doi: 10.1016/j.watres.2013.05.031.

168 Boyd, J.M., Hrudey, S.E., Li, X.-F., and Richardson, S.D. (2011). Solid-phase extraction and high-performance liquid chromatography mass spectrometry analysis of nitrosamines in treated drinking water and wastewater. *TrAC Trends Anal. Chem.* 30: 1410–1421. doi: 10.1016/j.trac.2011.06.009.

169 Bian, Y., Zhang, Y., Zhou, Y., Li, G., and Feng, X. (2020). Progress in the pretreatment and analysis of N-nitrosamines: an update since 2010. *Crit. Rev. Food Sci. Nutr.* 1–35. doi: 10.1080/10408398.2020.1803790.

170 Yoon, S., Nakada, N., and Tanaka, H. (2012). A new method for quantifying N-nitrosamines in wastewater samples by gas chromatography – triple quadrupole mass spectrometry. *Talanta* 97: 256–261. doi: 10.1016/j.talanta.2012.04.027.

171 Kadmi, Y., Favier, L., Soutrel, I., Lemasle, M., and Wolbert, D. (2014). Ultratrace-level determination of N-Nitrosodimethylamine, N-Nitrosodiethylamine, and N-Nitrosomorpholine in waters by solid-phase extraction followed by liquid chromatography-tandem mass spectrometry. *Cent. Eur. J. Chem.* 12: 928–936. doi: 10.2478/s11532-014-0537-z.

172 Lee, J.-H., Lee, S.-U., and Oh, J.-E. (2013). Analysis of nine nitrosamines in water by combining automated solid-phase extraction with high-performance liquid chromatography-atmospheric pressure chemical ionisation tandem mass spectrometry. *Int. J. Environ. Anal. Chem.* 93: 1261–1273. doi: 10.1080/03067319.2013.803281.

173 Raksit, A. and Johri, S. (2001). Determination of N-Nitrosodimethylamine in environmental aqueous samples by isotope-dilution GC/MS–SIM. *J. AOAC Int.* 84: 1413–1419. doi: 10.1093/jaoac/84.5.1413.

174 Mitch, W.A., Gerecke, A.C., and Sedlak, D.L. (2003). A N-Nitrosodimethylamine (NDMA) precursor analysis for chlorination of water and wastewater. *Water Res.* 37: 3733–3741. doi: 10.1016/S0043-1354(03)00289-6.

175 Grebel, J.E., Young, C.C., and (Mel) Suffet, I.H. (2006). Solid-phase microextraction of N-nitrosamines. *J. Chromatogr. A* 1117: 11–18. doi: 10.1016/j.chroma.2006.03.044.

176 Hung, H.-W., Lin, T.-F., Chiu, C.-H., Chang, Y.-C., and Hsieh, T.-Y. (2010). Trace analysis of N-nitrosamines in water using solid-phase microextraction coupled with gas chromatograph–tandem mass spectrometry. *Water Air Soil Pollut.* 213: 459–469. doi: 10.1007/s11270-010-0398-9.

177 Fan, -C.-C. and Lin, T.-F. (2018). N-nitrosamines in drinking water and beer: detection and risk assessment. *Chemosphere* 200: 48–56. doi: 10.1016/j.chemosphere.2018.02.025.

178 Fu, S.-C., Tzing, S.-H., Chen, H.-C., Wang, Y.-C., and Ding, W.-H. (2012). Dispersive micro-solid phase extraction combined with gas chromatography–chemical ionization mass spectrometry for the determination of N-nitrosamines in swimming pool water samples. *Anal. Bioanal. Chem.* 402: 2209–2216. doi: 10.1007/s00216-011-5681-0.

179 Spahr, S., Cirpka, O., Von Gunten, U., and Hofstetter, T.B. (2017). Formation of N-nitrosodimethylamine during chloramination of secondary and tertiary amines: role of molecular oxygen and radical intermediates. *Environ. Sci. Technol.* 51(1): 280–290. doi: 10.1021/acs.est.6b04780.

180 Fujioka, T., Takeuchi, H., Tanaka, H., Nghiem, L.D., Ishida, K.P., and Kodamatani, H. (2016). A rapid and reliable technique for N-nitrosodimethylamine analysis in reclaimed water by HPLC-photochemical reaction-chemiluminescence. *Chemosphere* 161: 104–111. doi: 10.1016/j.chemosphere.2016.06.094.

181 Reyes-Contreras, C., Domínguez, C., and Bayona, J.M. (2012). Determination of nitrosamines and caffeine metabolites in wastewaters using gas chromatography mass spectrometry and ionic liquid stationary phases. *J. Chromatogr. A* 1261: 164–170. doi: 10.1016/j.chroma.2012.05.082.

182 Liu, X., Cen, K., Zhao, L., Chen, Y., Lun, Z., Wu, X., Tian, Z., Shi, J., Zhu, X., Wei, J., Wang, N., and Chen, W. (2016). N-nitrodimethlyamine in natural and drinking water of high cancer incidence regions of Guangdong, China. *Appl. Geochem.* 74: 157–164. doi: 10.1016/j.apgeochem.2016.10.001.

183 Kim, G.-A., Son, H.-J., Kim, C.-W., and Kim, S.-H. (2013). Nitrosamine occurrence at Korean surface water using an analytical method based on GC/LRMS. *Environ. Monit. Assess.* 185: 1657–1669. doi: 10.1007/s10661-012-2658-1.

184 Plumlee, M.H., López-Mesas, M., Heidlberger, A., Ishida, K.P., and Reinhard, M. (2008). N-nitrosodimethylamine (NDMA) removal by reverse osmosis and UV treatment and analysis via LC–MS/MS. *Water Res.* 42: 347–355. doi: 10.1016/j.watres.2007.07.022.

185 Luo, Q., Bei, E., Liu, C., Deng, Y.-L., Miao, Y., Qiu, Y., Lu, W.-Q., Chen, C., and Zeng, Q. (2020). Spatial, temporal variability and carcinogenic health risk assessment of nitrosamines in a drinking water system in China. *Sci. Total Environ.* 736: 139695. doi: 10.1016/j.scitotenv.2020.139695.

186 Maqbool, T., Zhang, J., Qin, Y., Ly, Q.V., Asif, M.B., Zhang, X., and Zhang, Z. (2020). Seasonal occurrence of N-nitrosamines and their association with dissolved organic matter in full-scale drinking water systems: determination by LC-MS and EEM-PARAFAC. *Water Res.* 183: 116096. doi: 10.1016/j.watres.2020.116096.

187 Galaon, T., Cruceru, L., Petre, J., Pascu, L., Iancu, V., and Niculescu, M. (2016). New LC-MS/MS method for the determination of eight nitrosamines in drinking water. *J. Environ. Prot. Ecol.* 17: 74–82.

188 Ngongang, A.D., Duy, S.V., and Sauvé, S. (2015). Analysis of nine N-nitrosamines using liquid chromatography-accurate mass high resolution-mass spectrometry on a Q-Exactive instrument 7: 5748–5759. DOI: 10.1039/C4AY02967D.

189 Xie, Y., Rashid, I., (Joe) Zhou, H., and Gammie, L. (2002). Acidic methanol methylation for HAA analysis: limitations and possible solutions. *J. AWWA* 94: 115–122. doi: 10.1002/j.1551-8833.2002.tb10235.x.

190 Rubio, F.J., Urbansky, E.T., and Magnuson, M.L. (2000). Avoiding pitfalls in the determination of halocarboxylic acids: the photochemistry of methylation. *J. Environ. Monit.* 2: 248–252. doi: 10.1039/B000674M.

191 Zhang, A., Chu, W., Bond, T., Wang, F., Pan, Y., Tong, J., and Zhu, H. (2020). Interference from haloacetamides during the determination of haloacetic acids using gas chromatography. *J. Chromatogr. A* 1612: 460652. doi: 10.1016/j.chroma.2019.460652.

192 Chuang, Y.-H., Shabani, F., Munoz, J., Aflaki, R., Hammond, S.D., and Mitch, W.A. (2019). Formation of N-nitrosamines during the analysis of municipal secondary biological nutrient removal process effluents by US EPA method 521. *Chemosphere* 221: 597–605. doi: 10.1016/j.chemosphere.2019.01.053.

193 Health Canada. (2008). Guidelines for Canadian drinking water quality: guideline technical document – haloacetic acids. Water, Air and Climate Change Bureau, Healthy Environmentsand Consumer Safety Branch, Ottawa, Ontario. Retrieved from https://www.canada.ca/en/health-canada/services/publications/healthy-living/guidelines-canadian-drinking-water-quality-guideline-technical-document-haloacetic-acids.html. Accessed 14 December 2021.

194 Indian Standard Drinking Water Specification (Second Revision). Bureau of Indian Standards, IS 10500 (May), 1–11, 2012. Retrieved from http://cgwb.gov.in/Documents/WQ-standards.pdf. Accessed in November 2020.

195 Modificació a la Norma Oficial Mexicana NOM-127-SSA1-1994, Salud ambiental. Agua para uso y consumo humano. Límites permisibles de calidad y tratamiento a que debe someterse el agua para su potabilización. November 22, 2020. Diario Oficial (primera secció).

196 Decree of Health Ministry (No. 458) Egyptian Standards for Drinking Water and Domestic Uses (in Arabic Lang).

197 Ministry of Health, Labour and Welfare of Japan. (2015). Diagram 13 – drinking water quality standards in Japan. Retrieved from https://bit.ly/3biXT3l. Accessed in March, 2020.

198 Ministry of Health. (2005). Drinking-water standards for New Zealand (Revised 2008). Wellington: NZ Ministry of Health. Retrieved from https://codehub.building.govt.nz/resources/drinking-water-standards-for-new-zealand-2005. Accessed 14 December 2021.

10

Microplastics

Marta Llorca and Marinella Farré

Department of Environmental Chemistry, IDAEA-CSIC, Barcelona, Spain

10.1 Overview of Micro- and Nanoplastics

10.1.1 Properties

Microplastics (MPLs) are those plastic pieces below 5 mm including nanoplastics (NPLs). They are solid particles composed of mixtures of polymers (the primary component of plastics) and functional additives that enhance polymer properties such as flexibility and durability (i.e. flame retardants, impact modifiers, antioxidants, among others) [1, 2]. In addition, they may also contain impurities from their manufacture. Those small plastics can be indirectly formed through wear and tear of larger pieces of plastic (i.e. synthetic textiles), or can be directly manufactured and added to products such as exfoliating beads in facial or body scrubs [1].

MPLs include a wide range of microparticles types (pellets, fragments, fibbers, films, foam, etc.), a wide range of sizes (from 5 mm (MPLs) to 1 nm (NPLs)) [3, 4], with a high variety of polymer types. The most used ones in the industry and day-to-day life include polyethylene (PE, high- and low-density), polypropylene (PP), polyvinyl chloride (PVC), polystyrene (PS, including expanded EPS), polyurethane (PUR), polyethylene terephthalate (PET), and polyamides (PA) [4–6]. The structure of those polymers will confer them different properties such as density, tensile strength, glass transition temperature, and resistance to chemical products [7, 8]. According to their structure, polymers are classified as linear, branched, and cross-linked, while according to their thermal properties, they are classified as thermoplastics, thermosetting, and elastomers [9].

10.1.2 Legislation

Different initiatives have been developed at global and regional scale. For example, the United Nations Environment Organization has launched a global campaign to eliminate primary sources of marine plastic litter to reduce the excessive usage of single-use plastic by 2022 [10]. At the European level, different research initiatives have been promoted

Analytical Methods for Environmental Contaminants of Emerging Concern, First Edition. Edited by Núria Fontanals and Rosa Maria Marcé.

through the Horizon 2020 program to reduce MPLs in the environment [11], while the Member States must monitor MPLs [12]. In addition, the EU has implemented the EU recommendation about the Integrated Coastal Zone Management (ICZM) [11] and the Marine Strategy Framework Directive (MSFD) [13], both of them devoted to protecting European coastal areas and marine waters [10]. Descriptor 10 of the MSFD establishes the properties and quantities of marine litter that do not cause harmful effects on marine media and coastal areas [13]. Finally, in January 2018 the EU adopted the European Strategy for Plastics, which includes new strategies for life cycle economy and life cycle assessment for plastics [14].

10.1.3 Origin and Distribution

Microplastics are classified as primary or secondary according to whether the particles are originally manufactured in this size (primary) or whether they are the result of the breakdown of larger items (secondary) [5]. For example, primary virgin resin pellets are used during plastic manufacture as well as basic resin "feedstock" transport prior to plastic products production. Other primary MPLs are used as industrial scrubbers, plastic powders for molding, and in cosmetic formulations such as micro-beads, among others [5]. Secondary MPLs are the result of fragmentation and weathering of larger plastic items during manufacture of different products or into the environment [5].

MPLs reach the environment from different sources. In the case of primary MPLs, these are released from sewage discharges and factories or lost in small quantities during their transport as virgin pellets [5]. On the contrary, the main distribution sources of secondary MPLs are difficult to identify since they depend on the distribution of macroplastics and the degradation processes once they reach the environment. In addition, depending on litter size, this will be influenced by weathering effects to different degrees [5].

In the case of riverine systems (including water and river sediment), the presence of MPLs is due to anthropogenic mechanisms through their discharge from direct source industries as well as waste water treatment plants (although water treatments efficiently remove 80–90% of MPLs by trapping them in sludge) [3, 12]. Those MPLs that pass through riverine system reach seas and oceans through river discharge. This is one of the main sources of MPLs in marine environments, together with direct disposal of larger plastics, among other minor sources. Once there, it is expected that the low-density polymers remain at the water surface while the high-density ones reach sediments as their final sink [6, 10]. Nonetheless, low-density polymers also can reach sediments since their physicochemical characteristics may change due to weathering effects or even be modulated by an eco-corona of water organisms that surrounds MPLs [10]. The main factors that influence MPLs transport in sediments are: (i) gravity-driven transport in sediment-laden flows; (ii) settling, or conveyance through biological processes, of material that was formerly floating on the surface or suspended in the water column; (iii) transport by thermohaline currents, either during settling or by reworking of deposited microplastics [10, 15].

Regarding land sediments, MPLs reach them by diverse physical, biological and anthropogenic mechanisms [16]. Those MPLs are detected in sediments including agricultural soils. In this last case, their presence is explained by the reuse of sludge from WWTP as a fertilizer and irrigation with wastewater, weathering, and disintegration of plasticulture over fields, the fragmentation of plastic litter and plastic items and sedimentation from flooding soils [17–20].

Finally, those MPLs that are widespread in the environment can be accumulated in animals through ingestion due to their small size and, eventually, can be consumed by humans [1, 21].

10.1.4 Occurrence in the Environment

Plastic pollution is one of the major environmental challenges generated by the unsustainable use and disposal of plastic products by human societies.

Nowadays, it is recognized as a global complex, multidimensional, and multi-sectorial problem with environmental, economic, public health, food safety and even cultural impacts [10]. During the last years, researchers from different areas have been identifying sources, quantities, and impacts of plastic pollution although the knowledge is still limited. However, it is known that their presence in the media and especially in oceans is increasing. Among them, micro and nanoplastics (MNPLs) are of special interest due to their small size but also because they can be another source of pollutants through the release of additives and plasticizers [10]. In addition, MPLs can accumulate organic and inorganic contaminants as well as pathogens from the environment (air, water, or particulate matter), being an important vector for transport of those co-contaminants to aquatic organisms [22–26]. Due to their small size, MNPLs can enter the marine food web by their ingestion by aquatic organisms such as plankton, benthic protozoa, and bacteria [10, 25, 27, 28].

10.1.4.1 Water Systems

- **Marine environments**

The abundance of MPLs reported in various studies is subject to the sampling methods, the size ranges studied, and the measurement units chosen to report (e.g. items/km^2, items/m^2, items/m^3, ng/L, etc.). These disagreements in the literature make it difficult to compare surface water data, whether in a marine or riverine environment [7].

The occurrence of plastic and specifically MPLs in seas and oceans has been evidenced in many studies [6, 29–33]. According to Koelmans et al. [34], the average "whole ocean" concentration of plastic would be equal to ~2 ng/L, where Atlantic beaches in the vicinity of industrial areas, urban areas, and/or cargo/port facilities have the highest accumulation [29], as well as seawaters of the northern hemisphere [7] (Table 10.1). In general, there is a high variability on a small scale depending on seasonality, hydrographical conditions, population density, and industrial activities [7]. Nonetheless, the dynamic nature of the North Atlantic and North Pacific oceans makes it difficult to understand their distribution in this area, which includes oceanic currents, gyres, and recurring natural events. All these processes are responsible for transporting smaller plastic particles from sources to remote areas [35]. In this context, complex models, which includes multifactorial parameters, are needed to explain the distribution of MNPLs [7].

In the semi-enclosed area of Mediterranean Sea, the number of items ranges from 1,300 to 1,250,000 items/km^2 in the central area [36–40], between 103,000 and 310,000 items/km^2 in the Ligurian Sea [41–43], between 62,000 and 135,000 items/km^2 in the Western area [44–49], and between 140,418 and $3.76 \cdot 10^{10}$ items/km^2 in the Aegean-Levantine Sea

Table 10.1 Average items detected in marine environments. Adapted from Llorca et al. [10].

Average items/km^2		Ref.	Average items/m^3		Ref.
Northwestern Atlantic	2,500	[109]	East China Sea	0–144	[110]
Beaches of Guanabara Bay, Southeast Brazil	12,000,000–1,300,000,000	[111]	Yangtze Estuary	0.167	[112]
North Pacific	105,100	[113]	North Yellow Sea, China (Asia)	545	[114]
Subtropical Ocean Gyres (Pacific)	678,000	[115]	East coast of Japan (Asia)	0.03–0.075	[116]
Portuguese Coast	332,000,000–362,000,000	[22]	Surface Marine Polar waters (Arctic)	0.34	[117]
South Korea	46,334,000,000	[22]	Marine Polar waters (Arctic)	2.86	[118]
Japanese Sea	1,720,000	[119]	Marine Polar waters (Arctic) Sea ice	38–234	[24]
Seto Inland Sea	76,000	[120]	Arctic Central Basin	0.7	[25]
India (Bay of Bengal)	100–20,000	[121]	South Funen Archipelago (Baltic Sea)	0.07	[122]
Aegean-Levantine Sea	140,418	[123]	Jade Bay, Southern North Sea	1,770,000	[124]
Aegean-Levantine Sea	37,600,000,000	[125]	Svalbard (Norway)	0.34	[126]
NE Levantine coast of Turkey	1,067,120	[127]	Swedish Coast	150–2,400	[128]
Israeli Mediterranean Sea	1,518,340	[114]	NW Europe	0–1.5	[23]
Mediterranean Sea (coast of Italy)	1,250,000	[129]	North East Atlantic Ocean	2.46	[34]
Western Mediterranean Sea (Bay of Calvi)	62,000	[130]	Guanabara Bay, Brazil	1.4–21.3	[34]
Aegean Sea	1.31–2.64	[131]	Cape Cod (Caribbean)	0.00006–0.00832	[5]
Ligurian/Sardinian Sea	310,000	[132]	N Gulf of Mexico	5.0–18.4	[133]
Ligurian Sea	103,000	[134]	NE Pacific Ocean and coastal British Columbia	8–9,180	[27]

(Continued)

Table 10.1 (Continued)

Average items/km^2		Ref.	Average items/m^3		Ref.
Western Mediterranean Sea (Ligurian Sea)	125,930	[135]	**NW Pacific**	0.13	[28]
Adriatic Sea	472,000	[136]	**SE coastline of South Africa**	257.9–3,308.0	[25]
W Mediterranean Sea – Adriatic Sea	400,000	[137]	**Australia (Southern coastline from Australia to Antarctica)**	0.031	[138]
W Mediterranean Sea	135,000	[139]	**W Sardinia (Mediterranean Sea)**	0.15	[138]
Mediterranean Sea	243,000	[140]			[141]
Central W Mediterranean	147,000	[142]			[25]
Central Mediterranean Sea	1,300	[106]			
NW Mediterranean Sea	116,000	[143]			
NW Mediterranean Sea	183,000	[40]			
Western Mediterranean Sea	69,161	[144]			
Western Mediterranean Sea	82,000	[145]			
Western Mediterranean Sea (Gulf of Lion)	112,000	[146]			

[50–53]. In contrast to areas of Atlantic, the values are much higher, ranging from 2,500 items/km^2 in the NW [54] to $1.3 \cdot 10^9$ items/km^2 in Beaches of Guanabara Bay (SW Brazil) [55] or up to $3.62 \cdot 10^8$ items/km^2 in Portuguese Coast [56]. In Pacific Sea, the values are at similar levels, with 105,100 items/km^2 in North Pacific [57], $1.72 \cdot 10^6$ items/km^2 in Japanese Sea [58], and $4.63 \cdot 10^{10}$ items/km^2 in South Korea [59].

- **Inland waters**

  Rivers are considered as essential transport routes for plastic in coastal and lake environments as mentioned before. Although most of the studies have been focused on the investigation of mesoplastics, the presence of MPLs is ubiquitous in freshwater systems with concentration comparable to or higher than those observed in marine environments [60]. For example, the abundance of MPLs in North American rivers and estuarine systems ranges from $7.3 \cdot 10^5$ to $6.7 \cdot 10^6$ items/km^2 [61] and from $5.5 \cdot 10^3$ to $3 \cdot 10^5$ items/km^2 [62], respectively, and up to 0.26 items/m^3 in a Brazilian estuarine system [63]. Other studies carried out in China (Asia) evidenced the presence of MPLs in a river highly stressed by industrialization, up to 2,933 items/m^3 [64], and up to 10,200 items/m^3 in an estuary [65]. In contrast, the values reported for 5 urban estuaries settled in South Africa are much lower, ranging from 2 to 48.7 particles/m^3 [66]. In the case of Europe, the levels are also much lower, with values up to 0.028 items/m^3 in estuaries from UK [67], or between 0.32 and 45 items/m^3 in European rivers from France and Austria [68–70], up to 892,777 items/km^2 in Germany [71], and between 1 and 7,523 ng/L in the south part of the Ebro river [72, 73].

10.1.4.2 Sediments

Plastic materials with a density over the seawater (1.02 g/cm^3) are expected to sink and accumulate on seafloor sediments while low-density materials initially tend to float on the surface or suspended in the water column [74]. In addition, the association of particles with organic material and organisms produces density modification, facilitating the sinking of plastic debris and MPLs (known as biofouling). In the case of MPLs, these have an exceptional mobility once in marine media, due to the combination of particle properties (e.g. density, chemical composition, shape) and external hydrodynamics, marine sedimentology as well as physical oceanographic conditions [74]. For example, recent studies pointed out that the particle shape and the biofouling are the main contributors to settling/re-suspension behavior of MPLs. For example, the main hypothesis is that floating fibers and threads ("one-dimensional" (1-D) particles) are the first to begin sinking, followed by 2-D films and flakes, and then 3-D fragments [74]. This hypothesis has been confirmed by different researchers. Sanchez-Vidal et al. [6] detected high amounts of microfibers in deep-sea sediments from the Cantabrian Sea, Black Sea, and Mediterranean Sea (including the Alboran Sea, Levantine Sea, and Cretan Sea). In another study, Woodall et al. [75] showed that the number of microfibers was higher in deep-sea sediments (up to four orders of magnitude) than in the sea surface of contaminated areas of the Atlantic Ocean, the Indian Ocean, and the Mediterranean Sea. Regarding polymer type, the main polymers that have been detected in coastal sediments and deep-sea sediments include natural and regenerated cellulose [6] as well as synthetic plastics including PS, PE, PP [6, 76, 77], acrylic and polyamide (including nylon) [6], and ethylene-vinyl alcohol copolymers [78].

In the specific case of the Mediterranean Sea, the main quantities of MPLs detected in sea sediments varies between 4 [79] and more than 2,000 items/kg sediment DW [76]. Focusing on Spanish Mediterranean Coast sediments, Filgueiras et al. [80] investigated surface sediments from Algeciras to Barcelona, including samples from Málaga, Castell de Ferro, Almería, Cartagena, Benidorm, Benicarló, Vallcarca, and Palma de Mallorca. The number of MPLs ranged from 45.9 ± 23.9 MPLs/kg DW in Palma de Mallorca to 280.3 ± 164.9 MPLs/kg DW in Málaga. In addition, the authors found that MPLs concentrations are largely not associated with local sources of contamination. This finding agrees with previous work carried out in the Western region comparing the charge of

MPLs in sediments from Cabrera, a Marine Protected Area in the Balearic Islands, with a highly populated and touristic area in Mallorca, where the authors detected higher number of MPLs in a protected area (up to 900 items/kg DW) than in a touristic one [81]. Similar results were also obtained in the Aeolian Archipelago's islands, Tyrrhenian Sea, where the recorded values in this marine protected area were similar to those recorded in harbor sites (between 151 and 679 items/kg DW) [82], or the results from the Natural Park of Telaščica bay (Adriatic Sea), with the number of MPLs ranging from 32 to 378 items/kg DW [83], much higher than the recorded values in the Grand Harbor of Malta (<12 items/kg DW) [79]. One of the most polluted areas of Mediterranean Sea has been the lagoon of Venice (Italy), where MPLs < 1 mm were detected in almost all samples and ranging from 672 to 2,175 items/kg DW [76], followed by the Maremma Regional Park in Tyrrhenian Sea (Italy), with recorded values between 45 and 1,069 items/kg DW [84]. In this last case, the amount of MPLs in this regional park is highly influenced by fluvial inputs, the contribution of Ombrone river, as well as the impact of materials derived from agricultural activities in the coastal areas [84].

Due to their easy accessibility, numerous studies focus mainly on sandy beaches, where wind, marine currents and tides play an important role in sediment distribution and their accumulation on the shoreline [7]. For example, in Europe the presence of MPLs has been reported to be up to 2,000 items/kg in beach sediments highly impacted by plastic contamination, dominated by PE and PP [7]. In beaches from the Ebro Delta (Catalan coastal area) values up to 422 items/kg DW have been recorded, where fibers are also the most abundant type of MPLs [85]. These results were similar to those observed in, for example, the Tunisian coastal area, with MPLs ranging from 141 to 461 items/kg DW and fibers as the most abundant plastic form [77].

Finally, it is necessary to study the abundance in sub-tidal sediments, particularly in the areas far from the continental margins [7, 75, 86]. Recently, some studies have reported the first data for polar regions, confirming MPL contamination in sediments in both the Arctic and Antarctic regions [87–89].

10.1.4.3 Biota

The presence of MPLs in biota and their potential entrance to food chain have been investigated during the last years and found to represent an emerging threat to ecosystems, especially marine ecosystems. The presence of MPLs has been reported worldwide in a wide range of marine species from different trophic levels. Special interest should be paid to filter-feeding organisms, such as mussels, because of their role in marine food web and commercial interest. Although several studies reported that filter-feeding organisms ingest and subsequently eliminate MPLs, the potential effects at organ and tissue level remain unclear, especially considering exposure to different MPLs sizes and concentrations [10]. The study carried out by Gonçalves et al. [90] evaluating the pathophysiological effects on the Mediterranean mussel (*Mytilus galloprovincialis*) exposed to spherical polystyrene MPLs showed that mussels ingest MPLs and pass them through the lumen of the gut and, finally, excrete them through feces without any histopathological damage in whole-body sections, regardless of the plastic size. However, in the study developed by Kinjo et al. [91], significant differences in the gut retention time and the long-term egestion were observed when exposing mussels to PS microspheres sized 1, 10, and 90 μm.

Focusing on works with species directly collected from the environment, MPLs have been found at benthic level in invertebrates such as crustaceans or bivalves [7]. As examples, plastic filaments ranging from 3 to 20 mm length have been reported in commercial crab *Lithodes santolla* from Chile [92], and microfibers in farmed and wild mussels in Nova Scotia [93]. In the pelagic area, where fish are considered an important source of MPLs for higher trophic-level organisms [7], plastic fragments, pellets, and films have been found in, for example, the gastrointestinal tract of mesopelagic fish from the Pacific Ocean [94, 95]. On higher trophic levels, MPLs transfer has been investigated and levels have been reported in predator fish and marine mammals [96], including seabirds and sea turtles [97–100].

10.2 Sample Preparation and Collection

10.2.1 Protocols for Collecting and Preparing Samples

The protocols for collecting and preparing samples differ depending on matrix but also on the scope of the study and the cost. Schirinzi [7] summarizes the main methods according to sample origin (water, sediment, or biota) and these are further developed in this section.

10.2.1.1 Water

Sampling surface water for the analysis of MPLs used to be carried out by means of nets. Net tows employed for sampling plankton, insects, or another biota at sea have been adapted for the sampling of plastic particles in different water layers [7]. For example, manta trawls sample the first 15–25 cm, while the neuston trawls sample up to 50 cm. Net tows are generally deployed from vessels, and they are usually towed for 15–30 minutes (rarely 60 minutes) at ~2 knots [7]. The net can be between 1 and 8 m in length, and the mesh size is 50–500 μm, with 300 or 333 μm typically being the most used mesh. When the frame is wholly immersed, using a flowmeter is possible to measure the filtered volume and estimate the sampled area [7]. In the case of the sampling of the deepest part of the water column, bongo nets, or in some cases multi-level net tows, are typically used to collect particles from the bottom. Other sampling devices are the bulk water pump and underway pump sampling [7].

10.2.1.2 Sediment

The sampling of MPLs in sediments requires different efforts and resources depending on the water depth. Grabs are normally used to sample surface and near-surface sediments from a bulk sample, whereas coring devices sample different depth strata providing information on MPLs distribution [7].

10.2.1.3 Biota

In the case of biota, different sampling strategies are followed depending on the specie [101]. For example, in the case of river fishes, electrical fishing is the most used technique, followed by nets and rod. In seas and oceans, the most used strategy is net fishing.

After that, some physical characteristics of the organisms are measured, such as size, age, development, behavior and feeding strategies, and their interaction with the environment and humans [102].

10.2.2 Sample Extraction and Clean-up

The analysis of MPLs in environmental samples is a multi-step process (Figure 10.1) including some additional steps to separate plastic particles from the matrix by filtering/sieving or density separation and further treatments involving chemical or biological agents to reduce organic matter [7].

10.2.2.1 Separation

In the case of liquid samples, filtration is the most used approach, by means of glass fiber filters (0.2–0.7 μm), cellulose nitrate, or cellulose acetate filters (5 μm), or sieves (50–250 μm mesh size) [103].

For sediment samples, sieving is used to isolate the larger MPLs (>1 mm), with mesh size of sieves ranging across different pore sizes and multi-tier sieving with a decreasing mesh size [7]. For smaller MPLs (<1 mm), density separation combined with filtration are the methodologies of choice. Density separation uses the buoyancy force of a saline

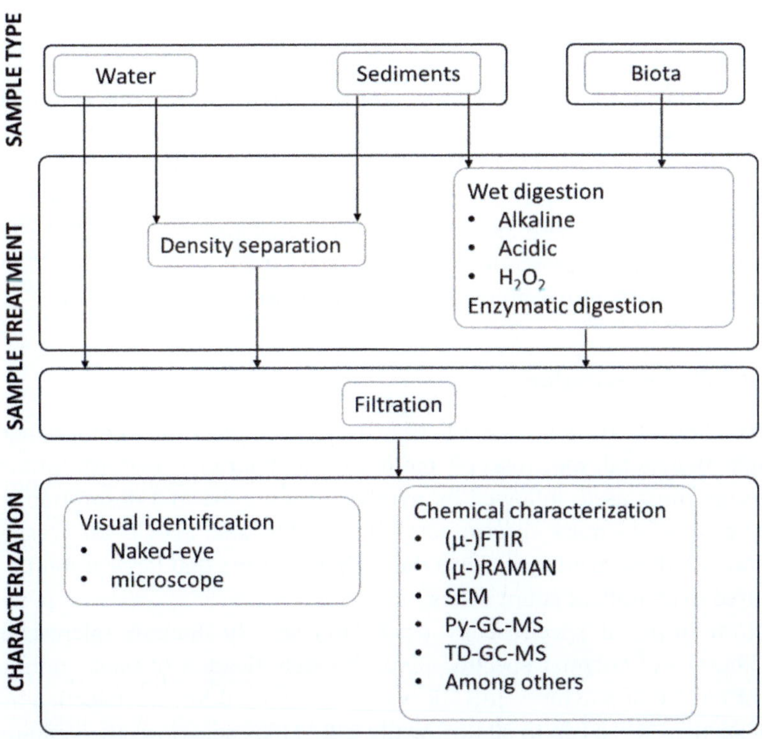

Figure 10.1 Scheme of the most commonly used methods for the analysis of MPLs.

solution (usually sodium chloride, sodium polytungstate, sodium iodide, and zinc chloride) with a density intermediate between that of plastic particles (0.8–1.70 g/cm^3) and that of organic/inorganic materials constituting sand or sediment, which is typically 2.65 g/cm^3 for some hours [7, 104]. After sedimentation, the supernatant where MPLs are assumed to be is filtered [7].

10.2.2.2 Matrix Removal by Digestion

The presence of organic materials (e.g. plankton, biofilms, rest of organisms) in the samples may interfere with the analysis. In such cases, biological or chemical digestion is needed to purify the sample matrix. This extraction is usually done by wet digestion of organic matter with H_2O_2 (especially when working with seawater and biota), alkaline or acidic attack [96], or enzymatic digestion (specially for biota) [28]. After digestion, the solutions can be directly filtered or extracted again by density separation.

10.3 Determination of MNPLs

For the characterization of MPLs in the environment, different physical and chemical methods are commonly used and combined to have the wide spectra of MPLs in one sample (Figure 10.1).

10.3.1 Physical Characterization

The physical characterization is used to classify MPLs according to their morphology (fragments, fibers, sheets, foam, among others), size, and color. This is usually done by means of visual identification (naked eye or by optical microscopes). For example, stereomicroscopes allow one to distinguish MPLs below 1 mm from other particles. For smaller particles, the most used technologies include scanning electron microscopy (SEM) alone or in combination with energy dispersive X-ray spectroscopy (SEM-EDS) and environmental scanning microscopy-energy dispersive X-ray spectroscopy (ESEM-EDS), which allows the elemental composition of polymers to be determined [7].

10.3.2 Chemical Characterization

The chemical characterization allows MPLs to be distinguished from other natural materials in the sample. In general, spectroscopic methods (e.g., Fourier-Transform Infrared and Raman) are the most used, followed by pyrolysis and/or thermal desorption gas chromatography coupled to mass spectrometry (Py-GC-MS and TD-GC-MS, respectively) [104]. However, these techniques have the main limitation that large amounts of sample are required to identify or count MPLs.

Fourier-Transform infrared spectroscopy ((μ-)FTIR) and (μ-)Raman microscopy, equipped with libraries of polymer spectra, allow the identification of most common polymers. Regarding size of polymers, μ-FTIR is commonly used for the detection and identification of smaller MPLs (up to 20 μm) while μ-RAMAN spectroscopy combined with imaging techniques offers the detection of MPLs up to 1 μm size [7, 10].

In the case of thermal analyses such as TD-GC-MS, measure changes in the physical and chemical properties of materials depending on their thermal stability. However, reference materials are needed to identify polymer types.

Finally, some recent works explore the capabilities of size exclusion chromatography coupled to high-resolution mass spectrometry (SEC-HRMS) for the identification and quantification of low molecular weight nano- and microplastics (between 500 and 3,000 Da) [72, 73, 105]. This methodology could be extended to other matrices, offering the possibility of detecting the smallest size of plastic litters, which are the ones identified as the most problematic from an ecotoxicological point of view [10].

10.4 Future Directions and Challenges

The main challenges regarding the investigation of microplastics are the standardization protocols for their analysis. This comprises sampling procedures as well as analytical procedures. In this context, it is necessary to develop suitable standardized strategies for sampling and analytical methodologies to compare data results [10]. An example of criteria could include: (1) sampling method strategy, (2) sample size, (3) sample processing and storage, (4) laboratory preparation, (5) clean air conditions, (6) negative controls, (7) positive controls, (8) target component, (9) sample (pre)treatment, and (10) polymer identification [106].

In terms of analysis, the best option is to participate in interlaboratory or proficiency test studies devoted to the analysis of MPLs in prepared samples [107, 108]. In addition, due to the variability of analytical procedures it is necessary to use Certified Reference Materials (CRM) to evaluate the robustness of the method. However, nowadays CRM for MPLs do not exist [10].

Finally, it is necessary to evaluate the environmental and human health risks due to the presence of MPLs associated to complex mixture of co-contaminants. This requires increasing the knowledge on exposure levels and effects of MPLs and associated contaminants such as lixiviates [10].

Acknowledgments

This work has been supported by DEGBIOPLAST project (RTI2018-097860-J-I00) from the Spanish Ministry of Science and Innovation and EUROqCHARM project (ID 101003805) Horizon 2020.

References

1 ECHA. (2020). Microplastics. European Chemical Agency. https://echa.europa.eu/hot-topics/microplastics.

2 polymerdatabase.com. (2019). Polymer Properties Database.

3 Corradini, F., Meza, P., Eguiluz, R., Casado, F., Huerta-Lwanga, E., and Geissen, V. (2019). Evidence of microplastic accumulation in agricultural soils from sewage sludge disposal. *Sci. Total Environ.* 671: 411–420. doi: 10.1016/j.scitotenv.2019.03.368.

4 Caldwell, J., Petri-Fink, A., Rothen-Rutishauser, B., and Lehner, R. (2019). Assessing meso- and microplastic pollution in the Ligurian and Tyrrhenian Seas. *Mar. Pollut. Bull.* 149. doi: 10.1016/j.marpolbul.2019.110572.

5 GESAMP. (2015). Sources, fate and effects of microplastics in the marine environment: a global assessment. In: *IMO/FAO/UNESCO-IOC/UNIDO/WMO/IAEA/UN/UNEP/UNDP Joint Group of Experts on the Scientific Aspects of Marine Environmental Protection* (P.J. Kershaw, ed.), 96.

6 Sanchez-Vidal, A., Thompson, R.C., Canals, M., and de Haan, W.P. (2018). The imprint of microfibres in southern European deep seas. *PLoS One* 13: e0207033. doi: 10.1371/journal.pone.0207033.

7 Schirinzi, G.F. (2020). *Chemical and Ecotoxicological Assessment of Microplastics and Emerging Risks in the Coastal Environments*. Universitat de Barcelona.

8 Brinson, H.F. and Brinson, L.C. (2015). Characteristics, applications and properties of polymers. In: *Polymer Engineering Science and Viscoelasticity*. Springer.

9 Peters, E.N., (2015). Thermoplastics, thermosets, and elastomers – descriptions and properties. In: Mechanical Engineers' Handbook. doi: 10.1002/9781118985960.meh109.

10 Llorca, M., Álvarez-Muñoz, D., Ábalos, M., Rodríguez-Mozaz, S., Santos, L.H.M.L.M., León, V.M., Campillo, J.A., Martínez-Gómez, C., Abad, E., and Farré, M. (2020). Microplastics in Mediterranean coastal area: toxicity and impact for the environment and human health. *TreAC Trends Environ. Anal. Chem.* 27: e00090. doi: 10.1016/j.teac.2020.e00090.

11 European Commission. Integrated coastal management. http://ec.europa.eu/environment/iczm/index_en.htm. Accessed 14 December 2021.

12 Li, X., Chen, L., Mei, Q., Dong, B., Dai, X., Ding, G., and Zeng, E.Y. (2018). Microplastics in sewage sludge from the wastewater treatment plants in China. *Water Res.* 142: 75–85. doi: 10.1016/j.watres.2018.05.034.

13 European Commission. (2008). Legislation: the marine directive.http://ec.europa.eu/environment/marine/eu-coast-and-marine-policy/marine-strategy-framework-directive/index_en.htm. (Accessed 14 December 2021).

14 European Commission. (2018). European strategy for plastics. http://ec.europa.eu/environment/waste/plastic_waste.htm. (Accessed 14 December 2021).

15 Kane, I.A. and Clare, M.A. (2019). Dispersion, accumulation, and the ultimate fate of microplastics in deep-marine environments: a review and future directions. *Front. Earth Sci.* 7.

16 Rillig, M.C., Ingraffia, R., and de Souza Machado, A.A. (2017). Microplastic incorporation into soil in agroecosystems. *Front. Plant Sci.* 8:: 1805.

17 Nizzetto, L., Langaas, S., and Futter, M. (2016). Pollution: do microplastics spill on to farm soils? *Nature* 537: 488.

18 Rochman, C.M. (2018). Microplastics research – from sink to source. *Science* 360: 28–29. doi: 10.1126/science.aar7734.

19 Bläsing, M. and Amelung, W. (2018). Plastics in soil: analytical methods and possible sources. *Sci. Total Enivron.* 612: 422–435.

20 Scheurer, M. and Bigalke, M. (2018). Microplastics in Swiss floodplain soils. *Environ. Sci. Technol.* 52: 3591–3598. doi: 10.1021/acs.est.7b06003.

21 EFSA. (2016). Panel on contaminants in the food chain – presence of microplastics and nanoplastics in food, with particular focus on seafood. *EFSA J.* 14: e04501. doi: 10.2903/j.efsa.2016.4501.

22 Cole, M., Lindeque, P., Halsband, C., and Galloway, T.S. (2011). Microplastics as contaminants in the marine environment: a review. *Mar. Pollut. Bull.* 62: 2588–2597.

23 Llorca, M., Farré, M., Karapanagioti, H.K., and Barceló, D. (2014). Levels and fate of perfluoroalkyl substances in beached plastic pellets and sediments collected from Greece. *Mar. Pollut. Bull.* 87: 286–291.

24 Rios, L.M., Moore, C., and Jones, P.R. (2007). Persistent organic pollutants carried by synthetic polymers in the ocean environment. *Mar. Pollut. Bull.* 54: 1230–1237. doi: 10.1016/j.marpolbul.2007.03.022.

25 Pittura, L., Avio, C.G., Giuliani, M.E., d'Errico, G., Keiter, S.H., Cormier, B., Gorbi, S., and Regoli, F. (2018). Microplastics as vehicles of environmental PAHs to marine organisms: combined chemical and physical hazards to the Mediterranean Mussels, Mytilus galloprovincialis. *Front. Mar. Sci.* 5: art. 103. doi: 10.3389/fmars.2018.00103.

26 Ashton, K., Holmes, L., and Turner, A. (2010). Association of metals with plastic production pellets in the marine environment. *Mar. Pollut. Bull.* 60: 2050–2055.

27 Wright, S.L., Thompson, R.C., and Galloway, T.S. (2013). The physical impacts of microplastics on marine organisms: a review. *Environ. Pollut.* 178: 483–492.

28 Cole, M., Webb, H., Lindeque, P.K., Fileman, E.S., Halsband, C., and Galloway, T.S. (2014). Isolation of microplastics in biota-rich seawater samples and marine organisms. *Sci. Rep.* 4: 4528.

29 Antunes, J., Frias, J., and Sobral, P. (2018). Microplastics on the Portuguese coast. *Mar. Pollut. Bull.* 131: 294–302.

30 León, V.M., García, I., González, E., Samper, R., Fernández-González, V., and Muniategui-Lorenzo, S. (2018). Potential transfer of organic pollutants from littoral plastics debris to the marine environment. *Environ. Pollut.* 236: 442–453.

31 León, V.M., García-Agüera, I., Moltó, V., Fernández-González, V., Llorca-Pérez, L., Andrade, J.M., Muniategui-Lorenzo, S., and Campillo, J.A. (2019). PAHs, pesticides, personal care products and plastic additives in plastic debris from Spanish Mediterranean beaches. *Sci. Total Environ.* 670: 672–684. doi: 10.1016/j.scitotenv.2019.03.216.

32 Lebreton, L.C.M., Greer, S.D., and Borrero, J.C. (2012). Numerical modelling of floating debris in the world's oceans. *Mar. Pollut. Bull.* 64: 653–661. doi: 10.1016/j.marpolbul.2011.10.027.

33 Constant, M., Kerhervé, P., Mino-Vercellio-Verollet, M., Dumontier, M., Sànchez Vidal, A., Canals, M., and Heussner, S. (2019). Beached microplastics in the Northwestern Mediterranean Sea. *Mar. Pollut. Bull.* 142: 263–273. doi: 10.1016/j.marpolbul.2019.03.032.

34 Koelmans, A.A., Bakir, A., Burton, G.A., and Janssen, C.R. (2016). Microplastic as a vector for chemicals in the aquatic environment: critical review and model-supported reinterpretation of empirical studies. *Environ. Sci. Technol.* 50: 3315–3326. doi: 10.1021/acs.est.5b06069.

35 Cózar, A., Martí, E., Duarte, C.M., García-de-lomas, J., Van Sebille, E., Ballatore, T.J., Eguíluz, V.M., González-Gordillo, J.I., Pedrotti, M.L., and Echevarría, F. (2017). The

Arctic Ocean as a dead end for floating plastics in the North Atlantic branch of the Thermohaline circulation. *Sci. Adv.* 3: e1600582. doi: 10.1126/sciadv.1600582.

36 Morris, R.J. (1980). Floating plastic debris in the Mediterranean. *Mar. Pollut. Bull.* 11(5): 125.

37 Ruiz-Orejón, L.F., Sardá, R., and Ramis-Pujol, J. (2016). Floating plastic debris in the Central and Western Mediterranean Sea. *Mar. Environ. Res.* 120: 136–144.

38 Suaria, G., Avio, C.G., Mineo, A., Lattin, G.L., Magaldi, M.G., Belmonte, G., Moore, C.J., Regoli, F., and Aliani, S. (2016). The Mediterranean plastic soup: synthetic polymers in Mediterranean surface waters. *Sci. Rep.* 6: 37551. doi: 10.1038/srep37551.

39 Cózar, A., Sanz-Martín, M., Martí, E., González-Gordillo, J.I., Ubeda, B., Gálvez, J.Á., Irigoien, X., and Duarte, C.M. (2015). Plastic accumulation in the Mediterranean Sea. *PLoS One* 10: e0121762. doi: 10.1371/journal.pone.0121762.

40 de Haan, W.P., Sanchez-Vidal, A., Canals, M., and Party, N.S.S. (2019). Floating microplastics and aggregate formation in the Western Mediterranean Sea. *Mar. Pollut. Bull.* 140: 523–535. doi: 10.1016/j.marpolbul.2019.01.053.

41 Fossi, M.C., Panti, C., Guerranti, C., Coppola, D., Giannetti, M., Marsili, L., and Minutoli, R. (2012). Are baleen whales exposed to the threat of microplastics? A case study of the Mediterranean fin whale (Balaenoptera physalus). *Mar. Pollut. Bull.* 64: 2374–2379. doi: 10.1016/j.marpolbul.2012.08.013.

42 Pedrotti, M.L., Bruzaud, S., Dumontet, B., Elineau, A., Petit, S., Grohens, Y., Voisin, P., Crebassa, J.-C., and Gorsky, G. Plastic fragments on the surface of Mediterranean waters. In: CIESM Workshop Monograph.

43 Pedrotti, M.L., Petit, S., Elineau, A., Bruzaud, S., Crebassa, J.-C., Dumontet, B., Martí, E., Gorsky, G., and Cózar, A. (2016). Changes in the floating plastic pollution of the Mediterranean Sea in relation to the distance to land. *PLoS One* 11: e0161581. doi: 10.1371/journal.pone.0161581.

44 Baini, M., Fossi, M.C., Galli, M., Caliani, I., Campani, T., Finoia, M.G., and Panti, C. (2018). Abundance and characterization of microplastics in the coastal waters of Tuscany (Italy): the application of the MSFD monitoring protocol in the Mediterranean Sea. *Mar. Pollut. Bull.* 133: 543–552. doi: 10.1016/j.marpolbul.2018.06.016.

45 Collignon, A., Hecq, J.-H., Galgani, F., Collard, F., and Goffart, A. (2014). Annual variation in neustonic micro-and meso-plastic particles and zooplankton in the Bay of Calvi (Mediterranean–Corsica). *Mar. Pollut. Bull.* 79: 293–298. doi: 10.1016/j.marpolbul.2013.11.023.

46 Collignon, A., Hecq, J.-H., Glagani, F., Voisin, P., Collard, F., and Goffart, A. (2012). Neustonic microplastic and zooplankton in the North Western Mediterranean Sea. *Mar. Pollut. Bull.* 64: 861–864. doi: 10.1016/j.marpolbul.2012.01.011.

47 Faure, F., Saini, C., Potter, G., Galgani, F., De Alencastro, L.F., and Hagmann, P. (2015). An evaluation of surface micro-and mesoplastic pollution in pelagic ecosystems of the Western Mediterranean Sea. *Environ. Sci. Pollut. Res.* 22: 12190–12197. doi: 10.1007/s11356-015-4453-3.

48 Fossi, M.C., Romeo, T., Baini, M., Panti, C., Marsili, L., Campani, T., Canese, S., Galgani, F., Druon, J.-N., and Airoldi, S. (2017). Plastic debris occurrence, convergence areas and fin whales feeding ground in the Mediterranean marine protected area pelagos sanctuary: a modeling approach. *Front. Mar. Sci.* 4: 167. doi: 10.3389/fmars.2017.00167.

49 Schmidt, N., Thibault, D., Galgani, F., Paluselli, A., and Sempéré, R. (2018). Occurrence of microplastics in surface waters of the Gulf of Lion (NW Mediterranean Sea). *Prog. Oceanogr.* 163: 214–220.

50 Gündoğdu, S. (2017). High level of micro-plastic pollution in the Iskenderun Bay NE Levantine coast of Turkey. *Su Ürünleri Dergisi* 34: 401–408.

51 Gündoğdu, S. and Çevik, C. (2017). Micro-and mesoplastics in Northeast Levantine coast of Turkey: the preliminary results from surface samples. *Mar. Pollut. Bull.* 118: 341–347. doi: 10.1016/j.marpolbul.2017.03.002.

52 Güven, O., Gökdağ, K., Jovanović, B., and Kıdeyş, A.E. (2017). Microplastic litter composition of the Turkish territorial waters of the Mediterranean Sea, and its occurrence in the gastrointestinal tract of fish. *Environ. Pollut.* 223: 286–294. doi: 10.1016/j.envpol.2017.01.025.

53 van der Hal, N., Ariel, A., and Angel, D.L. (2017). Exceptionally high abundances of microplastics in the oligotrophic Israeli Mediterranean coastal waters. *Mar. Pollut. Bull.* 116: 151–155. doi: 10.1016/j.marpolbul.2016.12.052.

54 Law, K.L., Morét-Ferguson, S., Maximenko, N.A., Proskurowski, G., Peacock, E.E., Hafner, J., and Reddy, C.M. (2010). Plastic accumulation in the North Atlantic subtropical gyre. *Science* 329: 1185–1188.

55 de Carvalho, D.G. and Neto, J.A.B. (2016). Microplastic pollution of the beaches of Guanabara Bay, Southeast Brazil. *Ocean Coast. Manage.* 128: 10–17. doi: 10.1016/j.ocecoaman.2016.04.009.

56 Antunes, J., Frias, J., Micaelo, A., and Sobral, P. (2013). Resin pellets from beaches of the Portuguese coast and adsorbed persistent organic pollutants. *Estuar. Coast. Shelf Sci.* 130: 62–69. doi: 10.1016/j.ecss.2013.06.016.

57 Eriksen, M., Lebreton, L.C., Carson, H.S., Thiel, M., Moore, C.J., Borerro, J.C., Galgani, F., Ryan, P.G., and Reisser, J. (2014). Plastic pollution in the world's oceans: more than 5 trillion plastic pieces weighing over 250,000 tons afloat at sea. *PLoS One* 9: e111913. doi: 10.1371/journal.pone.0111913.

58 Isobe, A., Uchida, K., Tokai, T., and Iwasaki, S. (2015). East Asian seas: a hot spot of pelagic microplastics. *Mar. Pollut. Bull.* 101: 618–623. doi: 10.1371/journal.pone.0111913.

59 Kim, I.-S., Chae, D.-H., Kim, S.-K., Choi, S., and Woo, S.-B. (2015). Factors influencing the spatial variation of microplastics on high-tidal coastal beaches in Korea. *Arch. Environ. Contam. Toxicol.* 69: 299–309. doi: 10.1007/s00244-015-0155-6.

60 Schirinzi, G.F., Köck-Schulmeyer, M., Cabrera, M., González-Fernández, D., Hanke, G., Farré, M., and Barceló, D. (2020). Riverine anthropogenic litter load to the Mediterranean Sea near the metropolitan area of Barcelona, Spain. *Sci. Total Environ.* 714: 136807. doi: 10.1016/j.scitotenv.2020.136807.

61 McCormick, A., Hoellein, T.J., Mason, S.A., Schluep, J., and Kelly, J.J. (2014). Microplastic is an abundant and distinct microbial habitat in an urban river. *Environ. Sci. Technol.* 48: 11863–11871. doi: 10.1021/es503610r.

62 Yonkos, L.T., Friedel, E.A., Perez-Reyes, A.C., Ghosal, S., and Arthur, C.D. (2014). Microplastics in four estuarine rivers in the Chesapeake Bay, USA. *Environ. Sci. Technol.* 48: 14195–14202. doi: 10.1021/es5036317.

63 Lima, A.R.A., Costa, M.F., and Barletta, M. (2014). Distribution patterns of microplastics within the plankton of a tropical estuary. *Environ. Res.* 132: 146–155. doi: 10.1016/j.envres.2014.03.031.

64 Wang, W., Ndungu, A.W., Li, Z., and Wang, J. (2017). Microplastics pollution in inland freshwaters of China: a case study in urban surface waters of Wuhan, China. *Sci. Total Env.* 575: 1369–1374. doi: 10.1016/j.scitotenv.2016.09.213.

65 Zhao, S., Zhu, L., Wang, T., and Li, D. (2014). Suspended microplastics in the surface water of the Yangtze Estuary System, China: first observations on occurrence, distribution. *Mar. Pollut. Bull.* 86: 562–568. doi: 10.1016/j.marpolbul.2014.06.032.

66 Naidoo, T., Glassom, D., and Smit, A.J. (2015). Plastic pollution in five urban estuaries of KwaZulu-Natal, South Africa. *Mar. Pollut. Bull.* 101: 473–480. doi: 10.1016/j.marpolbul.2015.09.044.

67 Sadri, S.S. and Thompson, R.C. (2014). On the quantity and composition of floating plastic debris entering and leaving the Tamar Estuary, Southwest England. *Mar. Pollut. Bull.* 81: 55–60. doi: 10.1016/j.marpolbul.2014.02.020.

68 Dris, R., Gasperi, J., Rocher, V., Saad, M., Renault, N., and Tassin, B. (2015). Microplastic contamination in an urban area: a case study in Greater Paris. *Environ. Chem. Lett.* 12: 592–599. doi: 10.1071/EN14167.

69 Wagner, M. and Lambert, S. (Eds.) (2018). *Freshwater Microplastics: Emerging Environmental Contaminants?* Cham: Springer Nature.

70 Lechner, A. and Ramler, D. (2015). The discharge of certain amounts of industrial microplastic from a production plant into the River Danube is permitted by the Austrian legislation. *Environ. Pollut.* 200: 159–160. doi: 10.1016/j.envpol.2015.02.019.

71 Mani, T., Hauk, A., Walter, U., and Burkhardt-Holm, P. (2015). Microplastics profile along the Rhine River. *Sci. Rep.* 5: 1–7. doi: 10.1038/srep17988.

72 Llorca, M., Vega-Herrera, A., Schirinzi, G., Savva, K., Abad, E., and Farré, M. (2021). Screening of suspected micro (nano) plastics in the Ebro Delta (Mediterranean Sea). *J. Hazard. Mater.* 404: 124022. doi: 10.1016/j.jhazmat.2020.124022.

73 Schirinzi, G.F., Llorca, M., Seró, R., Moyano, E., Barceló, D., Abad, E., and Farré, M. (2019). Trace analysis of polystyrene microplastics in natural waters. *Chemosphere* 236: 124321. doi: 10.1016/j.chemosphere.2019.07.052.

74 Chubarenko, I., Esiukova, E., Bagaev, A., Isachenko, I., Demchenko, N., Zobkov, M., Efimova, I., Bagaeva, M., and Khatmullina, L. (2018). Behavior of microplastics in coastal zones. In: *Microplastic Contamination in Aquatic Environments*, 1st edn (E. Zeng, ed.). Amsterdam: Elsevier.

75 Woodall, L.C., Sanchez-Vidal, A., Canals, M., Paterson, G.L., Coppock, R., Sleight, V., Calafat, A., Rogers, A.D., Narayanaswamy, B.E., and Thompson, R.C. (2014). The deep sea is a major sink for microplastic debris. *R. Soc. Open Sci.* 1: 140317. doi: 10.1098/rsos.140317.

76 Vianello, A., Boldrin, A., Guerriero, P., Moschino, V., Rella, R., Sturaro, A., and Da Ros, L. (2013). Microplastic particles in sediments of Lagoon of Venice, Italy: first observations on occurrence, spatial patterns and identification. *Estuar. Coast. Shelf Sci.* 130: 54–61. doi: 10.1016/j.ecss.2013.03.022.

77 Abidli, S., Antunes, J.C., Ferreira, J.L., Lahbib, Y., Sobral, P., and Trigui El Menif, N. (2018). Microplastics in sediments from the littoral zone of the north Tunisian coast (Mediterranean Sea). *Estuar. Coast. Shelf Sci.* 205: 1–9. doi: 10.1016/j.ecss.2018.03.006.

78 Mistri, M., Infantini, V., Scoponi, M., Granata, T., Moruzzi, L., Massara, F., De Donati, M., and Munari, C. (2017). Small plastic debris in sediments from the Central Adriatic Sea: types, occurrence and distribution. *Mar. Pollut. Bull.* 124: 435–440. doi: 10.1016/j.marpolbul.2017.07.063.

79 Romeo, T., D'Alessandro, M., Esposito, V., Scotti, G., Berto, D., Formalewicz, M., Noventa, S., Giuliani, S., Macchia, S., Sartori, D., Mazzola, A., Andaloro, F., Giacobbe, S., Deidun, A., and Renzi, M. (2015). Environmental quality assessment of Grand Harbour (Valletta, Maltese Islands): a case study of a busy harbour in the Central Mediterranean sea. *Environ. Monit. Assess.* 187: 747. doi: 10.1007/s10661-015-4950-3.

80 Filgueiras, A.V., Gago, J., Campillo, J.A., and León, V.M. (2019). Microplastic distribution in surface sediments along the Spanish Mediterranean continental shelf. *Environ. Sci. Pollut. Res.* 26: 21264–21273. doi: 10.1007/s11356-019-05341-5.

81 Alomar, C., Estarellas, F., and Deudero, S. (2016). Microplastics in the Mediterranean Sea: deposition in coastal shallow sediments, spatial variation and preferential grain size. *Mar. Environ. Res.* 115: 1–10. doi: 10.1016/j.marenvres.2016.01.005.

82 Fastelli, P., Blašković, A., Bernardi, G., Romeo, T., Čižmek, H., Andaloro, F., Russo, G.F., Guerranti, C., and Renzi, M. (2016). Plastic litter in sediments from a marine area likely to become protected (Aeolian Archipelago's islands, Tyrrhenian sea). *Mar. Pollut. Bull.* 113: 526–529. doi: 10.1016/j.marpolbul.2016.08.054.

83 Blašković, A., Fastelli, P., Čižmek, H., Guerranti, C., and Renzi, M. (2017). Plastic litter in sediments from the Croatian marine protected area of the natural park of Telaščica bay (Adriatic Sea). *Mar. Pollut. Bull.* 114: 583–586. doi: 10.1016/j.marpolbul.2016.09.018.

84 Guerranti, C., Cannas, S., Scopetani, C., Fastelli, P., Cincinelli, A., and Renzi, M. (2017). Plastic litter in aquatic environments of Maremma Regional Park (Tyrrhenian Sea, Italy): contribution by the Ombrone river and levels in marine sediments. *Mar. Pollut. Bull.* 117: 366–370. doi: 10.1016/j.marpolbul.2017.02.021.

85 Simon-Sánchez, L., Grelaud, M., Garcia-Orellana, J., and Ziveri, P. (2019). River Deltas as hotspots of microplastic accumulation: the case study of the Ebro River (NW Mediterranean). *Sci. Total Environ.* 687: 1186–1196. doi: 10.1016/j.scitotenv.2019.06.168.

86 Peng, X., Chen, M., Chen, S., Dasgupta, S., Xu, H., Ta, K., Du, M., Li, J., Guo, Z., and Bai, S. (2018). Microplastics contaminate the deepest part of the world's ocean. *Geochem. Perspect. Lett.* 9: 1–5. doi: 10.7185/geochemlet.1829.

87 Bergmann, M., Wirzberger, V., Krumpen, T., Lorenz, C., Primpke, S., Tekman, M.B., and Gerdts, G. (2017). High quantities of microplastic in Arctic deep-sea sediments from the HAUSGARTEN observatory. *Environ. Sci. Technol.* 51: 11000–11010. doi: 10.1021/acs.est.7b03331.

88 Reed, S., Clark, M., Thompson, R., and Hughes, K.A. (2018). Microplastics in marine sediments near Rothera research station, Antarctica. *Mar. Pollut. Bull.* 133: 460–463. doi: 10.1016/j.marpolbul.2018.05.068.

89 Munari, C., Infantini, V., Scoponi, M., Rastelli, E., Corinaldesi, C., and Mistri, M. (2017). Microplastics in the sediments of Terra Nova Bay (Ross Sea, Antarctica). *Mar. Pollut. Bull.* 122: 161–165. doi: 10.1016/j.marpolbul.2017.06.039.

90 Gonçalves, C., Martins, M., Sobral, P., Costa, P.M., and Costa, M.H. (2019). An assessment of the ability to ingest and excrete microplastics by filter-feeders: a case study with the Mediterranean mussel. *Environ. Pollut.* 245: 600–606. doi: 10.1016/j.envpol.2018.11.038.

91 Kinjo, A., Mizukawa, K., Takada, H., and Inoue, K. (2019). Size-dependent elimination of ingested microplastics in the Mediterranean mussel Mytilus galloprovincialis. *Mar. Pollut. Bull.* 149: 110512. doi: 10.1016/j.marpolbul.2019.110512.

92 Andrade, C. and Ovando, F., (Year) of Conference). First record of microplastics in stomach content of the southern king crab lithodes santolla (Anomura: Lithodidadae), Nassau Bay, Cape Horn, Chile. Universidad de Magallanes.

93 Mathalon, A. and Hill, P. (2014). Microplastic fibers in the intertidal ecosystem surrounding Halifax Harbor, Nova Scotia. *Mar. Pollut. Bull.* 81: 69–79. doi: 10.1016/j.marpolbul.2014.02.018.

94 Davison, P. and Asch, R.G. (2011). Plastic ingestion by mesopelagic fishes in the North Pacific Subtropical Gyre. *Mar. Ecol. Prog. Ser.* 432: 173–180. doi: 10.3354/meps09142.

95 Choy, C.A. and Drazen, J.C. (2013). Plastic for dinner? Observations of frequent debris ingestion by pelagic predatory fishes from the central North Pacific. *Mar. Ecol. Prog. Ser.* 485: 155–163. doi: 10.3354/meps10342.

96 Schirinzi, G.F., Pedà, C., Battaglia, P., Laface, F., Galli, M., Baini, M., Consoli, P., Scotti, G., Esposito, V., and Faggio, C. (2020). A new digestion approach for the extraction of microplastics from gastrointestinal tracts (GITs) of the common dolphinfish (Coryphaena hippurus) from the western Mediterranean Sea. *J. Hazard. Mater.* 397: 122794. doi: 10.1016/j.jhazmat.2020.122794.

97 Nelms, S.E., Galloway, T.S., Godley, B.J., Jarvis, D.S., and Lindeque, P.K. (2018). Investigating microplastic trophic transfer in marine top predators. *Environ. Pollut.* 238: 999–1007. doi: 10.1016/j.envpol.2018.02.016.

98 Burkhardt-Holm, P. and N'Guyen, A. (2019). Ingestion of microplastics by fish and other prey organisms of cetaceans, exemplified for two large baleen whale species. *Mar. Pollut. Bull.* 144: 224–234. doi: 10.1016/j.marpolbul.2019.04.068.

99 Provencher, J.F., Vermaire, J.C., Avery-Gomm, S., Braune, B.M., and Mallory, M.L. (2018). Garbage in guano? Microplastic debris found in faecal precursors of seabirds known to ingest plastics. *Sci. Total Environ.* 644: 1477–1484. doi: 10.1016/j.scitotenv.2018.07.101.

100 Caron, A.G.M., Thomas, C.R., Berry, K.L.E., Motti, C.A., Ariel, E., and Brodie, J.E. (2018). Ingestion of microplastic debris by green sea turtles (Chelonia mydas) in the Great Barrier Reef: validation of a sequential extraction protocol. *Mar. Pollut. Bull.* 127: 743–751. doi: 10.1016/j.marpolbul.2017.12.062.

101 Lusher, A.L., Welden, N.A., Sobral, P., and Cole, M. (2017). Sampling, isolating and identifying microplastics ingested by fish and invertebrates. *Anal. Methods* 9: 1346–1360. doi: 10.1039/C6AY02415G.

102 Michida, Y., Chavanich, S., Chiba, S., Cordova, M.R., Cozsar Cabanas, A., Glagani, F., Hagmann, P., Hinata, H., Isobe, A., and Kershaw, P. (2019). Guidelines for harmonizing ocean surface microplastic monitoring methods. Version 1.1.

103 Cutroneo, L., Reboa, A., Besio, G., Borgogno, F., Canesi, L., Canuto, S., Dara, M., Enrile, F., Forioso, I., and Greco, G. (2020). Microplastics in seawater: sampling strategies, laboratory methodologies, and identification techniques applied to port environment. *Environ. Sci. Pollut. Res.* 27: 8938–8952. doi: 10.1007/s11356-020-07783-8.

104 Pico, Y., Alfarhan, A., and Barcelo, D. (2019). Nano- and microplastic analysis: focus on their occurrence in freshwater ecosystems and remediation technologies. *TrAC Trends Anal. Chem.* 113: 409–425. doi: 10.1016/j.trac.2018.08.022.

105 Vega-Herrera, A., Llorca, M., Savva, K., León, V.M., Abad, E., and Farré, M. (2021). Screening and quantification of Micro(Nano)plastics and plastic additives in the seawater of Mar Menor Lagoon. *Front. Mar. Sci. accepted*. doi: 10.3389/fmars.2021.697424.

106 Hermsen, E., Mintenig, S.M., Besseling, E., and Koelmans, A.A. (2018). Quality criteria for the analysis of microplastic in Biota samples: a critical review. *Environ. Sci. Technol.* 52(18): 10230–10240. doi: 10.1021/acs.est.8b01611.

107 Isobe, A., Buenaventura, N.T., Chastain, S., Chavanich, S., Cózar, A., DeLorenzo, M., Hagmann, P., Hinata, H., Kozlovskii, N., Lusher, A.L., Martí, E., Michida, Y., Mu, J., Ohno, M., Potter, G., Ross, P.S., Sagawa, N., Shim, W.J., Song, Y.K., Takada, H., Tokai, T., Torii, T., Uchida, K., Vassillenko, K., Viyakarn, V., and Zhang, W. (2019). An interlaboratory comparison exercise for the determination of microplastics in standard sample bottles. *Mar. Pollut. Bull.* 146: 831–837. doi: 10.1016/j.marpolbul.2019.07.033.

108 QUASIMEME. QUASIMEME laboratory performance studies. http://www.quasimeme.org/links. Accessed 14 December 2021.

109 Carpenter, E.J., Anderson, S.J., Harvey, G.R., Miklas, H.P., and Peck, B.B. (1972). Polystyrene spherules in coastal waters. *Science* 178: 749–750. doi: 10.1126/science.178.4062.749.

110 Costa, M.F. and Barletta, M. (2015). Microplastics in coastal and marine environments of the western tropical and sub-tropical Atlantic Ocean. *Environ. Sci. Process. Impacts* 17: 1868–1879. doi: 10.1039/c5em00158g.

111 Wang, F., Wong, C.S., Chen, D., Lu, X., Wang, F., and Zeng, E.Y. (2018). Interaction of toxic chemicals with microplastics: a critical review. *Water Res.* 139: 208–219. doi: 10.1016/j.watres.2018.04.003.

112 Monteiro, R.C.P., Ivar Do Sul, J.A., and Costa, M.F. (2018). Plastic pollution in islands of the Atlantic Ocean. *Environ. Pollut.* 238: 103–110. doi: 10.1016/j.envpol.2018.01.096.

113 Bradney, L., Wijesekara, H., Palansooriya, K.N., Obadamudalige, N., Bolan, N.S., Ok, Y.S., Rinklebe, J., Kim, K.H., and Kirkham, M.B. (2019). Particulate plastics as a vector for toxic trace-element uptake by aquatic and terrestrial organisms and human health risk. *Environ. Int.* 131: 104937. doi: 10.1016/j.envint.2019.104937.

114 Waller, C.L., Griffiths, H.J., Waluda, C.M., Thorpe, S.E., Loaiza, I., Moreno, B., Pacherres, C.O., and Hughes, K.A. (2017). Microplastics in the Antarctic marine system: an emerging area of research. *Sci. Total Environ.* 598: 220–227. doi: 10.1016/j.scitotenv.2017.03.283.

115 Fendall, L.S. and Sewell, M.A. (2009). Contributing to marine pollution by washing your face: microplastics in facial cleansers. *Mar. Pollut. Bull.* 58: 1225–1228. doi: 10.1016/j.marpolbul.2009.04.025.

116 Cincinelli, A., Martellini, T., Guerranti, C., Scopetani, C., Chelazzi, D., and Giarrizzo, T. (2019). A potpourri of microplastics in the sea surface and water column of the Mediterranean Sea. *TrAC Trend Anal. Chem.* 110: 321–326. doi: 10.1016/j.trac.2018.10.026.

117 Harrison, J.P., Sapp, M., Schratzberger, M., and Osborn, A.M. (2011). Interactions between microorganisms and marine microplastics: a call for research. *Mar. Technol. Soc. J.* 45: 12–20. doi: 10.4031/MTSJ.45.2.2.

118 Lusher, A.L., Hernandez-Milian, G., O'Brien, J., Berrow, S., O'Connor, I., and Officer, R. (2015). Microplastic and macroplastic ingestion by a deep diving, oceanic cetacean: the true's beaked whale Mesoplodon mirus. *Environ. Pollut.* 199: 185–191. doi: 10.1016/j.envpol.2015.01.023.

119 Da Costa, J.P., Duarte, A.C., and Rocha-Santos, T.A. (2017). *Microplastics–Occurrence, Fate and Behaviour in the Environment.* In: Comprehensive Analytical Chemistry, 75. Amsterdam: Elsevier.

120 van Wijnen, J., Ragas, A.M., and Kroeze, C. (2019). Modelling global river export of microplastics to the marine environment: sources and future trends. *Sci. Total Env.* 673: 392–401. doi: 10.1016/j.scitotenv.2019.04.078.

121 Agamuthu, P., Mehran, S.B., Norkhairah, A., and Norkhairiyah, A. (2019). Marine debris: a review of impacts and global initiatives. *Waste Manag. Res.* 37: 987–1002. doi: 10.1177/0734242X19845041.

122 Andrady, A.L. (2011). Microplastics in the marine environment. *Mar. Pollut. Bull.* 62: 1596–1605. doi: 10.1016/j.marpolbul.2011.05.030.

123 Shahul Hamid, F., Bhatti, M.S., Anuar, N., Anuar, N., Mohan, P., and Periathamby, A. (2018). Worldwide distribution and abundance of microplastic: how dire is the situation? *Waste Manag. Res.* 36: 873–897. doi: 10.1177/0734242X18785730.

124 Dubaish, F. and Liebezeit, G. (2013). Suspended microplastics and black carbon particles in the Jade system, southern North Sea. *Water Air Soil Pollut.* 224: 1352. doi: 10.1007/s11270-012-1352-9.

125 Silva, A.B., Bastos, A.S., Justino, C.I.L., Da Costa, J.P., Duarte, A.C., and Rocha-Santos, T.A.P. (2018). Microplastics in the environment: challenges in analytical chemistry – a review. *Anal. Chim. Acta* 1017: 1–19. doi: 10.1016/j.aca.2018.02.043.

126 Jakimska, A., Huerta, B., Bargańska, Ż., Kot-Wasik, A., Rodríguez-Mozaz, S., and Barceló, D. (2013). Development of a liquid chromatography–tandem mass spectrometry procedure for determination of endocrine disrupting compounds in fish from Mediterranean rivers. *J. Chromatogr. A* 1306: 44–58. doi: 10.1016/j.chroma.2013.07.050.

127 Gago, J., Carretero, O., Filgueiras, A.V., and Viñas, L. (2018). Synthetic microfibers in the marine environment: a review on their occurrence in seawater and sediments. *Mar. Pollut. Bull.* 127: 365–376. doi: 10.1016/j.marpolbul.2017.11.070.

128 Sanchez-Vidal, A., Llorca, M., Farré, M., Canals, M., Barceló, D., Puig, P., and Calafat, A. (2015). Delivery of unprecedented amounts of perfluoroalkyl substances towards the deep-sea. *Sci. Total Environ.* 526: 41–48. doi: 10.1016/j.scitotenv.2015.04.080.

129 Fahrenfeld, N.L., Arbuckle-Keil, G., Naderi Beni, N., and Bartelt-Hunt, S.L. (2019). Source tracking microplastics in the freshwater environment. *TrAC Trends Anal. Chem.* 112: 248–254. doi: 10.1016/j.trac.2018.11.030.

130 Li, J., Liu, H., and Paul Chen, J. (2018). Microplastics in freshwater systems: a review on occurrence, environmental effects, and methods for microplastics detection. *Water Res.* 137: 362–374. doi: 10.1016/j.watres.2017.12.056.

131 Horton, A.A., Walton, A., Spurgeon, D.J., Lahive, E., and Svendsen, C. (2017). Microplastics in freshwater and terrestrial environments: evaluating the current understanding to identify the knowledge gaps and future research priorities. *Sci. Total Environ.* 586: 127–141. doi: 10.1016/j.scitotenv.2017.01.190.

132 Anbumani, S. and Kakkar, P. (2018). Ecotoxicological effects of microplastics on biota: a review. *Environ. Sci. Pollut. Res.* 25: 14373–14396. doi: 10.1007/s11356-018-1999-x.

133 Bowmer, T. and Kershaw, P., (2010). Proceedings of the GESAMP International Workshop on Microplastic Particles as a Vector in Transporting Persistent, Bio-accumulating and Toxic Substances in the Ocean, 28–30th June 2010, UNESCO-IOC, Paris. GESAMP.

134 Prata, J.C., Da Costa, J.P., Lopes, I., Duarte, A.C., and Rocha-Santos, T. (2019). Effects of microplastics on microalgae populations: a critical review. *Sci. Total Environ.* 665: 400–405. doi: 10.1016/j.scitotenv.2019.02.132.

135 Tourinho, P.S., Kočí, V., Loureiro, S., and van Gestel, C.A.M. (2019). Partitioning of chemical contaminants to microplastics: sorption mechanisms, environmental distribution and effects on toxicity and bioaccumulation. *Environ. Pollut.* 252: 1246–1256. doi: 10.1016/j.envpol.2019.06.030.

136 Jeong, J. and Choi, J. (2019). Adverse outcome pathways potentially related to hazard identification of microplastics based on toxicity mechanisms. *Chemosphere* 231: 249–255. doi: 10.1016/j.chemosphere.2019.05.003.

137 Guzzetti, E., Sureda, A., Tejada, S., and Faggio, C. (2018). Microplastic in marine organism: environmental and toxicological effects. *Environ. Toxicol. Pharmacol.* 64: 164–171. doi: 10.1016/j.etap.2018.10.009.

138 Setälä, O., Fleming-Lehtinen, V., and Lehtiniemi, M. (2014). Ingestion and transfer of microplastics in the planktonic food web. *Environ. Pollut.* 185: 77–83. doi: 10.1016/j.envpol.2013.10.013.

139 Toussaint, B., Raffael, B., Angers-Loustau, A., Gilliland, D., Kestens, V., Petrillo, M., Rio-Echevarria, I.M., and Van Den Eede, G. (2019). Review of micro- and nanoplastic contamination in the food chain. *Food Additives and Contaminants – Part A Chemistry, Analysis, Control, Exposure and Risk Assessment* 36: 639–673. doi: 10.1080/19440049.2019.1583381.

140 Hantoro, I., Löhr, A.J., Van Belleghem, F.G.A.J., Widianarko, B., and Ragas, A.M.J. (2019). Microplastics in coastal areas and seafood: implications for food safety. *Food Additives and Contaminants – Part A Chemistry, Analysis, Control, Exposure and Risk Assessment* 36(5): 674–711. doi: 10.1080/19440049.2019.1585581.

141 Browne, M.A., Dissanayake, A., Galloway, T.S., Lowe, D.M., and Thompson, R.C. (2008). Ingested microscopic plastic translocates to the circulatory system of the mussel, Mytilus edulis (L.). *Environ. Sci. Technol.* 42: 5026–5031. doi: 10.1021/es800249a.

142 Smith, M., Love, D.C., Rochman, C.M., and Neff, R.A. (2018). Microplastics in seafood and the implications for human health. *Curr. Environ. Health Rep.* 5(3): 375–386. doi: 10.1007/s40572-018-0206-z.

143 Hong, S.H., Shim, W.J., and Hong, L. (2017). Methods of analysing chemicals associated with microplastics: a review. *Anal. Methods* 9: 1361–1368. doi: 10.1039/c6ay02971j.

144 Fu, Z. and Wang, J. (2019). Current practices and future perspectives of microplastic pollution in freshwater ecosystems in China. *Sci. Total Environ.* 691: 697–712. doi: 10.1016/j.scitotenv.2019.07.167.

145 Bryant, J.A., Clemente, T.M., Viviani, D.A., Fong, A.A., Thomas, K.A., Kemp, P., Karl, D.M., White, A.E., and DeLong, E.F. (2016). Diversity and activity of communities inhabiting plastic Debris in the North Pacific Gyre. *mSystems* 1: e00024-16. doi: 10.1128/mSystems.00024-16.

146 Thiel, M., Luna-Jorquera, G., Álvarez-Varas, R., Gallardo, C., Hinojosa, I.A., Luna, N., Miranda-Urbina, D., Morales, N., Ory, N., Pacheco, A.S., Portflitt-Toro, M., and Zavalaga, C. (2018). Impacts of marine plastic pollution from continental coasts to subtropical gyres-fish, seabirds, and other vertebrates in the SE Pacific. *Front. Mar. Sci.* 5. doi: 10.3389/fmars.2018.00238.

Index

a

AA *see* annual average
acesulfame (ACE) 151
  chromatography 173–175
  detection 175–177
  figures of merit 177–179
  frequently used SPE cartridges 169–170
  legislation 154
  liquid chromatography 174
  metabolites and/or transformation products 155–157
  MRM transitions 177
  NORMAN PNECs 155
  occurrence 157–167
  properties 153–154
acetic anhydride with zinc powder 320
acetochlor 3, 15, 17
acetonitrile (ACN)
  artificial sweeteners 173, 178
  high production volume chemicals 233, 247, 251–252
  personal care products 107
  pesticides 15, 17
  pharmaceuticals 46, 55 *see also* disinfection byproducts
acetylsalicylic acid 39
ACMs *see* alicyclic musks
ACQUITY UPLC BEH C18 columns 55, 328
ADBI *see* celestolide
ADONA *see* 4, 8-dioxa-3 H-perfluorononanoic acid
AFFFs *see* aqueous film-forming foams
afidopyropen 17
aggregation-induced emission (AIE) 207
aglycon hesperetin dihydrochalcone 156
AHMI *see* phantolide
AHTN *see* tonalide
air samples
  high production volume chemicals 231–232, 238–244
  musk fragrances 269–271
  pesticides 8–9, 19
air-assisted-DLLME 106
alachlor 3
D-alanine 154
aldrin 8–9
alicyclic musks (ACMs) 266, 280
alitame (ALI) 151, 153
  figures of merit 177–179
  liquid chromatography 174
  occurrence 164, 167
  properties 153
alkaline hydrogenation of neohesperidin 154
N-alkyl FOSAs/-FOSEs, chromatography-mass spectrometry 205
N-alkyl perfluoroalkane sulfonamides 191
N-alkyl perfluoroalkane sulphonamido ethanols (FASEs) 188, 191, 195
N-alkyl polyflouoroalkyl phosphates 191
alkyl-bonded silicas and polymers 171
allergens *see* fragrance allergens
17-alpha-ethynylestradiol 39, 52
ambrettolide 265
amidosulfonic acid 156–157
amino methylphosphonic acid (AMPA) 3, 7

Analytical Methods for Environmental Contaminants of Emerging Concern, First Edition. Edited by Núria Fontanals and Rosa Maria Marcé.

ammonium acetate (NH_4AC) 55, 173
ammonium formate (NH_4CO_2) 171–172
analgesic/anti-inflammatories 39
annual average (AA) 2, 191
antianxiety medications 48
antibiotics 39, 41, 48, 50
anticancer drugs 43
antidepressants 39, 43, 49
antiepileptics 39, 40, 49
anti-inflammatories 39, 42–43, 47, 51–52, 54–55 *see also* non-steroidal anti-inflammatory drugs
antimicrobials (AMX) 55
antiparasites 43
APCI *see* atmospheric pressure chemical ionization
app-based biosensors 207–208
APPI *see* atmospheric pressure photoionization
aquatic environments and personal care product pathways 73–107
  fabric phase sorptive extraction 99–100
  fragrance allergen/siloxane analytical methods 96–97
  magnetic solid phase extraction 99
  miniaturized SBSE approaches 101
  molecularly imprinted solid phase extraction 99
  plasticizer analytical methods 89–95
  preservative analytical methods 84–88
  rotating-disk sorptive extraction 101–102
  solid phase extraction 75–99
  solid phase microextraction 102–104
  sorbent-based methods pathways 75–104
  stir-bar sorptive extraction 100–102
  UV filter analytical methods 76–83
  vacuum-assisted sorbent extraction 102
aqueous film-forming foams (AFFFs) 187, 190
  chromatography-mass spectrometry 201, 203–204
  metabolites and/or transformation products 192
  occurrence 193
  properties 190
artificial sweeteners (ASs) 151–186
  chromatography 172–175
  detection 175–177
  determination 172–180
  figures of merit 177–179
  frequently used SPE cartridges 169
  future directions and challenges 180
  hints and tips 178–180
  instrumental method 172–177
  legislation and environmental risk assessment 154–155
  metabolites and/or transformation products 155–157
  NORMAN PNECs 155
  occurrence 157–167
  overview 151–167
  properties 151–154
  protocols for collecting/preparing sampling 167
  sample extraction and clean-up, artificial sweeteners 167–172
  sample preparation and collection 167–172
arylcycloalkylamines 129, 138–139
ASP *see* aspartame
aspartame (ASP) 151–152, 154
  chromatography 173–175
  detection 175–177
  figures of merit 177–179
  frequently used SPE cartridges 169–170
  hints and tips 178
  legislation 154–155
  liquid chromatography 174
  metabolites and/or transformation products 156
  MRM transitions 177
  NORMAN PNECs 155
  occurrence 161–164
aspartic acid 154, 156
astkCARETM reagents 207
Athena C_{18}-WP columns 173, 175
atmospheric pressure (AP)GC-QTOF 100
atmospheric pressure chemical ionization (APCI)
  disinfection byproducts 330
  high production volume chemicals 205, 252
  perfluorinated substances 201, 205–206

pharmaceuticals 55, 59
atmospheric pressure photoionization (APPI) 55, 59, 201, 205–206
atrazine 6
atrazine-desethyl (DEA) 3
atrazine-desisopropyl (DIA) 3
ATTI *see* traseolide
azinphos-ethyl 7–8

b

Bakerbond SDB-1 316
BAM *see* 2,6-dichlorobenzamide
BAμE coatings 101
2,4-D-based herbicides 7
BBzP *see* butyl benzyl phthalate
BCAA *see* bromochloacetic acid
benzenesulfonamides (BSAs)
  determination 251–252
  properties 223–225
  sample extraction and clean-up 237, 238, 244, 247
benzodiazepines 129
benzophenone-type UV filters 105
benzothiazoles (BTHs)
  determination 251–252
  legislation 227
  occurrence 229–231
  properties 223–225
  sample extraction and clean-up 237, 244, 247
benzotriale UV filters 100
benzotriazole ultraviolet stabilizers (BUVs)
  determination 251
  metabolites and/or TPs 227
  occurrence 229, 231
  properties 223–225
  sample extraction and clean-up 233, 238, 244, 247
benzotriazoles (BTRs)
  determination 251–252
  legislation 227
  metabolites and/or transformation products 227
  occurrence 228–231
  properties 223–225
  sample extraction and clean-up 233, 237–238, 247
17-beta-estradiol 39
beta-agonists 51–52
β-blockers 41, 51–52
BHA *see* 2-tert-butyl-4-methoxyphenol
BHT *see* butylhydroxyanisole; 2,6-di-tert-butyl-4-methylphenol
Białk-Bielińska, A. 47–48
bioconcentration factors (BCFs) 269–270
biofouling 358
biosensors, perfluorinated substances 206–207
biota samples
  high production volume chemicals 232–233, 247–250
  microplastics 359–361
  musk fragrances 269–271, 276–279
  pesticides 18
Bon Elut-PPL cartridges 317
borosilicate glass 313–314
boscalid diuron pesticides 3, 4
bromochloacetic acid (BCAA) 288, 310, 319
bromodichloromethane (BDCM) 298, 307
bromoform (TBM) 289, 298, 307 *see also* THM4
BSAa *see* benzenesulfonamides
BSTFA *see* *N*,*O*-bis(trimethylsilyl) trifluoroacetamide
BTHs *see* benzothiazoles
BTRs *see* benzotriazoles
butyl benzyl phthalate (BBzP) 226
butylhydroxyanisole (BHT), personal care products 73
2,6-di-tert-butyl-4-methylphenol (BHT) 226–231, 233, 247

c

C8-C14 PFCAs 190–191
C18 cartridges, perfluorinated substances 200–201
C18 coated open systems, personal care products 103
C18 columns 55, 321
C18 sorbent, pesticides 17, 20–22
C18-microspheres, personal care products 98
C18-WP columns 173, 175
CAA *see* monochloroacetic acid
CAC *see* chitosan activated carbon
candle soot (CS) 104
cannabinoids 128

capillary liquid chromatography 201
Captiva SPE modifications 200–201
carbamazepine 39, 40
carbofuran 9
carbon nanomaterials 11, 99, 103
Carbotrap samplers 17, 20
Carbowax/DVB 317–318
carboxy-metabolites of NSAIDs 54–55
cashmeran (DPMI) 264, 267–268, 270–271
cathinones 128
cationic PFAss 187, 190
CC-MS/MS, perfluorinated substances 201
CECs *see* contaminants of emerging concern
celestolide (ADBI) 263–264, 266–269
Certified Reference materials (CRMs) 363
CF *see* cyclophosphamide
Che, J.S. 276
Chelon labrosus (grey mullet liver) 204
chemical characterization of microplastics 362–363
chemical ionization (CI) 205–206, 329
chitosan activated carbon (CAC) 98
chlordane 8–9
chloridazon 3
chlorinated polyfluorinated ether sulfonic acid (Cl-PFESA) 194–196, 204
chlorination, artificial sweeteners 157
4-chloro-4-deoxygalactose (4-GC) 156
chloroform (TCM)
  determination 320–331, 334
  legislation 298, 307
  occurrence 310
  properties 289
  protocols for collecting/preparing samples 314–315 *see also* THM4
9-chlorohexadeca fluoro-3-oxanonane-sulfonate (9Cl-PF3ONS) 193
9-chlorohexadecafluoro-3-oxanonane-1-sulfonate (F53B major) 187, 190, 193–194
chloronitromethane (CNM) 309
chloronitromethane (CNM) Poly-Sery PWAX SPE cartridge 170
11-chlororeicosafluoro-3-oxaundecane-1-sulfonic acid (F53B minor) 187, 190, 193–194
chloro-sulfonic acid 151, 154
chlorothalonil 3, 11
chlorpyrifos 3–4, 6–9, 20, 23
chromatography
  artificial sweeteners 172–175
  disinfection byproducts 320–328
  gel permeation chromatography 247, 279
  musk fragrances 279–280
  new psychoactive substances 140
  reversed-phase 107 *see also* gas chromatography; liquid chromatography
chromatography-mass spectrometry 201–206
CI *see* chemical ionization
CID *see* collision-induced dissociation
ciprofloxacin (CIP) 39, 41
citric acid 47
civetone 265
Clean Water Act (CWA) 2
clofibric acid 39
Cl-PFESA *see* chlorinated polyfluorinated ether sulfonic acid
CNM *see* chloronitromethane
CNW Poly-Sery PWAX SPE cartridge 170
CNW Poly-SeryPSD 170
collision-induced dissociation (CID) 204–205
Commission on Narcotic Drugs 131
common household products 73–74
Compound Discoverer 3.0 software 23
contaminants of emerging concern (CECs) 72, 108
cosmetics 73–74
criterion maximum concentration (CMC) 2
CRMs *see* Certified Reference materials
CS *see* candle soot
current-use pesticides (CUPs) 9
CWA *see* Clean Water Act
cyanopropyl groups 98
cyclamate (CYCL)
  chromatography 173–175
  figures of merit 177–179
  frequently used SPE cartridges 169–170
  liquid chromatography 174
  metabolites and/or transformation products 156
  MRM transitions 177
  NORMAN PNECs 155

occurrence 157–167
properties 151–152, 154
cyclic perfluoroalkyl sulfonic acids (cyclic PFSAs) 189, 193–194
cyclohexanol 156
cyclohexanone 156
cyclohexylamine 156
cyclophosphamide (CF) 43
cypermethrin 9

d

daily care products 73–74
daily tolerable intake (DTI) 191
DBAA *see* dibromoacetic acid
DBAN *see* dibromoacetonitrile
DCAA *see* dichloroacetic acid
DCAL *see* dichloroacetaldehyde
DCAN *see* dichloroacetonitrile
DCM *see* dichloromethane
DCNM *see* dichloronitromethane
DDE *see* dichloro-diphenyl-dichloroethane
DDT *see* dichloro-diphenyl-trichloroethane
DEA *see* atrazine-desethyl
deep eutectic solvent (DES)-based polymeric hydrogel 11
deep eutectic solvent-based polymeric hydrogel (DES)-DLLME 105
deglycosylation of NHDC 156
DEHP *see* di-(2-ethylhexyl)phthalate
DEHT *see* bis(2-ethylhexyl)teraphthalate
deltamethrin 9
desethylatrazine 6
desphenyl chloridazon 3
detection
artificial sweeteners 175–177
disinfection byproducts 328–330
musk fragrances 280
new psychoactive substances 140–142
pesticides 22–24
determination
artificial sweeteners 168–169, 172–180
disinfection byproducts 320–335
high production volume chemicals 251–253
microplastics 362–363
musk fragrances 279–280
new psychoactive substances 139–143
personal care products 107–108
pesticides 20–25
PFASs 201–210
pharmaceuticals 51–59
DGT *see* diffusive gradients in thin films
DIA *see* atrazine-desisopropyl
diazinon 6–8
dibromoacetic acid (DBAA) 288, 307
dibromoacetonitrile (DBAN) 311, 314–315
dibromochloromethane (DBCM) 289–298, 307
dichlofenthion 7
dichloro-diphenyl-dichloroethane (DDE) 2–3
dichloro-diphenyl-trichloroethane (DDT) 2–3, 9
dichloroacetaldehyde (DCAL) 310–311
dichloroacetic acid (DCAA) 288, 307, 310
dichloroacetonitrile (DCAN) 307–308, 311, 314–315
2,6-dichlorobenzamide (BAM) 3
1,6-dichloro-1,6-dideoxyfructose (1,6-DCF) 156
dichlorodiphenyldichloroethane (DDD) 8–9
dichloromethane (DCM) 233, 238, 244, 247
dichloronitromethane (DCNM) 309
2,4-dichlorophenoxyacetic acid (2,4-D) 5
dichobenil 3
diclofenac (DIC) 39, 42, 48
DiDP 226
dieldrin 8–9
diethyl phthalate (DEP) 226
differential mobility spectrum (DMS)-MS 204–205
diffusive gradients in thin films (DGT) 270–271
dihydroxyacetyl sulfamate 157
di-iso-nonyl phthalate (DiNP) 226, 229–230
diketopiperazine (DKP)
detection 176–177
figures of merit 177–179
liquid chromatography 174
metabolites and/or transformation products 156
MRM transitions 177
occurrence 164–165
dimethoate 7

dimethyl phthalate (DMP) 223, 226
dimethyl sulfate 54
2,4-D dimethylamine salts (2,4-D DMA) 5
3,3-dimethylbutyl moiety 154
N,N-dimethylsulfamide (DMS) 3
dimethyl(3,3,3-trifluoropropyl) silyldiethylamine (DIMETRIS) 52
diode array detector (DAD) 105, 107
4,8-dioxa-3 H-perfluorononanoic acid (ADONA) 187, 190
diPAPs *see* polyfuoroalkyl phosphate diesters
Directive 98/83/EC 2
Directive 2000/60/EC 72–73
Directive 2008/105/EC 155
disinfection byproducts
  chromatography 320–328
  detection 328–330
  determination 320–335
  figures of merit 330–333
  future directions and challenges 335
  hints and tips 333–335
  instrumental method 320–330
  linearity 330
  precision and accuracy 331–332
  properties 288–306
  protocols for collecting/preparing samples 313–314
  sample extraction and clean-up 314–319
  sensitivity 332–333
disinfection byproducts (DBP) 287–351
  legislation 298, 307–308
  metabolites and/or transformation products 308–310
  occurrence 310–313
  overview 287–313
  properties 288–306
  sample preparation and collection 313–319
dispersive liquid-liquid microextraction (DLLME)
  high production volume chemicals 237
  musk fragrances 271, 279
  perfluorinated substances 199
  personal care products 104–106
dispersive solid phase extraction (d-SPE)
  high production volume chemicals 247
  musk fragrances 271
  new psychoactive substances 139
  perfluorinated substances 200–201
  pesticides 17
  pharmaceuticals 47, 51
ditridecyl pthalate (DTDP) 223
diuron 3, 4, 7–8
divinylbenzene (DVB) 237, 276, 317–318
DKP *see* diketopiperazine
DMP *see* dimethyl phthalate
DMS *see* N,N-dimethylsulfamide
DPMI *see* cashmeran
Drinking Water Directive 98/83/EC 2
drinking water treatment plants (DWTPs) 309–313
2,4-DTBP 226
DTDP *see* ditridecyl pthalate
DTI *see* daily tolerable intake
dust samples
  high production volume chemicals 231–232, 238–244
  perfluorinated camples 198
DVB/CAR/PDMS fiber 317–318

e

Early Warning Advisory (EWA) 128–131
ECHA *see* European Chemicals Agency
EDs *see* endocrine disruptors
EDTA-McIlvaine buffer 47
EE *see* electroenhanced
EFSA *see* European Food Safety Authority
2-EHMC *see* 2-ethylhexylmethoxycinnamate
EI *see* electron ionization
electroenhanced (EE-)SPME 103
electron capture detection (ECD) 22, 52, 319
electron ionization (EI)
  artificial sweeteners 176
  disinfection byproducts 329
  musk fragrances 280
  perfluorinated substances 201, 204–206
  pesticides 22
electrospray ionization (ESI)
  artificial sweeteners 170, 173, 176, 177
  disinfection byproducts 330
  HESI 22, 252
  high production volume chemicals 252
  musk fragrances 280
  pesticides 22

pharmaceuticals 55, 59
ElutPlexa SPE cartridges 169–170
EMR *see* enhanced matrix removal
enantiomers
new psychoactive substances 140
pesticides 17, 22
endocrine disruptors (EDs) 54–55, 75
endosulfan 3–4, 6–7, 20, 23
enhanced matrix removal (EMR) 200–201
Env + SPE cartridges 169–170, 199–200
Envi-Carb cartridges 199–201
environmental occurrence
disinfection byproducts 310–313
high production volume chemicals 228–231
microplastics 355–360
musk fragrances 268–270
perfluorinated substances 193–197
personal care products 73–74
pesticides 5–9
pharmaceuticals 39–43
Environmental Protection Agency *see* United States Environmental Protection Agency
Environmental Protection Agency (EPA) 2
environmental quality standards (EQS) 2, 20, 155, 191
Environmental Risk Assessment (ERA)
artificial sweeteners 154–155
pharmaceuticals 45
EOF *see* extractable organic fluorine
EPA *see* United States Environmental Protection Agency
EQS *see* environmental quality standards
ESA *see* ethanesulfonic acid
ESI *see* electrospray ionization
esterase 151, 156
estriol 39
estron (E1) 43
estrone 39
EtAc *see* ethyl acetate
ethanesulfonic acid (ESA) 3
ethion 7
ethyl acetate (EtAc) 233, 244, 247
N-ethyl perfluorooctane sulfonamidoethanol (EtFOSE) 192, 205–206
N-ethyl perfluorooctanesulfonamide (EtFOSA) 194
2-ethylhexylmethoxycinnamate (2-EHMC), personal care products 73
di-(2-ethylhexyl)phthalate (DEHP) 72–73, 99, 226, 228–230, 252
bis(2-ethylhexyl)teraphthalate (DEHT) 229–230, 252
European Chemicals Agency (ECHA) 191, 267
European Environment Agency (EEA) 6
European Food Safety Authority (EFSA) 191
European Union (EU)
artificial sweeteners 155
high production volume chemicals 226, 227
microplastics 353–354
musk fragrances 267
perfluorinated substances 191
personal care products 72–73
pesticides 2, 6, 9
pharmaceuticals 39, 44–45, 59
European Watch List (WL) 44–45, 73
EWA *see* Early Warning Advisory
exaltolide 265
exaltone 265
extractable organic fluorine (EOF) 210

f

F53B major *see* 9-chlorohexadecafluoro-3-oxanonane-1-sulfonate
F53B minor *see* 11-chlororeicosafluoro-3-oxaundecane-1-sulfonic acid
FA *see* formic acid
fabric phase sorptive extraction (FPSE) 99–100
famoxadone 3–5, 20–21, 23
FASAs *see* perfluoroalkane sulphonamides
FASEs *see* N-alkyl perfluoroalkane sulphonamido ethanols
fat-soluble UV filters 71
Fe_3O_4 nanoparticles 98, 105
fenbendazole (FEN) 43
fenobucarb 9
filtration of samples, new psychoactive substances 133–134
fish samples

pesticides 7–8 *see also* water samples
flame ionization detector (FID) 22, 104, 105
floating organic drop (SFO), personal care products 106, 107
florisil 172, 247, 279
fluopyram 6
fluorotelomer acrylates (FTAs) 189, 191, 193
fluorotelomer alcohols (FTOHs) 189–193, 195, 205–206
fluorotelomer alkyl betaine (FTAB) 192, 193–196
fluorotelomer carboxylic acids (FTCAs) 189, 201–203
fluorotelomer iodides (FTIs) 189, 191
fluorotelomer methacrylates (FTMAs) 189
fluorotelomer olefins (FTOs) 189
  chromatography-mass spectrometry 205–206
  metabolites and/or transformation products 191
  occurrence 193
fluorotelomer sulfonic acids (FTSAs) 189, 201–203
fluorotelomer unsaturated carboxylic acids (FTUCAs) 189
fluoxatine (FLX) 43
FLX *see* fluoxatine
focused ultrasonic solid-liquid extraction (FULSE) 47
formic acid (FA) 55, 59, 328
  artificial sweeteners 173
FOSAs *see* perfluorooctane sulfonamides
FOSES *see* perfluorooctane sulfonamidoethanols
FOUEA *see* 2H-perfluoro-2-decenoic acid
Fourier Transform infrared spectroscopy (μ-FTR) 362
FPSE *see* fabric phase sorptive extraction
fragrance allergens
  analytical methods in aqeous samples 96–97
  liquid-based extraction techniques 106, 107
  single-drop microextraction 106
  solid phase microextraction 102
fragrances 72
freezing samples, new psychoactive substances 133
FTAB *see* fluorotelomer alkyl betaine
FTAs *see* fluorotelomer acrylates
FTCAs *see* fluorotelomer carboxylic acids
FTIs *see* fluorotelomer iodides
FTMAs *see* fluorotelomer methacrylates
FTOHs *see* fluorotelomer alcohols
FTOs *see* fluorotelomer olefins
FTSAs *see* fluorotelomer sulfonic acids
FULSe *see* focused ultrasonic solid-liquid extraction
fungicides 1, 3, 6, 9, 17, 21–23

g

GAC *see* green analytical chemistry
galaxolide (HHCB) 263, 264
  determination 280
  metabolites and/or transformation products 267
  occurrence 268–270
  properties 266–267
  protocols for collecting/preparing samples 270–271
  sample extraction and clean-up 276
galaxolidone (HHCB-lactone) 264, 267–269, 279–280
GAPs *see* Global Atmospheric Passive Sampling
gas chromatography (GC)
  artificial sweeteners 172–173, 176
  disinfection byproducts 317–320, 322–327, 329
  high production volume chemicals 237, 251
  musk fragrances 276, 279–280
  new psychoactive substances 140
  perfluorinated substances 201, 205–208
  personal care products 75
  pesticides 10–11, 20, 22
  pharmaceuticals 47, 51–55
gas chromatography (GC)-APPI-HRMS, perfluorinated substances 205–206
gas chromatography (GC)-ECD, disinfection byproducts 320, 328, 332
gas chromatography (GC)-(EI)QqQ, high production volume chemicals 252

gas chromatography (GC)-FID 104, 105
gas chromatography (GC)-MS/MS
  artificial sweeteners 175
  disinfection byproducts 332
gas chromatography (GC)-MS/MS(QqQ), perfluorinated substances 205–206
gas chromatography (GC)-QTOF, perfluorinated substances 205–206
gas chromatography-mass spectrometry (GC-MS)
  disinfection byproducts 320, 328–329, 335
  perfluorinated substances 201
  personal care products 101–102, 107–108
  pesticides 21
  pharmaceuticals 51–55
gas chromatography-mass spectrometry (GC-MS) analysis, personal care products 104
GCB *see* graphitized carbon black
gel permeation chromatography (GPC) 247, 279
Generally Recognized As Safe (GRAS) 154
GenX 187, 188, 190, 193–196, 203–204
glass storage of pesticides 9
glass-fibre filters (GFF) 198, 231–232
Global Atmospheric Passive Sampling (GAPs) 195
globalide 266
glyphosate 3
GPC *see* gel permeation chromatography
grab sampling 16
graphene sponge (GS)-SPE 98
graphite carbon SPE cartridges 200–201
graphitized carbon black (GCB) 17
GRAS *see* Generally Recognized As Safe
green analytical chemistry (GAC) 98

h

HAA5 307, 310
HAA9 310
HAAMEs 335
habanolide 266
haloacetamides (HAMs) 288
  determination 320–322, 326, 328–330
  hints and tips 335
  legislation 307–308
  main physical-chemical properties 301–303
  metabolites and/or transformation products 309
  occurrence 311
  precision and accuracy 332
  protocols for collecting/preparing samples 314
  sample extraction and clean-up 315
  sensitivity 333
haloacetic acids (HAAs)
  determination 320–321, 328, 330
  hints and tips 335
  linearity 330
  main physical-chemical properties 290–291
  metabolites and/or transformation products 309
  occurrence 310–312
  parametric and guideline values in drinking water 308
  precision and accuracy 332
  properties 288
  protocols for collecting/preparing samples 314
  sample extraction and clean-up 316
  sensitivity 332
haloacetonitriles (HANs) 288
  determination 320, 322–325, 328
  legislation 307–308
  occurrence 311
  precision and accuracy 331
  properties 298–300
  protocols for collecting/preparing samples 314
  sample extraction and clean-up 315–318
halobenzoquinones (HBQs)
  determination 320–321, 328, 330
  disinfection byproducts 317
  legislation 307–308
  metabolites and/or transformation products 309
  occurrence 311
  properties 288, 294–298
  protocols for collecting/preparing samples 313–314
haloectaldehydes (HALs) 288
  determination 320, 322–323

main physical-chemical
properties 292–293
occurrence 310–311
properties 288
protocols for collecting/preparing
samples 314
sample extraction and clean-up 315–317
sensitivity 333
haloketones 316
halonitromethanes (HNMs)
determination 320–321, 325–326,
328–329
legislation 307–308
main physical-chemical
properties 304–305
metabolites and/or transformation
products 309
occurrence 312
precision and accuracy 331–332
properties 298
protocols for collecting/preparing
samples 314
sample extraction and
clean-up 315–318
sensitivity 333
HALs *see* haloectaldehydes
HAMs *see* haloacetamides
HANs *see* haloacetonitriles
HBQs *see* halobenzoquinones
heated ESI (HESI) 22, 252
heptafluorobutyrylimidazole (HFBI) 320
herbicides 1, 3, 6–7, 9, 17, 21–23
HESI *see* heated ESI
hesperetin dihydrochalcone
4′-β-d-glucoside 156
heveltolide 266
hexachlorobenzene (HCB) 8–9
hexachlorocyclohexane 6
hexane on-cell PLE clean-up 247
hexane/acetone mixtures 247
HFPO-DA *see* GenX
HHCB *see* galaxolide
HHCB-lactone *see* galaxolidone
high production volume chemicals
(HPVCs) 223–262
air samples 231–232, 238–244
biota samples 232–233, 247–250
determination 251–253
dust samples 231–232, 238–244
figures of merit 253
future directions and
challenges 253–254
hints and tips 253
legislation 226–227
metabolites and/or transformation
products 227–228
occurrence 228–231
overview 223–231
properties 223, 225–226
protocols for collecting/preparing
samples 231–233
sample extraction and clean-up 233–250
sample preparation and
collection 231–250
sediment samples 232, 244–247
sludge samples 232, 244–247
soil samples 232, 244–247
water samples 231
high-performance liquid chromatography
(HPLC) 20–22, 24
artificial sweeteners 176–178
disinfection byproducts 321
new psychoactive substances 140
perfluorinated substances 201
personal care products 98
pharmaceuticals 55
high-performance liquid chromatography
(HPLC)-(ESI)MS/MS, perfluorinated
substances 203–204
high-performance liquid chromatography
(HPLC)-HRMS, new psychoactive
substances 139
high-performance liquid chromatography
(HPLC)-MS/MS
artificial sweeteners 177–178
new psychoactive substances 139
perfluorinated substances 203–204
high-performance liquid chromatography
(HPLC)-TOF-MS 24
high-resolution mass spectrometry
(HRMS) 1
artificial sweeteners 176
disinfection byproducts 328
new psychoactive substances 141–142
perfluorinated substances 203–206
personal care products 108

pesticides 11, 23–24
HILIC *see* hydrophilic interaction liquid chromatography
HKUST-1 MOF 98
HLB *see* N-vinylpyrrolidone-DVB sorbent
HNMs *see* halonitromethanes
hormones 43, 48–49
HPVCs *see* high production volume chemicals
HRMS *see* high-resolution mass spectrometry
HR-WAX SPE cartridge 169–170
HR-X SPE cartridge 169–170
HS-SPME
disinfection byproducts 317–318, 321, 331–333
musk fragrances 276
perfluorinated substances 205–206
hybrid LTQ-Orbitrap 204–205
hybrid mass analyzers, new psychoactive substances 141–142
hydrophilic balanced monomers, new psychoactive substances 136
hydrophilic interaction liquid chromatography (HILIC)
artificial sweeteners 173, 175, 178
pesticides 20–22
pharmaceuticals 55
hydroxy acesulfame 157
hydroxyl radicals 157
hydroxy-metabolites of NSAIDs 54–55
3-(3-hydroxy-4-methoxyphenyl)propionic acid 156
Hyper Sep Retain PEP 171

i

ibuprofen 39, 42
IC *see* ionic chromatography
ICZM *see* Integrated Coastal Zone Management
identification points (IPs) 59
ifosfamide (IF) 43
ILIS *see* isotopically labelled internal standards
imazalil 7–8
in syringe μ-SPE-HPLC-PDA methodology 99
indomethacin 39
inland water systems, microplastics 357–358
insecticides 1, 4, 6–9, 21–23, 105
instrumental method
artificial sweeteners 172–177
disinfection byproducts 320–330
new psychoactive substances 140–142
perfluorinated substances 201–207
Integrated Coastal Zone Management (ICZM) 353–354
internal standards (IS) 136, 177
iodine-containing trihalomethanes (I-THMs) 298, 310, 317–318
ion trap (IT) 55, 280
ionic chromatography (IC) 22, 314, 321, 327
ionic chromatography (IC)-MS 319, 330
ionic chromatography (IC)-MS/MS 332
ionic liquid (IL) 101, 103, 105, 107
ionic liquid (IL)-DLLME, microextraction liquid phase approaches 105
ionic liquid (IL)-HS-SDME, musk fragrances 271
ionic liquid (IL)-USAEME, personal care products 107
ion-pair extraction (IPE) 199
IPs *see* identification points
IRMS *see* isotope ratio mass spectrometer
ISO 5667-3 guide 167
ISO Standard on Water Quality – Sampling 5667 9
iso-acesulfame 157
ISOLUTE ENV + cartridges 199–200
isoproturon 6
isotope ratio mass spectrometer (IRMS), pesticides 23
isotopically labelled internal standards (ILIS) 55
I-THMs *see* iodine-containing trihalomethanes

k

ketamine analogues 129, 138–139
ketoprofen (KET) 39, 42

l

large volume injection (LVI) 279–280
LC-MS *see* liquid chromatographic-mass spectrometry
legislation
disinfection byproducts 298, 307–308

high production volume chemicals 226–227
microplastics 353–354
musk fragrances 267
perfluorinated substances 190–191
personal care products 72–73
pharmaceuticals 44–45
limits of quantification (LOQs)
artificial sweeteners 171–172
disinfection byproducts 332–333
high production volume chemicals 244
perfluorinated substances 205, 208–209
pesticides 11, 23–24
lindane 7–9
lipid regulators 49
lipophilic balanced monomers, new psychoactive substances 136
liquid chromatographic-mass spectrometry (LC-MS)
artificial sweeteners 176
disinfection byproducts 319, 330, 332, 335
perfluorinated substances 211
personal care products 107–108
pesticides 10–11, 21–22
pharmaceuticals 56–59
liquid chromatography (LC)
artificial sweeteners 171–174
disinfection byproducts 280, 314, 321, 327–328
high production volume chemicals 251
HILIC 20–22, 55, 173, 175, 178
new psychoactive substances 140
perfluorinated substances 201–203, 207–208
pesticides 20–23
pharmaceuticals 47, 56–59
liquid chromatography (LC)/ESI-QTOF-MS, musk fragrances 280
liquid chromatography (LC)/UPLC, pharmaceuticals 55
liquid chromatography (LC)-DAD, microextraction liquid phase approaches 105
liquid chromatography (LC)-ESI-MS/MS, artificial sweeteners 178
liquid chromatography (LC)-(ESI)QqQ, high production volume chemicals 252
liquid chromatography (LC)-(HESI) Q-Orbitrap 252
liquid chromatography (LC)-MS/MS
artificial sweeteners 171–172, 177
disinfection byproducts 332
musk fragrances 280
new psychoactive substances 140–142
perfluorinated substances 201, 203–204
pharmaceuticals 59
liquid chromatography (LC)-Orbitrap analyzers, high production volume chemicals 252
liquid chromatography (LC)-Orbitrap-MS analyzers, pesticides 11
liquid-phase microextraction (LPME) 11
liquid-liquid extraction (LLE)
disinfection byproducts 314–315, 319–320, 331–333, 335
high production volume chemicals 233
musk fragrances 271
new psychoactive substances 135–136
perfluorinated substances 199–200
personal care products 104–107
pesticides 10–11
liquid-liquid microextraction (LLME) 106, 315–316, 320, 331–334
liquid-phase microextraction (LPME)
high production volume chemicals 237
pesticides 11
Lithodes santolla (commercial crab) 360
LODs
artificial sweeteners 165–166, 168–169
perfluorinated substances 203, 205–209
personal care products 98, 99, 101–107
LOQs *see* limits of quantification
LTQ-Orbitrap 204–205
Luna C18 columns 328
Luna HILIC columns 175
Luna Omega Polar C18 321
LVI *see* large volume injection

m

macrocyclic musk fragrances (MCMs) 98, 265–266, 268, 276, 280
MAE *see* microwave-assisted extraction
magnesium sulphate ($MgSO_4$) 17, 334
magnetic solid phase extraction (MSPE) 99
magnetism-enhanced monolith-(ME-MB-IT/SPME), personal care products 103

marine environments 204, 355–357, 359, 360 *see also* water samples
Marine Strategy Framework Directive (MSFD) 353–354
mass spectrometry (MS)
  artificial sweeteners 157, 176
  high production volume chemicals 252
  new psychoactive substances 140–141
  pesticides 11, 23–24
  pharmaceuticals 47 *see also* high-resolution mass spectrometry; tandem mass spectrometry
matrix effects, new psychoactive substances 142
matrix removal by digestion, microplastics 362
Maximal Environmental Concentrations (MECs) 39, 45
MBTFA *see* *N*-methyl-bis(trifluoroacetamide)
MCMs *see* macrocyclic musk fragrances
MCPA *see* 2-methyl-4-chlorophenoxyacetic acid
MDLs *see* method detection limits
mecoprop 3–4
N-MeFOSA *see* N-methyl perfluorooctanesulfonamide
MeFOSE *see* 3M N-methyl perfluorooctane sulfonamideoethanol
membrane devices 10, 17, 20, 103–104, 231
membrane-assisted sorptive extraction (MASE) 271
MET *see* methothrexate
metabolites
  artificial sweeteners 155–157
  disinfection byproducts (DBP) 308–310
  high production volume chemicals 227–228
  musk fragrances 267–268
  new psychoactive substances 131–132
  perfluorinated substances 191–192
  personal care products 73
  pesticides 3–4
  pharmaceuticals 37–39, 40–43, 54–55
metal nanoparticles 98, 102–105
metalaxyl 6
metal-organic frameworks (MOFs) 98
methanol:water extractions 15, 17, 276
methanol (MeOH)
  artificial sweeteners 156, 173, 178
  disinfection byproducts 316–319, 328
  high production volume chemicals 231, 233, 238, 244, 251–252
  perfluorinated substances 199–200
  personal care products 107
  pharmaceuticals 55
methlydesphenyl chloridazon 3
method detection limits (MDLs)
  disinfection byproducts 333
  musk fragrances 270, 276
method quantitation limits (MQLs) 208–209, 252
methothrexate (MET) 43
3M N-methyl perfluorooctane sulfonamideoethanol (MeFOSE) 192
N-methyl perfluorooctanesulfonamide (MeFOSA) 194, 205–206
N-methyl-bis(trifluoroacetamide) (MBTFA) 52
2-methyl-4-chlorophenoxyacetic acid (MCPA) 6
methyl-tert-butyl ether (MTBE) 200, 314–317, 334
N-methyl-N(tert-butyldimethylsilyl)-trifluoroacetamide (MTBSTFA) 320
N-methyl-*N*-(trimethylsilyl) trifluoroacetamide (MSTFA) 52
metolachlor 3
metopropol (MTP) 41
metronidazol (METR) 41
$MgSO_4$ *see* magnesium sulphate
microextraction liquid phase approaches, personal care products 104–107
microextraction SPE (μ-SPE) 98
μ-LC, new psychoactive substances 140
μ-SPE-HPLC-PDA methodology 98
microplastics 353–374
  biota 359–361
  biota samples 360–361
  chemical characterization 362–363
  determination 362–363
  future directions and challenges 363
  inland water systems 357–358
  legislation 353–354
  marine environments 355–357
  matrix removal by digestion 362
  occurrence 355–360

origin and distribution 354–355
overview 353–360
physical characterization 362
properties 353
protocols for collecting/preparing samples 360–361
sample extraction and clean-up 361–362
sample preparation and collection 360–362
sediment samples 358–360
separation 361–362
water samples 355–358, 360
microwave-assisted (MA)-DLLME 106
microwave-assisted extraction (MAE)
high production volume chemicals 238, 244, 247
new psychoactive substances 139
pharmaceuticals 47
microwave-assisted (MA)-HS-SPME 276
microw-wave assisted headspace solid phase microextraction (MA-HS-SPME) 276
MID *see* multiple ion detection
miniaturized SBSE approaches 101
miniscale-LLE (mLLE), personal care products 104
mirex 7
Mixed-Mode LC (MMLC) 55
MK *see* musk ketone
MM *see* musk moskene
Model MC-400 Multi-corer 198–199
MOFs *see* metal-organic frameworks
molecularly imprinted polymers (MIPs) 46, 99, 171, 207
molecularly imprinted solid phase extraction (MISPE) 99
monobromoacetic acid (BAA) 307
monochloroacetic acid (CAA) 307, 332
monoPAPs *see* polyfluroalkyl phosphate monoesters
MQLs *see* method quantitation limits
MRM *see* multiple reaction monitoring
MS/MS *see* tandem mass spectrometry
MSA-DLLME (magnetic stirring) 106
MSFD *see* Marine Strategy Framework Directive
MSPE *see* magnetic solid phase extraction
MSTFA *see* *N*-methyl-*N*-(trimethylsilyl) trifluoroacetamide
MT *see* musk tibetene
MT, legislation 267
MTBE *see* methyl-tert-butyl ether
MTBSTFA *see* N-methyl-N(tert-butyldimethylsilyl)-trifluoroacetamide
MTP *see* metopropol
multiple ion detection (MID) 23
multiple reaction monitoring (MRM)
artificial sweeteners 176–178, 180
new psychoactive substances 140–141
perfluorinated substances 207–208
pesticides 22
pharmaceuticals 59
multiresidue analysis, personal care products 100
multiwalled carbon nanotubes (MWCNTs) 99, 103
muscone 266
musk ambrette (MA) 263–264, 266–267
musk fragrances 263–286
air samples 269–271
biota samples 269–271
chromatography 279–280
detection 280
determination 279–280
future directions and challenges 280
legislation 267
metabolites and/or transformation products 267–268
occurrence 268–270
overview 263–270
properties 263–267
protocols for collecting/preparing samples 270–271
sample extraction and clean-up 271–279
sample preparation and collection 270–279
sediment samples 269, 271, 276
single-drop microextraction 106
sludge samples 271, 276
soil samples 269, 271, 276
surface water samples 269
wastewater samples 268
water samples 270–275

musk ketone (MK)
legislation 267
metabolites and/or transformation products 267–268
properties 263–264, 266
sample extraction and clean-up 276
musk MC4 266
musk moskene (MM) 265
musk tibetene (MT) 265
musk xylene (MX)
legislation 267
metabolites and/or transformation products 267–268
properties 263, 265–266
musk-NN 266
MWCNTs *see* multiwalled carbon nanotubes
MX *see* musk xylene
Mytilus galloprovincalis (mussel) 204
Mytilus galloprovincialis (Mediterranean mussel) 359

n

naproxen (NPX) 39, 43, 48
National Priorities List (NPL) 2
NCI *see* negative chemical ionization
NDEA *see* *N*-nitrosodiethylamine
NDMA *see* *N*-nitrosodimethylamine
NDPA *see* *N*-nitrosodipropylamine
NDPhA *see* *N*-nitrosodiphenylamine
needle trap (NT) microextraction 271
negative chemical ionization (NCI) 320
neohesperidin dihydrochalcone (NHDC)
artificial sweeteners 154, 156
chromatography 173–175
detection 176–177
figures of merit 177–179
frequently used SPE cartridges 169
liquid chromatography 174
metabolites and/or transformation products 156
MRM transitions 177
NORMAN PNECs 155
occurrence 162–164, 167
properties 151, 153
neonicotinoids 4, 6–8
neotam (NEO)
detection 176–177
figures of merit 177–179
frequently used SPE cartridges 169
legislation 154
liquid chromatography 174
MRM transitions 177
NORMAN PNECs 155
occurrence 163–164, 167
properties 151, 153–154
neutral PFASs (nPFASs) 193, 199–200
new psychoactive substances (NPS) 127–149
analytical methods from 2015 to present 137–138
chromatographic separation 140
detection 140–142
determination 139–143
extraction procedures and clean-up 134–136
figures of merit 142
future directions and challenges 143–144
hints and tips 142–143
instrumental method 140–142
market categories 128–129
market, dynamics and international control 130–131
metabolites and/or transformation products 131–132
occurrence 132–133
other environmental matrices 136–139
overview 127–133
properties 127–130
protocols for collecting/preparing samples 133–134
sample preparation and collection 133–139
UNODC-Early Warning Advisory 128–131
urban wastewater 133–136
NHDC *see* neohesperidin dihydrochalcone
nitro musks (NMs) 264–265, 267–268, 269–270, 276, 280
nitrosamines 288, 306, 308, 313–314, 329
N-nitrosodiethanolamine 313

N-nitrosodiethylamine (NDEA) 308, 313, 319
N-nitrosodimethylamine (NDMA)
  determination 320, 329, 332–333
  future directions and challenges 335
  legislation 308
  metabolites and/or transformation products 309–310
  occurence 312–313
  properties 298
  sample extraction and clean-up 319
N-nitrosodiphenylamine (NDPhA) 308, 313, 320
N-nitrosodipropylamine (NDPA) 308, 319
N-nitrosodipyrrolidine (NPYR) 308, 319
N-nitrosopiperidine (NPIP) 319
non-agricultural use of pesticides 6
non-polar FPSE coatings 100
non-steroidal anti-inflammatory drugs (NSAIDs) 39, 42–43, 47, 49, 51–52, 54–55
NORMAN Ecotoxicology Database 155
NPIP *see* *N*-nitrosopiperidine
NPS *see* new psychoactive substances
NPX *see* naproxen
NPYR *see* *N*-nitrosodipyrrolidine
N-vinylpyrrolidone-DVB sorbent 75

o

OA *see* oxanilic acid
Oasis HLB
  artificial sweeteners 169–170, 172
  disinfection byproducts 316–319
  high production volume chemicals 233, 237, 244
  perfluorinated substances 200
  pesticides 11, 17
  pharmaceuticals 46
Oasis MAX
  artificial sweeteners 169–170
  pharmaceuticals 46
Oasis MCX, high production volume chemicals 244
Oasis WAX
  artificial sweeteners 170
  perfluorinated substances 200
  pharmaceuticals 46
octhilinone pesticides 3, 4
octocrylene (OCR) 100
oestrogens 49, 51–52
opioids 129
Orbitrap analyzers
  high production volume chemicals 252
  pesticides 23 *see also* LC-Orbitrap-MS analyzers
Orbitrap-MS analyzers
  perfluorinated substances 203
  pesticides 11
Organisation for Economic Cooperation and Deveopment (OECD) 223
organochlorine pesticides (OCPs) 7–9, 20, 23
organophosphate esters (OPEs)
  determination 251–252
  legislation 227
  occurrence 228–229
  properties 224, 226
  sample extraction and clean-up 237, 244, 247
organophosphate insecticides 6
organophosphorus 7–8
Ostro SPE modifications 200–201
oxaloacetate 156
oxanilic acid (OA) 3
ozonation experiments 156
ozone radicals 157

p

P3MHpA *see* perfluoro-3-methylheptanoic acid
PAE metabolites (PAEMs) 251–253
PAN *see* polyacrylonitrile
parabens 71–75, 98–105
paracetamol 39
passive sampling
  perfluorinated substances 195, 198
  personal care products 103
  pesticides 10, 17, 20
  pharmaceuticals 46
  TWA 103
PCMs *see* polycyclic musks
PCPs *see* personal care products
PDMS *see* polydimethylsiloxane
PECs *see* predicted environmental concentration

penconazole 3, 4
O-(2,3,4,5,6-pentafluorobenzyl) hydroxylamine (PFBHA) 318–319
pentafluoropropionic anhydride (PFPAA) 52
perfluoralkyl amides 192
perfluorinated substances 187–222
- biosensors 206–207
- chromatography-mass spectrometry 201–206
- dust samples 198
- figures of merit 207–209
- future directions and challenges 210–212
- hints and tips 209–210
- legislation 190–191
- metabolites and/or transformation products 191–192
- occurrence 193–197
- overview 187–197
- properties 190
- protocols for collecting and preparing samples 197–199
- sample extraction and clean-up 199–201
- sample preparation and collection 197–201
- sediment samples 198
- sludge samples 196

perfluoroalkane sulphonamides (FASAs) 188, 195
perfluoroalky carboxylic acids (PFCAs) 187–188
- determination 201–203, 205
- legislation 190–191
- metabolites and/or transformation products 191–192
- occurrence 193–196

perfluoroalkyl acids (PFAAs) 187, 195–196, 201
perfluoroalkyl iodides (PFAIs) 188
perfluoroalkyl phosphinic acids (PFPiAs) 188, 192
perfluoroalkyl phosphonic acids (PFPAs) 187–188, 190
- chromatography-mass spectrometry 201–203
- metabolites and/or transformation products 192
- occurrence 193–194
- sample extraction and clean-up 200–201

perfluoroalkys sulfonic acids (PFSAs) 187–188, 191–195, 200–201
2H-perfluoro-2-decenoic acid (FOUEA) 193
perfluoroether carboxylic acids (PFECAs) 188, 193–194, 201–203
- *see also* GenX

perfluoroether sulfonic acids (PFESAs) 188, 193–196
perfluoroethylcyclohexanesulfonate (PFECHS) 197
perfluorohexane sulfonate (PFHxS) 190–191
perfluoro-3-methylheptanoic acid (P3MHpA) 193
perfluorooctane sulfonamides (FOSAs) 193–194, 196, 200–201, 205–206
perfluorooctane sulfonamidoethanols (FOSEs) 192–193, 205–206
perfluorooctane sulfonic acids (PFOSs) 187
- determination 203–205, 207
- future directions and challenges 211
- legislation 190–191
- occurrence 193–197

perfluorooctane sulfonyl fluoride (PFOSF) 190–191
perfluorooctane sulphonamido ethanol-based phosphate esters (SAmPAPs) 188, 192, 195–196
perfluorooctanoic acids (PFOAs) 187
- biosensors 207
- chromatography-mass spectrometry 203–206
- future directions and challenges 211
- legislation 191
- metabolites and/or transformation products 192
- occurrence 193–196
- properties 190

perfluoropentanoic acid (PFPeA) 193, 195–196
permanganate (Mn(VII)) 157
permethrins 9
peroxone treatment 157
Persistent Organic Pollutants (POPs) 9, 190–191

personal care products (PCPs) 71–126
  determination 107–108
  future directions and challenges 108
  legislation 72–73
  liquid-based extraction 104–107
  metabolites and/or transformation products 73
  microextraction liquid phase approaches 104–107
  occurrence 73–74
  overview 71–74
  pathways to aquatic environment 73–74
  properties 71–72
  sample preparation in aquatic environments 74–107
  sorbent-based methods 75–104
pesticides 1–36
  air matrices 19
  biota 18
  chromatography 20–21
  detection 22–24
  determination 20–25
  figures of merit 24
  future directions and challenges 25–26
  hints and tips 24–25
  instrumental method 20–24
  metabolites and/or PTs 3–4
  occurrence in the environment 5–9
  overview 1–9
  pharmaceuticals 54–55
  properties 1
  protocols for collecting and preparing samples 9–10
  sample extraction and clean-up 10–20
  sample preparation and collection 9–10
  soil samples 7–8, 15–17, 21
  water samples 12–14
PET nanofiber 98
PFAAs *see* perfluoroalkyl acids
PFAIs *see* perfluoroalkyl iodides
PFASs *see* polyfluoroalkyl substances
PFBHA *see* O-(2,3,4,5,6-pentafluorobenzyl) hydroxylamine
PFCAs *see* perfluoroalky carboxylic acids
PFECAs *see* perfluoroether carboxylic acids
PFECHS *see* perfluoroethylcyclohexanesulfonate
PFESAs *see* perfluoroether sulfonic acids
PFHxS *see* perfluorohexane sulfonate
PFOA *see* perfluorooctanoic acid
PFOSF *see* perfluorooctane sulfonyl fluoride
PFOSs *see* perfluorooctane sulfonic acids
PFPAA *see* pentafluoropropionic anhydride
PFPAs *see* perfluoroalkyl phosphonic acids
PFPeA *see* perfluoropentanoic acid
PFPiAs *see* perfluoroalkyl phosphinic acids
PFSAs *see* perfluoroalkyl sulfonic acids
phantolide (AHMI) 264, 267–269
pharmaceuticals 37–70
  determination 51–59
  future trends 60
  legislation 44–45
  metabolites and/or transformation products 37–43
  occurrence 39–43
  overview 37–45
  properties 37–38
  sampling and sample preparation 46–51
  selected transformation products 40–43
  solid samples 46–50
  water samples 46–47, 51
phenethylamines 128
phenolic preservatives, single-drop microextraction 106
phenols, pesticides 11
phenoxy acid pesticides 17
phenylalanine 154, 156
1,2-phenylenediamine 320
phenylketonuria 154
phenylureas 17
phloroglucinol 156
Phree SPE modifications 200–201
phthalate esters (PAEs)
  determination 251–252
  legislation 226
  metabolites and/or transformation products 227
  occurence 230
  PAEM metabolites 224, 251–253
  properties 223–224
  sample extraction and clean-up 233, 237, 244, 247
phthalates
  legislation 72–73
  liquid-based extraction techniques 107
  metabolites pathways 227–228
  microextraction liquid phase approaches 104–106

properties 72
solid phase extraction 98–99
solid phase microextraction 102–104
stir-bar sorptive extraction 100
physical characterization of microplastics 362
piperazines 129, 138–139
plant-based substances 129
plasticizers 72, 89–95, 102, 107
see also phthalates
PLE *see* pressurized liquid extraction
Plexa PAX, artificial sweeteners 169–170
PNEC *see* Predicted No Effect Concentration values
Polar Organix Chemical Inegrative Sampler (POCIS)
high production volume chemicals 231
perfluorinated substances 198
personal care products 103–104
pesticides 10
pharmaceuticals 46
polyacrylonitrile (PAN) 171
polycyclic musks (PCMs) 264
future directions and challenges 280
occurrence 269–270
properties 266–267
sample extraction and clean-up 276
polydimethylsiloxane (PDMS) 100–101, 237, 276
polydimethylsiloxane/divinylbenzene (PDMS/DVB) 237, 276
polyfluoroalkyl substances (PFASs) 187, 190
biosensors 207–208
determination 201–210
figures of merit 207–209
future directions and challenges 210–212
hints and tips 209–210
occurrence 193–197
properties 190
protocols for collecting and preparing samples 198–199
sample extraction and clean-up 199–200
sample preparation and collection 197–198
smarthphone app-based biosensors 207–208
polyfluroalkyl phosphate monoesters (monoPAPs) 189
polyfuoroalkyl phosphate diesters (diPAPs) 189, 192, 201–203
polymer-based SPE sorbents, artificial sweeteners 178
polymeric ionic liquids (PIL)-based materials 103
polymeric reversed-phase sorbents 75
polypropylene (PP) containers 198–199
polytetrafluoroethylene (PTFE) 9, 313–314
polyurethane foam filter (PUF)
high production volume chemicals 231–232
perfluorinated substances 198–200
pesticides 17, 20
POPs *see* Persistent Organic Pollutants
positive chemical ionization (PCI) 329
predicted environmental concentration (PECs), pharmaceuticals 45
Predicted No Effect Concentration (PNEC) values 155
preservatives
analytical methods in aqueous samples 84–95
determination 107
liquid-based extraction techniques 104–107
properties 71–72
solid phase microextraction 102
pressurized liquid extraction (PLE) 200
artificial sweeteners 168, 172
high production volume chemicals 238, 244, 247
musk fragrances 276, 279
perfluorinated substances 199–200
pesticides 15, 17
primary secondary amine (PSA) 17, 247
prochloraz 7
programmable temperature vaporization (PTV) 279–280, 321, 334
prometon 6
properties
disinfection byproducts 288–306
high production volume chemicals 223, 225–226
microplastics 353
musk fragrances 263–267

  new psychoactive substances 127–130
  perfluorinated substances 190
  personal care products 71–72
  pesticides 1
  pharmaceuticals 38
propiconazole 9
propisochlor 15, 17
protocols for collecting/preparing samples
  artificial sweeteners 167
  disinfection byproducts 313–314
  high production volume chemicals 231–233
  microplastics 360–361
  musk fragrances 270–271
  new psychoactive substances 133–134
  perfluorinated substances 197–199
  pesticides 9–10
PSA *see* primary secondary amine
PTFE *see* polytetrafluoroethylene
PTV *see* programmable temperature vaporization
PUF *see* polyurethane foam filter
pydiflumetofen enantiomers 17

q

QFF *see* quartz-fibre filters
Q-Orbitrap analyzers, high production volume chemicals 252
QqQ *see* triple quadrupole analyzers
Quadrupole Time of Flight (QTOF) systems
  artificial sweeteners 176
  high production volume chemicals 252
  new psychoactive substances 141
  perfluorinated substances 205–206
  personal care products 100, 108
  pesticides 23
Quadrupole Time-of-Flight mass spectrometry (QTOF-MS), perfluorinated substances 203
quadrupole-linear ion trap mass spectrometer (QTRAP) 23, 252
quality assurance/quality control (QA/QC) 142
quality control (QC) samples 142, 209
quartz-fibre filters (QFF) 198, 231–232, 244
QuEChERS method
  high production volume chemicals 247
  musk fragrances 276, 279
  new psychoactive substances 139
  perfluorinated substances 200–201
  pesticides 11, 15, 17
quinolone antibiotics 48

r

rainbow trout and perfluorinated substances 192
(μ-)RAMAN spectroscopy 362
RDSE *see* rotating-disk sorptive extraction
redox states of halobenzoquinones 298
Registration, Evaluation, Authorization and Restriction of Chemicals (REACH Regulation) 226–227, 267
relative standard deviation (RSD) 205–206, 208–209, 231–232
restricted access materials (RAMs) 171
retention time (RT) 140–142
reversed phase (RP) polymers 136
reversed-phase chromatography 107
Risk Quotient (RQ) 45
River Basin Management Plants (RBMPs) 6
romandolide 266
rotating-disk sorptive extraction (RDSE) 101–102
RP column, artificial sweeteners 173, 175
RP-LC
  artificial sweeteners 172–173
  new psychoactive substances 140
  pharmaceuticals 55
RPLC-ESI(-)-MS/MS, artificial sweeteners 178
RPLC-MS/MS, artificial sweeteners 178
RP-LC-MS/MS, artificial sweeteners 173
RQ *see* Risk Quotient
RT *see* retention time

s

saccharine (SAC)
  chromatography 173–175
  detection 176–177
  figures of merit 177–179
  frequently used SPE cartridges 169–170
  hints and tips 178
  legislation 154
  liquid chromatography 174
  metabolites and/or transformation products 156
  MRM transitions 177

NORMAN PNECs 155
occurrence 157–167
properties 151–152, 154
Safe Drinking Water Act (SDWA) 2
SAmPAPs *see* perfluorooctane sulphonamido ethanol-based phosphate esters
sample extraction and clean-up
artificial sweeteners 167–172
disinfection byproducts 314–319
high production volume chemicals 233–250
microplastics 361–362
musk fragrances 271–279
new psychoactive substances 134–136
perfluorinated substances 199–201
sample preparation and collection
artificial sweeteners 167–172
disinfection byproducts 313–319
high production volume chemicals 231–250
microplastics 360–362
musk fragrances 270–279
new psychoactive substances 133–139
perfluorinated substances 197–201
personal care products (PCPs) 74–107
pesticides 9–10
pharmaceuticals 46–51
SArAs *see* synthetic aromatic antioxidants
SBDLME *see* stir-bar dispersive liquid microextraction
SBSDME *see* stir-bar sorptive-dispersive microextraction
SBSE *see* stir-bar sorptive extraction
Scientific Committee on Cosmetic Products and Non-Food Products (SCCS) 72
SCRAs *see* synthetic cannabinoid receptor agonists
SDME *see* single-drop microextraction
SDWA *see* Safe Drinking Water Act
SEC-HRMS *see* size exclusion chromatography coupled to high-resolution mass spectrometry
sediment samples
high production volume chemicals 232, 244–247
microplastics 358–360
musk fragrances 269, 271, 276
perfluorinated camples 198
pesticides 7–8 *see also* soil samples
selected ion monitoring (SIM) 207–208, 320
selected reaction monitoring (SRM)
disinfection byproducts 328–330
pesticides 22
pharmaceuticals 59
selected-ion-storage (SIS) 328–329
self-dope polythiophene (STP) 103
semipermeable membrane devices (SPMDs) 10, 17, 20, 103–104, 231
semi-polar compounds, pharmaceuticals 54–55
semivolatile organic compounds (SVOCs) 232, 238, 244
separation
microplastics 361–362 *see also* chromatography
sewage treatment plants (STPs) 228–229
SFC *see* supercritical fluid chromatography
SFO *see* floating organic drop
short-chain PFASs 209–210
SIL *see* stable isotopic labelling
silica, artificial sweeteners 172
silica SPE cartridges 244, 247
silica/alumina columns 279
silica-based MCM-41 mesoporous materials 98
silica-based SPE sorbents 178
SILISs *see* stable isotope labelled internal standards
siloxanes 96–97
legislation 72–73
liquid-based extraction techniques 106–107
properties 72
solid phase microextraction 102
silylation 52, 175
SIM *see* selected ion monitoring
simazine 6
single-drop microextraction (SDME)
disinfection byproducts 320
musk fragrances 271
personal care products 104, 106
SIP *see* solvent-impregnated polyurethane foam samplers
SIS *see* selected-ion-storage
size exclusion chromatography coupled to high-resolution mass spectrometry (SEC-HRMS) 363
SLE *see* solid-liquid extraction

sludge samples
  high production volume chemicals 232, 244–247
  musk fragrances 268, 271, 276
  perfluorinated substances 196
smarthphone app-based biosensors 207, 208
SMIP *see* surface molecularly imprinted polymers
soil samples
  high production volume chemicals 232, 244–247
  musk fragrances 269, 271, 276
  pesticides 7–8, 15–17, 21
solid phase extraction (SPE)
  artificial sweeteners 168–172, 178
  disinfection byproducts 316–317, 320, 331–333, 335
  high production volume chemicals 233, 247
  musk fragrances 271
  new psychoactive substances 133–135, 136
  perfluorinated substances 198–201
  personal care products 75–99
  pesticides 10–11, 17, 20
  pharmaceuticals 46–47, 51, 54 *see also* dispersive solid phase extraction
solid phase extraction (SPE)-LC-APCI/MS-MS, artificial sweeteners 171
solid phase microextraction (SPME)
  arrow 276, 279
  artificial sweeteners 168, 171
  disinfection byproducts 319
  high production volume chemicals 237
  perfluorinated substances 199, 205–206
  personal care products 102–104
  pesticides 11, 17, 23
  thin film 103–104, 171
solid samples
  new psychoactive substances 138–139
  pharmaceuticals 46–50 *see also* sediment samples; sludge samples; soil samples
solid-liquid extraction (SLE) 200
  high production volume chemicals 247
  perfluorinated substances 200
  pesticides 15–17
solvent-impregnated polyurethane (SIP) foam samplers 195, 198–200
sorbent-based methods, personal care products 75–104
Soxhlet extractions 198, 200, 276
SPE *see* solid phase extraction
SPMDs *see* semipermeable membrane devices
SPME *see* solid phase microextraction
SRM *see* selected reaction monitoring
stable isotope labelled internal standards (SILISs) 52, 54
stable isotopic labelling (SIL), personal care products 106
steroid hormones 43
stir-bar sorptive-dispersive microextraction (SBSDME) 101
stir-bar dispersive liquid microextraction (SBDLME) 101
stir-bar sorptive extraction (SBSE)
  musk fragrances 271, 276, 279
  personal care products 99–102
  pesticides 11, 17
stir-bar sorptive extraction (SBSE)-LC-MS/MS, personal care products 100–101
STP *see* self-dope polythiophene
STPs *see* sewage treatment plants
Strata-X
  artificial sweeteners 169–170
  perfluorinated substances 199
  pesticides 11
  pharmaceuticals 46
Substances of Very High Concern (SVHCs) 190–191, 226, 227
sucralose (SUC)
  chromatography 173–175
  detection 175–176
  figures of merit 177–179
  frequently used SPE cartridges 169–170
  hints and tips 178
  legislation 154–155
  liquid chromatography 174
  metabolites and/or transformation products 155–157
  MRM transitions 177
  NORMAN PNECs 155
  occurrence 157–167
  properties 151–152, 154
sulfamethazine 39
sulfamethoxazole (SMX) 39, 41

2-sulfamoylbenzoic acid 156
sulfate and UV radiation of acesulfame 157
2-sulfobenzoic acid 156
sulfonic acids
  amidosulfonic acid 156–157
  FTSAs 189, 201–203
  pesticides 11
  PFESAs 188, 193–196
  PFOSs 187, 190–191, 193–197, 203–205, 207, 211
  PFSAs 187–188, 191–195, 200–201
supercritical fluid chromatography (SFC), pesticides 22
surface molecularly imprinted polymers (SMIP) 99
Surface Water European Watch List (WL) 73
surface water samples 269 *see also* water samples
suspect screening 141
SVHCs *see* Substances of Very High Concern
SVOCs *see* semivolatile organic compounds
synthetic aromatic antioxidants (SArAs) 226
synthetic cannabinoid receptor agonists (SCRAs) 129–131
synthetic cannabinoids 128
synthetic cathinones 128
synthetic opioids 129
synthetic phenolic antioxidants (SPAs) 224–227, 230, 247, 251, 252

t

tamoxifen (TAM) 43
tandem mass spectrometry (MS/MS)
  artificial sweeteners 157, 176–177
  disinfection byproducts 320
  high production volume chemicals 244, 252
  new psychoactive substances 141–142
  perfluorinated substances 204–205, 207–208
  personal care products 107–108
  pesticides 11
  pharmaceuticals 55, 59 *see also* gas chromatography-MS/MS
targeted screening 141
targets/new psychoactive substances 131–134, 136
TBAL *see* tribromoacetaldehyde
TBM *see* bromoform
TCAA *see* trichloroacetic acid
TCAL *see* trichloroacetaldehyde
TCM *see* chloroform
TCNM *see* tribromonitromethane
TD *see* terbuthylazine-desethyl; thermal desorption
TD-GC-MS, microplastics 363
TED *see* terbumeton-desethyl
Tenax TA samplers 17, 20
terbumeton-desethyl (TED) 3
terbuthylazine-desethyl (TD) 3
terbuthylazine-2-hydroxy (T2H) 3
terbutryn 3, 4
2-tert-butyl-4-methoxyphenol (BHA) 231, 233, 251–252
TF *see* thin film
TFAA *see* trifluoroacetic anhydride
thermal desorption (TD) 100, 238
thiamethoxam 6
thibetolide 265
thin film (TF)-SPME 103–104, 171
THM4 (trihalomethanes, chlorine/bromine-containing regulated) 298, 307–310, 314–319, 330
THMs *see* trihalomethanes
Thunnus albacares (yellowfin tune muscle tissue) 204
Time-of-Flight mass spectrometry (TOF-MS), disinfection byproducts 328
Time-of-Flight (TOF) systems 23, 24 *see also* Quadrupole Time of Flight systems
time-weighted average (TWA) passive sampling 103
TMCS *see* trimethylchlorosilane
tobacco-specific nitrosamines 313
tolerable weekly intake (TWI) 191
toluene 151, 154
tolylfluanid 3
tonalide (AHTN) 264
  detection 280
  determination 279–280
  legislation 267
  metabolites and/or transformation products 267
  occurrence 268–270

properties 266–267
sample extraction and clean-up 276
total oxidisable precursors assay (TOPA) 211
TRA *see* tramadol
tramadol (TRA) 43
Tran, N.H. 170, 173, 175
trans-cyclohexane-1,2-diol 156
transformation products (TPs)
artificial sweeteners 155–157
disinfection byproducts 308–310
disinfection byproducts (DBP) 308–310
high production volume chemicals 227–228
musk fragrances 267–268
new psychoactive substances 131–132
perfluorinated substances 191–192
personal care products 73
pesticides 1, 2, 3–4, 6, 11, 24
pharmaceuticals 37–39, 40–43, 45
traseolide (ATTI) 264, 267–269
triazines 3, 6–8, 17
tribromoacetaldehyde (TBAL) 311, 316
tribromonitromethane (TCNM) 309, 312, 314, 318, 333
trichloroacetaldehyde (TCAL) 307, 310–311, 314–315
trichloroacetic acid (TCAA) 288, 307, 310
trichloroacetonitrile (TCAN) 311, 313–315, 320, 321, 331, 334
triclocarban 39
triclosan (TCS) 39, 71–72
trifluoroacetic anhydride (TFAA) 52
trihalogenated HALs 316–317
trihalomethanes (THMs) 288
determination 320, 322–324
hints and tips 334
legislation 308
main physical-chemical properties 289
metabolites and/or transformation products 308
occurrence 310–312
parametric and guideline values in drinking water 307
precision and accuracy 331
protocols for collecting/preparing samples 313–314
sample extraction and clean-up 316–318
sensitivity 332
trimethoprim 39
trimethylchlorosilane (TMCS) 52
trimethylsilyldiazomethane 52
N,O-bis(trimethylsilyl)trifluoroacetamide (BSTFA) 52
triple quadrupole (QqQ) analyzers
artificial sweeteners 176, 178, 180
high production volume chemicals 252
musk fragrances 280
new psychoactive substances 140
pesticides 22–23
pharmaceuticals 55
tryptamines 129
TTR isomers 225, 227–231, 233, 251–252
TWA *see* time-weighted average passive sampling
TWI *see* tolerable weekly intake
two-dimensional gas chromatography (GC x GC) 279–280

u

UAE *see* ultrasound-assisted extraction
UA-LLME and UA-DLLME (ultrasound) 106
UDMH *see* unsymmetrical dimethylhydrazine
UHPLC *see* ultra-high-performance liquid chromatography
UHPLC-HRMS, perfluorinated substances 203–204
UHPLC-MS/MS, high production volume chemicals 252
ultra performance liquid chromatography (UPLC)
disinfection byproducts 321
perfluorinated substances 201
pharmaceuticals 55
ultra-high-performance liquid chromatography (UHPLC)
new psychoactive substances 140
perfluorinated substances 201
pesticides 20–23
pharmaceuticals 55

ultra-performance convergence chromatography (UPC^2) 199, 203
ultra-short-chain PFASs 203
ultra-sonic-assisted extraction (USE) 238, 244, 247
ultrasound-assisted emulsification microextraction (USAEME) 104, 106–107
ultrasound-assisted extraction (UAE)
  musk fragrances 276
  new psychoactive substances 139
  pharmaceuticals 47
United States Environmental Protection Agency (USEPA)
  disinfection byproducts 308, 311, 314–315, 319, 335
  perfluorinated substances 191
  pesticides 2
  pharmaceuticals 46–47, 51
Unmwelt Bundesamt (German Environmental Agency) 39
UNODC-Early Warning Advisory 128–129, 130–131
unsymmetrical dimethylhydrazine (UDMH) 312
UPC^2 *see* ultra-performance convergence chromatography
UPLC-MS/MS, perfluorinated substances 208–209
UPLC-QqQ-MS/MS, perfluorinated substances 204
Uptisphere Strategy C18 columns 328
urban wastewater
  new psychoactive substances 133–136 *see also* water samples
USAEME *see* ultrasound-assisted emulsification microextraction
USE *see* ultra-sonic-assisted extraction
USEPA *see* United States Environmental Protection Agency
UV filters 76–83
  fabric phase sorptive extraction 100
  high production volume chemicals 252
  liquid-based extraction techniques 104–107
  occurrence 73–74
  properties 71
  solid phase extraction 75, 98
  solid phase microextraction 102–104
  stir-bar sorptive extraction 100–102
UV irradiation, artificial sweeteners 156–157
UV/DAD (diode array detector) 104–105
UV-Vis spectra 107

V

vacuum-assisted sorbent extraction (VASE) 102
VASE *see* vacuum-assisted sorbent extraction
venlafaxine (VNF) 43
veterinary antianxiety medications 48
veterinary antibiotics 48, 50
N-vinylpyrrolidone-DVB sorbent (HLB) 169–170
  artificial sweeteners 171–172
  disinfection byproducts 317
  high production volume chemicals 244, 247
  perfluorinated substances 199–201
  personal care products 75, 101–103
  pharmaceuticals 46–47 *see also* Oasis HLB
VNF *see* venlafaxine

W

wastewater samples
  musk fragrances 268 *see also* water samples
wastewater treatment plants (WWTPs)
  artificial sweeteners 157, 166–167
  disinfection byproducts 313
  musk fragrances 266–267, 268
  new psychoactive substances 131, 132
  perfluorinated substances 193–198, 203–204
  personal care products 73–75
Water Framework Directive (WFD) 2
water samples
  high production volume chemicals 231, 233–237
  microplastics 355–358, 360
  musk fragrances 270–275
  pesticides 7–8, 12–14
  pharmaceuticals 46–47, 51 *see also* disinfection byproducts
water-soluble UV filters 71

WAX cartridges
  artificial sweeteners 169–172
  perfluorinated substances 200–201
  pharmaceuticals 46
WFD *see* Water Framework Directive
World Health Organization (WHO) 155–156, 226, 307–308
WWTPs *see* wastewater treatment plants

X
XAD resin and PUF 198, 199–200
XDB-C18, pesticides 20–22

Z
Zn(II) based-MOF 98
Zn-Zn nanoparticles 104
Zorbax Eclipse XDB-C_8 173
zwitterionic PFASs 187, 190